Crisis in the Workplace: Occupational Disease and Injury

Crisis in the Workplace: Occupational Disease and Injury
A Report to the Ford Foundation

Nicholas Askounes Ashford
Center for Policy Alternatives
Massachusetts Institute of Technology

The MIT Press
Cambridge, Massachusetts, and London, England

This book was set in IBM Composer Press Roman
by Techdata Associates,
printed and bound
by The Colonial Press Inc.
in the United States of America

Library of Congress Cataloging in Publication Data

Ashford, Nicholas A
Crisis in the workplace.

Bibliography: p.
Includes index.
1. Industrial hygiene–United States. 2. Industrial safety–United States. 3. United States. Occupational Safety and Health Administration. I. Ford Foundation. II. Title.
HD7654.A77 614.8'52 75-28424
ISBN 0-262-01045-3

Στήν 'Ανδρονίκη

Contents

Preface

In May 1972 I began what was to be a short study of occupational health and safety in the United States for the Ford Foundation. The purpose of the study was to identify the important problems and actors in the field and to recommend what action, if any, private philanthropy should take in the area. After three months of intensive interviews with representatives of management, labor, government, and public-interest groups, and with occupational safety and health professionals, it became apparent that much more investigation and analysis would be required than had originally been envisioned. It also became clear that the experience of other countries had particular relevance for the United States. Consequently, I spent ten weeks during the summer of 1972 interviewing people involved with the problems of job safety and health in eight Western European countries, the World Health Organization, and the International Labour Organization. In the end, my "short" study required two years for its completion.

This resulting report is an analysis of the technical, legal, political, and economic problems in occupational health and safety. In a sense, then, this is actually four reports, reflecting four points of view on the subject that decision-makers might take depending on their backgrounds. Chapter 3 explains the problems from a scientific point of view. Chapters 4-6 deal with the legal and political aspects of the implementation of the Occupational Safety and Health Act of 1970. Chapters 7-9 examine the economic issues and the problems of manpower development. Chapter 10 describes some of the important activities of the private (i.e., nongovernmental) actors. Each approach is important and utilizes a different language to describe the issues. Separate chapters are devoted to a discussion of the foreign experience (Chapter 11) and to the special problems of agricultural workers (Chapter 12).

It is not expected that every reader will want to read the entire report. A summary may be found in Chapter 1, while a general introduction is provided in Chapter 2. Chapter 13 represents an attempt to bring together the many themes—legal, economic, scientific, and political—that will determine the future of occupational safety and health in the United States.

The original report submitted to the Ford Foundation included a 600-page set of appendices containing descriptions and analyses of the activities of the various groups active in the field: professional societies, private organizations, labor unions, insurance carriers, management groups, consulting firms, and public-interest groups. Detailed descriptions of activities in eight Western European countries were also included. These appendices were intended to serve as important background for the analysis of the chapters on key activities and the foreign experience and to illuminate what innovative programs exist and what areas are receiving insufficient attention from the private groups concerned with workplace health and safety problems. Because of the already voluminous nature of the report, it was decided that the appendices should not be published at this time. Queries regarding their availability should be directed to the author.

The first draft of this report was submitted in May 1974, and the material should be considered current as of that date. A revision of the manuscript was completed in January 1975, and an attempt has been made since to update some important areas, including changes occurring as late as June 1975. As this work goes to press, the policy analysis and recommendations are still on point.

The report could not have been completed without the aid of my writing staff. Although the conception, ideas, and research that comprise the report are essentially my own, the staff carried out a significant portion of the actual writing and made important additions to the basic work. I am, of course, also indebted to the many people who consented to be interviewed; they provided valuable materials and a refreshing amount of candor.

Special recognition goes to Sheldon Friedman for his extensive contributions in Chapter 7, Chapter 9, and Section 3.4, and for conceiving and writing Section 4.5 with the assistance of Joseph Hadzima. I would especially like to thank Judith Katz for her extensive contributions in Chapters 9, 10, and 11 and for her assistance in the final preparation and organization of the report. The author is grateful to Joseph Hadzima for his significant contributions to Chapters 4, 5, 6, and 10; Les Boden for his extensive work in Chapter 8 and assistance in Chapter 7; and Jacqueline Cooke for her efforts in Chapter 12. Contributions were made to Chapter 2 by Charles Edmundson, to Chapter 3 by Elizabeth Barrett Levy and Kenneth Wong, and to Chapter 6 by Mariette Howell. Numerous people typed the manuscript, and I am indebted to every one.

Many useful additions were provided by readers of the first draft. These readers were Peter S. Barth of the University of Connecticut, Miles O. Colwell of the Aluminum Company of America, Bert Cottine of the Health Research Group, George Hagglund of the University of Wisconsin School for Workers, Harriet Hardy, Lorin Kerr of the American Public Health Association, Seymour Lusterman of The Conference Board, Warren R. Muir of the Council on Environmental Quality, Mary-Win O'Brien of the Department of Labor, George Perkel of the Textile Workers of America, George Taylor of the AFL-CIO, and Anne Trebilcock of the United Auto Workers.

Additional thanks go to Timothy F. Cleary of the Occupational Safety and Health Review Commission, Daniel Krivit of the Subcommittee on Manpower, Compensation and Safety, U.S. House of Representatives, Benjamin Mintz of the Department of Labor, and Ray Randlett of the Department of Labor for their contributions to the updating of Chapters 4, 5, and 6.

Lastly, I wish to thank Basil Whiting of the Ford Foundation, the Center for Policy Alternatives at M.I.T., and especially its director, Dr. J. Herbert Hollomon, for their support and patience during the time this report was in preparation.

Nicholas A. Ashford
Cambridge, Massachusetts

Crisis in the Workplace: Occupational Disease and Injury

1 Summary

1.1 INTRODUCTION

This first chapter is a summary of a two-year analysis of the technical, legal, political, and economic problems related to health and safety in the work environment.* *Health, work,* and *environment* rank among the most important areas of social concern today, and the point where these concerns converge—the workplace—has become a microcosm of national conflict. Not only are issues directly related to the long-neglected field of occupational health and safety receiving increased attention, but the nation is beginning to perceive that some of its larger concerns may be centrally connected to the workplace. For example, the growing debate over health care and preventive medicine cannot proceed far without encountering the probable occupational causation of a surprising proportion of disease. Problems of industrial productivity, work alienation, and management-labor disputes relate increasingly to occupational health and safety. In addition, the monetary costs of job injury and disease are beginning to be more fully realized and deserve closer attention during a period of inflation and of materials and energy shortages.

This modest rebirth of interest in occupational health and safety was stimulated in part by the recent passage of federal legislation—the Coal Mine Health and Safety Act of 1969 and the Occupational Safety and Health Act (OSHAct) of 1970. But the legislation was itself a response to several issues that had surfaced during the 1960s or before, that had grown in immediacy during the late sixties, and that have not yet been adequately addressed by the programs deriving from the new legislation:

- Probably the most important single factor was the increase in the reported injury rate in industry, an increase of nearly 29% from 1961 to 1970. The National Safety Council estimated 14,000 deaths due to accidents on the job and 2.2 million disabling injuries annually (with the latter statistic probably underestimated by a factor of five). A number of spectacular disasters—largely in mining—punctuated this trend.
- A second and potentially more important factor is occupational disease, a subject hotly debated by management, labor, and government—and one that is for the most part *not* reflected in the above *injury* statistics. The last decade has seen both new and newly acknowledged evidence for the occupational origin of much disease. Coal miners' pneumoconiosis, asbestos-caused cancer, beryllium disease, and the recently discovered vinyl chloride-caused liver cancer are the most familiar examples. Such problems appear to be more far-reaching, however. While statistical evidence for

*This study was commissioned by the Ford Foundation in May 1972. The Foundation's charge to the author was to "examine the nature and severity of occupational health problems in the United States, concentrating on the domestic policy context but including consideration of relevant foreign experience, all with the purpose of recommending appropriate courses of action." The views expressed are, of course, those of the author and not necessarily those of the Ford Foundation.

occupational disease is difficult to obtain, the Department of Health, Education, and Welfare estimates 390,000 new cases of occupational disease each year and as many as 100,000 deaths.

• A third factor is the rapid rate of technological change. The United States is the most technologically advanced nation in the world, and in the last one to three decades American ingenuity and industry has introduced into the workplace at an increasing rate new chemicals, new productive processes, and new forms of stress associated with heat, vibration, and noise. Relatively few of these developments have been examined for their impact on the human organism.

• A fourth, related factor was the rise of the environmental movement with its concern for the effect of toxins on ecological systems. This movement raised the consciousness of millions of citizens to the dangers of industrial pollution to the *general* environment. And from there it was only a step to an increased awareness by some that pollution almost always begins in a workplace and that exposures of people to toxic substances is often many times more serious in *workplace* environments than elsewhere.

• Finally, the changing character of the workforce has also contributed to the new importance of job health and safety issues. The increased educational levels of the labor force, the rise in wages in the manufacturing sectors, and the consequent shifting of worker demands has slowly placed job safety and health higher in the priorities of workers and their representatives. On the management side, some industries are seeking a better understanding of the worker and his or her needs in an effort to cope with the relative decrease in American industrial productivity—and this has brought about some increase in job safety and health consciousness.

The passing of federal legislation was a natural response to the pressures resulting from these developments. It must be recognized, however, that here, as with the National Environmental Policy Act or the proposed Toxic Substances Control Act, we face difficulties inherent in any attempt to use law as the predominant mechanism for the social control of science and technology. There is no philosophical reason why a society should not choose to be as protective of people in their working environment as it is with regard to the food they eat or the nonwork environment to which they are exposed. But the goals of the new OSHAct cannot be attained—indeed the Act cannot even be adequately implemented—without concurrent attention to the fundamental conflicts that exist between people who play various roles in the institutions of the society. Law enforcement, while necessary, is not sufficient; other activities must also be undertaken to supplement the law and to deal with systemic inadequacies in the economic, political, scientific, and technological realms.

There are at least four kinds of conflicts characterizing health and safety in the work environment. These are discussed in detail in later sections and will be reviewed only briefly here:

The first type of conflict is the clashing of different self-interests that is characteristic of management-labor relations on many issues. In unionized situations, this often adversarial process occurs in the form of discussions, negotiations, grievance procedures, collective bargaining, and strikes. Fundamentally, the basic conflict of self-interests stems from management's desire to keep costs down and to maintain control of the workplace versus workers' desire to gain the largest possible package of wages and benefits, job security, and control. While the system of industrial relations is complex and numerous exceptions exist, there are three characteristics of this relationship that are especially important for an analysis of occupational health and safety:

- By and large, management is responsible in both tradition and law for providing a "safe" workplace. While labor does participate in joint safety committees, these are often not of great importance, and such participation does not erode labor's basic position that management is responsible and government should ensure that management lives up to its responsibilities. Hence, until recently, health and safety were not central issues in collective bargaining.
- Even when health and safety were at issue, the worker was traditionally aware that improvements could be attained only by trading off some portion of his wage and benefit demands for them. In extreme cases, employers could maintain that costly health and safety improvements would render the facility unprofitable and cause it to be closed, depriving workers of their jobs.
- Both labor and management have difficulty balancing the concrete and immediate costs of health and safety improvements against their often indeterminate and long-term benefits. Short-term and known considerations usually win out. This often means that actions are taken to limit injuries, which are dramatic and whose costs are reflected in immediately perceived pain and in workmen's compensation premiums, but that improvements relating to health are limited. The impact of health hazards is often deferred for many years, and employers dislike "internalizing" costs whose benefits—the possible absence of diseases at some future time when employees may be working elsewhere or retired—do not appear to accrue sufficiently to the employer.

The Occupational Safety and Health Act of 1970 is slowly serving to raise the consciousness of both management and labor.* The mandate to comply with health and safety standards is causing management to internalize costs, in much the same way that workmen's compensation legislation did. The critical difference is that the emphasis is now on *prevention* related to both health and safety rather than on treatment or compensation for injury, and management and labor now argue about the nature of safe and healthful work conditions. These new conflicts thus center around issues that, if resolved, are more likely to improve workers' health and safe-

*In coal mining, the Coal Mine Health and Safety Act of 1969 is also changing management and labor attitudes.

ty (and productivity in the long run) than the resolution of conflicts over who will pay for the harm.

The second kind of conflict derives from an insufficient data base regarding the nature and severity of health hazards. A good example is the controversy over the adoption of standards for 14 carcinogens by the Occupational Safety and Health Administration (OSHA) in the Department of Labor. Establishing "safe levels" for toxic materials on the data presently available is difficult enough; the chronic nature of many diseases, including cancer, compounds the problem by making the unequivocal establishment of causation difficult. In the case of cancer, in particular, we have not yet settled the debate over whether there is *any* "safe level" for exposure to a carcinogen or whether we should adopt the concept of a "zero threshold," i.e., no permitted exposure. The lack of a firm data base on the effects of exposure to toxic materials explains in part why only three permanent health standards—those for asbestos, the 14 carcinogens, and vinyl chloride—have been promulgated by the Department of Labor.

This problem of inadequate knowledge raises the further question of the burden of proof required to make legal-political decisions. The standard of proof required for acceptance of a scientific thesis is much higher than that required by the law. When it comes to safeguarding rights under the law, a "scintilla of evidence" may justify legal sanctions, controls, and even the establishment of liability. The law thus attempts to make the best decisions on the information that exists. Absolute certainty or even consensus is not always required. What is required is an opportunity for all the available evidence to come in and for all parties to be heard. This pervasive principle in our system of jurisprudence is not fully understood by "pure scientists" (who sometimes view the law as "unscientific"); nor is it easily accepted by parties adversely affected by the result of a legal proceeding. In such situations, demands are sometimes made for "scientific proof" that is not available. And, as we shall see, it is very unlikely that sufficient "scientific proof" will *ever* be available—or is even *possible*—with respect to some occupational health issues.

A third kind of conflict relates to differences in perception of what is just or fair in public policy and what are the appropriate limits to public policy. The controversies over the Delaney Amendment on the testing of food additives and the proposed Toxic Substances Control Act hinge on the question, "Do we prohibit use or consumption of a material until it is proven safe, or do we allow its use until it is proven harmful?" Either course involves costs and risks to someone, and honest people can differ on where fairness lies.

The question of public policy with respect to the general environment as compared to the work environment is complicated by differences in the degree and selectivity of the risk posed. Chemicals that degenerate in the general environment or are used in diluted form may present low risks to the general population and confer important benefits (e.g., certain pesticides). Similarly, materials such as asbestos may save lives as fire retardants or brake linings. In such cases, the general risk may be low and almost randomly distributed, and there is thus a certain equity in the use of these substances that gives us comfort. But if selected groups of chemical, agricultural, or asbestos workers are exposed to severe health risks, there is a strong argument that these situations are not equitable, even if more lives are saved than are lost by the continued use of these materials. The nonrandom selection of those who bear these extra risks deserves special attention by both public and private decision-makers.

But what kind of attention? How protective should the government be of its citizens and how much responsibility should be assigned to the individual? To the firm? For a number of reasons suggested above and detailed below, it is clear that a strong governmental role in seeking improved job environments is needed *and* that increased efforts by individual workers, their representatives, employers, and numerous private forces are called for. All of this, however, will involve continued controversy over what is fair and appropriate.

The fourth kind of conflict relates to the fact that various institutions, forces, and mechanisms in our technological society are not connected very well. For example, those people and institutions concerned with problems in the *general* environment are not yet interacting adequately with those concerned with the *work* environment. A rather vivid example of this exists again in the handling of pesticides. "General" environmentalists succeeded in eliminating DDT for use as a pesticide because it posed serious ecological dangers to wildlife and to the consumer. These efforts resulted in the use by farmers of parathion, which deteriorates in the environment much faster than DDT. However, parathion is very much more harmful than DDT for the people who must handle it. Because mechanisms did not exist to coordinate the resolution of problems in the work environment with those of the general environment, the work environment may have been made more dangerous.

There are other persuasive arguments for closer collaboration between those concerned with these two environments: (1) Very often persons who work in industrial plants also live near those plants and are thus in a position of "double jeopardy." (2) A health danger to the general population may be recognized as a result of illness among persons working with high concentrations of previously unknown hazards. (3) Efforts to clean up the factory may result in polluting the general environment, and, conversely, efforts to "contain the pollution" may mean increased pollution within the plant. The problems of the general and workplace environments are thus intimately connected and must be attacked together.

A second serious discontinuity is in the field of health services where *preventive* medicine is not tied sufficiently to the *treatment* establishment. Occupational health problems can be most effectively dealt with through the practice of preventive medicine, and preventive medicine is grossly underdeveloped in the United States. In fact, many professionals concerned with medical-care delivery systems will define preventive medicine as merely early detection of disease, rather than an elimination of those hazards that can ultimately cause disease. The line is not a thin one; the distinction is very important, especially with diseases whose progress is difficult to reverse, such as cancer. The growing interest in Health Maintenance Organizations (such as the Kaiser-Permanente Plan) reflects merely the importance of early detection of disease. Preventive medicine needs a different kind of mechanism for its institutionalization—one which emphasizes the reduction of exposure to potentially harmful substances.

A third example relates to the different languages used and different problem areas dealt with by the various professional interests in the society. General practitioners know little of occupational medicine. The industrial hygienist views preventive medicine differently than does the occupational physician. The personnel director may not want workers to talk directly with the plant doctor or industrial hygienist because management-labor relations are his "problem area." And finally, the safety professional does not talk the same language as the health professional even though it is futile to draw a distinction between safety and health in many instances.

Consider the example of noise. Noise has been recognized as a safety problem by safety professionals, and they have sought to reduce the noise level of both the workplace and the general environment to minimize hearing loss. But the issue of noise as a stressor and a cocausative factor of disease is of little, if any, interest to the safety professional. Each professional thus chooses to deal with only some aspects of a problem, and words like "noise" mean different things to different professionals.

The problems in the work environment are complex and demand both long- and short-term transdisciplinary approaches for their solution. The issues go far beyond the medical and technical problems of disease and injury. Hopefully this report will serve to persuade decision-makers that it is futile to place sole reliance on any single policy instrument to reduce occupational injury and disease—be it research and medicine, the law and the regulatory process, or market incentives.

The remainder of this summary elaborates on the matters touched on above. It reviews the nature and dimensions of occupational health and safety problems; the informational, economic, and manpower implications of these problems; activities of management, labor, and public-interest groups; relevant foreign experience; and the design of national policy in this area.

1.2 THE NATURE AND DIMENSIONS OF OCCUPATIONAL HEALTH AND SAFETY PROBLEMS

1.2.1 Safety versus Health

Safety hazards are those aspects of work environments which can cause burns, electrical shock, cuts, bruises, sprains, broken bones, or the loss of limbs, eyesight, or hearing. In general, the harm is usually of an immediate and sometimes violent nature, is very often associated with industrial equipment or the physical environment, and often involves an employment task that requires care and training. I have noted above the alarming increase in such injuries over the past decade.

Safety professionals insist that the term "safety" is all-encompassing and includes health—but it is fair to say that their concern is more with the explosive nature of chemicals than with their toxicology and more with the effects of noise on hearing than with its role as a stressor and cocausative factor of disease. However, to the extent that the causes of accidents are recognized by the state workmen's compensation system, they do become a concern for the firm, its insurance carrier, and its safety professionals.

Typical *health* hazards include toxic and carcinogenic chemicals and dusts, often in combination with noise, heat, and other forms of stress. Other health hazards include physical and biological agents. The interaction of health hazards and the human organism can occur either through the senses, by absorption through the skin, by intake into the digestive tract via the mouth, or by inhalation into the lungs. The results of these interactions can be respiratory disease, heart disease, cancer, neurological disorder, systemic poisoning, or a shortening of life expectancy due to general physiological deterioration. The disease or sickness can be acute or chronic, can require a long latency period even if the original exposure is brief, and can be difficult or impossible to diagnose early or with certainty. (It should also be noted that disease can give rise to accidents.)

The real world, unfortunately, does not offer isolated hazards. Chemical, physical, biological, and stress hazards are often found in combination, and their effects can be not merely additive but intensified (synergistic). Carbon monoxide and heat, amphetamines and overcrowding, asbestos and smoking, and promoters of cancer are all examples of agents whose effects can be synergistic. Most such combination effects are probably still to be recognized, and this recognition often occurs only after adverse effects are accidentally encountered (as in the case of barbiturates and alcohol).

Unlike safety hazards, the effects of health hazards may be slow, cumulative, irreversible, and complicated by nonoccupational factors. While an unguarded blade in a rotary saw may present a severe and imminent danger, it is often difficult to perceive the severity or imminent danger contained in a brief exposure to a potential carcinogen that can take years to cause a tumor or death. However, the probability of dying from cancer may be just as high as that of having an accident with the saw.

The failure of the nation's injury reporting system and workmen's compensation system to include occupational disease adequately has contributed to the failure of society to recognize the severity of occupational health hazards. The very nature of the differences between health and safety hazards has resulted in a pervasive safety bias that has affected legislation, the setting of standards, enforcement, manpower development, employer and employee education, and technology development. This relative overemphasis on safety has fostered complacency and has thus prevented much needed progress in the more neglected area of occupational health.

1.2.2 Disease and Occupational Causation

The Public Health Service estimates that 390,000 new cases of occupational disease appear annually. Epidemiological analyses of excess mortality among workers in several industries suggest that as many as 100,000 deaths occur each year as a result of occupational disease. Evidence from the insurance industry indicates that the *excess* risk of death, in actuarial terms, is substantial among certain occupational groups. The excess risk is on the order of magnitude of all the other risks of death we must face—in other words, the probability of death (given age, race, sex, etc.) facing members of certain hazardous occupations is in some cases greater than twice the norm for a matched cohort drawn from the general population.

Occupational factors may very well play a far more significant role than is presently realized in the causation of the major diseases and health problems that confront us. Two million people die every year in the United States. Heart disease, the leading cause of death (accounting for 38.7% or about 750,000 deaths), is only 25% "explained" by known physiological and environmental factors, such as excess weight, hypertension, serum cholesterol, and cigarette smoking. An unknown but quite possibly substantial proportion of the 75% of heart disease risk that is presently unaccounted for could be related to work and its attendant hazards, particularly stress.

Cancer is the second leading cause of death in the United States today, with an annual toll of 300,000. The incidence of cancer has risen rapidly with industrialization: in 1900, 3.7% of all deaths were attributable to cancer, but by 1968 the proportion of deaths from this cause was 16.5%. In part, improved diagnosis and longer life expectancy are responsible for this dramatic increase. Nevertheless, there are indications that the true incidence of cancer has been on the upswing.

Research in the United Kingdom indicates that more than 80% of cancer is of environmental origin and therefore, theoretically, preventable. This conclusion is based upon the observation that the variation in cancer incidence rates among geographic and occupational environments is enormous. Domestically, an HEW Task Force on Research Planning in Environmental Health Science reported that

> there is abundant evidence that the great majority of malignant neoplasms—probably over 90 percent of the total—are induced, maintained or promoted by specific environmental factors. Many of the known environmental causes of cancer are physical and chemical agents that directly concern the environmental health professions. *Carcinogenesis must therefore be regarded as one of the most significant potential consequences of environmental contamination.* (Emphasis added)

Although 80 or 90% of cancer could be environmentally caused, it is not presently known how much is occupationally related. There seems to be a general consensus among cancer researchers and environmentalists that probably one-half of all cases are complicated by occupational factors. The experiences of chemists, asbestos workers, underground uranium miners, and, most recently, rubber workers handling vinyl chloride, amply document the proposition that "excess" cancer of various types is indeed occupationally related.

Since the production of petrochemicals in the United States has doubled every five years since the end of World War II, the incidence of cancer characterized by long latency periods may be expected to rise significantly over the next twenty to thirty years. This expectation is reflected in the considerable legislative activity directed toward the regulation of chemicals—and especially of chemicals known to be carcinogenic.

Chronic diseases of the respiratory system have been reported with increasing frequency in the United States and have become major causes of death and disability. Chronic bronchitis and emphysema are the fastest-growing diseases in the country, doubling every five years since World War II, and account for the second highest number of disabilities under social security.

Much of the respiratory disease that plagues workers is known to be job-related. Both "specific" diseases such as coal workers' pneumoconiosis and general diseases such as emphysema and chronic bronchitis can be of occupational origin.

Even if only 5% of heart disease, cancer, and respiratory disease is in fact occupationally related, the number of deaths that could thereby be classified as "job-related" would exceed 60,000. Thus, the Public Health Service estimate of "as many as 100,000" may not be unreasonable. A recent study conducted for NIOSH by the University of Washington reported that 31% of over 1100 medical conditions found in 908 participants were of probable occupational origin, with an additional 10% manifesting "suggestive history." The probable incidence of occupational disease was 28.4 per hundred workers. Only 2% of these illnesses were reported on the employer's log required by OSHA, and only 3% were found in workmen's compensation records.

The risk of occupational illness is not shared equally by all members of the labor force. Miners, construction and transportation workers, and blue-collar and lower-level supervisory personnel in manufacturing industries experience the bulk of both occupational disease and injury. Further, the job illness and injury rate in agriculture is exceeded only by those in mining and construction.

Occupational health problems are not restricted, however, to the industrial or agricultural worker. They affect white-collar workers and corporate executives as well. Dentists are being studied for the possible effects of X-radiation, mercury, and anesthetics on their having the highest rate of suicide of any professional group and excess diseases of the nervous system, leukemia, and lymphatic malignancies. Operating-room nursing personnel suffer several times the miscarriage rates of other nurses and give birth to a larger proportion of children with congenital deformities. Cosmetologists (beauticians) display excess cancer and respiratory and cardiac disease. Administrators are far more likely to develop coronary disease than are scientists and engineers.

These and numerous other examples suggest that presently available statistics seriously understate the enormity of America's occupational health and safety problem and the degree to which the nation's general health problems may be occupational in origin.

1.3 THE OCCUPATIONAL SAFETY AND HEALTH ACT OF 1970

1.3.1 Provisions

The Occupational Safety and Health Act of 1970 imposes on virtually every employer in the private sector an unprecedented *general duty* to "furnish to each of his employees employment and a place of employment which are free from recognized hazards that are causing or are likely to cause death or serious physical harm to his employees." In addition, each employer must comply with occupational safety and health *standards* promulgated and enforced under the Act by the newly established Occupational Safety and Health Administration (OSHA) in the Department of Labor. OSHA has the authority to enter any workplace without advance notice and to propose penalties upon discovery of violations.

The Occupational Safety and Health Review Commission was also established under the Act as an independent quasi-judicial review board consisting of three members appointed to six-year terms by the President. The Review Commission rules upon all enforcement actions of the Occupational Safety and Health Administration. Under certain circumstances the decision of the Review Commission is reviewable in the courts.

To assist the Secretary of Labor in enforcing the Act, Congress established the National Institute for Occupational Safety and Health (NIOSH), a research body, located within the Department of Health, Education, and Welfare, which is responsible for developing and recommending occupational safety and health standards to OSHA. The agency is specifically required to publish a list of all known toxic substances and the concentrations at which these substances exhibit toxicity effects. In order to accomplish these tasks, NIOSH is authorized to conduct research and undertake experimental programs in occupational safety and health.

In addition to the three agencies described above, the Act established a National Advisory Committee on Occupational Safety and Health (NACOSH) consisting of "representatives of management, labor, occupational safety and occupational health professions, and of the public." NACOSH is required to advise, consult with, and make recommendations to the Secretary of Labor and the Secretary of Health,

Education, and Welfare on matters relating to the administration of the Act. The Secretary of Labor may also appoint other advisory committees to aid in the setting of standards.

The Act contains provisions that allow workers to become involved in the administrative and enforcement activities of the Act even though no enforceable duties are imposed on them. Perhaps the most important employee right is that of requesting an OSHA inspection. Any employee or employee representative may notify the Occupational Safety and Health Administration of violations of standards or of the general duty obligation by his employer. The employee's name must be withheld from his employer if he so requests. In addition, the Act gives employees the so-called walkaround right, enabling their representative to accompany the federal inspector on his inspection of the workplace.

An important and far-reaching provision of the Act allows the individual states to regain local authority over occupational health and safety by submitting state plans that are, and will continue to be, at least as effective as the federal program.

1.3.2 Implementation at the Federal Level

The OSHAct has failed thus far to live up to its potential for reducing job injury and disease, for fostering the internalization of the social costs of health and safety hazards, for encouraging technological innovation, or for stimulating job redesign. OSHA has had little measurable impact in reducing injuries and deaths, and the problems in the health area have become even more serious since 1970.

After five years, OSHA has adopted only three new health standard packages—for asbestos, for 14 carcinogens, and for vinyl chloride—all of which were adopted only after considerable pressure and counterpressure on the part of unions, public-interest groups, and employer groups. The lack of a firm data base, strong commercial interests, and continued pressure by public-interest groups and unions indicate that the adoption of health standards will continue to be a highly adversary and time-consuming process.

Inadequate levels of compliance manpower has meant that the probability of a firm's being inspected is relatively low. Inadequate policy direction and inadequate training of compliance officers has meant the perpetuation of a severe safety bias and the overlooking of many serious health hazards. Moreover, even if employers are cited, the fines are often far too low to serve as an economic incentive to improve working conditions. On the other hand, it is possible that the issuing of a citation itself, rather than fines, provides a more effective incentive when nonserious violations are cited. There is little indication, however, that OSHA's enforcement activities have resulted in significantly improved health and safety conditions.

The potential strength of the OSHAct's "general duty" obligation lies in its imposition of a general responsibility upon the employer to take affirmative action in improving workplace conditions. The applicability of the employer's general duty even in the absence of injury can in principle encourage general preventive responses to injury and illness. However, OSHA continues to underemphasize the employer's general duty obligation, and it thus leaves workers virtually unprotected in cases where there is reason to believe that excessive exposure to a given agent may be harmful, but where no standards have been established.

The OSHAct, is, however, providing an important stimulus to labor unions in collective-bargaining activities with regard to job health and safety. Contractual embodiment of OSHAct provisions in collective-bargaining agreements and their implementation through revitalized national and local union safety committees of greatly expanded scope and function are likely to be of considerable importance in implementing the OSHAct in the years ahead.

The implications of the OSHAct for other areas of law remain to be seen. The relationship between the OSHAct, existing labor law, arbitration, and collective bargaining is rapidly changing. The need for industry to share information and technology to provide a safe and healthful workplace, for instance, will have to be reconciled with our antitrust policy. And the OSHAct could have significant implications for employees' rights to refuse hazardous work without penalty, both in arbitration and in proceedings before the National Labor Relations Board or the courts.

In sum, neither the Occupational Safety and Health Administration nor the National Institute for Occupational Safety and Health has formulated a coordinated and visible occupational health and safety policy. Both agencies are doing only a fraction of what needs to be done or what they could do with their *present* resources. The activities of research, enforcement, and manpower training are not coordinated. Recent changes of leadership at both agencies, however, hold out once more the hope for innovative reform.

1.3.3 The Role of the States

In addition to the difficulties encountered at the federal level, state preemption of federal occupational health and safety authority poses serious problems. Such decentralization of control will necessarily result in unequal protection of employees and has the potential of decreasing the overall level of protection. OSHA will be very hard pressed to administer and evaluate many completely different programs. The concept of the general duty clause will be destroyed as many state judicial systems and enforcement agencies attempt to apply their version of a general duty requirement, if they have one, to occupational health problems. A worker's health will thus be dependent upon the state in which he is employed.

The states' previously poor record in occupational health and safety enforcement does not suggest that sufficiently increased worker protection will result from the return of the programs to the states, even though the total occupational health and safety compliance force will be increased somewhat. State takeover could severely limit the ability of labor unions to act in a watchdog role and hamper their advocacy efforts to improve working conditions. Manufacturing concerns with operations in several states may find it difficult to conform uniformly to different state requirements, and management strategy will consequently become more complicated. Thus, the introduction of state plans raises both old and new issues within a different framework. Recently, however, several large industrial states have withdrawn their state plans and have thus reversed the earlier trend toward returning authority to the states.

1.4 THE INFORMATION PROBLEM

In almost every public policy arena complaints are common that knowledge and its transmission as useful information are inadequate and imperfect. Indeed, it is a cliché to note that the task of the policy-maker (and, as noted above, of the legal system) is to make decisions and keep the society functioning in the face of inadequate information. The field of occupational health and safety is no exception; problems of knowledge and information are pervasive. However, while some of the most important knowledge gaps may never be filled, many of these problems seem likely to yield to the resources and approaches that private philanthropy can bring to bear.

It is helpful to consider these problems in terms of the generation of knowledge, its dissemination, and its utilization.

1.4.1 The Generation of Knowledge

The systematic researching and establishing of "safe levels" of exposure for every chemical product in commercial use is simply impossible in any practical sense. First, there are as many as 12,000 materials of *known* toxicity in commercial use today. Second, as noted above, most human exposure to chemicals is to mixtures, which may lead to multiple etiology (causation) of disease or to synergistic effects. Research suggests that 5% of the possible pairs of known toxic materials behave with adverse synergistic effects, meaning that there are millions of such pairs to investigate. (Multiple combinations increase the numbers astronomically.) Third, the time lag between exposure of humans to occupational health hazards and the appearance or diagnosis of chronic disease is sometimes two decades or more, meaning that research, to be timely, must be undertaken with existing populations that have been exposed for long periods of time (a form of epidemiological study) or with lower orders of life. Finally, the rate of increase of new substances for which standards must be set is much greater than the rate at which we are able to improve our standard-setting or enforcement ability.

These considerations have important implications for research strategy:

- While screening procedures for chemicals must be further developed, any attempt to develop rigorous scientific "proof" of the validity of exposure limits for all such substances is misguided. Such efforts must be targeted in rifle-shot fashion at situations posing the clearest or greatest potential danger. However, many dangerous substances will be recognized much too late—as was the case with vinyl chloride.
- A high priority should be given to epidemiological studies—the analysis of morbidity patterns among various occupations and their correlation with exposure to possible health hazards. The purpose here would be to uncover what types of past and present exposures have had adverse effects. Foreign experience ought to be constantly monitored.
- Relatively high priority should also be given studies attempting to determine patterns of interaction between exposure to chemical substances or dusts and noise, heat, or stress.
- Technological research on new processes involving closed systems should be a first priority item for long-term solutions to the problem of exposure to chemicals.

Federal leadership in elaborating and funding a research strategy based on these principles is badly needed.

1.4.2 The Dissemination of Information

Even if adequate information could somehow be generated, we would still be faced with a host of problems regarding the dissemination of information. Information does little good if it is not made widely available to decision-makers and to parties who bear the consequences of the decision-makers' actions—and if it does not reach them in a form they find useful. These needs are discussed further in a later section. Here we note the special problem of inequality of access to information.

Inequality of access—for example, between management and labor or between large firms and small ones—creates incentives for special interests to withhold or distort potentially damaging (or beneficial) information. Differential access converts information into a bargaining advantage for the more knowledgeable party, and compounds the difficulties of public and private decision-makers who must evaluate the merits of a bewildering variety of conflicting claims. Thus the phenomenon of differential access to information transforms the problem of improving our understanding from the purely scientific and technical realm of information generation, dissemination, and utilization, to the "political" arena. Moreover, the pervasiveness of patent rights and licensing arrangements, and the proprietary nature of much of the information about newly introduced substances and processes, means that the problem of differential access to information concerning workplace hazards is endemic to our economic system, not a mere superficial imperfection.

While such problems are serious, much progress can be made by well-known techniques of training, technical assistance, and publication.

1.4.3 The Utilization of Information

In large measure, on-the-job hazards are attributable to a lack of utilization of information that already exists. A large body of knowledge has already been generated and fairly widely disseminated that holds promise of easing occupational health and safety problems, but decision-makers have been painfully slow in utilizing it.

A tragic example of this situation is provided by the case of America's uranium miners, an excess number of whom have died or will soon die prematurely from lung cancer of occupational origin. It has long been known that exposure to airborne radiation carries greatly increased risk of lung cancer, and that the airborne radiation to which this nation's underground uranium miners were being exposed was well in excess of levels proven to be unsafe. It has, furthermore, long been known in Europe that this hazard can be substantially alleviated by the simple expedient of installing adequate ventilation in uranium mines. Failure of public and private decision-makers to utilize this widely disseminated information has exacted a terrible toll among America's uranium miners.

Another example is the adoption of substitute materials for asbestos. The debate surrounding the industrial use of asbestos proceeded largely on the assumption that asbestos was an essential raw material without close substitutes for many uses, such as in shipbuilding. Given this assumption, the terms of the debate were confined to the choice of an appropriate asbestos exposure standard and the costs of meeting it in terms of filtration and ventilation equipment and the like. Yet, while this controversy raged, the Swedish shipbuilding industry had for some time been using what are thought to be safer substitute materials in place of asbestos. These materials completely eliminate the necessity of exposing shipyard workers to asbestos, at least in the construction of new ships. It seems clear that we have many lessons to learn from other countries with regard to superior technologies for improving the level of occupational health and safety. An important objective of public policy must, therefore, be to facilitate the process of diffusion of technological information among different nations.

Increasing the effective utilization of available information calls for a strategy combining enforcement, public awareness, information and technical assistance, and active concern on the part of those directly affected by workplace hazards.

1.5 ECONOMIC CONSIDERATIONS

1.5.1 The Overall Costs of Occupational Health and Safety Hazards

The total cost of occupational hazards—in terms of lost wages, medical expenses, insurance claims, production delays, lost time of coworkers, and equipment damage—was estimated by the National Safety Council at $9.3 billion during 1971—nearly 1% of the GNP. This figure, moreover, is likely to be a gross understatement of even the direct costs to the GNP of both occupational injuries and illness. An estimated 25 million workdays were lost through absenteeism during 1972. This reported figure is equivalent to a loss of 100,000 man-years of work. It has been estimated that reduction of one day per year in the annual rate of absenteeism among the U.S. labor force would add $10 billion to the GNP. Much disease resulting in absenteeism is probably occupationally related, although it is not reported or recognized as such. Thus, if *unreported* occupational illness is responsible for even one additional day of absence per worker per year, inclusion of the hidden costs of such occupational hazards would result in a doubling of present National Safety Council estimates of the cost of occupational illness and injury. (Many of the other costs of chronic occupational illness—early death or retirement, reduced efficiency, family and community problems—are also not reflected in the Council's estimates.)

1.5.2 The Market Paradigm and the Role of Incentives

Occupational disease and injury are part of the human and social costs of production. As such, the level of workplace-related injury and disease is influenced by the economic forces that come to bear on firms and their employees. Companies, for example, must produce an adequate rate of return if they are to remain economically viable, and competition among firms results in constant pressure to reduce costs—a socially desirable objective in itself. Unfortunately, it is also the case that many ways of reducing costs also increase workplace hazards.

Many policy-makers are accustomed to thinking about the problem of providing a safe and healthful workplace in terms of the market paradigm. According to this view, the reason that the free market results in too much workplace injury and illness is because market "signals" or incentives are "wrong" from society's point of view. Employers, for instance, are not held financially accountable for the full human and social consequences of their failure to provide safe and healthful working conditions. A disproportionate share of the damage associated with occupational illness and injury thus befalls working people, their families, and society at large, without ever directly entering a corporate profit-and-loss statement. As a result, business has insufficient incentive to improve job health and safety, and the total costs of production enter neither the price nor the wage equations. The problem of public policy then becomes one of finding ways to make the market more effective in obtaining the socially desired level of job-related hazard. This approach suggests that public policy should be geared toward intervention in the market system to make it function in such a way that all prices reflect true social costs and all "externalities" are "internalized."

The workmen's compensation system is an example of an attempt to "internalize" costs after the fact—i.e., after the harm has been done. While the system has been largely inadequate, the shifting of the cost of injury from the individual employee to the employer gave rise to the adage "safety pays." The complete adage might better be put, "safety pays when you're forced to bear the cost of an unsafe workplace." Job safety became a matter of self-interest to both management and labor when reducing accidents on the job could reduce insurance premiums. (It should be noted again that this system does not deal adequately with occupational health problems.)*

The market paradigm is helpful in determining the usefulness of market incentives for dealing with occupational health problems. The imperfections in the market approach are inherent and severe:

- The deficiencies in the knowledge of the nature and severity of health hazards are the most serious imperfections, since market analysis assumes adequate knowledge of costs and benefits. The problem of chronic disease that manifests its harm (costs) far into the future presents a "discounting" problem for even the most far-sighted management. This discounting problem for future harm (costs) is paralleled by a discounting problem for future benefits as well. Even if a firm thought it profitable in terms of conserving its scarce human resources to reduce chronic disease, the uncertainty of future benefits would make it less likely that action would be taken.
- There are serious reasons for questioning the notion that the existing level of workplace hazards represents working people's free-market "choice" regarding the assumption of job-related risk. Beyond important informational problems, a wide variety of other forces—including social, cultural, psychological, and environmental factors—influence workers' decisions regarding the assumption of job-related risks. An inability to assess or relate to low-probability, large-harm contingencies is a behavioral trait common to many, if not most, individuals. Further, many workers are socialized to accept the hazardous nature of certain jobs and are convinced of the necessity of performing them in order to earn their livelihood. The Puritan work ethic and the notion that work has to be hard and unpleasant contributes toward the willingness to accept hazardous conditions.

The pervasiveness of these market imperfections with regard to occupational health and safety emphasizes the importance of social policy and private initiative in forcing employers to internalize more of the total costs associated with occupational health and safety hazards. The workmen's compensation system rests on incentives (premium payments) that are keyed to the immediate and relatively undeniable nature of *injuries*; such a post hoc approach is not sufficient when dealing with the more complex matter of *disease*—especially those diseases that take de-

*Workmen's compensation, in paying for treatment, rehabilitation, or harm, is probably not nearly as cost-effective as prevention. However, since prevention costs must be faced immediately, whereas treatment and compensation costs may be deferred—or even avoided entirely—the higher dollar costs of delayed action may be discounted to a relatively lower present value than prevention costs.

cades to manifest their presence. Thus, in the absence of some form of compulsion through OSHA-like standards or through collective-bargaining agreements, it is unrealistic to expect individual employers to assume what they view to be competitively disadvantageous costs.

1.5.3 Other Economic Considerations

Attempts to diminish occupational hazards are often complicated by other kinds of economic problems, not necessarily identifiable as "market imperfections." Safe conditions are often more expensive to achieve for smaller firms than for larger ones since the cheapest control technologies often require a minimum size that is larger than the operations of many production units. Also, these smaller firms may often find it difficult to locate and pay for the expert advice or information they may require.

It has been widely noted that the most profitable firms in an industry are typically also the safest. This may, of course, be explained by many factors, such as better management (for both production and safety), the utilization of longer time horizons over which to evaluate the investment in good safety practices, and better access to information and expertise. The converse of this observation is that many firms with clear safety and health problems—often the ones with the worst problems—may also be economically marginal. Such firms are often found in highly competitive industries and are under constant threat of failure. When these marginal firms have larger competitors with substantial reserves, or foreign competitors not forced to internalize the costs of health and safety, they may be faced with a choice between hazardous working conditions and bankruptcy.

The costs of providing a safe and healthful workplace *may* initially be high, especially if modification of existing facilities occurs, and the marginal producer *may* be forced out of business. But it is not clear that the smaller or marginal firm will *always* be forced out of business. The smaller firm's relatively smaller investment in capital equipment, closer employer-employee relations, and greater ease of relocation and job redesign may offset the advantages of economies of scale and large reserve funds of the larger firm. Furthermore, in many cases, the workplace can be made safer and healthier by education and management-labor cooperation, with little capital expenditure for new equipment or expensive substitute materials. In such cases, the size or marginal nature of the firm need not be a reason for ignoring workplace hazards.

Critical differences in incentives may exist at various levels within a firm. Top management, for instance, may believe in principle that it is consistent with the long-run economic interest of the firm to adopt good health and safety practices. Nonetheless, it may at the same time judge the short-run performance of its middle management by production criteria that result in disease and injury being higher than it intends.

In the short run, the costs of complying with the OSHAct will be high and unequal in different industries, and severe pressures to resist the enforcement of the legislation will arise. However, if enforcement persists and other activities to facilitate adjustment are pursued aggressively, the longer run should see an encouragement of new technology, substitute materials, and redesigned jobs that should result in higher total productivity, decreased worker absenteeism, and improved job health and safety.

1.5.4 International Competitive Implications

Although the percentage of the U.S. Gross National Product that depends on foreign trade is much smaller than that for Western European countries, more concern for what the enforcement of occupational health and safety legislation might mean to business is expressed in the United States. The National Association of Manufacturers, for instance, has expressed special concern for the effects of occupational health and safety legislation on America's international competitive posture.

Ironically, if imposing strict occupational health and safety legislation raises the price of American products (e.g., asbestos) to the point where we import the product from abroad, we are then in fact exporting disease and injury. Further, multinational companies may locate a production site in a particular country because the occupational health and safety of a worker is less emphasized there and hence the product of his work is cheaper. We are thus exporting jobs, albeit dangerous ones. (We could, of course, levy an import tax on all unsafely produced goods.) This is precisely why equity demands that international standards and work practices for occupational health and safety be established. Europe is moving in that direction. The United States is unlikely to agree to such standards and in some instances may seek to keep its own standards below those of other advanced nations. In these cases, it is not a competitive *disadvantage* the United States is seeking to avoid, but rather a competitive *advantage* the United States wishes to capture—at the expense of its workers. More production of unsafely produced products means more injuries and disease at home.

Despite all the voiced concern, it is not clear that environmental controls will severely affect America's competitive position. A report for the National Commission on Materials Policy from the National Academies of Science and Engineering examines the impact of general environmental controls:

> There are only a few econometric estimates of the *impact of environmental control on the short-term balance of payments and level of aggregate income.* One, using an econometric model of the U.S. economy and projected future American EC measures (assuming other nations do *not* follow suit), emerges with the following results: Over the 1972-80 period, pollution control and compensatory macroeconomic policy measures are expected to raise the U.S. unemployment rate by 0.3%, raise the average annual rate of inflation by 0.26%, reduce fixed investment not related to pollution control by an average of $2.3 billion annually, and exert *a*

> *negative impact on the U.S. balance of trade in the amount of $1.9 billion* (Council on Environmental Quality, 1972). If, on the other hand, pollution control costs are 50% higher than current estimates, or standards are raised correspondingly, the average annual 1980 negative trade balance effect is estimated to be in the neighborhood of $3.2 billion or 6-7% of current level of exports.
>
> The meager empirical evidence available at this time suggests the balance of payments, employment, and national income effects resulting from domestic environmental controls are likely to be relatively small for the United States. This is true even if the trend toward indirect subsidization is reversed toward a full "polluter pays" principle. Thus a tentative conclusion is that the *aggregate impact* of domestic environmental controls should not be a matter of deep concern for policymakers. Of more crucial importance are the short-term impacts on particular firms and industries. Substantial additional research is needed on these questions. (Emphasis added)

Moreover, the President's Science Advisory Committee takes into account the expectation that foreign rivals would be adopting similar regulating controls in stating:

> The United States has consistently been a net exporter of chemical products. It is also an exporter of technology and research and development but with repatriation of profits through multinational firms. The regulatory process does not seem to have been a major force in this trend. Rather, it simply reflects business opportunities and the regulations prevailing in other nations.
>
> The Panel is persuaded that much or all of the industrialized world is moving towards a pattern of regulation which resembles that of the U.S.

Thus the notion that the United States would suffer a severe competitive disadvantage internationally if it enforces strict environmental controls in the occupational health and safety area is open to question. Such enforcement may, however, change the mix of small and large firms in a hazardous industry, with some smaller and marginal firms disappearing. It would seem, then, that competition in international markets does not appear to create an economic imperative for relaxing the provisions of occupational health and safety legislation. What seems to emerge is an imperative to adopt uniform international standards and practices and to devise special forms of assistance to smaller or marginal producers.

1.6 MANPOWER IMPLICATIONS

The field of occupational health and safety requires expertise in many disciplines in order to respond to the congressional mandate of assuring safe and healthful working conditions. In order to develop a coordinated manpower policy, an understanding of the breadth and depth of the needed expertise is essential.

Manpower needs can be classified into four categories, each of which includes a number of different types of professionals:

1. Enforcement personnel are needed to ensure compliance with the OSHAct in both safety and health by inspecting workplaces and citing violations of standards.

2. Qualified researchers, including chemists, physicians, environmental physiologists, epidemiologists, toxicologists, industrial hygienists, and engineers, are needed to identify potential occupational health and safety hazards, to assess their effects upon those who work and have previously worked under exposure to these hazards, and to determine the requirements for assuring a safe and healthful working environment.
3. Personnel who provide health and environmental services are needed to aid in developing and implementing techniques, methods, and programs for the prevention and treatment of occupational disease and injury. This group includes occupational physicians, industrial nurses, safety professionals, industrial hygienists, and technicians, as well as executives and managers in business, labor unions, and government.
4. Personnel are needed to teach the skills necessary for performing the tasks within each of the above-mentioned groups.

The manpower resources presently provided by the health and safety professionals, government, industry, and labor are not nearly sufficient to come to grips with the expanding problems of occupational hazards or to meet the mandate of the OSHAct. The medical profession is oriented to treatment, not prevention, and it does little to promote occupational medicine. Furthermore, the occupational physician is not so highly regarded or remunerated as other medical specialists, and he generally has insufficient training for the task. While nursing is potentially a great source of expertise for the small firm, the quality of nursing education in these matters is too low presently to be of immediate help. There are not enough industrial hygienists to meet the present technical need, and the market has not yet responded by recognizing the need.

The total shortage of environmental and health services manpower has been estimated at 57,000 (including occupational physicians, nurses, industrial hygienists, safety professionals, technicians, and paraprofessionals). These estimates are necessarily subjective in that they depend on the evaluator's perception of how safe or healthful the workplace should be. There are several clearly definable (though less clearly measurable) benchmarks that require manpower

- to bring all firms up to the "best practice" level (i.e., to the safest firms in each industry)
- to bring all firms into compliance with the present (and future) requirements of the OSHAct through adherence to existing (and expected future) standards
- to match the best foreign practices
- to achieve a certain target level of safety and health that goes beyond the specific standards issued under the Act and complies with a "general duty" obligation
- to achieve a proper balance between safety and health and between different industries.

It is clear that the nation's present job health and safety manpower levels meet *none* of these benchmarks and that manpower development will remain the critical limiting resource factor in improving workplace conditions for the American worker.

The problems of insufficient numbers and inadequate training are compounded by a poor distribution of existing manpower. The safety bias that pervades this field is especially serious in the case of manpower and is manifested in a severe imbalance in the mix of safety versus health professionals and in the types of facilities available for their training. This imbalance is in large measure attributable to the rational response of firms and insurance companies to economic forces and government regulations. Workmen's compensation, insofar as it acts as an incentive at all, primarily influences firms' demands for *safety* as opposed to *health* manpower, as few diseases are recognized as being of occupational origin for compensation purposes. The OSHAct with its emphasis on safety has, if anything, worsened the manpower imbalance, as firms demand additional *safety* personnel to comply with the Act and government demands additional *safety* personnel to enforce it.

Labor faces a serious shortage of the skilled health and safety manpower that it should employ if it is to perform the essential function of serving as a check on management. In view of its limited resources, organized labor is in great need of manpower on a part-time or consulting basis, especially for the purpose of conducting training programs that can begin to provide working people with the self-help skills needed to monitor their own job environment.

Industry faces a variety of disincentives to the development and utilization of occupational health and safety manpower. Firms are reluctant to pay training costs for health and safety professionals, because the professional, once trained, can sell his services to another employer and also because, even while he is employed by the firm that paid for his training, the firm does not capture the full social value of his services. The result is that firms devote insufficient resources to the training of health and safety manpower. Moreover, the perceived interest of the firm may diverge from the public interest, as industry's safety bias illustrates. Without a change in present public policy, there is little basis for hope that the market will induce private industry to utilize properly the health and safety manpower it employs.

Small firms face special manpower problems. Often the large fixed cost of setting up a full-time in-house manpower capability is prohibitive for the smaller firm. Yet the market has proven extremely sluggish in providing the kind of ancillary consulting services that could be a useful substitute for full-time staff. Labor markets generally respond slowly to new challenges such as that provided by the OSHAct; but the uncertainty as to how seriously occupational health and safety standards will be enforced also accounts for the slowness of the market to respond.

It has already been mentioned that the activities of government have served to worsen imbalances in the supply and demand for health and safety manpower, directly through insufficient government demand for occupational *health* manpower and indirectly through enforcing incentives on employers that make them respond in a similar manner. In addition, government has also failed to promote the training and development of health and safety manpower; in fact, it has made decisions resulting in severe cutbacks in existing manpower training programs.

It is not clear that the market demand for occupational health and safety manpower has increased sufficiently to justify extensive expenditures in manpower training. Indications are that the demand for occupational health professionals has increased greatly recently, but the supply still remains relatively small. Safety professionals are very often "created" as needed by conferring the title on already employed engineers. In any event, the great uncertainty in the future of occupational health and safety in the United States is a disincentive to new entrants into the field. But if we anticipate that the demand for these professionals will increase rapidly, training them now will avoid years of delay.

In sum, then, there is a serious shortage of qualified occupational health and safety manpower in terms of what needs to be done, but this shortage is not sufficiently recognized in the market: insufficient resources are devoted to the training of such manpower; the cost-effectiveness of presently committed training resources is low; and present government policies are either inadequate or exacerbate these problems.

1.7 ACTIVITIES OF MANAGEMENT, LABOR, AND PUBLIC-INTEREST GROUPS

Recent years have seen a modest increase in the activities of labor, management, and public-interest groups regarding occupational health and safety issues. These activities include legal action, lobbying and advocacy, educational and information services, and technical assistance. In addition, there are important changes occurring in labor-management relations on these matters.

1.7.1 Legal Action

Legal action taken by public-interest groups, unions, and employer groups has played an important role in the implementation of the OSHAct. Many of the positive steps taken by OSHA have been in response to this legal action or fear of future legal action. Indeed, for all of the health standards thus far adopted through the hearing procedure, petitions or litigation have been filed by unions, public-interest groups, or employer groups. Legal action was utilized in different cases for the adoption of temporary emergency standards or for the adoption or revision of permanent standards.

Legal action has been aimed not only at the issuance of health standards but also at the policies of the Department of Labor in implementing the OSHAct. As mentioned earlier, unions are greatly concerned about the return of occupational health and safety enforcement to the states. A suit by the Industrial Union Department (AFL-CIO) and the Illinois AFL-CIO seeks to prohibit OSHA from approving plans that do not meet the statutory requirements of the OSHAct.* The suit alleges that OSHA is illegally allowing states to operate occupational health and safety programs using OSHA-approved plans that are based on unsecured promises of future state performance and that do not meet the "at least as effective as" requirement of the Act (particularly for development and enforcement of standards and for adequate funding). The suit also challenges the adequacy of the personnel qualification requirements of the states.

In general, labor unions have been reluctant to pursue litigation against individual employers because, historically, they have not had much success in the courts. The availability of such alternatives as the filing of OSHA complaints and Review Commission briefs, contract bargaining, and the threat of litigation have also served to minimize the incidence of court action. It should also be pointed out that the expense of court action, while high for labor unions that have their own legal departments, is prohibitive for unorganized workers.

The impact of court action and legal assistance on the establishment of health standards, on the testing of various provisions of the OSHAct (particularly with regard to employee rights), and on the determination of liability regarding occupational disease, has been very strong. The continuation of such legal action seems to be necessary for improving worker health and safety since it appears that the OSHAct is being implemented reluctantly by the present administration.

Legal assistance remains relatively unavailable to unorganized workers. To date, Ralph Nader's Health Research Group and the Environmental Defense Fund have provided valuable legal assistance to unions and to their members in filing complaints, petitions, and lawsuits and in assessing the legal aspects of occupational health issues. Some unions and public-interest groups are also educating union members on their legal rights under the Act, as will be discussed below. However, while the unorganized worker does benefit generally from the precedents set by union lawsuits, his individual problems rarely reach either OSHA or the courts. Indeed, the unorganized worker is unlikely to know of the existence of the OSHAct, let alone have information about the legal assistance available from public-interest groups.

1.7.2 Lobbying and Advocacy

While legal action has been crucial in forcing the implementation of the OSHAct, lobbying, testifying, advocacy, and political activities have also helped to shape the course of federal and state occupational health and safety efforts. Various groups

*The District Court for the District of Columbia dismissed the suit, upholding the challenged state plan's regulations; the case is now on appeal.

representing management, labor, the public interest, and health and safety professionals have brought pressure to bear from all directions on the federal government through legislative and administrative agency testimony and publicity efforts.

On the management side, the U.S. Chamber of Commerce has testified in support of the removal of first-instance citations and sanctions, removal of the prohibition of advance notice of inspections, and decentralization of the federal authority through state takeover. In testimony on the proposed Toxic Substances Control Act, the Manufacturing Chemists Association opposed mandatory premarket testing and clearance of new chemical substances, which would have indirect effects on workers manufacturing these products. In January 1973 the National Association of Manufacturers recommended adoption of numerous amendments to the Act to provide on-site consultation, advance notice of inspection, and the abolition of mandatory first-instance sanctions. These changes would remove provisions of these acts that industry considers ill-advised or unfair. At the same time, these changes could drastically slow or weaken enforcement efforts or divert already inadequate enforcement manpower into other activities.

The outrage of small businessmen at OSHA "harassment" did result in the passage by Congress of an amendment that would have had the effect of exempting small businesses (with less than fifteen employees) from key provisions of the OSHAct, but President Nixon vetoed the appropriations bill to which it was attached. The House again sought to exempt small businesses in the fiscal 1975 appropriations bill; but the Senate took no such action, and OSHA appropriations were ultimately provided without the exemption amendment.

The labor movement, whose lobbying efforts contributed significantly to the passage of the OSHAct in 1970, has continued to speak out forcefully regarding its subsequent implementation. In the five years since the OSHAct was passed, the labor movement, spearheaded by the AFL-CIO and its Industrial Union Department (IUD), has become increasingly disillusioned with the implementation of the Act and the government's inadequate commitment to protecting worker health. Instead of experiencing the collaboration with the Department of Labor that they had expected, AFL-CIO officials found themselves thrust into an adversary posture. As a result, the AFL-CIO has spent a great deal of its occupational health and safety effort in lobbying against a weakening of the federal health and safety program and in playing a watchdog role over the Labor Department. The IUD has applied steady pressure for recognition of workers' rights under the law; and in its statements to the Department of Labor and its frequent press conferences, the IUD has strongly opposed state takeover of occupational health and safety, on-site consultation, and the low funding and manpower provided both OSHA and NIOSH.

The AFL-CIO and the IUD have been joined by several large, politically sensitive unions on job health and safety issues. The United Steelworkers of America has vigorously opposed return of enforcement to the states. The United Mine Workers has lobbied strongly to correct "fiscally discriminatory and punitive administrative decisions" regarding benefits under the Black Lung Program. The Oil, Chemical, and Atomic Workers Union has been perhaps most active in both advocacy and lobbying on issues directly affecting workers: it has criticized the Labor Department for denying to workers the information found by OSHA inspectors, and it has led the way in forcing the adoption of carcinogen standards.

While some professional societies have hesitated to become involved in advocacy activity, others have lent their expertise both in support of and in opposition to federal efforts. The American Conference of Governmental Industrial Hygienists has urged OSHA to make distinctions between known human carcinogens and those found in animals in the establishment of standards. Both the American Society of Safety Engineers and the American Industrial Hygiene Association have expressed concern over the number and competency of occupational health and safety personnel associated with the implementation of the OSHAct and state plans under the Act. The Industrial Medical Association indicated in testimony its support of a strong federal role with a broader role for NIOSH in the standard-setting process. One of the most active societies, the American Public Health Association, has exhibited a high level of concern over workers' health and actively supported the OCAW strike of the Shell Oil Company over health and safety issues. The newly formed and well-balanced Society for Occupational and Environmental Health promises to be a valuable group whose viewpoints may influence future governmental activity in occupational and environmental health.

Management and labor have always been active in lobbying and advocacy efforts. However, management has had much greater access to expert testimony on technical, medical, or scientific matters to present before congressional committees or administrative agencies. The small but effective group of experts made available to labor by some public-interest and professional groups in the past few years remains inadequate to counterbalance industry expertise. The continued voicing of all points of view is necessary for a resolution of occupational health and safety issues. It is clear that the interests of working people have not been adequately represented by the scientific and medical communities before legislative and administrative tribunals; and since many decisions turn on technical arguments, scientific-medical expertise should be as available to labor as it is to management.

1.7.3 Educational and Information Services

As noted above, the dissemination of information about workplace hazards is crucial to improving the work environment. This includes information on how hazards can be recognized, monitored, and controlled and on how the OSHAct can be used to alleviate occupational health and safety problems.

Educational and information services have increased since the passage of the

OSHAct. Management groups, trade associations, and insurance carriers have been the main providers of information and education for employers on their responsibilities under the OSHAct. The Conference Board and the Industrial Health Foundation, for instance, disseminate the results of their research findings among their membership, which is comprised primarily of companies. And in 1971 the National Association of Manufacturers sponsored a nationwide closed-circuit teleconference on the OSHAct's requirements that reached more than 10,000 executives. Inasmuch as a large number of professionals are employed by industry, information and educational activities directed toward them also benefit the employer.

With respect to labor unions and workers generally, the information and training gap is vast and of extreme importance. (Some workers are not even told the identity of the materials they work with.) Unions and a few public-interest groups are attempting to provide information and training opportunities, and these can be categorized into the following areas: (1) education of workers concerning the law and workers' legal rights under the OSHAct; (2) dissemination of technical information on occupational disease and potential workplace hazards; (3) training in the identification and monitoring of workplace hazards and the establishment of worker self-help programs; and (4) the establishment of ongoing national or regional training centers and programs.

1.7.4 Technical Assistance

In addition to information and educational services, both management and labor are in need of technical and medical services and assistance. Employers presently receive technical consultative services on occupational health and safety problems from three groups: private consulting firms, trade associations, and insurance carriers. In general, these services are designed to minimize the financial burden on employers in their efforts to comply with the OSHAct. Large corporations are often able to obtain sufficient technical services from in-house safety, industrial hygiene, and medical personnel. Law firms are also a potential source of both legal and technical advice, but their services are usually sought only after an OSHA inspection has occurred. The smallest firms, however, do not have adequate access to either in-house or consultant expertise.

The larger insurance carriers generally provide employers with assistance in the recognition and evaluation of risks and in loss prevention and reduction. Because insurance services and policies are most responsive to the level of workmen's compensation claims, where emphasis is on injury rather than disease, extensive occupational *health* services are generally not provided by carriers. In fact, the insurance industry is opposed to changes in workmen's compensation laws that would extend coverage to occupational diseases.

Technical assistance to workers, while on the increase, is still less available than it is for management. Of seven major unions studied for this report, only one large union—the United Mine Workers*—has full-time medical expertise on its staff, and only three large unions—the United Auto Workers, United Rubber Workers, and OCAW—have full-time industrial hygiene or scientific expertise. Needless to say, these individuals can only provide a modicum of assistance to local unions and workers or to joint labor-management health and safety committees. Generally, health activist groups have not been able to provide a great deal of technical assistance to or through unions. The difficulty in providing such services lies in the workers' lack of responsibility for and control of the workplace, workers' traditional acceptance of hazardous conditions and their fears of job loss, and the inadequate resources that can be brought to bear by such groups.

To a great extent, occupational health needs and services are neglected because occupational health is not a part of the general health care that working people receive. Physicians in general practice have little interest or training in occupational medicine. Isolation of occupational from general health care partly explains the underdevelopment of occupational medicine in the United States, although some would argue that bringing both together would cause concern for occupational disease to disappear altogether. It is interesting to note that many Eastern European countries deliver *all* health care at the factory site, and indications are that occupational medicine in Poland, at least, is more developed than in the United Kingdom, where medicine is also socialized but not delivered at the workplace.

1.7.5 Labor-Management Relations

In addition to increased activities in information dissemination and technical assistance, changes are occurring in labor-management *interactions* concerning the settlement of safety and health disputes. Historically, organized labor has not emphasized health and safety in collective bargaining for several reasons. Worker concern with inflation and economic problems has often preempted health and safety considerations. The complex nature of the collective-bargaining process and the often coincident distress of a strike have limited the number of issues that can be realistically negotiated. Most importantly, the fear that jobs would be lost if strict safety and health standards were introduced, or if a worker filed a complaint with the state authorities, has often deterred workers or unions from raising this issue except in extremely serious situations. Further, it was not until 1966 that the National Labor Relations Board established the principle that health and safety issues are mandatory subjects for bargaining. With the advent of the OSHAct, labor hoped for a strong federal role, which has not so far been forthcoming. As a result, there is

*The reader is reminded that coal mining is covered by the Coal Mine Health and Safety Act of 1969 and not by the OSHAct, although research to further the objectives of both Acts is done by NIOSH.

now a trend toward contract bargaining as a major union mechanism for achieving improved health and safety conditions.

Strikes and the threat of strikes were powerful weapons in 1973 collective-bargaining negotiations. The Oil, Chemical, and Atomic Workers Union obtained its first success on safety and health provisions when management realized the seriousness of the union's threats to strike over these issues. After deciding not to come to agreement with OCAW on its contract proposals, the Shell Oil Company was faced with a four-month strike and a worldwide consumer boycott campaign. Wildcat strikes and threatened formal strikes played an important role in several contractual breakthroughs by the United Automobile Workers on health and safety issues. Although not related to master contract bargaining, a strike by members of a Textile Workers local over a specific nerve poison found in their Columbus Coated Fabrics plant resulted in a negotiated occupational health program. The United Rubber Workers and several rubber companies have agreed to sponsor jointly occupational health research at two public health schools.

Collective bargaining has the potential to go far beyond the mandates of the OSHAct by obligating employers to interact closely with workers rather than merely complying with loosely enforced and inadequate government standards. The negotiation process enables different local and industrywide needs to be met, particularly where hazards are extensive. Further, it may move the responsibility for occupational health and safety out of the sole hands of management and thus encourage the participation of workers in the process of controlling technology in the workplace. To the extent that collective bargaining changes ways of doing business with regard to health and safety on the job, indirect benefits may accrue to unorganized workers, although very slowly.

It should be noted that, not only has occupational health and safety been one of the main bargaining issues for the United Farm Workers, it has also been a main *organizing* issue. It remains to be seen whether occupational health and safety in general will give rise to new organizing efforts on the part of the industrial unions.

The discussion in this section has centered around organized group activity. Unfortunately, neither the small firm nor the unorganized worker has received the assistance that accrues to the larger firm or to organized labor. Indications are that the smaller firms and unorganized workers, which represent the majority of firms and workers in the United States, incur a disproportionately higher amount of occupational disease and injury. National policy must ultimately address their problems and provide them with the information, educational services, legal aid, and technical assistance that they both badly need.

1.8 THE FOREIGN EXPERIENCE

There are inherent difficulties in making comparisons between national systems, due in part to historical differences in social, cultural, and political values. A comparative discussion is further complicated by difficult-to-know subtleties of each system. For example, although a country may have very protective legislation regarding worker health and safety, it is necessary to know how extensively the legislation is enforced, what the qualifications and roles of the inspectors are, and how effective compliance is at the local level. The effectiveness of policies, then, is difficult to measure. Yet there are clear and significant differences in legislation and enforcement and in patterns of health and safety concern and activity in the European countries that are worthy of consideration by U.S. policy-makers.

No attempt is made here to represent the situation in Europe as ideal. There are problems in each of these countries, both in terms of the amount of job injury and disease and in terms of systemic problems preventing the further reduction of injury and disease. (For example, lack of a reliable scientific data base and lack of services and protection in small firms are problems afflicting many countries.) My purpose is to draw the best examples from European occupational health and safety approaches to show that there are ways in which other productivity-oriented economies pay more attention to worker protection, and especially to occupational health problems, than does the United States.

In many European countries, the heads of the federal agencies and departments responsible for occupational health and safety have considerably more technical and medical expertise than is the case in the United States. The European counterpart of the National Institute for Occupational Safety and Health is at a much higher organizational level within the government, or else is an independent organization. Furthermore, European institutes have greater access to competent researchers, and the research they do is more relevant to critical occupational health and safety problems.

European governments consult more with the employer, and first-instance citations are infrequent. Warning and improvement notices are the most-used enforcement mechanism. Standards are generally regarded as guidelines, and the inspector has wide discretionary power. This flexible approach to enforcement is possible *only* because it stands at the peak of a well-developed infrastructure of personnel, attitudes, values, and intraplant health and safety mechanisms. In many European nations, for instance, inspectors are more specialized and better trained, and physicians are very often part of the inspection system. Further, there are legal requirements for occupational health services in the workplace or, in the case of small firms, for participation in a joint medical service.

The differences in the way Western European countries address occupational health and safety problems suggest the following reforms for the United States:

- The chief administrators of NIOSH and OSHA should be competent health and safety professionals as well as good administrators.
- Mechanisms need to be encouraged which facilitate closer cooperation between NIOSH and OSHA in policy formulation.

- NIOSH should be either located higher within the HEW hierarchy, combined with the National Institute of Environmental Health Sciences, or removed entirely from HEW.
- Occupational health and safety research needs serious reconsideration in order to make research more relevant to critical occupational health and safety problems.
- The government should provide more information to employers, employees, and the public, but not in the form of consultative services that would detract from the presently weak enforcement capability.
- More emphasis should be placed on enforcement of health standards and the general duty clause to improve health and safety conditions.
- Better training of physicians and other health professionals in the recognition of occupational diseases and in the importance of prevention strategies should be encouraged by means of federal support.
- More and better training of inspectors, especially in occupational health, is desperately needed.
- Serious consideration should be given to mandating in-house occupational health services or, in the case of small firms, participation in a joint medical service.
- Occupational disease needs to be fully reported, through a national registry, by employers and private physicians.
- Occupational disease needs to be fully recognized for compensation purposes (this should include *any sickness caused by exposure to toxic materials).*
- Serious consideration should be given to the mandating of joint occupational health and safety committees for firms, whether organized or not.
- More equal representation of management and labor should be attempted in the staffing of the government and private agencies responsible for health and safety.
- The problems of mental health, ergonomic factors, synergistic effects, and carcinogenesis should receive more attention.
- Mechanisms for the international exchange of scientific-medical information and personnel should be developed further and fully utilized.
- Attempts at the adoption of international standards should receive full attention.

Many of these reforms are possible within the present U.S. system. However, they require a firm national commitment to improve job health and safety—and such a commitment does not appear to be forthcoming as of this writing.

It should be obvious to the reader that more effective international exchange of scientific-medical information and expertise related to occupational health and safety is a necessity. The amount of research that needs to be done—in both basic science and epidemiology—is so vast that coordination on an international scale is required. It should not have been necessary for the United States to rediscover "black lung" or asbestosis. Foreign information officially unrecognized is foreign information ignored. It has been noted earlier that environmental factors may be responsible for as much as 90% of all human cancers; the striking variation in cancer rates in different areas of the world provides a challenge and an opportunity to identify the environmental factors that must be responsible.

1.9 THE DESIGN OF POLICY

I have sought throughout this work to demonstrate that occupational health and safety problems will not yield significantly to simplistic strategies based on single approaches. "Enforcing the law," as is argued by some, will not suffice; nor will "research" to determine "safe levels" of exposure or "hypersusceptibility."* Rather, the design of policy to bring about improvements in occupational environments must rely upon four sets of policy instruments: (1) the law, (2) market incentives, (3) the generation, dissemination, and utilization of knowledge, and (4) the development of personnel in the various professions, in labor unions, in management, and in government with the requisite knowledge of the issues.

While the policy instruments cited above must be further developed and aggressively wielded, they will nonetheless be insufficient even when working in concert. In addition, workers themselves must assume a greater role in self-monitoring the quality of their own work environments. It is true that both law and tradition place ultimate responsibility for job health and safety on the employer. But full reliance on law or market incentives to encourage fulfillment of this obligation is impossible in any practical sense. While some progress can be expected through the courts and hearings—and in laboratories and classrooms—much more progress is likely if workers and their representatives are enabled to function responsibly and effectively in fruitful interaction with, and as a necessary check on, employers.

This latter conclusion is based on the fact that too little is now known—or even, at least in the short run, knowable—about health hazards for scientific proof and definitive standards to be widely forthcoming. Rather, most decisions regarding workplace environmental quality must be worked out in other forums on the basis of existing, often underutilized knowledge. In part these forums will be the courts or administrative agencies, and in part they will be the arenas of politics and public opinion. In the main, however, these issues need to be addressed in the day-to-day relationship between informed employers and employees—and in a deepening of the trend toward placing a higher priority on job environment issues in collective bargaining.

Governmental reforms suggested by European experience were presented in the previous section. In addition, the indiscriminate return of authority for occupational health and safety to the states should be curtailed.

Recommendations for the private sector fall into two general categories:

*Contrary to popular belief, the best evidence does not support the thesis that the worker who is "accident-prone" or "hypersusceptible" to disease accounts for a significant amount of job-related injury or disease. However, the continuing adherence to this belief prevents more progressive and innovative measures from being taken to reduce job illness and injury.

*Activities that can have a direct and early impact on occupational health and safety problems**

1. The education of workers
   - by employers
   - by universities with worker education programs or by other institutes and centers offering long- and short-term training in occupational health and safety, collective bargaining, and industrial relations
   - by health activist and public-interest groups to train workers or their representatives in their legal rights under the OSHAct, in recognizing occupational hazards, and in utilizing self-help techniques
2. The support of information dissemination activities and the development of new information sources
   - by encouraging the transfer of technical and medical information between employer and employees
   - by encouraging efforts to develop and disseminate readable materials for small employers and unorganized workers
   - by supporting joint development of audiovisual materials for use by labor, management, and public-interest groups
   - by supporting information clearinghouses of a neutral nature or designed to meet the information needs of particular groups
   - by establishing mechanisms to ensure the rapid international transfer of information
3. Cooperative efforts in the rapid development and transfer of new technology designed to assure safer and healthier working conditions
4. Encouragement of professional manpower development
   - by maintaining the viability of occupational health and safety programs in universities where they already exist but where support has been withdrawn by the government
   - by establishing or supporting college programs for training multidisciplinary professionals in combined occupational health and safety curricula
   - by supporting the development and teaching of specific occupational health and safety courses in the medical, nursing, public health, law, and business schools
5. Increased utilization of occupational health and safety manpower by employers and unions
6. The support of individuals or groups within the professions who can awaken professional and public-interest groups to these problems
7. The sponsoring of conferences on occupational health and safety (there is a need for such conferences within various interest groups—labor, management, profession-

*The mechanisms for supporting these activities can include direct funding, the establishment of academic chairs in occupational health and safety, scholarship aid for students, and the support of internships in labor unions and industry.

al groups, etc.–and also for a variety of conferences to which all interested parties are invited)
8. The support of selective research to demonstrate the seriousness of combined effects of hazards (synergism) and the effects of stress as a cocausative factor in occupational disease
9. The support of health activist and environmental and public-interest groups in the use of legal action to force implementation of the OSHAct
10. The support of the above-mentioned activities on an immediate but short-range basis while serious policy questions are examined

Policy analysis of occupational and environmental safety and health issues to be done over the longer term *
1. The monitoring of public policy and the performance of government agencies, including governmentally produced data on the nature and extent of job health and safety problems and on efforts to alleviate these problems
2. Mechanisms for linking approaches to problems of the general environment with those of the work environment, and problems of occupational health with those of general health-care delivery (the design of toxic substances control legislation and national health insurance are important considerations)
3. Alternative institutional frameworks for implementing the OSHAct, delivering occupational health and safety services, educating employers and employees, and acquiring and disseminating information
4. Examination of the changing disease and injury profile in the United States, and the setting of priorities for curative or preventive action
5. Research policy: recognizing that there is a serious lack of a firm data base and that all the research that needs to be done cannot be done, it is important that the research that is conducted be carefully selected according to a conscious policy
6. Development of alternative technologies to decrease occupational exposure to harmful materials (including studies to test the feasibility of using closed systems for chemical processing and of using less harmful substitute materials)
7. Assessing the potential of job and workplace redesign for reducing workplace injury and disease (both ergonomic and motivational factors should be included in such studies)
8. Development of simple measurement and control techniques which can be used in the workplace by workers, their representatives, and first-line or middle-management personnel
9. Manpower development: the appropriate numbers, training, distribution, and mix of occupational health and safety professionals should be examined along with policies that determine the supply and demand for such personnel

*Such policy analysis should be conducted at universities, research institutes, and independent centers. It should be multidisciplinary and should engage health and safety professionals, lawyers, economists, engineers, scientists, and persons with business, managerial, and labor experience.

10. Economic policy: the impact of occupational health and safety legislation on GNP, on the mix of large and small firms in various industries, on employment, and on international competition should be assessed
11. Examination of foreign experience in occupational health and safety and of its relevance to the United States
12. Formulation and assessment of alternative coordinated policies for funding activity by private philanthropy and by other institutional elements.

2 Introduction

2.1 THE WORKPLACE–A MICROCOSM OF NATIONAL CONFLICT

In our system of government, the passing of federal legislation was the natural response to the expanding problems in occupational health and safety. However, it must be recognized that here, as with the National Environmental Policy Act or the proposed Toxic Substances Control Act, we are faced with the inherent difficulties encountered when the law is used as the *predominant* mechanism for the social control of science and technology. We have been recently reminded that ours is a government of laws and not of men. However, the law, and especially the Occupational Safety and Health Act, cannot be successfully implemented if we continue to ignore the fundamental conflicts and tensions which exist between various groups of people and between various institutions in our society.

These tensions and frustrations deserve special attention if we are to establish coordinated mechanisms and policies to solve these very difficult problems. One might be tempted to believe that the approach to reducing occupational injury and disease need not be any more complicated than the approach to controlling polio. I shall attempt to show why this is not so and to explain why the issue of health and safety in the work environment has raised the tempers of industry, government, and labor alike.

There is no philosophical reason why society should not choose to be as protective of man in his working environment as it is with regard to the food he eats or the nonwork environment to which he is exposed. One of the reasons why the issue of health and safety on the job appears to be different is that the mechanisms for allocating (internalizing) the costs and assessing the benefits of job health and safety are undeveloped or not well understood.

The remainder of this section is devoted to a discussion of five kinds of conflicts and tensions which characterize the work environment: (1) conflicts caused simply by different self-interests, such as are characteristic of management-labor relations on many issues; (2) conflicts which arise because of a lack of knowledge and the absence of a sufficient data base; (3) conflicts due to differences in the perception of what is just or fair; (4) conflicts which have their roots in the nonconnectedness or dispersion of important institutions in our society; and (5) conflicts due to the different languages which are used and the different problems which are dealt with by various professional interests in society.

2.1.1 Conflicts Reflecting Different Self-Interests

In the United States, management and labor characteristically settle their disputes in an adversary manner, either formally through collective bargaining or informally through grievance procedures, discussions, and negotiations–often conducted on an individual basis. Whatever the mechanisms, management (and very often labor) has a certain "package" of economic benefits in mind, consisting of wages plus fringe benefits with which it is willing to part.

Leaving aside for the moment the effect of workers' health and safety on productivity, the issue has been very much one of the self-interests of the worker, to be balanced against his wages, pension, etc. That is, the worker could traditionally demand safer working conditions only if he were willing to trade off some of the other benefits in his total package. Partly because job health and safety were difficult to quantify in monetary terms, they received the lowest priority in worker demands. Even in dangerous jobs, the introduction of hazardous duty pay was not based on any assessment of potential social harm, but rather on what the "market" would bear. Thus the costs of health and safety in the workplace were not internalized in either the price of the employer's output or in the wages paid to employees.

The workmen's compensation system did attempt to "internalize" costs after the fact, i.e., after the harm had been done. But, as will be shown in Chapter 8, the system has been largely inadequate and has failed to cover occupational disease. Nonetheless, the shifting of costs from employee to employer gave rise to the adage "safety pays." The complete adage might better be put, "safety pays when you are forced to bear the cost of an unsafe workplace." Job health and safety becomes the self-interest of both management and labor when reducing accidents on the job can reduce insurance premiums.

Workmen's compensation liability also served to fortify the business assumption that management, not labor, is responsible for, and hence should have control of, the workplace. This turned out to be a mixed blessing. Work alienation and workers' desire to control their own workplace environment provide a constant source of conflict. Indeed, a major union—the Oil, Chemical, and Atomic Workers Union—recently struck over the issue of control of health and safety issues in the workplace.

Where the harm is not recognized by the workmen's compensation system—as is the case for most occupational disease—the self-interests of management and labor are competitive. The issue becomes one of increased productivity versus improved workers' health, safety, and job satisfaction. In the long run, it is probably true that safety and health do pay, i.e., higher productivity will result from healthier and safer workplaces. However, two characteristics of the firm militate against this long-run choice being made: (1) managers are judged in the short run, and (2) the management of smaller firms does not devote the minimum "critical level" of either expertise or effort to solving the technical problems of health and safety. Thus, the operationally perceived self-interests of management and labor differ and contribute to their adversary postures.

When *perceived* self-interests differ, the solutions offered for the resolution of conflict sometimes only postpone the tensions. Increasing the wages of workers alienated from their jobs only temporarily solves the conflict. The problems of job health and safety often stem from a misperception of issues by both labor and management.

When *actual* self-interests of management and labor differ, conflict naturally ensues, but compromise and negotiation can sometimes lead to real solutions.

The Occupational Safety and Health Act of 1970 is slowly raising the consciousness of both management and labor. The costs of compliance with the new health and safety standards imposed on management are causing an internalization of costs, in much the same way that workmen's compensation legislation did. The critical difference is that emphasis is now on *prevention* rather than treatment or compensation, and management and labor now argue about what are safe and healthy work conditions. These new conflicts center around issues which, if resolved, are more likely to improve workers' health and safety (and productivity in the long run) than resolution of the conflict over who will pay for the harm.

The adversary posture in which labor and management find themselves gives rise to accusations that "labor is using health and safety issues to extort higher wages and benefits from management" and that "management seeks every way to free-ride on workers' health and safety." While examples can be found to support both these charges in specific instances, the adversary nature of the conflict between articulated self-interests obscures other conflicts whose resolution or appreciation may lead more effectively to improvement of the work environment. These other conflicts are discussed in the remainder of this section.

2.1.2 Conflicts Caused by a Lack of Knowledge

A good example of a conflict caused by lack of knowledge is the controversy over the adoption of standards for 14 carcinogens by OSHA, the Occupational Safety and Health Administration in the Department of Labor. Establishing "safe levels" for toxic materials on the data presently available is difficult enough; the chronic nature of many diseases, including cancer, compounds the problem by making the unequivocal establishment of causation difficult. In the case of cancer, in particular, there is still a raging controversy over whether there is *any* safe level for exposure to a carcinogen. The lack of a firm data base with regard to exposure to toxic materials explains in part why only three new health standards—for asbestos, for fourteen carcinogens, and for vinyl chloride—have been adopted by the Department of Labor.

Independent scientists taking the most protective view are automatically viewed as "quarterbacking for labor," while company scientists are expected to take the least protective view. In fact, the division of views does fall roughly this way, but it is also true that a broad range of opinions are represented in each camp.

There exist inherent difficulties in using law as the predominant mechanism for the social control of science and technology. One crucial problem has to do with the nature of the burden of proof required to make legal-political decisions. The standard of proof required for acceptance of a scientific thesis is much higher than that required by the law. When it comes to safeguarding rights generally under the law, a "scintilla of evidence" may justify legal sanctions, control, and even the establishment of liability. This pervasive principle in our system of jurisprudence is not fully understood by "pure scientists," who sometimes view the law as "unscientific." The law attempts to make the best decisions on the information that exists. Absolute certainty or even consensus is not always required. What is required is an opportunity for all the evidence to come in and for all parties to be heard. Contradictory or poorly substantiated evidence results in the continuation of controversy long after a standard has been set or a "decision" made.

Another area in which lack of knowledge is crucial is in the area of heart disease. In the recently released report by HEW entitled *Work in America*,[1]* it is reported that

> research findings suggest that [diet, exercise, medical care, and genetic inheritance] may account for only about 25% of the risk factors in heart disease, the major cause of death. That is, if cholesterol, blood pressure, smoking, glucose level, serum uric acid, and so forth, were perfectly controlled, only about one-fourth of coronary heart disease could be controlled. Although research on this problem has not led to conclusive answers, it appears that [the] work role, work conditions, and other social factors may contribute heavily to this "unexplained" 75% of risk factors.

An attempt by the National Commission on Workmen's Compensation (established by the OSHAct) to have heart disease recognized as a work-related compensable disease failed because the compensation system was not meant to be a proxy for social security. I submit that if we knew to what extent and in what cases heart disease is caused directly by the work environment, the outcome would have been different.

Finally, with regard to the information problem, a crucial policy question arises: Should we put our efforts into trying to improve the data base in cases where new technologies and materials entering the workplace will cause more new problems than we can ever hope to solve? Specifically with regard to toxic materials, if 500 new chemicals enter the work environment annually, and if the number of combinations of chemical and physical hazards (compounded by the element of stress) increases factorially, would it not be better to put our efforts into instituting engineering controls for all substances—not because all substances are harmful, but because it would be more economical to redesign chemical processes than to do research on all materials? This question needs to be addressed.

2.1.3 Conflicts Due to Differences in the Perception of What Is Just or Fair

The raging controversies over the Delaney Amendment concerning the testing of food additives or the Toxic Substances Control Act hinge on the following ques-

*References can be found at the back of the chapter.

tion: Should we prohibit use or consumption of a material until it is proven safe, or should we allow its use until it is proven harmful? The main issue becomes one of where the burden of proof should lie—and honest men can differ on this question.

An analysis of the risk situation in the work environment indicates a difference from the situation in the general environment. A failure to appreciate this difference accounts for much conflict which is incorrectly interpreted as resulting from a conflict of interest between employer and worker.

If society collectively decides to tolerate a certain sulfur content in its heating fuels to save costs and is willing to suffer an increase in respiratory disease as a result, there is a certain equity which gives us comfort. All citizens save money, and the risk of disease (within an urban area) is randomly distributed, more or less. But if the society decides to use asbestos wherever it will save money, and a selected group of asbestos workers pay for it with their lives and health, there is a strong argument that the situation is not equitable, even if more lives are saved than lost. For example, by the use of asbestos brake linings more lives may be saved for the public on the highways than are lost by workers in the plant. The nonrandom selection of those who bear the risk, however, deserves special attention. This distinction is often lost. Whether it is fair for the lives of selected employees to be sacrificed for the betterment of society and whether employees can rationally assume the risk of hazardous employment are questions that need to be addressed.

Finally, honest men will differ and argue on how much control a government should exercise to protect its citizens. Should seat belts be mandatory? Should workers themselves decide on what a particular risk on the job is worth to them? One could argue that, ideally, individuals should make these decisions—but the underlying assumptions are that all pertinent information is known, that the labor market is efficient, and that man is economically rational. Unfortunately, the situation is hardly that in most instances.

2.1.4 Conflicts Due to the Unconnectedness of Important Institutions and Agencies

The fourth kind of conflict comes from the fact that the various institutions, forces, and mechanisms in our technological society are not connected very well. For example, people concerned with problems in the *general* environment are not interacting or connected strongly with people concerned with the *work* environment. The Environmental Protection Agency, for example, generally functions in a well-defined area that is quite different and apart from the Occupational Safety and Health Administration, although recently they have been forced to coordinate some of their policies—as, for example, in the cases of pesticide and noise standards.

A vivid example of what can happen when general environmental concerns are not connected with those of the work environment occurred in the case of pesticides. The general environmentalists succeeded in eliminating the use of DDT as a pesticide because it was of some harm to wildlife and to the consumer. Their efforts resulted in the large-scale use of a substitute, parathion, which deteriorates in the environment much faster than DDT. Food then arrives at the consumer's table as a safer item. However, parathion is very much more harmful than DDT for the people who must handle it. Because mechanisms did not exist to coordinate the resolution of problems in the work environment with those of the general environment, the work environment may have been made more dangerous.

A second "unconnection" has to do with something of which U.S. health officials are well aware–the fact that *preventive* medicine is not tied to the *treatment* establishment. In fact, if you talk to people involved with medical-care delivery, many of them will define preventive medicine as an early detection of disease, rather than as an elimination of those hazards that can ultimately cause disease. And the line is not a thin one–the distinction is very important, especially with diseases whose progress is difficult to reverse after onset, such as cancer. The rising interest in Health Maintenance Organizations (such as the Kaiser-Permanente Plan) reflects, at last, a realization that preventive medicine needs a different kind of mechanism for its institutionalization.

A third important unconnection is that between the Occupational Health and Safety Administration (OSHA) in the Department of Labor and the National Institute for Occupational Safety and Health (NIOSH) in HEW. Both agencies are concerned with occupational health and safety, but the enforcement function and priority setting in OSHA are barely connected to the research and manpower development mandated for NIOSH. In fact, OSHA speaks of an increased health orientation, while the Office of Management and Budget has all but eliminated NIOSH's budget for occupational health manpower development. As we shall see, we pay a heavy price for this unconnection.

2.1.5 Conflicts Due to Unidisciplinary Approaches to Problems

We come finally to the kind of conflict that is caused by the different languages used and different problem areas dealt with by the various professional interests in the society. The safety professional does not talk the same language as the health professional although, in fact, it is futile to draw a distinction between safety and health in many instances. Consider the example of noise. Noise has been recognized as a safety problem by safety professionals, and they have sought to reduce the noise level of both the workplace and the general environment in order to minimize hearing loss. But the issue of noise as a stressor and a cocausative factor of disease is of little, if any, interest to the safety professional. Perhaps it is of little interest to the toxicologist as well. Each professional chooses not to deal with some aspects of the total problem, and words like "noise" mean different things to different professionals.

General practitioners know little of occupational medicine. The industrial hygienist views preventive medicine differently than the occupational physician. And finally, the personnel director may not think it wise for workers to talk to the plant doctor or industrial hygienist because management-labor relations are his "problem area."

The problems in the work environment are complex and demand a multidisciplinary approach, both long- and short-term, for their solution.

2.2 TOWARD A DEFINITION OF OCCUPATIONAL HEALTH AND SAFETY

Since the workplace does represent a microcosm of national conflict, the problem of occupational health and safety must be understood beyond in a broader sense than simply the elimination of disease and injury related to the work environment. The objectives of occupational health have been defined more broadly by the International Labour Organization and the World Health Organization as

> the promotion and maintenance of the highest degree of physical, mental, and social well-being of workers in all occupations; the prevention among workers of departures from health caused by their working conditions; the protection of workers in their employment from risks resulting from factors adverse to health; the placing and maintenance of the worker in an occupational environment adapted to his physiological and psychological condition.[2]

In the broadest sense, then, occupational health and safety must encompass not only the problems of *preventing* physical disease and injury, stress, psychological disorders, and work alienation, but also the *promotion* of mental well-being and job satisfaction.

The professionals, institutions, and laws related to job health in our society have been historically quite separated from those related to job safety. For example, the prevention and the treatment of disease and injury have been unconnected. The physical health establishment has been isolated from that of mental health, and both concentrate on problems separated from the concerns of the social psychologist. While *acute* disease and injury are prevalent in the workplace and their treatment has become institutionalized, work-related *chronic* health problems such as respiratory and heart disease, as well as exposure to carcinogens or other toxic substances, present particularly difficult problems which require different kinds of solutions. In a sense, then, the occupational health and safety "field" is not a field at all operationally, but rather a collection of activities that are at present unconnected in any meaningful way.

Implicit in a broader definition is the observation that occupational health and safety might be better referred to as "occupational *politics*, health, and safety" because many of the difficulties cannot be explained, and many of the solutions cannot be found, in technological and medical practices alone.

In this treatise I hope to review the problems of the work environment in the unified and connected way which is a prerequisite for any real understanding or progress in this very troubled area.

2.3 THE REASONS FOR THE RECENT INTEREST IN OCCUPATIONAL HEALTH AND SAFETY

While the subject of occupational health and safety is not new, a tremendous rebirth of interest in the area has been stimulated by the passing of recent federal legislation—the Coal Mine Health and Safety Act of 1969 and the Occupational Safety and Health Act of 1970. This legislation has directly caused increased activity; but the legislation was itself a response to a concern for occupational health and safety issues brought about by several increasingly important factors.

Probably the most important single factor was the observed increase in the industrial accident rate, which rose nearly 29% from 1961 to 1970.[3] The National Safety Council presently estimates 14,000 deaths due to accidents on the job and 2.2 million disabling injuries annually. The latter statistic is probably underestimated by a factor of five. For the most part, these statistics do not reflect occupational disease. The safety record of American industry did not go unnoticed by labor, by the insurance companies, or by industry itself, and Congress was moved into action in the late 1960s.

However, statistics alone rarely cause congressional action. In the coal mining industry, the frequent occurrence of mining disasters involving a large number of sudden deaths, and the dangerous and unhealthy state of the mines generally, provided the driving force for the passage of the Coal Mine Health and Safety Act of 1969. In one West Virginia county there were eleven major mine disasters in twenty years.[4] The final straw came with the Farmington, West Virginia, coal mine explosion on November 20, 1968, in which 78 miners died.[5] The coal mine legislation and the congressional momentum thus generated helped pave the way for job health and safety legislation applicable to the rest of American industry, including agriculture. The Occupational Safety and Health Act (OSHAct) was passed one year later in 1970.

While accidents on the job produced the initial pressure for federal action, other important concerns, seemingly unconnected with job safety, provided additional persuasive power. These other concerns are steadily increasing in their importance relative to the traditional concern for accidents on the job.

The most important of these concerns, which has caused the most vocal and bitter management-labor-government disputes, is the problem of occupational disease. Both new evidence and newly acknowledged evidence for the occupational origin of much disease, riding on the crest of an increased environmental and health consciousness on the part of the public, have thrust the issue of job health into the forefront of activity and debate. Black lung, asbestosis, asbestos-caused cancer, and beryllium disease are the most familiar examples of occupational illness. The problems appear to be much more far-reaching, however. While statistical evidence for occupational disease is difficult to obtain, HEW estimates 390,000 new cases of occupational disease each year, and as many as 100,000 deaths.[6]

The slowness of the medical-scientific and political communities to recognize the occupational origin of much disease was paralleled in the state workmen's compensation system. An increased consciousness of the inequalities and inadequacies of the system with regard to injury slowly brought about a reexamination of the system which included attempts (though meager) to raise the issues of compensability for occupational disease and the legal requirements for a determination of the occupational origin of disease.

The changing character of the workforce has also contributed to increasing the importance of job health and safety issues. Rising wages in the manufacturing industries and the consequent rethinking of worker goals has slowly elevated the priority of job health and safety in contract negotiations and other union activities. And the closer examination of the American worker by some industries, in an effort to understand the relative decrease in American industrial productivity, has also brought about some increase in job health and safety consciousness.

Finally, impetus for federal legislation resulted from an awareness that state programs in occupational health and safety were unable to handle the rapidly expanding problems with which they were confronted. As a result of these numerous concerns, then, the passing of far-reaching federal legislation was inevitable.

2.4 A BRIEF HISTORY OF HEALTH AND SAFETY LEGISLATION

2.4.1 State Occupational Safety and Health Legislation

General awareness of the need for legislation protecting the health and safety of the worker first arose in the United States during the nation's great period of industrial expansion in the last third of the nineteenth century, as the development of new industries and methods of production brought radical changes and new hazards to the working environment. The Massachusetts Department of Factory Inspection, created in 1867, was the first state agency to deal with job safety, while in 1877 Massachusetts also enacted the first worker safety law, which required the guarding of hazardous spinning machinery in textile plants.[7] Other states gradually followed this lead, and by 1900 most of the heavily industrialized states had at least some rudimentary form of legislation requiring employers to reduce or eliminate certain workplace hazards.[8]

All through this initial period, however, no legislation was developed for helping workers once they had been injured in job-related accidents. Such workers could seek compensation only by suing their employers for recovery of damages under common-law rules of liability, which required them to prove that their injuries were due specifically to employer negligence. Of course, their chances of success under this slow and costly system were limited—the more so as the courts were biased in favor of employers,[9] who were able to resort to three common-law defenses: (1) the doctrine of contributory negligence, used when it could be shown that an accident was in part the result of a worker's own carelessness; (2) the fellow servant rule, used in those instances in which a fellow employee's negligence had contributed to an accident; (3) the theory of assumption of risk, based on the premise that in accepting a particular job, a worker also implicitly accepts whatever risks are connected with it.[10]

Growing dissatisfaction with the unfairness and impracticality of this system eventually led to efforts at reform,* and in 1902 Maryland passed an act providing for a cooperative accident insurance fund, which represented the first legislation in the United States embodying the compensation principle in any degree. Even though the law applied only to fatal accidents, and provided meager benefits at that, it was soon declared unconstitutional by the courts on the grounds that the procedure for deciding compensation denied the employer the right of trial by jury.[12] Then in 1908, under the prodding of President Theodore Roosevelt, Congress passed a compensation act covering certain federal employees, thus providing a new impetus to state efforts in the same direction. In 1910 New York ventured to adopt a wide-ranging workmen's compensation act which was compulsory for certain especially hazardous jobs and optional for others, but it too was declared unconstitutional by the courts on the grounds of deprivation of property without due process of law.[13] This last decision set the stage for a major confrontation between the proponents and opponents of compensation legislation. In the words of the *Compendium* of the National Commission on State Workmen's Compensation Laws,

> This decision was met with an explosion of criticism from all sides. Theodore Roosevelt was so angry that he openly advocated the passage of laws which would permit the recall of judicial decisions. While even the supporters of compensation legislation considered this measure too extreme, fear of its passage prompted many conservatives to support compensation legislation by more traditional means.[14]

The result was that many state courts began adopting a more liberal attitude toward compensation. In 1911 New Jersey passed the first state workmen's compensation law to be upheld, and other states quickly followed this lead. By 1920 a total of

*However, there is good evidence that the probability of high payments under the common law was growing over time. A recently completed study indicates that defendants were more and more regularly losing negligence suits during the period from 1875 to 1905.[11] Even though the study focuses on negligence suits in general, work-related negligence suits were included. Litigation was growing rapidly in the years after 1875, and more plaintiffs were recovering damages. These two facts indicate that employers were increasingly vulnerable, and it was therefore in their economic interest to lobby for the passage of compensation legislation which established payments.

forty states, along with the District of Columbia and the territories of Alaska and Hawaii, had passed workmen's compensation laws, while in 1948 Mississippi became the forty-eighth state to enact such legislation.[15]

One important result of the initial opposition of the courts to compensation was that the fear of unconstitutionality caused state legislatures to limit the scope of their legislation.[16] Thus, the coverage provided by the new laws that appeared after 1911 tended to be fragmentary and elective rather than comprehensive and compulsory. In the words of Abner Brodie, writing in the Wisconsin Law Review,

> Coverage of the early laws was limited; even when elective, most acts applied only to specified hazardous industries. None covered all classes of employees. Agricultural workers, domestic help, and casual workers were most commonly excluded. Only a few acts applied to public employment. In general, compensation laws limited indemnity benefits to maximum total amounts, even for permanent disability or death. Cash benefits were usually stated as a percent of wages at the time of injury, 50 percent being the most common, although a few acts provided for about two-thirds of wages, subject to statutory maximum compensation ranging from $10 weekly in several states up to $15. Several states made no provision at all for medical benefits. Where provided they were limited in duration or amount or both.[17]

Another limitation of these early laws was that none of them expressly covered occupational diseases: "Statutes which provided compensation for 'injury' were frequently interpreted to include disability from disease, but those acts which limited compensability to 'injury by accident' excluded occupational disease."[18]

Once efforts in the area of workmen's compensation bore fruit, public interest and concern over the general issue of worker safety declined markedly. There followed a long period of relative inactivity which lasted into the 1960s, and "though some state agencies did good, enduring work in the fields of boiler and elevator safety, the general picture was discouraging."[19] As *Bitter Wages*, the 1973 Nader Study Group Report on Disease and Injury on the Job, puts it,

> A hard look at the various laws designed to promote job safety and the compensation of workers disabled at work would have uncovered numerous shortcomings. Nonetheless, the problems of occupational safety and health languished in the limbo of neglect for more than half a century. The muckraking era in American journalism, essentially the work of intellectual reformers, failed to generate a grass-roots constituency and passed with the advent of World War I. The assumption prevailed that the new workmen's compensation laws were causing accident rates to decline. Nobody paid heed to the inadequacy of existing state health and safety laws and their non-enforcement. Nor was there any attempt to arouse among the workers themselves an awareness of the full dimensions of the health and safety issue, nor a sense of responsibility for their own self-protection and survival. The labor movement was gradually building political muscle, but it did not press for the reduction of the frequency and severity of work accidents and diseases.[20]

The result of this long-standing neglect was that even by the end of the 1960s occupational safety and health legislation in most states remained inadequate. The situation with regard to workmen's compensation was actually worse than ever, since benefit levels generally had not kept pace with increasing wage levels and the rising cost of living, while approximately 20% of all American workers lacked any compensation coverage whatsoever.[21] State job safety laws tended to be poorly enforced and also left many workers, particularly those engaged in agriculture, unprotected.[22]

State occupational health legislation was particularly inadequate, and state occupational health programs were generally weak, understaffed, and poorly coordinated and organized. In a review of state occupational health legislation as of December 31, 1968, Andrew Mosey and Louise Ede of BOSH (the predecessor of NIOSH) found that although all states were concerned about protecting the health of their workers, there was "no uniformity in the laws and rules and regulations issued to achieve this end, either in language or scope of legislation or in currency of standards, resulting in an inevitable disparity in effectiveness of protection afforded the worker."[23] Some of Mosey and Ede's specific findings were the following:

1. Although the fifty states plus the District of Columbia and Puerto Rico had statutory authority (express or implied) to establish an occupational health entity, only twenty had done so by statute.
2. From one to five state agencies had the power to advise, inspect, investigate, adopt, promulgate, and/or enforce in the area of occupational health; the right to enter for purposes of inspection or investigation was usually delegated to labor departments, industrial commissions, and/or state mining inspectors, "but not usually to health departments except in the area of sanitation or other 'public health' problems."
3. While reporting of occupational disease was required in all states, the District of Columbia, and Puerto Rico, either by statute or by rules and regulations for health control or workmen's compensation purposes, there was no uniformity in the system of reporting: some states made this the responsibility of the individual physician by requiring the reporting of specified occupational diseases, while other states placed the responsibility on the employer, tying this in with certain other reporting required by the workmen's compensation law.
4. Health services were required by statute in twenty-six states, with thirteen requiring periodic physical examinations and twenty requiring the provision of first-aid facilities.
5. While nationally recognized governmental or nongovernmental standards were referenced either in statutes or in rules and regulations in all but four states, "these references generally were not up-to-date, referring to standards from one to more than twenty years old; in eight of the states the referenced standards were suggested as guides."[24]

Bitter Wages provides additional detail about the weaknesses of state occupational health efforts through the end of the 1960s. In particular, it notes the fractionalization of authority over occupational health among various departments within state governments, the scarcity of health personnel in state occupational health programs, and the poor record of the states with regard to research into, and control of, industrial diseases.[25] This general pattern of neglect on the part of the states eventually led to federal action and the passage of the Occupational Safety and Health Act of 1970.

2.4.2 Federal Occupational Safety and Health Legislation Prior to the OSHAct

Federal involvement in the area of occupational safety and health began in the 1890s, with the passage of rudimentary legislation governing coal mine and railroad safety.[26] In 1910 the Federal Bureau of Mines was established in the Department of the Interior to deal with mine safety in general, but initially its powers were severely limited. The bureau first received authority to conduct inspections and investigations in 1941, but it was not allowed to promulgate and enforce safety and health standards until 1952.[27]

In 1914 the Office of Industrial Hygiene and Sanitation was organized within the Public Health Service's Division of Scientific Research.[28] * Two years later Congress passed the Federal Employees Compensation Act, which provided compensation for federal employees injured while on the job.[30] Under the Walsh-Healey Public Contracts Act of 1936, employers with government contracts exceeding $10,000 were required to comply with certain safety and health standards promulgated under the Act,[31] although the law's effectiveness was undermined by a lack of enforcement.[32]

During the 1930s and 1940s both the federal government and the states reduced their occupational safety and health efforts as the country faced more immediate problems:**

> Both the depression and World War II interfered with overall activity in the health and safety field. Labor was concerned with wages and establishment of union rights; management was in no position to expend time or money on non-profit-making endeavors during the lean years. Winning the war took precedence over other considerations in the early forties.[33]

*Reorganizations of the Public Health Service in 1937 combined the Office of Industrial Hygiene and Sanitation and the Office of Dermatosis Investigations into the Division of Industrial Hygiene. This Division was placed in the National Institutes of Health in 1939. In 1953 HEW was established, and the Division of Industrial Hygiene became the Occupational Health Program in HEW. The Occupational Health Program was designated in 1968 as the Bureau of Occupational Safety and Health. In 1969 the Bureau began to carry out HEW's responsibilities (except for black lung benefits) under the federal Coal Mine Health and Safety Act of 1969. When the Occupational Safety and Health Act of 1970 became law, HEW reorganized the Bureau as NIOSH within the Health Services and Mental Health Administration. A reorganization effective July 1, 1973, transferred NIOSH to HEW's Center for Disease Control in Atlanta.[29]

**However, it was pointed out to the author that public funds were used under President Roosevelt to expand occupational health in the Public Health Service and many states started their efforts at this time. (Harriet Hardy, private communication)

Neglect of the job safety and health problem continued in the postwar era. In 1951 Senator Hubert Humphrey of Minnesota proposed an Accident Prevention Act which would have established a bureau in the Department of Labor to develop safety (but not health) standards, along with an independent board to enforce them; but, like several other job safety bills introduced over the next dozen years, it failed to attract enough support and was forgotten.[34]

Eventually this state of affairs began to change:

> By the sixties a new factor entered the picture—environmental control. As the country became conscious of air and water pollution, a parallel concern about the polluted working environment developed. This coincided with several other elements—sharp increase in industrial accidents starting in 1958, research into the danger of new chemicals and other physical agents, and general prosperity which enabled unions to work for non-wage demands and management to pay for needed controls. The groundwork for comprehensive legislation was laid.[35]

In 1965 the McNamara-O'Hara Public Service Contract Act was passed, extending to suppliers of government services the same protection as had been embodied in the Walsh-Healey Act. The Metal and Non-metallic Mine Safety Act of 1966 provided for inspections and for certain health and safety standards in the minerals industry.[36] Finally, the end of the decade was marked by the enactment of two federal occupational safety and health laws of unprecedented scope: the Coal Mine Health and Safety Act of 1969 and the Occupational Safety and Health Act of 1970. The Federal Coal Mine Health and Safety Act set mandatory health and safety standards for underground mines (with the Bureau of Mines exercising enforcement authority), appropriated aid to the states for the establishment or improvement of local protection and compensation programs, and provided for federal funding of research into coal workers' pneumoconiosis, better known under the name of "black lung." The Act also guaranteed compensation to miners disabled by pneumoconiosis or to their widows.[37] (In May 1973 the Department of the Interior transferred responsibility for enforcing this Act and the Metal and Non-metallic Mine Safety Act from the Bureau of Mines to a newly created Mining Enforcement and Safety Administration (MESA). The Steelworkers and United Mine Workers opposed this move, demanding instead that responsibility for administering and enforcing the two Acts be transferred entirely away from the Interior Department, which they considered "industry-oriented," to the Department of Labor.)[38]

2.4.3 A Brief Legislative History of the OSHAct*

On January 23, 1968, President Lyndon B. Johnson proposed in his Manpower Message to Congress "the Nation's first comprehensive Occupational Health and

*A useful account of the background and legislative history of the OSHAct can be found in *The Job Safety and Health Act of 1970*, a Bureau of National Affairs operations manual. An anec-

Safety Program,"[39] declaring that "It must be our goal to protect every one of America's 75 million workers while they are on the job."[40] Shortly thereafter the administration submitted to the 90th Congress an occupational safety and health bill, which was introduced in the House by Congressman James O'Hara (D-Michigan) and in the Senate by Senator Ralph Yarborough (D-Texas).

This bill contained three major provisions. First, the Secretary of Health, Education, and Welfare was directed to conduct an extensive program of research as a basis for developing comprehensive safety and health standards. Second, the Secretary of Labor was empowered to set and enforce such standards; and to impose strong legal sanctions, including civil and criminal penalties, for violations; he was also authorized to conduct inspections of workplaces subject to the provisions of the act and to shut down immediately machinery or processes presenting an imminent danger to workers. And third, federal assistance was provided to the states to develop and strengthen their own occupational safety and health programs.[41,42]

According to the Nader report *Bitter Wages*, business interests, led by the U.S. Chamber of Commerce, the National Association of Manufacturers, the American Iron and Steel Institute, and the Manufacturing Chemists Association, strenuously opposed the bill.[43] The report also claims that "union leaders saw no reason to exert themselves" on behalf of the measure.[44] In any case, the bill did not reach the floor of either the House or the Senate.

In early 1969, with the advent of the 91st Congress, the Democrats once again introduced occupational safety and health bills (H.R. 3809 and S. 2193) similar to the legislation that had been stalled during the previous year. On the Republican side, President Nixon called for "proposals to help guarantee the health and safety of workers" in his Domestic Programs and Policies Message to Congress on April 14, 1969, while some four months later Congressman William H. Ayres (R-Ohio) and Senator Jacob K. Javits (R-New York) introduced in the House and Senate two Nixon administration occupational safety and health bills (H.R. 13373 and S. 2788). (Figure 2.1 charts the history of the various Democratic and Republican occupational safety and health bills introduced during the 91st Congress.)

The administration legislation differed from that of the Democrats chiefly in that it vested authority for setting and enforcing standards in a new National Occupational Safety and Health Board (whose members were to be appointed by the President) rather than in the Secretary of Labor. The Nader report mentions earlier claims that the administration bill "featured vague language and the complete absence of mandatory protection for workers," and "would have foisted on the workers a cumbersome bureaucratic machine that could choose to do absolutely nothing."[45]

---

dotal critique of the same background and history is given in *Bitter Wages*, Ralph Nader's Study Group Report on Disease and Injury on the Job (Grossman, New York, 1973). The present account draws heavily on both these sources.

	January 1969	August 1969	March 1970
House	H.R. 3809 [O'Hara (D–Mich.)] **Provisions** Secretary of HEW to conduct occupational safety and health research and develop criteria for standards. Secretary of Labor to set and enforce standards. Federal assistance to states to develop and strengthen their own programs.	H.R. 13373 [Ayres (R–Ohio)] **Provisions** New and independent National Occupational Safety and Health Board (with members appointed by the President) to set and enforce standards. Secretary of Labor given discretion to conduct inspections and required to petition NOSH Board to determine if a violation has occurred.	H.R. 16785 [Daniels (D–N.J.)] **Provisions** Contains provisions of H.R. 3809 plus a general duty clause. Numerous worker rights. Enforcement through administrative citation.
Senate	**May 1969** S. 2193 [Williams (D–N.J.)] **Provisions** Similar to H.R. 3809.	**August 1969** S. 2788 [Javits (R–N.Y.)] **Provisions** Senate version of H.R. 13373.	

Figure 2.1 History of Democratic and Republican occupational safety and health bills during the 91st Congress.

September 1970	November 1970	December 1970
H.R. 19200 [Steiger (R–Wis.)] **Provisions** Revised version of H.R. 13373. Standards to be set by an independent National Occupational Safety and Health Board. Enforcement by an independent Occupational Safety and Health Appeals Commission. Labor Department given authority to initiate complaints and make inspections.	House approves H.R. 19200 over H.R. 16875.	Joint Senate-House Conference Committee Bill (signed into law as the Occupational Safety and Health Act of 1970 by President Nixon on December 20, 1970). **Provisions** Secretary of Labor to set and enforce standards. Independent Occupational Safety and Health Review Commission to review enforcement cases. National Institute for Occupational Safety and Health to be established in the Department of Health, Education, and Welfare.
September 1970 S. 4404 [Dominick (R–Colo.)] **Provisions** Senate version of H.R. 19200.	**November 1970** Senate approves S. 2193 (as reported from committee, this bill is similar to H.R. 16785). Final version of S. 2193 includes amendments [Javits (R-N.Y.)] for an independent Occupational Safety and Health Review Commission and for a National Institute in HEW to conduct research and formulate recommended standards.	

With legislation now in from both parties, in 1969 and 1970 both the House and Senate Labor Subcommittees conducted intensive hearings on occupational safety and health. This time around, management representatives threw their support behind the Nixon administration's proposals, while organized labor strongly supported H.R. 3809.[46]

After a difficult period of conflict and compromise, the House in November 1970 approved H.R. 19200, a revised administration bill cosponsored by Congressman William A. Steiger (R-Wisconsin) and Robert Sikes (D-Florida); it contained more provisions for worker protection (like those found in the Daniels bill) than did the previous administration proposals, but it divided authority further. Standards were to be set by an independent National Occupational Safety and Health Board (under formal rule-making procedures), but would be enforced by an independent Occupational Safety and Health Appeals Commission. The Labor Department was also to share enforcement responsibility, with authority to initiate complaints and make inspections. In adopting the Steiger proposal, the House passed over H.R. 16785, a bill sponsored by Congressman Dominick V. Daniels (D-New Jersey) which had been approved in June by the House Labor Committee. The Daniels bill was stronger than H.R. 3809 in that it included guarantees of worker involvement in efforts to safeguard the workplace.

On its side, the Senate in November 1970 approved a compromise version of S. 2193, the Democratic bill originally introduced in early 1969, which incorporated many of the provisions of H.R. 16785. In its final form, the bill included an amendment by Senator Javits creating an independent three-member Occupational Safety and Health Review Commission to exercise final administrative review of enforcement cases, although both standard-setting authority and initial enforcement authority were to remain with the Secretary of Labor. The bill also included a Javits proposal for elevating to National Institute status the former Bureau of Occupational Safety and Health in HEW, to conduct research and to formulate recommended occupational health and safety standards for the Labor Department. In adopting S. 2193, the Senate passed over S. 4404, a bill introduced by Senator Peter H. Dominick (R-Colorado) which was the Senate equivalent of H.R. 19200, the Steiger proposal approved by the House. Thus, despite the fact that both bills in their final versions were the result of compromise, the Senate occupational safety and health bill tended more to reflect Democratic and organized labor thinking, while the House bill inclined in the direction of Republican and management views.

In December 1970 a joint Senate and House conference committee hammered out a final version of the two conflicting bills. During the crucial period, according to the Nader report, "the pressure from the unions bore down on the issue of who should set standards. Before the conference had begun, the union representatives . . . had decided that what they wanted above all else was that the Department of Labor, and not a board, set the standards. In order to reach this goal, the unions were willing to make certain concessions, which they conveyed to Senator Williams and Representative Perkins (both Democratic members of the conference committee)."[47] As a result, the committee adopted the Senate bill's provisions giving the Secretary of Labor standard-setting and initial enforcement authority, and an Occupational Safety and Health Review Commission authority to exercise final administrative review of enforcement cases. On the other hand, in the face of House opposition, the Senate dropped its demand that the Secretary of Labor be given administrative authority to shut down plants or processes in "imminent danger" situations. The committee also retained Senator Javits's proposal for a National Institute for Occupational Safety and Health and settled conflicts over the walkaround right, deciding that an employee representative had the right to accompany a federal inspector whether or not the employer decided to exercise that same right for himself. And finally, the committee reached agreement on a general duty clause for employers and employees, on provisions for investigation and record-keeping, and on numerous other lesser details. On December 17, 1970, the conference committee version of the bill was passed by both the House and Senate, and on December 29, 1970, President Nixon signed the Occupational Safety and Health Act into law.

2.5 ATTEMPTS TO BRING OCCUPATIONAL HEALTH AND SAFETY PROBLEMS TO THE ATTENTION OF THE PUBLIC

In 1965, before public awareness of the problems of occupational safety and health had been aroused, a study of critical occupational health needs was developed internally for the Public Health Service by the National Advisory Environmental Health Committee.[48] The study was stimulated not by inadequate health standards, shocking statistics, or immediate trauma, but rather by a realization that most environmental research was being done on air and water pollution, and by a belief that more research was needed on the occupational environment.[49] The report, entitled *Protecting the Health of Eighty Million Americans: A National Goal for Occupational Health* and commonly known as the Frye Report, was submitted to and endorsed by four representatives of labor, management, and the environmental-medical establishment. However, the Public Health Service did not take up the challenge contained in its findings, and the report went unnoticed by Congress and the public. The report is good reading today, however, because it contains, not only a valuable assessment of occupational health problems, but also an estimate of what occupational disease is costing the American economy and a suggested plan of action.

While the Frye Report was the only significant policy document on occupational health and safety produced in the 1960s, the early 1970s saw four popularized attempts to bring the issue of job health and safety to the attention of the American public, three written by members of the labor movement and one by a Ralph Nader task force. Significantly, while there were great differences among these books, each of them emphasized occupational health problems.*

The Nader task force report *Bitter Wages* (Grossman, New York, 1973), written by Joseph Page and Mary-Win O'Brien, has received the most public attention. This anecdotal account of occupational health and safety problems, politics, and history leading up to the passing of the OSHAct weaves a picture of conspiracy and malfeasance on the part of industry and a picture of neglect and disinterest on the part of labor and government. The book succeeds in doing what was intended—to raise the temper of the reader in either concurrence or disbelief. *Bitter Wages* imparts the notion that there is no more time to remain uncommitted. Finally, a worker-student alliance is seen as a possible mechanism for creating the conditions for necessary workplace improvement. The report accurately conveys the kinds of tensions that underlie the occupational health and safety problems.

The American Worker: An Endangered Species (Ballantine Books, New York, 1972) was written by Franklin Wallick, managing editor of the weekly United Auto Workers' newspaper and a union employee for twenty years. The book is a popularized overview of the field of occupational health and safety addressed both to the uninformed layman and also to the average worker. In his attempt to get occupational health and safety and the new law publicized, Wallick uses an anecdotal style to deal with a little bit of everything in the field: standard-setting, determination of threshold limit values for exposure to toxic materials, occupational medicine, government policy, hazard surveillance, and worker training.

Wallick is in many instances a realist in his criticisms. He recognizes that, historically, labor has been as negligent in addressing workplace perils as have industry and government. Union leaders, for the most part, will continue to be unresponsive, like most institutions, until workers apply the necessary pressure for occupational health and safety to become a first-rate priority. If Wallick appears a bit tough on his own ranks, he is equally, if not more, critical of government's attitudes and participation in the field. He attacks the Nixon administration for strangling the OSHAct by failing to provide adequate funds, manpower, and motivation, and he praises the Johnson administration (under Labor Secretary Wirtz) for being instrumental in initiating action on worker health and safety.

In answering the question of the underlying cause of the occupational health malaise, Wallick differs significantly from Nader:

*Since the preparation of this section, two more relevant popular presentations have been published: Rachel Scott's *Muscle and Blood* (Dutton, New York, 1974) and Paul Brodeur's *Expendable Americans* (Viking, New York, 1974). In addition, there is J. Davitt McAteer's *Coal Mine Health and Safety: The Case of West Virginia* (Praeger, New York, 1973), which, as its title implies, deals specifically with problems in the coal mining industry.

One mistake in the otherwise excellent Nader *Task Force on Job Safety and Health* is the simplistic notion that some kind of conspiracy lurks behind the failure of unions, of employers, of government, and of science. I question any conspiracy. Ineptitude, yes. But conspiracy, no. The problem, as I see it, is a failure to see the whole picture, a failure to get a real handle on ways to grapple—or, to use a contemporary phrase, a lack of consciousness about the dimensions of the problem.[50]

Wallick is likewise candid about what he terms the "worker-student alliance." He states that "there are few notions which contain more wishful thinking than the perennial notion that workers and students can rise up in an alliance that will remake the world." The hostilities unions feel toward the outsider, often people with motivations not akin to their own, stem from a resentment of "elitism" and its blinders:

What the "elitists" lack is an appreciation of the native intelligence and finesse of a lot of union leaders who may lack sophisticated polish, but have a keen sense of timing, mood, and what is possible. Unions are political institutions and must be understood in that context.[51]

Wallick is, however, extremely optimistic, despite his candor about the shortcomings of labor and the naiveté of students. His enthusiasm arises from an understanding of the potential of the American worker to become a prime mover in workplace health and safety. His encouragement and advice to the average worker is perhaps the greatest achievement of *The American Worker.* In the book, Wallick provides the worker with valuable information, of both a technical and political nature, with which to begin the battle for job health and safety.

Peril on the Job: A Study of Hazards in the Chemical Industry (Public Affairs Press, Washington, D.C., 1970) was written by Ray Davidson, editor of the Oil, Chemical, and Atomic Workers' newspaper. The book's style is basically anecdotal, almost half of it being devoted to detailed descriptions of and quotations from interviews with workers in the chemical- and oil-related industries. Indeed, Davidson clearly states that his book "is an attempt to portray industrial health and safety conditions as they actually exist. The matter has been approached from its underside, from the human side where the suffering occurs, rather than from the overview of theory or broad statistics."[52]

But *Peril on the Job* is not solely a collection of horror stories detailing the pervasiveness of hazards to workers across a wide range of chemical processes and substances. It also presents a strong case for the argument that industry's occupational health malaise results from conflicts of interest that are inherent in the management system. For example, the fact that management operates "by the dollar yardstick," with a drive toward maximum production, is in conflict with the need to shut down equipment for preventive maintenance. The need of the middle-level supervisor to maintain a high production record if he is to be promoted, or the need of the line foreman to maintain his level of prestige and social status despite the fact that he is aware of hazards but has no authority to correct them, is inherently in conflict with the improvement of health and safety conditions. Managers who demand that machines operate in excess of capacity come into conflict with design engineers and maintenance crews. Top-level supervisors who work solely in the front office cannot be fully aware of working conditions, and, Davidson maintains, "the conflict of interest between working conditions and profit maximization takes an even more direct form when, as is the case in many plants, the safety director has other managerial duties."[53] He sums the argument up succinctly: "At every supervisory level, the overwhelming pressure is to maximize production. In order to do this, risks are taken on health and safety precautions whenever there is conflict."[54]

Davidson's view that the management system is in a state of conflict extends to the company doctor's role as well:

The M.D. is in a peculiar position. If he tells a worker that his illness is job-caused, the employee may sue the company for workmen's compensation and subpoena the doctor to testify. At best, the typical medical man dislikes appearing on a witness stand. It must be even more difficult when he is put in the position of testifying against the company which has paid his fee. This is a conflict of interest which perhaps would be eliminated by the establishment of socialized medicine. Or, as a more conservative step, a system for paying doctors other than by company check might be devised.[55]

The merits of *Peril on the Job* lie in the second half of the book, where the author demonstrates his understanding of the functioning of bureaucracies and points out where the weaknesses and conflicts lie which prevent workplace health from being a first-rate priority. Davidson's analogy is a good one: "The left hand may dispense healing balm while the right hand is committing manslaughter and the higher authority at the head is out of contact with both."

Jeanne M. Stellman, another member of OCAW, and Dr. Susan M. Daum have coauthored *Work Is Dangerous to Your Health* (Vintage Books, New York, 1973), a valuable handbook for workers in which occupational health hazards and what the worker can do about them are explained in clear and concise language. The book is an important aid both in recognizing hazards and in taking appropriate actions to correct them. Except for an inadequate coverage of chemical carcinogens, the book provides most of what workers need to know about job health and safety. The handbook is not an attempt to agitate workers "to cause trouble for their employer," but, rather, provides a mechanism for workers to make informed judgments about the risks they assume at their place of employment.

The HEW report *Work in America* (The MIT Press, Cambridge, Mass., 1973), which deals with the broader problems of work, provides a valuable integration of the issue of occupational health and safety with other quality-of-work issues. Poor job health and safety may be as much a determinant of work alienation as monotonous, meaningless tasks—and vice versa.

Aside from books dealing specifically with occupational health and safety issues, many periodicals have carried articles covering special occupational health and safety topics. For example, disease caused by exposure to asbestos has been discussed in the *Saturday Review*[56] and in a series of five articles by Paul Brodeur in the *New Yorker.*[57] Articles dealing with the OSHAct appear regularly in law reviews, business journals, *Fortune* magazine, and trade journals. Finally, the amount of coverage given job health and safety in the *New York Times, Washington Post*, and the *Wall Street Journal* has increased markedly in the two years of preparation of this report.

Finally, it should be mentioned that at least three blue-ribbon studies were undertaken in the early 1970s which have relevance for occupational health problems: *Man's Health and the Environment–Some Research Needs*, a report of the HEW Task Force on Research Planning in Environmental Health Science (March 1970); *Man, Materials, and Environment*, a report for the National Commission on Materials Policy by the National Academy of Sciences and the National Academy of Engineering (March 1973); and *Chemicals and Health*, a report of the Panel on Chemicals and Health of the President's Science Advisory Committee (Science and Technology Policy Office, National Science Foundation, September 1973). Unfortunately, none of these reports approaches the evaluation of occupational health problems contained in the Frye Report.

The 1970 HEW study devotes some attention to problems of the work environment, but the latter two studies ignore the workplace for all practical purposes. The studies provide some valuable insights, but they are deficient in four critical ways: (1) they accept too easily the idea that a finite threshold for cancer may exist; (2) they believe that screening can be performed to detect workers who are hypersusceptible to industrial toxins; (3) they believe that research can be done to provide a sufficient data base for the necessary decisions to be made; and (4) they ignore or underemphasize the equity question in the difference between selective victimization in the workplace (or in poor or highly polluted areas) and the random risk present with general environmental pollution.

This report to the Ford Foundation deliberately avoids an anecdotal approach to the problems of occupational health and safety. Its purpose is to provide an in-depth analysis of the technical, legal, economic, and political issues in occupational health and safety, with recommendations for action. It will hopefully be read by those who must make decisions for industry, labor, government, and the public sector. I have found it necessary to explode some myths–e.g., the accident proneness or hypersusceptibility of workers as an explanation for most injuries and disease, the ability of market mechanisms to solve occupational health and safety problems, and the ability of basic research to come up with all the answers necessary for well-informed decisions to be made. Hopefully, this report will not be quoted nine years from now as "an interesting report, forgotten and ignored."

2.6 QUESTIONS THIS REPORT ATTEMPTS TO ANSWER

The following list of questions is not meant to be all-inclusive, but to be representative of the multifaceted problems that need to be addressed in occupational health and safety.

2.6.1 Analytical Questions

- How serious, and how pervasive, are injury and disease in the workplace? What is the occupational disease and injury profile?
- Do white-collar workers and corporate executives have occupational health and safety problems, or are the problems restricted to the industrial worker?
- To what extent are costs presently being borne by the worker, by industry and the employer, and by the society generally?
- How has the new federal legislation shifted the allocation of these costs?
- Are good health and safety policies profitable in the short run? in the long run?
- How would full enforcement of occupational health and safety standards affect the mix of small and large firms? the displacement of workers?
- How do alternative health and safety policies affect the competitive position of the United States in the world market?
- To what extent is selective enforcement and voluntary compliance effective?
- As legislation, how good is the Occupational Safety and Health Act?
- What are the apparent occupational health and safety objectives of the present administration? How do they differ from congressional intent vis-à-vis the Occupational Safety and Health Act?
- To what extent is the Act being implemented?
- Apart from the administration of the Act, what legal action is being taken?
- What are the present and projected manpower needs? What manpower policies have been adopted by management, labor, government, and educational institutions?
- What are the present and projected research and technology needs? What policies have been adopted by management, government, research institutions, and industry?
- To what extent are health and safety issues related to worker alienation and job satisfaction?
- To what extent are health and safety issues becoming important in collective bargaining?
- How serious is the information-transfer problem? To what extent is ignorance a factor in an unsafe and unhealthy work environment?
- To what extent is worker carelessness a significant factor in occupational injury and disease?

• What contributions have management, labor, the government, and educational and research institutions made in the field?
• How serious is the technology-transfer problem, both vertically within the society and horizontally at the national and international level?

2.6.2 Policy Questions

• How protective of the worker should society be?
• Should the Occupational Safety and Health Act be amended?
• What are the priority areas for action, and what is the most effective way of using available funds to bring about change?
• What manpower policies should be adopted?
• What research should be done?
• What technology needs to be developed?
• What can be done to increase information transfer?
• What can be done to increase the transfer of technology?
• What can be done to make educational and research institutions more aware of and responsive to needs in the area?

2.6.3 Questions Concerning the Role of the Private Sector

• Why should the private sector do anything?
• What research should be supported?
• What educational and training programs should be supported?
• What information-dissemination activities should be supported?
• What public-interest legal activities should be encouraged?
• Can private philanthropy serve a function as a neutral forum for bringing together people and resolving conflict?

REFERENCES

1. *Work in America*, Report of a Special Task Force to the Secretary of Health, Education, and Welfare (MIT Press, Cambridge, Mass., 1973), p. 79.

2. World Health Organization, WHO Technical Report Series, No. 66 (1963), p. 4.

3. Hearings on Small Business and the Occupational Safety and Health Act of 1970 Before the Subcommittee on Environmental Problems Affecting Small Business of the H.R. Select Committee on Small Business, 92d Congress, 2d Session, June 22, 1972.

4. "Coal Mining: The Way to Dusty Death," *The Sciences*, Vol. 11, No. 7 (August-September 1971), p. 30.

5. *Ibid.*

6. *The President's Report on Occupational Safety and Health*, G.P.O. Document No. 2915-0011 (May 1972).

7. Roland P. Blake, ed., *Industrial Safety* (Prentice-Hall, Englewood Cliffs, N.J., 1963), pp. 13, 22.

8. *Ibid.*, p. 13.

9. National Commission on State Workmen's Compensation Laws, *Compendium on Workmen's Compensation*, G.P.O. Document No. 496-632 (1973), p. 12.

10. Blake, *Industrial Safety*, p. 14.

11. National Commission, *Compendium*, pp. 17-18.

12. M. J. Fox, Jr., and Gary S. Nelson, "A Brief History of Safety Legislation and Institutions of the United States and Texas," *Business Studies* (North Texas State University), Vol. 11, No. 2 (fall 1972), p. 49.

13. Richard Posner, "A Theory of Negligence," *Journal of Legal Studies*, Vol. 1, p. 29.

14. National Commission, *Compendium*, p. 18.

15. Blake, *Industrial Safety*, p. 15.

16. National Commission, *Compendium*, p. 18.

17. Abner Brodie, "The Adequacy of Workmen's Compensation as Social Insurance: A Review of Developments and Proposals," *Wisconsin Law Review* (January 1963), p. 57, quoted in National Commission, *Compendium*, p. 18.

18. National Commission, *Compendium*, p. 18.

19. Joseph A. Page and Mary-Win O'Brien, *Bitter Wages* (Grossman, New York, 1973), p. 58.

20. *Ibid.*, p. 59.

21. Senate Labor and Public Welfare Committee Report No. 91-1282 on S. 2193, 91st Congress, 2d Session (October 6, 1970), in *The Job Safety and Health Act of 1970* (the Bureau of National Affairs, Inc., Washington, D.C., 1971), pp. 239-240.

22. Page and O'Brien, *Bitter Wages*, p. 87.

23. Andrew D. Hosey and Louise Ede, *A Review of State Occupational Health Legislation*, Bureau of Occupational Safety and Health, HEW (January 1970), pp. 10-13.

24. *Ibid.*, pp. 11-13.

25. Page and O'Brien, *Bitter Wages*, pp. 80-85.

26. *Job Safety and Health Act of 1970*, p. 14.

27. Page and O'Brien, *Bitter Wages*, p. 87.

28. *Slow Progress Likely in Development of Standards for Toxic Substances and Harmful Physical Agents Found in Workplaces*, Report to the Committee on Labor and Public Welfare, U.S. Senate, by the Comptroller General of the United States (September 28, 1973), p. 91.

29. *Ibid.*, p. 7.

30. *Ibid.*, p. 109.

31. Fox and Nelson, "Brief History," p. 53.

32. *Job Safety and Health Act of 1970*, p. 15.

33. *Ibid.*

34. Page and O'Brien, *Bitter Wages*, p. 64.

35. *Job Safety and Health Act of 1970*, p. 15.

36. Fox and Nelson, "Brief History," p. 54.

37. "Coal Mining: The Way to Dusty Death," p. 30.

38. *Occupational Safety and Health Reporter* (the Bureau of National Affairs, Inc.), Vol. 2, No. 49 (May 10, 1973), p. 1441; Vol. 3, No. 1 (June 7, 1973), pp. 28, 30.

39. *Job Safety and Health Act of 1970*, p. 16.

40. Page and O'Brien, *Bitter Wages*, p. 137.

41. *Ibid.*, p. 140.

42. *Job Safety and Health Act of 1970*, p. 16.

43. Page and O'Brien, *Bitter Wages*, p. 141.

44. *Ibid.*

45. *Ibid.*, pp. 172-173.

46. *Job Safety and Health Act of 1970*, p. 17.

47. Page and O'Brien, *Bitter Wages*, p. 178.

48. *Protecting the Health of Eighty Million Americans: A National Goal for Occupational Health*, Special Report to the Surgeon General of the United States Public Health Service, HEW (November 19, 1965).

49. George R. Taylor, AFL-CIO, private communication.

50. Franklin Wallick, *The American Worker: An Endangered Species* (Ballantine Books, New York, 1972), pp. 160-161.

51. *Ibid.*, p. 143.

52. Ray Davidson, *Peril on the Job: A Study of Hazards in the Chemical Industries* (Public Affairs Press, Washington, D.C., 1970), p. 4.

53. *Ibid.*, p. 129.

54. *Ibid.*, p. 112.

55. *Ibid.*, p. 142.

56. Bruce Porter, "An Asbestos Town Struggles with a Killer," *Saturday Review of the Society* (March 1973), pp. 26-31.

57. Paul Brodeur, "Annals of Industry," *New Yorker* (October 29-November 26, 1973).

3 The Nature and Dimensions of Occupational Health and Safety Problems

3.1 INTRODUCTION

In this chapter a brief description of typical occupational health and safety problems will be given, along with a discussion of (1) injury and disease statistics, (2) the state of the art for research into the prevention and causation of injury and disease, and (3) expected new hazards and trends for the future. The chapter is not intended to be a definitive treatise on the scientific-medical aspects of occupational health and industrial safety. The reader is referred to the literature for detailed treatment of these subjects.[1]

While, as we shall see, it is sometimes impossible or futile to draw a distinction between health and safety, there are operationally significant distinctions reflecting (1) the time frame in which the harm occurs or is recognized, (2) the cause of the harm, (3) the kind of professionals dealing with the problems, and (4) the extent to which the harm can be quantified monetarily and incorporated into the operating costs of the firm or business. For the purposes of analysis, then, safety and health are first discussed separately.

3.2 SAFETY HAZARDS

3.2.1 Definition

Safety hazards are commonly recognized to arise in those work environments which can cause burns, electrical shock, cuts, bruises, sprains, broken bones, or the loss of limbs, eyesight, or hearing. It can be said generally that the harm is usually of an immediate and sometimes violent nature, is very often associated with industrial equipment or the physical environment, and often involves an employment task which requires care and training.

Safety professionals insist that the term "safety" is all-encompassing and includes health—but their concern is more with the explosive nature of chemicals than with their toxicology, and with the effects of noise on hearing rather than with noise as a stressor and cocausative factor of disease. To the extent that the causes of accidents are recognized by the state workmen's compensation system, they become a concern for the firm, its insurance carrier, and its safety professionals.

3.2.2 Specific Safety Hazards

Almost every work environment involves one or several safety hazards, as evidenced by the fact that there are 2.5 million work-related injuries each year.[2] The following represent some of the major safety hazards presently existing (see also Section 6.3.2 for a discussion of research priorities for safety).

Fires and Explosions

Chemicals and pressurized containers or transmission lines are the sources of many fires and explosions which cause damage to both workers and property. Thus, the interests of workers and employers are probably more consonant here than they are with regard to many other hazards which affect workers alone. Fire insurance provides a fairly healthy incentive system through the insurance carrier's inspection force, although industrial fires and explosions remain a serious problem in terms of both human and financial loss.

Electrocution

Electrical hazards and possible resulting electrocution constitute perhaps the most pervasive safety hazard next to noise. It has often been pointed out that the 110-volt system for *domestic* electricity provides a safer situation than the 220-volt system used in Europe. However, much of the equipment used in the American industrial workplace requires 220 or more volts. Electrical hazards can often be inexpensively eliminated by proper grounding or shielding and are an almost guaranteed item to be checked for by OSHA inspectors.

Dangerous Stationary Machines

All machines have moving parts which, if not shielded, may cause injury. Typical of machines used in various phases of industry are those capable of sawing, molding, rolling, cutting, mixing, flattening, bending, grinding, or just moving semifinished parts to another part of the factory. Generally, these are very large and have rapidly moving parts which carry great momentum. Additionally, they are all incapable of differentiating hands or other parts of the human body from their proper working material and can cut, mutilate, or in some other way injure the worker.

Operations Requiring Eye Protection

Industrial machines that cut metal or wood also create airborne particles of the substance being cut. Grinding and other operations which mechanically abrade also produce flying chips or dust particles. These airborne particles can get into the eyes and inflict injury, the severity of which may depend on the size and shape of the particle as well as on the manner in which it is projected into the eye. Machines creating this sort of hazard can be properly shielded or ventilated, but often are not.

Liquid aerosols can be equally dangerous to the sensitive membranes of the eyes. Also, frequent injuries occur from splashes, spills, and mists, especially from strong acids or bases. Other liquids, too—for example, certain organic solvents—can hurt the eye when protective glasses are not worn. Fires and explosions also require eye protection.

Moving and Lifting Equipment, and Falls

Workers may be injured by moving or lifting equipment where there are no guardrails, warning bells, horns, or aisle markers. As with any moving-vehicle accident, the heavier the machine and the faster it is moving, the more damage it is likely to cause. The high injury rate in stevedoring is due to these kinds of hazards. Many of the machines used by industry today have large, heavy parts engaged in a rapid, repetitive motion. These machines are usually not designed with the operator's protection in mind and are often associated with crushing and mangling accidents to the hand, arm, or foot. Falls from farm equipment cause frequent injuries. In many plants, falls are a frequent cause of injury. In most cases, falls could rather easily be avoided by providing guardrails and warning markers. Safety lines, belts, and nets would also prevent serious injury in the event of a fall.

Noise

Excessive noise presents at least three kinds of safety hazards. A sudden loud noise can create enough pressure in the ear to damage or rupture the eardrum. Constant loud noise causes eventual hearing loss.* In addition, the presence of loud background noise (even at 90 dBA) can increase the likelihood of accident by reducing a worker's visual acuity,[3] by making it impossible for him to hear warning signals, or by dulling his other senses. There is evidence indicating that workers exposed to high noise levels have higher accident rates than people working in a quiet environment.[4] Noise as a stressor and direct or cocausative factor in disease is discussed in Section 3.3.2.

Heat Stress

Under conditions of extreme heat or cold, changes of the body's core temperature can occur, thereby generating considerable physical stress. The human body can tolerate considerable temperature variation before irreversible harmful health effects occur, but discomfort, cramps, and psychological effects may occur long before there is any major change in the core temperature of the body. Working in conditions of heat stress, especially while performing intense physical work, may decrease job performance, accuracy, concentration, and certain physiological responses. This, then, increases the possibilities of accidents occurring. Moreover, workers exposed to high ambient temperatures are often involved in activities for which the accident rate may already be high, such as firefighters, furnace men, forge-hammer workers, foundry workers, steelworkers, glassblowers, iron smelters, etc.

*The proposed OSHA noise standard—with exposure allowed at 90 dBA for eight hours—affords inadequate protection. A large percentage (~20%) of those exposed (for 40 years) are expected to suffer measurable hearing loss. As of this writing the Environmental Protection Agency is challenging the proposed OSHA standard, and hearings are under way to help resolve the issue.

Vibration

NIOSH investigators recently estimated that "as many as eight million people in the United States may be exposed to industrial vibration."[5] Transportation equipment provides the largest single source of exposure to occupational vibration. Construction workers and agricultural workers who operate heavy equipment, truck and forklift drivers, railroad workers, and intercity bus drivers are exposed to varying but often considerable amounts of vibration in the course of their jobs. The second largest occupational group subject to vibration is foundry workers. Forty percent of the nearly 500,000 workers in this industry are exposed to vibration on the job. Other exposures to vibration result from the use of pneumatic tools in drilling, riveting, mining, and tunneling. When a worker is exposed to repeated impact blows or tool vibrations, he generally braces himself so as to absorb the energy. The muscular activity accompanying this bracing is usually associated with an increase in the metabolic load and presents a physiological stress. This stress, in turn, can lead to a slowing of reflex activity and will certainly increase fatigue. The likelihood of an accident is accordingly increased. There is also a strong relationship between vibration and visual acuity which can affect both the rate of accidents and the extent of quality control problems in industry.

Ergonomic Factors

Industrial processes are often designed to accomplish a repeated task as rapidly as possible so as to produce the largest possible number of finished products per unit time. The actual positioning and comfort of the workers who must operate the machines are rarely considered in the design stage. Yet certain aspects of the workplace, such as lighting, accessibility of the control panel, and the shape and size of tools, are intimately related to the efficiency and well-being of the worker. These factors comprise the study of ergonomics, which includes both physiological and psychological stresses of the worker's activity that can result in fatigue, unnecessary errors, and accidents. Tasks as well as equipment can be designed so that physiological and psychological stresses are at a minimum and job performance and safety are at a maximum.

Combined Hazards

Individual hazards are rarely found alone, and hazards often combine in a synergistic way. Thus, noise reduces visual acuity and probably increases the likelihood of an accident.[6] Carbon monoxide is more toxic when heat stress is present.[7]

3.2.3 Severity

Safety hazards present several levels of danger: (1) immediate and violent harm, as exemplified by someone operating a punch press who loses his hand if he fails to move it away from the work surface quickly enough, (2) slowly acting, but severe harm nonetheless, as exemplified by the worker exposed to noise each day until he has damaged his hearing irreversibly, and (3) damage caused by a combination of hazards, none of which might have been severe by itself but which act together to cause damage of a severe nature either in the short run or at a much later time. The minor cuts and bruises–and the "near misses"–do not show up in accident statistics or workmen's compensation records. Only injuries of a "severe" nature are recognized by the system.

The Coal Mine Health and Safety Act applies specifically to the coal mining industry, while the OSHAct is meant to have a much wider application. However, the Occupational Safety and Health Administration has launched a Target Industry Program to focus attention on five industries with high rates of job-related injuries. They are listed below with their injury frequency rates (to be compared with the national average of 14.8 disabling injuries per million man-hours worked):[8]

- longshoring–69.9
- roofing and sheet metal–43.0
- meat and meat products–38.5
- mobile homes and miscellaneous transportation equipment–37.6
- lumber and wood products–36.1.

3.3 HEALTH HAZARDS

3.3.1 Definition and Types of Occupational Disease

The objectives of occupational health have been defined by a joint ILO-WHO committee as

> the promotion and maintenance of the highest degree of physical, mental, and social well-being of workers in all occupations; the prevention among workers of departures from health caused by their working conditions; the protection of workers in their employment from risks resulting from factors adverse to health; the placing and maintenance of the worker in an occupational environment adapted to his physiological and psychological condition.[9]

In occupational health, there is concern for the preservation of the worker's health both in the course of his work and after he has left his work, either for the day or for retirement. In safety, the concern is primarily focused on the time the worker is actually working. Since occupational health hazards (1) are much more difficult to recognize than safety hazards and may act very slowly on the body, (2) can cause harm which compounds the effects of nonoccupational health hazards (such as smoking), (3) are of many different kinds, all of which can act simultaneously on the body, and (4) can manifest themselves in sickness while the worker is no longer at his job, the task of preserving the health of workers must necessarily be more complicated and difficult than that of assuring workers' safety. (See Section

8.7 for a discussion of workmen's compensation and the special problems of disease.)

Specific occupational health hazards can be classified into four general types:

- physical–noise, heat, vibration, and radiation
- chemical–dusts, poisonous fumes, and gases, toxic metals and chemicals, and carcinogens
- biological–bacteria, fungi, and insects
- stress–including that caused by physical, chemical, and ergonomic factors, and psychological factors such as disciplinary pressure on the job.

At the present time, most of the effort in occupational health is spent in the abatement of physical and chemical hazards, partly because these hazards or their effects are most easily detected or measured by the senses or by analytical instrumentation.

In order to determine if a health hazard exists, it is important to recognize that the occupational health of workers is dependent on the interaction of three factors: (1) the *environmental factor*, (2) the *human factor*, and (3) the *characteristics of the job task*. Environmental monitoring, such as the measurement of noise or dust levels, addresses itself to the first factor. Biological monitoring is concerned with changes in human physiology (although canaries are used in the mines to detect the presence of methane, because the lethal dose is so much lower for canaries than for humans). Ergonomics, or human factors engineering, and the behavioral sciences address the third factor, although all too frequently they ignore stress and the psychosocial aspects of work.

All three factors must be monitored or measured for a proper analysis of hazards on the job. Environmental factors are usually easier to measure, although the results of biological monitoring, e.g., measuring the amount of mercury in the urine or blood instead of in the plant, may be a more reliable indicator of danger. There are problems with mandatory biological monitoring, however: (1) waiting to take corrective action until materials have been detected in the body assures that action will come only after some harm has occurred; (2) since people may vary in their work habits and metabolism, a large enough sample of workers must be monitored to get reliable data; (3) the medical examination of workers is fraught with problems in labor-management relations; and (4) there is a constitutional question of the invasion of privacy.* Finally, the characteristics of the job task are immensely complex and require considerable multidisciplinary expertise for their full appreciation.

*Another kind of monitoring can be accomplished by epidemiology, or the study of excess disease or sickness in selected groups. Epidemiology is useful in detecting the presence of health problems without relying on the identification of a specific hazard. Certain jobs can thus be singled out as dangerous even if the hazards involved are difficult to identify. For example, rubberworkers, who work with as many as 3000 chemicals,[10] show a higher incidence of leukemia than the general population, though the specific causative agents have not been identified.[11]

After the detection of hazards by environmental or biological monitoring, the *control* of hazards can also be accomplished in a number of ways:

- Problems can be attacked at the source—e.g., noisy equipment can be replaced, or a closed system designed to prevent the emission of toxic materials, or alternative methods of production and substitute materials, can be used.
- Equipment and processes can be modified—e.g., leaks repaired, additions made, or better maintenance provided.
- The environment around the source of the hazard can be controlled—e.g., by ventilation, acoustical enclosures, or relocation.
- Personal protective devices can be provided for the worker—e.g., respirators, eye goggles, and ear plugs.
- Jobs may be redesigned and work practices may be altered—e.g., working with toxic materials under stressful conditions can be eliminated.
- The complete removal of workers from the hazardous area or a shortening of their exposure time can be effected.

It may not always be clear which method of control is most suitable for a given hazardous situation. Ear protection coupled with short time on the job may be the logical solution for airplane mechanics, while acoustical baffling or relocation of equipment may be better for impact noise. Closed systems may be the ideal solution for chemicals, while dust control might best be effected in some cases by ventilation and the removal of the worker from the proximity of the operation. What is important is that the worker's health be a determinant of the "most suitable" method.

The interaction of health hazards and the human organism can occur through the senses, as a result of absorption through the skin, by intake into the digestive tract via the mouth, or by inhalation into the lungs. The results of such interactions can include respiratory disease, heart disease, cancer, neurological disorders, a large number of systemic poisonings, and a general shortening of life expectancy due to general physiological deterioration. Disease can also give rise to accidents. The disease or sickness can be acute or chronic, can require a long latency period to appear (even if the exposure occurs over a short period of time), and can be difficult or impossible to diagnose with certainty.

To help focus our consideration of occupational health hazards, some specific hazards are discussed in the following section. The magnitude of the problem is more fully discussed in Section 3.4.

3.3.2 Specific Health Hazards

Noise

In addition to contributing to hearing loss and accidents, noise has a potentially serious effect as a stressor and cocausative factor in disease. Noise triggers changes in cardiovascular, endocrine, and neurologic function that are suggestive of a stress reaction.[12] That is, noise induces physiological and psychological adjustments of the body and mind similar to the response that one would have under an alert condition. It is this state of *continuous* alertness and tension which is harmful to health. Animal studies indicated that noise exposure can induce cardiovascular disease, hypertension, gastrointestinal ailments, reproductive dysfunctioning, and even tissue damage of the kidneys and liver. Human epidemiological study also shows that cardiovascular irregularity, neurological disorder, irritability, and social conflict are more frequently encountered in high-noise-level trades. The relative importance of noise as a stressor and its interaction with other stressors are still unknown.

The most serious effect of noise as a stressor may be its interaction with chemical or physical hazards. Noise and overcrowding have been shown to increase amphetamine toxicity in laboratory animals.[13] Workers in noisy assembly lines who take pep pills may be seriously harming themselves. The NIOSH-recommended exposure limit of 90 dBA for eight hours is "primarily designed to conserve hearing. For other disturbances connected with noise such as stress-related illness and performance losses, there is *insufficient or inconclusive evidence upon which to base a standard.*"[14] This should in no way be interpreted as a dismissal of the effects of noise as a stressor. In fact, NIOSH's behavioral research team wishes to pursue this research.[15]

Some Harmful Dusts

Many occupational poisonings that affect the internal organs of the body occur as a result of breathing airborne materials. Such materials, whether generated by grinding, crushing, or some other mechanical means of attrition, are categorized as dusts. Dusts can be inorganic or organic in nature, and the extent of biological damage done when they are inhaled depends on the size and density of the particles, the concentration of dust particles in the breathing zone, the composition of the dust itself, the length of exposure, the rate and depth of breathing during the exposure, the solubility of the dust, and the rate of elimination from the body.

A few of the more common dust-related diseases are silicosis (from the inhalation of silica), asbestosis (from the inhalation of asbestos particles), coal workers' pneumoconiosis (from the inhalation of coal dust), berylliosis (from the inhalation of beryllium dusts), and byssinosis (from the inhalation of cotton dust).

Silica Exposure to silica is found in such industries as mining, quarrying, stonecutting, founding, and glass manufacturing. Silicosis manifests itself in shortness of breath on exertion, coughing, chest pain, and weakness associated with pulmonary fibrosis and obstructive emphysema. Tuberculosis may be a complication. After development of symptoms, survival time is not likely to be very long, but symptoms may develop only after many years of silica dust inhalation. The population at risk is reported to be 1.1 million.[16]

Asbestos Asbestos is a general term encompassing several fibrous mineral silicates including chrysotile, amosite, anthophyllite, and crocidolite. The hazard occurs in the industries involved in the mining of asbestos, the handling and crushing of asbestos, and the manufacture and use of asbestos products. Prolonged inhalation of asbestos particles may result in severe pulmonary fibrosis and respiratory disability. The progress of asbestosis is insidious as long as exposure continues. Some coughing, expectoration, anorexia, and loss of weight occurs. The disease continues to progress even after exposure is discontinued, and the fibrosis extends into almost all portions of the lung tissue. A causal relationship between asbestos and cancer of the lung, pleura, and intestine has also been established.[17] The population at risk is reported to be 350,000.[18]

Beryllium Beryllium is a light, stable metal with a high melting point and a very high strength-to-weight ratio. It is both extremely hard and ductile. Because of these properties, it has found wide application in industry. Acute beryllium disease is found in workers who are (or have been) employed in industrial plants extracting beryllium from the ore, in metallurgical industries, in ceramics industries, and in the flourescent lamp industry. Sometimes, granulomatous lesions of the skin, liver, kidney, and other organs may appear, but the most striking damage is to the lungs. In both the acute and the chronic forms of the disease, symptoms include coughing, dyspnea, anorexia, and weight loss. In the acute form, symptoms of pneumonitis may appear within a few hours or weeks. The disease symptoms then progress rapidly. Recovery in about six months with the use of oxygen, steroids, and antibiotics is possible. Chronic berylliosis is usually more delayed in onset, sometimes developing years after exposure to beryllium, and the occurrence of symptoms is usually precipitated by some other stress. Death in beryllium disease cases is usually due to some complication, such as congestive heart disease. Evidence exists to indicate a higher death rate from lung cancer among people who have had beryllium disease, as well as increased cancer of the liver, gall bladder, and bile ducts.[19] The population at risk is reported to be 30,000.[20]

Cotton Dust Cotton dust is found throughout the textile industry and affects cotton, flax, and soft hemp workers. The greatest exposure to cotton dust generally occurs during the carding of cotton—an operation that includes combing out cotton fibers and setting them parallel by drawing them between toothed cylinders. Also, machinery throughout the cotton mill must be cleaned frequently, and much exposure to cotton dust occurs then. Materials adhering to cotton fibers, such as unwanted leaf, stem, and boll trash, elicit an allergic reaction or response when inhaled, giving rise to the disease known as byssinosis. It generally advances through three recognizable stages. The first is commonly called "Monday fever" and begins only after exposure of more than ten years. Workers develop an irritating cough, tightness of the chest, and breathlessness. The attacks usually come on Monday, or after a period of absence from the plant, and last only a day or so; hence, the name "Monday fever." In the second stage, these symptoms extend over more days in the week and finally become permanent. If exposure to cotton dust is ended at this stage, recovery can occur. In the third stage, the tightness of the chest and labored breathing become so distressing the worker must leave the industry. The disease can progress then to chronic bronchitis and emphysema. The population at risk is reported to be 800,000.[21]

Some Toxic Gases, Metals, and Chemicals

A toxic material, of which NIOSH has listed over 12,000,[22] has been defined as

> one which demonstrates the potential to induce cancer;* to produce long-term disease or bodily injury; to affect health adversely; to produce acute discomfort; or to endanger the life of man or animals through exposure via the respiratory tract, skin, eye, mouth, or other routes.[23]

The potential harm from chemical exposures occurs both in the chemical industry itself, which converts raw materials into various chemicals, and then in the manufacturing and other industries which utilize these chemicals. Harmful effects include local and systemic damage. Local effects usually include injury to the skin, eyes, and mucous membranes of the upper respiratory tract. Systemic effects are manifest when the chemical enters the body via the respiratory or gastrointestinal tract and is absorbed and distributed to the internal organs.

Poisoning by toxic gases is frequent. Some are colorless and odorless, such as carbon monoxide; but others are easily identified by the senses. Sulfur dioxide, nitrogen oxides, chlorine, ammonia, acrolein, and alkaline mists are responsible for a high incidence of acute respiratory disease, and prolonged exposure to them leads to chronic obstructive pulmonary disease.[24] Ozone is primarily an irritant to the eyes and mucous membranes. Inhalation of ozone may produce pulmonary edema and hemorrhage if the exposure is severe enough. Less serious exposures produce headache, malaise, shortness of breath, and fatigue.

Industrial solvents are encountered in many working processes. Exposure to toxic doses of these substances may cause diseases such as anemia (from benzene), liver and kidney damage (from carbon tetrachloride), neurological disorders, arteriosclerosis, and heart disease (from carbon disulfide).[25] By way of example, consider the variety of acute and chronic health effects that can be produced by typical exposures to benzene, phenols, and sulfuric acid:

1. Benzene is an irritant of the skin, the eyes, and the mucous membranes of the

*For the purposes of much of the following discussion, toxic materials will be distinguished from carcinogens (cancer-causing agents). See Section 3.5.2.

respiratory tract. Contact with the skin can produce erythema, vesiculation, and dermatitis. Inhalation of high concentrations of benzene vapor initially produces symptoms of central nervous system depression, such as headache, fatigue, and dizziness. Convulsions and death may follow. Low-level chronic exposure commonly produces blood changes, such as leukemia and leukopenia. There may be bone marrow effects as well. Symptoms include headache, dizziness, weakness, breathlessness, bleeding from mucous membranes, proneness to infection, and easy bruising. Furthermore, there is strong evidence that benzene is carcinogenic; for example, leukemia is noticed with higher than expected frequency in benzene workers.
2. Phenols are primary irritants and are highly corrosive. Chronic poisoning may produce symptoms of headache, fatigue, nervousness, insomnia, paresthesias, and weight loss. Renal and hepatic damage may also occur. Acute exposures are characterized by central nervous system disorders, and pneumonia may follow.
3. Sulfuric acid is very widely used in industry. Its mist is a respiratory irritant, and the liquid form can produce severe burns and ulceration.

Toxic metals that are serious health hazards include lead, mercury, cadmium, nickel, gold, and beryllium. Most of these produce very serious systemic effects upon chronic exposure, often leading to a shortened life expectancy, and are lethal upon acute exposure. Lead poisoning usually results from the inhalation of lead dust or fumes. Symptoms include abdominal pain, headache, weakness, loss of appetite, nausea, vomiting, and weight loss. Serious exposures may result in lead palsy and lead encephalopathy. Mercury poisoning can occur from the inhalation of vapor or from percutaneous absorption. Acute exposure to high concentrations may produce abdominal pain, vomiting, diarrhea, pneumonitis, renal damage, and circulatory or respiratory failure. Inorganic mercurials may produce tremors, emotional instability (possibly a factor in the high suicide rate found among dentists), renal damage, restriction of visual fields, and digestive disturbances upon excessive exposure. Chronic exposure to organic mercurials can produce ataxia, tremor, emotional instability, dysarthria, paresthesias, and impaired hearing.

Occupational skin disease (dermatitis) is among the most easily recognized and commonly reported occupational diseases. Though usually not life-threatening, absorption of toxic materials through the skin may produce extremely serious systemic damage to vital organs or the nervous system. For example, four drops of parathion absorbed through the skin may be lethal (see Chapter 12 on the problems of agricultural workers).

Toxic materials, in addition to being poisonous in the ways described above, may also be carcinogenic, mutagenic, or teratogenic. (A mutagenic material is one capable of causing a genetic change in the chromasomal makeup of an offspring. A teratogenic material is one capable of causing birth defects.) Organic compounds, acquired or synthesized mainly from petrochemicals, may be highly carcinogenic, and inorganic chemicals are also not free from complicity. An increased incidence of cancer with exposure to airborne arsenic at levels of 0.2 mg As/cubic meter has been reported. (The population at risk is reported to be 1,500,000.)[26] The effects of the gas sulfur dioxide are enhanced by airborne particulate matter, and sulfur dioxide may promote the carcinogenic action of other airborne substances. (The population at risk is reported to be 500,000.)[27]

Aside from their toxic, carcinogenic, mutagenic, or teratogenic effects, toxic materials may be responsible for a great many accidents. Regular exposure to low levels of anesthetic gases in the hospital operating room may slow reaction times by 50% and generally dull the senses of the operating personnel.[28]

Carcinogens

Cancer is a widespread cause of death and disability in this country today. There is abundant evidence that the great majority of malignant neoplasms—probably over 90% of the total—are induced, maintained, or promoted by specific environmental factors.[29] It is difficult to isolate the causative agents in most cases, much less prove that a cancer has been induced by an occupational hazard. However, epidemiologic studies are finding higher than average rates of specific kinds of cancer in certain industries and are thus able to trace the origin of cancer to specific agents. The first cancer attributed to occupational exposure was discovered in 1775 by the British surgeon Percival Potts among London chimney sweeps who exhibited a high rate of scrotal cancer, presumably due to exposure to coal tar in chimneys.[30]

Some of the elements and substances which are recognized as carcinogenic are tar, dye stuffs, paraffin, aniline, chromates, Xrays, sunlight, arsenic, asbestos, vinyl chloride, and other hydrocarbon derivatives. Occupational cancer may follow chronic inflammation which has been produced by continuous and long-term exposure to an irritating or toxic substance. Most often, cancer is found in the skin, scrotum, urinary bladder, or lungs. It can appear wherever the carcinogenic agent comes in contact with the body or at sites other than at the point of contact. For example, tar cancer is common in fishermen. The tar is used to keep nets from rotting, and a fisherman will often hold the needle in his mouth while mending his nets. Cancer of the lip has been found in fishermen in a frequency greater than that for the average population. On the other hand, inhaling vinyl chloride causes cancer in the liver or brain.

Not only do asbestos workers suffer from excess lung cancer and pleural cancer (mesothelioma), they also exhibit a high incidence of cancer of the stomach, colon, and rectum.[31] Coke-oven workers as a group are two and a half times as likely to die from lung cancer as steelworkers who do not work in coke plants. They are also seven and a half times more likely to die from kidney cancer than would normally be the case.[32] Dye workers have a high incidence of papillomas and carcinomas of the urinary bladder. These have been traced to the prolonged inhalation of certain aromatic amines, including aniline, beta-naphthylamine, and benzidine.

New chemicals are being introduced into industry at an alarming rate—500 per year—which is considerably faster than would allow for the proper experimentation and testing of these materials for carcinogenic properties.[33]

Heat Stress

Heat stress in the workplace is commonly encountered in steel mills, foundries, deep and surface mines, the glass industry, chemical plants, and construction sites. However, it is also well known that the performance of sedentary production workers and even office employees are greatly affected by the temperature of the room in which they work.

The effects of heat on human beings and on machines are closely analogous. Engineers have known for centuries that machines can only work within a certain temperature range and that to keep the machine working, the heat generated and received by the machine must be dissipated at exactly the same rate. The human subject behaves similarly. He must be able to dissipate the metabolic heat he generates and the convective or radiated heat he receives to keep his body temperature close to 98.6°F. Otherwise, he will not be able to work as well as he should and may be unable to work at all. The only significant built-in heat dissipation mechanism that a man has is the evaporative cooling mechanism of sweating. Thus, what an industrial hygienist must do is try to figure out how much heat the worker can dissipate under a given set of conditions. He might try to increase the worker's heat dissipation capacity by ventilation, meanwhile decreasing the heat passed on to the worker by insulating hot objects and shielding the radiant heat from hot surfaces. If the balance cannot be made, then heat stress becomes evident and a severe hazard exists.

The immediate effect of heat stress may be physiological or psychological. It has been noted that heat stress accelerates fatigue, causes sensory-motor coordination to deteriorate, decreases learning capacity, and increases the rejection rate in production lines. People also become more irritable, so that anger and other emotional states arise more easily. When the stress is more intense, systemic failure may develop which could be fatal. Commonly encountered heat stress episodes include heat stroke, a rapidly fatal phenomenon due to failure of the body's thermoregulatory mechanism; heat exhaustion, in which people become extremely weak and may even faint due to excessive losses of body water through sweating; and heat cramps, in which a person develops painful skeletal muscle spasms due to excessive loss of body salt content through sweating. Excessive sweating can also induce skin rashes and subsequent secondary skin infections.

Chronic heat stress has been found to be deleterious to the cardiovascular system. Since the physiological effect of heat stress is an increase of blood flow to the body's periphery for sweating and heat dissipation, it is understandable that heat stress imposes a burden to this system. Epidemiological studies indicated that lost time due to sickness is 63% higher in coal miners working at 80°F than at 70°F, and that the death rate is 35% higher for coal miners working in hotter mines.

Other studies also indicated more frequent heart defects and arteriosclerosis in steel- and glassworkers. Workers in heat stress environments are customarily self-selective in physique and have more tolerance for acute effects of heat stress; this means that although the epidemiological studies are indicative, they do not sufficiently emphasize the chronic effects of heat stress on a general cross section of workers.

The combined effect of given levels of heat stress and carbon monoxide is known to be more deleterious to health than exposure to either of them separately at the same levels. It is well known that carbon monoxide ties up hemoglobin, thus reducing the oxygen-carrying capacity of the blood. As explained before, heat stress imposes a circulatory burden. Thus, the two agents, through different mechanisms, are attacking the same system, and this is possibly why the synergism exists. Since both heat and carbon monoxide are the products of industrial combustion, simultaneous exposure to the two agents is common.

The difficulty involved in setting policy for dealing with heat stress rests in the uncertainty surrounding its combined effects with other forms of stress and its subsequent chronic effects. As the current OSHA standard for heat stress is the absence of discomfort and short-term physiological unbalance, the protection offered is limited. The NIOSH criteria document recognizes, though, that available data "indicate the complexity of multiple stresses. They raise fundamental questions as to the validity of [threshold limit values] based on single stress experiments for toxic substances in the presence of heat."[34] Further study is required if other effects which may be more significant are to be used as the criteria for a standard.

Vibration

The preliminary nature of research on vibration prevents the drawing of firm conclusions regarding the adverse consequences for workers' health of exposure to this widespread occupational hazard. Researchers have found that rhesus monkeys subjected to vibration under laboratory test conditions have developed anemia and lesions in the gastrointestinal tract resulting in internal hemorrhaging.[35]

Another researcher, referring to the implication of vibration exposure for human health, has pointed out that

> at resonant frequencies [when the body is in tune with the source of vibration] the gut mass—that entire area between the pelvis and the rib cage—tends to move up and down and create pressure on the heart and lung area. This is a very stressful situation. Such movement can cause lesions in some organs and blood vessels.
>
> We've also found that resonance of the respiratory system can make voice communications very difficult, because there is an oscillating sound superimposed on the vocal sounds. The eyeball also has a distinct resonance when a person is subjected to vibration, and this can affect vision somewhat.[36]

Thus, vibration can be a serious form of stress, which may mean that this hazard has far broader adverse implications for physical and mental health than are presently understood. Vibration has also been causally linked to low back pain, a pervasive complaint among truck and bus drivers, as well as members of other occupational groups.

Radiation

Ultraviolet (peak around 3000 Å) Exposure to high intensities of ultraviolet light is common to welders, glassblowers, steel-mill workers, foundry workers, electricians, etc. The effects occur mainly in the skin, resulting in actinic skin, but the eyes can become severely burned if inadequate precautions are taken. Skin cancer has been associated with prolonged exposures to ultraviolet radiation.

Infrared The action here is thermal; therefore, the body is warned to remove itself from the exposure. However, the eye may be burned by levels of infrared radiation which do not burn the skin. Workers exposed to this form of radiation include blacksmiths, furnace and foundry workers, glassblowers, kiln operators, welders, and steel-mill workers.

Ionizing radiation This type of radiation includes both electrically charged or neutral particles and electromagnetic radiations that interact harmfully with body tissue. Corpuscular radiations include alpha particles, beta particles, neutrons, and protons. Electromagnetic wave radiations include gamma and Xrays. These radiations all differ in their penetrating powers and in the extent of the damage they cause as they move through tissue. The ionizing radiations are produced both naturally, by radioactive decay processes, and artificially, by such devices as accelerators.

The biological effects of ionizing radiation may be either somatic or genetic. Somatic effects include radiodermatitis, chronic or acute radiation syndrome, skin cancer, leukemia, cataracts, sterility, and life shortening. The more severe somatic effects are not common in industry, and usually result only from accidents.

Genetic effects are not well understood. Most evidence is from animal studies or is epidemiologic in nature (e.g., studies of offspring of people exposed to radiation in Hiroshima and Nagasaki). It has been shown that ionizing radiation is capable of producing mutations. The long-term effect of this is dependent upon whether the mutant is able to survive, whether the mutant is able to reproduce, and whether the mutation itself can be passed on.

Ergonomic Factors

Occupational stress is a general term used to refer to responses to stimuli at work that can cause a disruption to the worker's physiological or psychological equilibrium. The stress can be physiological in origin, i.e., caused by physical or chemical hazards, or it can be psychological, i.e., caused by psychological or psychosocial factors.

The reduction of stress and fatigue is crucial to the control of both accidents and disease. For example, there are strong indications that stress plays a major role in heart disease, so that redesigning jobs may be an important means of controlling this ailment, as well as other degenerative diseases.[37] A recent Japanese study on the relationship between life-style, work, and sudden death showed a sharply defined course of susceptibility: executives die from heart failure six times more frequently than office workers, and there is a strong peak in the mortality curves for men in their twenties and thirties.[38]

Throughout this discussion of the various health hazards, the importance of stress (particularly heat and noise stress) as a cocausative factor in toxicity has been emphasized. It is quite possible that this effect may be more important than the direct effects of stress at low levels.

Finally, it should be noted that such aspects of job stress as information overload, role conflict and ambiguity, and other factors involved in the organization of work may well be important determinants of workers' mental health.

Combined Hazards

Hazards, whether chemical, physical, biological, or stress, often combine in such a way that their effects are not merely additive but synergistic. (Occasionally, two hazards may behave antagonistically, i.e., they may lower each other's deleterious effects; for example, tobacco and alcohol work oppositely in their effects on the norepinephrine level.) Carbon monoxide and heat, amphetamines and overcrowding, asbestos and smoking, are examples of agents with synergistic effects. The real world does not contain isolated hazards, yet combination effects are still only rarely studied. Because there are 12,000 listed toxic materials in the United States, most synergisms are likely to be found only after their effects have been observed (as in the case of barbiturates and alcohol).

3.3.3 Severity

Unlike safety hazards, health hazards are often slow-acting, cumulative, irreversible, and complicated by nonoccupational factors. An unguarded blade in a rotary saw presents an obviously severe and imminent danger, but it is often difficult to comprehend the severity or imminent danger of a brief exposure to a potential carcinogen that might take years to cause a tumor or death. However, the "chance" of dying from cancer may be just as high as that of having an accident with the saw. The failure of the injury reporting system and the workmen's compensation system to recognize occupational disease has contributed to the failure of society to realize the severity of occupational health hazards.

3.4 THE MAGNITUDE OF THE PROBLEM AND THE COSTS OF OCCUPATIONAL INJURY AND DISEASE

The present costs of occupational injury and disease are reflected in (1) the numbers of reported job-related injuries, diseases, disabilities, and deaths; (2) lost workdays; (3) the population exposed to potential hazards; (4) workmen's compensation costs; (5) workers' perception of the magnitude of health risks facing them; and (6) suggestive evidence that the magnitude of occupational health and safety problems may in fact be much larger than currently available statistics indicate. To the extent that job disease and injury is treated under existing general health programs, private and public, the costs are not immediately identifiable as being occupationally related. In addition, there are a host of costs not "internalized" in any accounting system.

3.4.1 The Reported Number of Injuries, Illnesses, Disabilities, and Deaths

Official statistics present a grim picture of the costs of occupational injury and disease in the United States. National Safety Council estimates indicate that 14,200 deaths and 2,300,000 disabling injuries (100,000 of which cause permanent disability) resulted from *accidents* at nonagricultural workplaces during 1971.[39] Since the working population of this country numbers roughly 80 million, this would mean that in 1971 more than 1 in 40 workers suffered a temporarily or permanently disabling injury at work. Moreover, these figures do not reflect occupational disease (see Section 3.4.6).

Estimates have been released by the Bureau of Labor Statistics for the rate of occupational injury and disease in the latter half of 1971. These data, the first available under the reporting requirements and system of data collection mandated by the OSHAct, indicate a rate of *recordable* occupational injuries and illnesses of 13.8 per 100 man-years worked. At this rate, it is possible that more than one employee in eight experienced a *recordable* occupational injury or illness during 1971.[40] Altogether, a total of 3.1 million occupational injuries and illnesses were recorded during the last six months of 1971 (exclusive of the public sector, mining, railroads, agriculture, and the self-employed).

Preliminary BLS estimates indicate that 5.6 million *recorded* job-related injuries and illnesses were sustained by American workers during 1972.[41] On the average, at least one out of every ten workers experienced a job-related injury or illness during 1972 in private nonfarm industries other than railroads and mining. Moreover, these estimates, shocking though they may be, seriously understate the magnitude of occupational *disease.*

The rate of industrial injuries has been showing an upward trend in recent years. Along with GNP, industrial accidents have been on the increase since 1958: according to former Secretary of Labor Wirtz, the rate of disabling work-related accidents, measured in terms of the number of disabling injuries per million man-hours worked, grew 20% in the decade following the postwar low year of 1958.[42] Over the ten-year period prior to the enactment of the OSHAct (1961-1970), there was a steady increase in the rate of industrial accidents in manufacturing, amounting to 29% over the entire interval.[43] The last survey conducted by the Bureau of Labor Statistics using the pre-OSHAct reporting methods confirmed the continuation of this upward trend. This survey, conducted in 1971, indicated a continuing increase in the rate of disabling work injuries in manufacturing to a level of 15 per million man-hours worked–the highest rate of industrial accidents in twenty years.[44]

3.4.2 Lost Workdays and Other Measurable Costs to GNP

Of the 5.6 million recorded job-related injuries and illnesses sustained by U.S. workers during 1972, 30% resulted in the loss of one or more days of work. An estimated 25 million workdays–excluding fatalities–were lost during 1972.[45] This reported figure is equivalent to a loss of 100,000 man-years of work. The *recorded* time lost as a result of occupational illness and injury is roughly of the same order of magnitude as time lost due to industrial strife.*

The total cost of occupational injuries–in terms of lost wages, medical expenses, insurance claims, production delays, lost time of coworkers, and equipment damage–was estimated by the National Safety Council at $9.3 billion during 1971, nearly 1% of GNP.[48] This figure, moreover, is likely to be a gross understatement of even the direct costs to GNP of occupational illness and injury. For example, it is possible that a considerable proportion of absenteeism due to sickness, presently attributed to nonoccupational causes, is in fact job-related. On average, each employee is absent from work roughly six days per year, according to the BLS.[49] It has been estimated that reduction of one day per year in the annual rate of absenteeism among the U.S. labor force would add $10 billion to GNP.[50] If hazardous and unpleasant working conditions such as excessive noise are responsible for even one of these six days of absence per year, inclusion of the hidden costs of this occupational hazard alone would result in a doubling of present National Safety Council estimates of the cost of occupational illness and injury. (For a fuller discussion of other components of the economic cost of occupational illness and injury, see Section 7.4.8.)

3.4.3 Population Exposed to Potential Hazards

Of course, the risk of occupational injury and illness is not shared equally by all members of the labor force. Miners, construction and transportation workers, and

*The time lost as a result of occupational injuries has been estimated by Jerome Gordon to be at least an order of magnitude greater than time lost due to industrial strife. According to Gordon, in one recent year, 255 million man-days were lost due to work injuries, as compared to 25.4 million man-days lost to strikes.[46] Data from Great Britain support Gordon's estimates: in 1967, 23 million workdays were lost as reported in the workmen's compensation system as compared with 3 million lost through strikes.[47]

blue-collar and lower-level supervisory personnel in manufacturing industries carry the brunt of the occupational injury and disease burden.

Occupational disease is not confined to the industrial worker. Dentists are being studied for the possible effects of X-radiation, mercury, and anesthetics on their having the highest rate of suicide of any professional group, as well as excess diseases of the nervous system, leukemia, and lymphatic malignancies.[51] Operating-room personnel suffer ten times the miscarriage rate as other nursing personnel and give birth to a larger proportion of children with congenital deformities.[52] Cosmetologists (beauticians) display excess cancer and respiratory and cardiac disease.[53] Administrators are far more likely to develop coronary disease than are scientists and engineers.[54]

Some of the most serious occupational health problems are quite general and widespread throughout industrial environments. Dr. William H. Stewart, former Surgeon General of the United States, reported to Congress in 1968 that 65% of industrial workers were "exposed to toxic materials or harmful physical agents such as severe noise or vibration." Of these, only 25% were "adequately safeguarded" by controls and protective devices. "The remaining workers were plainly unprotected or working in conditions which needed immediate attention."[55] In a recent survey of places of employment in the Chicago area, 72.8% of plants visited were found to have one or more employees subject to occupational health hazards.[56]

OSHA chose to concentrate its early health enforcement efforts on five major health hazards: asbestos, cotton dust, silica, lead, and carbon monoxide. These substances were chosen for inclusion in the Target Health Hazards Program for three reasons: (1) they were deemed serious health hazards by NIOSH; (2) they are substances which can effectively be measured and monitored; and (3) large numbers of workers are exposed to them (asbestos, over 200,000; cotton dust, 800,000; silica, 1.1 million; lead, 1.6 million; carbon monoxide, unknown, but high).[57] A high probability of death or serious impairment of health exists for these substances. For example, it has been estimated that of the 500,000 workers now or previously exposed to asbestos, approximately 100,000 will die of lung cancer, 35,000 of pleural cancer (mesothelioma), and about 35,000 of asbestosis (scarring of the lungs).[58] (The pleural cancer death rate in the general working population would be 50 per 500,000.)[59]

Other groups of workers also suffer disproportionately higher injury and disease. Of the 6000 men who have been uranium miners, an estimated 600-1100 will die during the next 20 years as a result of radiation exposure, principally from lung cancer.[60] Coke-oven workers as a group are two and a half times as likely to die from lung cancer as steelworkers who do not work in coke plants, and are seven and a half times as likely to die from kidney cancer. For topside (of the coke oven) workers with at least five years on the job, the lung cancer rate is ten times the expected–15 deaths out of 132 workers.[61] Dr. Joseph Wagoner, head of the NIOSH epidemiology team, has stated that no matter what industrial group he chooses to look at, his batting average is always high. Industrial workers suffer a disproportionately higher occupational disease rate than the general population.[62]

The incidence of occupational injuries and illness varies widely among industries. The first data collected by the Bureau of Labor Statistics under the new OSHAct reporting requirements–encompassing the last half of 1971–are illustrative in this regard. The rate of recordable injuries and illness per 100 man-years worked ranged from a low of 218 in finance, insurance, and real estate to a high of 2215 in contract construction.[63] Though contract construction accounts for only 6% of private nonagricultural employment, 19% of recorded fatalities (nearly 800) and close to 12% of recordable injuries and illnesses (over 360,000) were sustained by construction workers.[64]

Preliminary statistics for 1972 likewise exhibit considerable interindustry variation in the rate of recorded occupational injury and disease. The number of recorded injuries and illnesses per 100 workers during 1972 by major industry division ranged from a high of 19.0 in contract construction to a low of 2.5 in finance, insurance, and real estate.[65]

Manufacturing, which accounts for approximately one-third of all employment in the private nonfarm sector, had almost half the total injuries and illnesses.[66] The incidence rate for manufacturing as a whole (15.5) was nearly as high as in contract construction. Among the 21 subdivisions within manufacturing, the following experienced particularly high rates of *recorded* occupational illness and injury:

- lumber and wood products (25.1)
- fabricated metal products (22.8)
- primary metal industries (20.9)
- furniture and fixtures (19.3)
- food and kindred products (19.)
- stone, clay, and glass products (18.7)
- rubber and plastic products (18.6)
- transportation equipment (18.1)
- machinery, except electrical (17.0)
- paper and allied products (16.0).[67]

The National Safety Council estimates for 1971 indicate that although farm employees make up only about 4.4% of the workforce, they suffer some 16% of the occupational deaths and 9% of the disabling injuries.[68] Since food may well form the next area of resource crisis, thus providing pressure against increased expenditures for agricultural workers' health and safety, the situation may worsen (see Chapter 12).

The rapid pace of technological change seems geared to ensure that the incidence of occupational health and safety problems will continue to rise if strict preventive measures are not taken. Several years ago during the OSHAct hearings, it was reported that there were already "6000 known toxins in industry about which we know little or nothing," and technological change continually augments this deadly list. Senator Harrison Williams, Jr., of New Jersey, head of the Senate Subcommittee on Labor and sponsor of the OSHAct, quoted an HEW publication which stated that "every 20 minutes, a new and potentially toxic chemical is introduced into industry."[69] In the words of the Assistant Surgeon General,

> Approximately one-half million chemicals are produced and used in this country; some 12,000 of which are in widespread use throughout industry. In addition, an estimated 3000 new chemicals are synthesized each year, with approximately 500 finding use in industry. Add the thousands of physical agents such as noise, ultraviolet radiation, etc., to which numerous workers are exposed and the potential health danger to workers is vast.[70]

During 1973 NIOSH listed 12,000 toxic substances in use in industry. Other technological changes, such as the introduction of high-speed steels and dry drilling, have caused increases in a range of occupational health and safety hazards in many industries, including higher levels of noise, dust, and heat.

3.4.4 Costs of Workmen's Compensation

In 1970 workmen's compensation cost employers $4.9 billion, or 1.13% of covered payrolls.[71] This proportion has risen steadily since 1959, when it was at a postwar low of 0.89%.

Workmen's compensation costs are often substantially greater in particularly hazardous industries. Daniel Berman cites the case of the pulpwood industry, in Mississippi and elsewhere, for which workmen's compensation costs range up to $26 per $100 of payroll.[72] Similarly, Fred Foulkes cites a case from the longshoring industry (traditionally an extremely hazardous activity) in which workmen's compensation costs were sufficiently different between two ports (Philadelphia and Hampton Roads) to influence economic decisions regarding the unloading of ships.[73]

Besides the direct costs of insurance premiums, the medical, legal, and administrative costs of the workmen's compensation system are substantial. In some cases, the present method of administering workmen's compensation actually results in increased pain and suffering for the victims of occupational illness and injury. For example, workers disabled on the job are often instructed by their lawyer to refrain from seeking proper medical attention pending the disposition of their workmen's compensation claim in order to improve their chances of receiving a larger settlement. On the other hand, insurance companies sometimes try to drag out the proceedings as long as possible in an effort to force the claimant to "give in" and obtain medical care. Dr. A. H. Chapman, a Kansas City physician and medical writer, had the following to say about administration of the workmen's compensation system:

I have had experience with the workmen's compensation courts, since every busy physician sees at least a few people who get injured on the job and then end up in the endless grind of the workmen's compensation courts. The workmen's compensation courts are a mess, a pernicious mess. They go on for years on a case, while the worker's condition deteriorates. I remember one lawyer who was defending a man with severe hysterical disabilities after an injury. He stressed that he wanted only an evaluation and letter from me. When I asked if he didn't think it would be a good idea to treat the man, he replied that *it would hurt the case if the man recovered.* The lawyers work on a 50/50 basis; they get half of whatever the patient is awarded.[74]

Proposed reforms are likely to result in considerable increase in the cost of the workmen's compensation system in the near future. The National Commission on State Workmen's Compensation Laws estimates that compliance with all of its recommendations will result in workmen's compensation cost increases ranging up to 65% in some states.[75] These figures, moreover, are likely to be a substantial underestimate, particularly if occupational *disease* becomes more widely compensable as a result of the commission's recommendations. (See Section 7.4.8 and Chapter 8 for a fuller discussion of the costs of the workmen's compensation system and its limitations as a method of internalizing the social costs of occupational illness and injury.)

3.4.5 Workers' Perception of the Magnitude of Their Occupational Health and Safety Problem

Survey of Working Conditions

Surveys conducted on behalf of the Department of Labor in 1969 and 1973 by a team from the University of Michigan's Institute for Social Research provide useful information regarding workers' perception of the magnitude of the occupational health and safety problems that they face. Workers surveyed—1533 in 1969 and 2157 in 1973—were chosen to be broadly representative of the U.S. labor force in a statistically meaningful sense. The results of the 1969 survey are listed in Table 3.1. If anything, the population interviewed overrepresented groups in the labor force less subject to occupational hazard, such as white-collar professions. It is therefore interesting and noteworthy that 38% of workers surveyed in 1969, and 41% of those surveyed in 1973, reported health and safety hazards on the job. Indeed, health and safety hazards were more frequently cited than any other labor standards problem area.[76] Of the workers interviewed, 13% in 1969 and 14% in 1973 reported having actually experienced a work-related illness or injury in 1969, and 42% (5.5% of all workers surveyed) had lost two or more weeks of work for this reason. Two-thirds of this latter group, in turn, reported difficulty in meeting their medical and living expenses while out of work due to occupational injury and illness.[77]

As part of the same surveys, workers were asked to state which job-related prob-

Table 3.1
Rankings of Labor Standards Problem Areas

Problem area	Percentage of All Workers Reporting One Problem or More in the Area[1]	Percentage of Those Reporting the Problem Who Regarded It as Sizeable or Great	Percentage of All Workers Considering Protection against the Problem as Very Important
Inadequate fringe benefits	39[2]	43	43
Health and safety hazards*	38	51	64
Transportation to and from work	35	39	39
Unpleasant physical working conditions*	33	38	51
Inconvenient or excessive hours	30	38	39
Inadequate income	26	62	68
Work-related illness or injury*	13	56	71
Unsteady employment	10	37	59
Occupational handicaps	9	39	39
Invasion of privacy by employer	8[3]	28	48
Mistreatment by employment agencies (within past 3 years)	7[4]	68	50
How democratically one's union is run	6[5]	58	56
How well one's union is managed	5[5]	58	56
Failure to receive wages or salary due (within past 3 years)	5[6]	43	68
Age discrimination	5	35	48
Inadequate expense coverage during work-related illness or injury (within past 3 years)*	4[7]	39	69
Race or national origin discrimination	3[8]	53	47
Sex discrimination	3[9]	42	38
Wage garnishment or assignment (within past 3 years)	2[10]	72	57

*Health- and safety-related items.
1. The percentages in this column are based on the full sample. Those in footnotes are based on all workers eligible to experience the problem.
2. 45% of wage and salary workers.
3. 9% of wage and salary workers.
4. 47% of those who had dealt with an agency in the past 3 years.
5. 18% of union members.
6. 6% of those who had worked for wages or salary at some time in the past 3 years.
7; 68% of those who had been away from work for 2 weeks or more due to a work-related illness or injury at some time in the past 3 years.
8. 17% of blacks.
9. 8% of women.
10. 2% of those who had worked for wages or salary at some time in the past 3 years.

Source: U.S. Department of Labor

lems were in need of attention. A large proportion of blue-collar unionists in the 1969 sample listed "good health and safety practices" as being "very important" in relation to other labor standards problem areas. The labor standards problem listed next often among this group as being "very important" was "contingency protection" (like workmen's compensation). In the nature of their responses, blue-collar unionists differed markedly from other groups such as white-collar workers, top management, and, interestingly, their own top union officials (see Table 3.2).

These surveys are also potentially capable of providing valuable information regarding the distribution of the burden of job-related hazards among occupations, industries, and demographic groups. Hopefully, such analysis will be forthcoming in the near future.

Allied Industrial Workers' Survey

A recent membership survey conducted by the research staff of the Allied Industrial Workers International Union (AIW) gives similar results.[78] The AIW is a medium-sized Milwaukee-based labor union representing production and maintenance workers in the metalworking industry. A questionnaire on job satisfaction was printed in the February 1973 issue of the *Allied Industrial Worker*, the union's journal. Asked to rate the importance of a variety of possible job improvements—(1) better health and safety practices, (2) better contingency protection (e.g., workmen's or unemployment compensation), (3) more pay, (4) improved fair employment practices (no discrimination on account of sex, race, or age), and (5) increased opportunity to do interesting and satisfying work—a higher proportion of respondents (79%) rated better health and safety practices as "very important" than any

Table 3.2
Ranking of Work Elements According to the Percentage of Respondents Who Felt the Need for Attention to Be "Very Important"

	Top Management	Top Union Officials	White-collar Workers	Blue-collar Unionists
Adequate income	1st	1st	2nd	Last
Opportunity to achieve and grow on the job	2nd	4th	1st	3rd
Good health and safety practices	3rd	2nd	4th	1st
Fair employment practices	4th	3rd	3rd	4th
Contingency protection	Last	Last	Last	2nd

Source: Neal Q. Herrick, "Institutional Attitudes Toward Human Fulfillment Through Work," draft for submittal to *Monthly Labor Review* (July 1971), Table 2, p. 29. Quoted as Table 14 in Daniel M. Berman, *Death on the Job: The Politics of Occupational Health in the United States* (unpublished manuscript, 1974), p. 62.

other listed job improvement.[79] Next came "better contingency protection," rated as "very important" by 71% of respondents.*

In the words of the AIW's research director, "In the face of considerable controversy over the issue of job safety and health as a major federal concern, it is interesting to note that the results of the present study indicate that union members do have a real concern about this issue."[80]

3.4.6 Suggestive Evidence on the True Magnitude and Social Cost of Occupational Safety and Health Problems

Official statistics barely represent the tip of the iceberg as far as the magnitude of the U.S. occupational health and safety problem is concerned. In a report done for the BLS to assess the adequacy of pre-OSHA occupational injury statistics, it was found that among an intensively studied sample of California firms, the ratio of serious injuries to reported disabling injuries was ten to one. This finding suggests that the true national level of "serious" but not "disabling" occupational injury is nearer to 25 million than 2.5 million per year.[81] Such chronic underreporting was found to be attributable in part to the widespread practice of shifting injured workers to "less demanding" jobs and other similar abuses, methods of avoidance, and lax reporting practices.[82] Sometimes a "less demanding job" meant lying in bed in the plant dispensary all day.

Previous official estimates presented by the National Safety Council and the Bureau of Labor Statistics, though clearly inadequate as indices of the true extent of occupational injury, have been even more seriously deficient with regard to the potentially far more important social problem of occupational *illness* and *disease.* In this respect, our data-gathering agencies have mirrored—and reinforced—the pervasive bias toward safety rather than health noted elsewhere throughout this report.

Prior to enactment of the OSHAct, there was no systematic attempt to collect data on the incidence of occupational illness and disease—as opposed to injury—in the United States.** One result of the OSHAct has been the initiation of new record-keeping and reporting practices which require firms to record each occurrence of occupational disease among their employees. Yet the results of the first eighteen months of data collection under the new statistical reporting systems indicated that OSHA had been seriously deficient in carrying out its congressional mandate to institute accurate reporting practices with regard to the incidence of occupational

*It is also of more than passing interest that "increased opportunity to do interesting and satisfying work" was rated "very important" by 68% of respondents.

**In 1965, 27,000 occupational diseases were reported in California, a state where such disease is reportable and where there are vigorous health and safety programs. If this rate holds nationally, then *there are every year at least 336,000 cases of occupational disease among the 75 million employed civilians in the United States.*[83]

disease. According to BLS statistics, occupational *illness* accounted for under 5% of the total incidence of occupational *injury* and *illness* during the period between July 1, 1971, and December 31, 1972. These estimates reveal that roughly 133,000 cases of occupational illness, a large proportion of which were of the relatively non-serious variety (dermatitis, for example, accounted for 33.9% of all recorded occupational illness), and only 600 deaths from occupational disease were recorded during the last six months of 1971.[84] Yet other estimates indicate that the true incidence of the more serious types of occupational disease is substantially greater. The Public Health Service estimate is 390,000 new cases of occupational disease annually, and an estimate of one million new cases caused by exposure to industrial pollutants in the workplace has also been made.[85] Epidemiological analyses of excess mortality among workers in several industries suggest that as many as 100,000 deaths occur each year as a result of occupational disease.[86] Thus, we are faced with estimates for death from occupational disease which range from about 1000 to 100,000 annually. The number is most certainly greater than 1000. Whether it is in fact 10,000 or 50,000 should not prevent the recognition that there is a serious problem.

Evidence from the insurance industry indicates that the excess risk of death, in actuarial terms, is substantial among certain occupational groups. This *excess* risk is on the order of magnitude of the normal risk of death; in other words, the probability of death (given age, race, sex, etc.) facing members of certain hazardous occupations is in some cases greater than twice the norm for a matched cohort drawn from the general population. For example, according to 1967 life tables, the probability of death for white males age 35 was 2 in 1000; for lumbermen, boilermakers, and structural-iron workers in this age-race-sex cohort, the probability of death was over 4 in 1000.[87]

Occupational factors may well play a far more significant role than is presently realized in the causation of the major diseases and health problems which face us. For example, heart disease, the leading cause of death in the United States (accounting for 38.7%, or about 750,000, of the 2,000,000 deaths reported annually),[88] is only 25% "explained" by known physiological and environmental factors, such as overweight, hypertension, serum cholesterol, and cigarette smoking.[89] An unknown, but quite possibly substantial, proportion of the 75% of heart disease risk which is presently unaccounted for could be related to work and its attendant hazards, particularly stress.

Cancer is the second leading cause of death in the United States today, with an annual toll of 300,000. The incidence of cancer has risen rapidly with industrialization: in 1900, 3.7% of deaths were due to cancer, but by 1968 the proportion of deaths from this cause was 16.5%.[90] In part, improved diagnosis and longer life expectancy due to decreasing prevalence of previous health "scourges" are responsible for this dramatic increase. Nevertheless, there are indications that the *true* incidence of cancer has in fact been on the upswing (see Figure 3.1).

Research in the United Kingdom indicates that over 80% of cancer is of environmental origin and is therefore, theoretically, preventable.[91] This conclusion is based upon the observation that the variation among geographic and occupational environments in cancer incidence rates is enormous. The HEW Task Force on Research Planning in Environmental Health Science reported that

> there is abundant evidence that the great majority of malignant neoplasms—probably over 90 percent of the total—are induced, maintained, or promoted by specific environmental factors. Many of the known environmental causes of cancer are physical and chemical agents that directly concern the environmental health professions. Carcinogenesis must therefore be regarded as one of the most significant potential consequences of environmental contamination.[92]

Of the 80 or 90% of cancer which could be environmentally caused, it is not presently known how much is occupationally related. The experience of chemists, asbestos workers, underground uranium miners, rubberworkers handling vinyl chloride, and other occupational groups amply documents the probability that "excess" cancer of various types is indeed occupationally related.*

Chronic diseases of the respiratory system have been reported with increasing frequency in the United States and have become a major cause of death and disability.[95] Chronic bronchitis and emphysema are the fastest-growing diseases in the United States. They have doubled every five years since World War II, and account for the second highest number of disabilities under social security.[96] Emphysema and chronic bronchitis comprise 75% and 19%, respectively, of all reported (non-cancer) respiratory disease;[97] they account for more days away from work and school than all other diseases combined, and they cause more than one billion days lost from work annually.[98] One insurance company gives the following assessment of the prevalence of chronic respiratory diseases:

> More than 30,000 deaths are attributed each year to the chronic respiratory diseases as the primary cause of death. . . . One inquiry into the presence and severity of chronic respiratory diseases in deaths of persons aged 35-74 indicated that *the prevalence of these diseases at time of death was more than nine times greater than*

*For example, C.E. Searle has noted that the incidence of bladder cancer among workers exposed to certain chemical substances ranges from 50 to 100%.[93]

Similarly, in an examination of the cause of death among 2152 chemists aged 20 to 64, it was found that there were 444 deaths from cancer, whereas 354 were expected—a finding which was highly statistically significant.[94] It can be hypothesized that this excess mortality due to cancer was linked to the carcinogenic nature of certain chemical agents to which these chemists were exposed during the course of their occupation.

In 1968 I was told by a statistician for the National Air Pollution Control Administration that he had observed that chemists as a group die five years sooner than physicists or mathematicians whose mortality and morbidity characteristics are essentially identical. These findings would seem to indict chemicals in general as a cause of death, and would seem to make the suggestion that only a few chemicals are doing the harm, improbable. (I was unable to locate published results of these findings.)

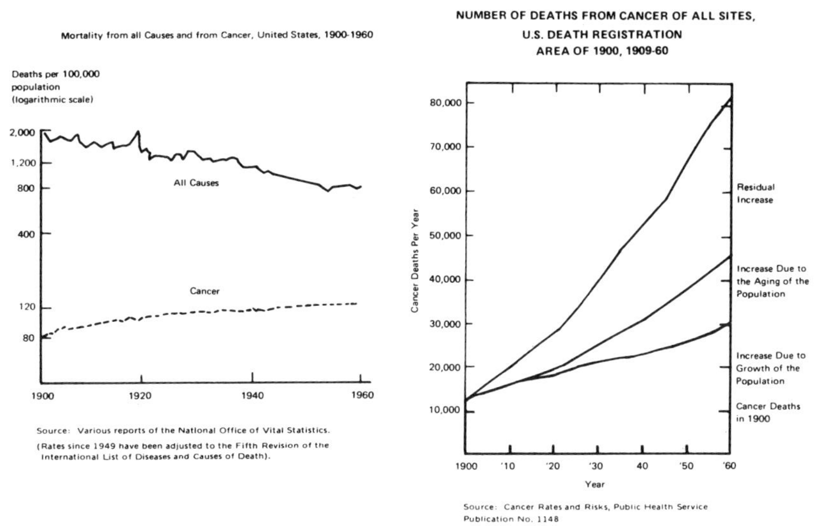

Figure 3.1 Incidence of cancer in the United States, 1900-1960. From *Chemicals and Health*, Report of the Panel on Chemicals and Health of the President's Science Advisory Committee (September 1973), p. 153.

that reported for these diseases as the underlying cause of death. The consensus is that existing coding practices have the effect of underestimating the contribution of chronic respiratory diseases to mortality in the United States.[99]

Thus, the estimate by the National Lung and Heart Institute Task Force Report on Respiratory Diseases that, excluding cancer and certain other respiratory diseases, 150,000 deaths annually occur as a result of respiratory disease, seems realistic.[100]

Much of the respiratory disease which plagues the worker is also known to be job-related. Both "specific" diseases, such as black lung and general emphysema, and chronic bronchitis can be of occupational origin. The Public Health Service estimates that *of the 390,000 new cases of occupational diseases which occur annually, a good percentage of them are respiratory disorders.*[101] The National Tuberculosis and Respiratory Disease Association (now the American Lung Association) has also noted the occupational link of respiratory disease:

According to Edward Baier, deputy director of NIOSH, the death rate from respiratory disease among coal workers is five times that of the general population of workers. Furthermore, although coal miners have a higher than average rate for occupational accidents, there are four deaths from coal miner's pneumonoconiosis for every one from accidents.[102]

The President's Science Advisory Committee conservatively estimates that at least 9000 deaths annually have occupational respiratory disease as a primary contributing cause. Their estimate for deaths linked to air pollution also numbers 9000.[103] (Note: the sum of these estimates for occupational and nonoccupational respiratory disease is 18,000. If, as has been suggested, an underreporting by a factor of nine occurs, the total estimate for deaths due to respiratory disease is 162,000—in good agreement with the 150,000 estimated by the National Lung and Heart Institute.)

Even if only 5% of heart disease, cancer, and respiratory disease were in fact occupationally related, the annual number of occupationally caused deaths would be 60,000. Thus the Public Health Service estimate of "as many as 100,000" may not be unreasonable.

Mental illness is another leading public health problem which may be far more extensively job-related than is presently realized. The contribution of occupational stress and hazardous, unpleasant, or monotonous working conditions to the promotion of mental illness is not well understood. At the present time, though, the incidence of mental health problems is known to be substantial among certain occupational groups.[104] Recent evidence that certain prevalent forms of mental illness, such as schizophrenia, may be caused by chemical agents adds another dimension, potentially of great importance, to the possible links between occupational factors and mental health problems.

Since the original submission of this report, a NIOSH-sponsored study has been completed by the University of Washington.[105] The study indicates that 31% of over 1100 medical conditions found in 908 participants were of probable occupational origin, with an additional 10% indicating "suggestive history." The probable incidence of occupational disease was 28.4 per 100 workers. Only 2% of this occupational illness had been reported on the employer's log (required by OSHA), and only 3% was found in workmen's compensation records.

The occupational disease profile found in the study (see Table 3.3) indicated that skin diseases, once thought to comprise the bulk of occupational disease, in fact accounted for only 18%. Hearing loss accounted for 28%, and the remaining 54% was accounted for by respiratory conditions, toxic effects, eye conditions, anemia, and diseases of the musculoskeletal and connective tissues. This survey confirmed what previous epidemiological studies had suggested: a large portion of serious disease is of occupational origin.

For all of these reasons, then, we can be certain that presently available statistics seriously understate the enormity of America's occupational health problem and the attendant social costs of hazardous working conditions. (For a fuller discussion of the nature of the social costs imposed by hazardous working conditions—costs which are not presently "internalized" in any accounting system—see Section 7.4.8.)

3.5 RESEARCH EFFORTS AND RESEARCH NEEDS

This section and the next are concerned with the present and future role of research in expanding our understanding of the causes of occupational injury and disease and improving the methods of prevention. This section outlines the present distribution of efforts, and discusses some recent attempts to assess what research is needed. Section 3.6 focuses on three controversial issues in order to emphasize the

Table 3.3
Medical Conditions by Occupational Relationship

Condition	Total	Probable Occupational Disease[1]	Can't Evaluate[2] and Doubtful[3]	Suggestive History[4]	Probable Nonoccupational Disease[5]
Anemia and other blood diseases	49	7	21	1	20
Conjunctivitis and other diseases of the eye	74	31	1	24	18
Refractive errors	84	0	1	0	83
Hearing loss, mild	86	68	12	0	6
Hearing loss, moderate	32	26	4	0	2
Hearing loss, advanced	4	2	2	0	0
Hearing loss, possible threshold	16	0	16	0	0
Hypertension	138	1	0	0	137
Heart disease	14	0	1	0	13
Other diseases of the circulatory system	3	0	0	0	3
Respiratory conditions excluding upper	289	49	76	11	153
Upper respiratory conditions	73	38	2	21	12
Skin conditions	128	64	4	35	25
Diseases of musculoskeletal and connective tissues	28	9	2	7	10
Symptoms and ill-defined conditions	17	1	6	1	9
Nonsymptomatic conditions	27	27	0	0	0
Toxic and other adverse effects	23	9	1	13	0
Low-grade toxic effects	14	13	1	0	0
Other conditions	17	1	1	0	15
Total	1116	346	151	113	506

1. Manifestations of disease are consistent with those known to result from excessive exposure to a given injurious agent; this injurious agent is present in the patient's working environment, and significant contact in the course of usual duties is likely.
2. An association between the disease and occupational factors has never been clearly established (e.g., cardiovascular disease, respiratory infections, mental and psychoneurotic episodes).
3. Manifestations of disease are not entirely consistent with those known to result from excessive exposure to an injurious agent *or* manifestations of illness are consistent with those known but no significant contact with the alleged injurious agent can be established.
4. Manifestations of disease as given in history are consistent with those known to result from excessive exposure to a given injurious agent; this exposure is taking place or has taken place; there is no objective evidence of disease at the time of the medical examination.
5. Manifestations of disease but no relationship to occupational exposures.

Source: D. P. Discher et al., *Pilot Study for Development of an Occupational Disease Surveillance Method*, HEW Publication No. (NIOSH) 75-162.

deficiencies in our present knowledge; these are the alleged phenomena of accident proneness and hypersusceptibility, the idea of a finite threshold for toxic materials and the applicability of this idea to carcinogens, and the role of stress in the etiology of occupational illness and disease.

A wide range of research is relevent to the understanding of occupational health and safety problems. Industrial pollution has consequences both inside and outside the industrial operation. Thus, the entire body of environmental health science and medicine and public health research is important for occupational health, and vice versa. This importance has four dimensions: (1) however serious the implications of pollutants are for public health, their effects are very often felt more strongly where the concentration is the highest—in the industrial setting; (2) very often persons who work in industrial plants also live near those plants and are thus put in a position of "double jeopardy"; (3) adverse health effects will often appear first as a result of occupational exposure to high concentrations of previously unknown hazards; and (4) efforts to clean up the factory may result in polluting the general environment, and, conversely, efforts to "contain the pollution" may mean increased pollution within the plant. The problems of the general and workplace environments are intimately connected and must be attacked together.

The effects of stress are increasingly implicated as a cause or cofactor in heart disease and other degenerative disease. Research to clarify the role of stress therefore has general as well as occupational health importance. Research into the causes of accidents, identification of allegedly "accident-prone" persons, and motivational studies likewise have far-reaching implications for both occupational and general safety.

3.5.1 Distribution of Research Efforts

Environmental health research is supported primarily by the government, either in the several agencies responsible for the various health problems or at the schools of public health and environmental medicine. Safety research is performed in a few scattered university schools for safety and departments of industrial engineering, and in private organizations such as the National Safety Council, the National Fire Protection Association, or the insurance companies.

The distribution of research efforts in environmental health among government agencies during fiscal 1972 was studied by the President's Science Advisory Committee, with the following results:

> There are four principal contributing agencies (EPA, AEC, HEW, and DOD). The major contributors in HEW are the National Cancer Institute, the National Institute of Environmental Health Sciences, National Institute of Occupational Safety and Health, and FDA. In addition, there is a residue of research scattered widely among a number of other agencies (USDA, USDI, NSF, NASA, and other National Institutes of Health). [See Table 3.4.]

Table 3.4
Total Level of Federal Effort in Environmental Health Research for Fiscal 1972 (Millions of Dollars)

Agency	Level of EH Research
AEC	40.1
EPA	16.8
DOD	23.4
HEW	23.4
NCI	(38.6)
NIEHS	(15.6)
FDA	(49.7)
NIOSH	(14.8)
Other	60.9
Total	215.1

Source: *Chemicals and Health*, Report of the Panel on Chemicals and Health of the President's Science Advisory Committee, Science and Technology Policy Office, National Science Foundation (September 1973).

The total level of investment in environmental health research (according to a broad, inclusive definition) for FY72 is $215 million. This includes research on infectious agents as well as chemical and physical ones. It also includes work dedicated to physical measurement and characterization of environmental agents as well as research in behalf of control.

If one excludes biological infectious agents, the total investment in this research area is $124 million. [This is summarized in Table 3.5.] Note that of the total of $124 million, nearly $46 million or 37% is spent supporting research on ionizing and electromagnetic radiation—mostly by AEC.

To determine the amount of effort aimed solely at an understanding of the effects of chemical and physical agents on biological organisms in a narrow sense, figures corresponding to certain categories have been extracted from these tables. These reflect the research done to promote understanding of absorption (or entry) of the agents into the organism, metabolism, distribution, and mechanisms of biological action, modification, interactions, and biological consequences or end-points. *It is these figures which come closest to a representation of the meaningful biological research effort in this field.* [See Table 3.6.]

From these figures one can conclude that the FDA and EPA are generally poorly equipped scientifically (16% of the total research effort). The AEC contributes 32% and NIH 28%. There is a heavy emphasis on radiation research—most of it on effects of ionizing radiation and most of this in the AEC. The DOD contributes less than 8%, and almost all of this is concerned with radiation and other physical agents.

In addition, the committee's report from which these figures were taken observed that the investment in research on mechanisms of biological action may be thought of as an index of the sophistication of this research effort. The total investment in

Table 3.5
Total Federal Research Effort (Broadly Defined) in Behalf of Physical and Chemical Agents for Fiscal 1972 (Millions of Dollars)

Agency	Physical	Chemical	Total
EPA	3.6	11.6	15.2
AEC	35.7	4.3	39.9
HEW	9.9	32.6	42.5
NCI	(0.9)	(12.7)	(13.7)
NIEHS	(1.6)	(11.6)	(13.2)
NIOSH	(5.4)	(5.9)	(11.3)
FDA	(2.0)	(2.3)	(4.3)
DOD	6.7	0.8	7.5
Other	5.7	12.9	18.6
Total	61.6	62.2	123.8

Source: *Chemicals and Health*, p. 88.

Table 3.6
Federal Research Effort Aimed at Understanding the Human Biological Effects of Physical and Chemical Environmental Agents*

Agency	Physical (Millions of $s)	Chemical (Millions of $s)	Total (Millions of $s)	Percent of Total
EPA	2.8	7.3	10.2	11.8
AEC	25.1	2.7	27.8	32.2
HEW	6.9	26.2	33.1	38.6
NCI	(0.8)	(11.6)	(12.4)	(14.5)
NIEHS	(1.6)	(10.0)	(11.5)	(13.4)
NIOSH	(2.7)	(2.8)	(5.6)	(6.5)
FDA	(1.8)	(1.8)	(3.6)	(4.2)
DOD	6.2	0.6	6.9	7.9
Other	4.9	3.2	8.1	9.5
Total	46.0	40.0	86.0	100.0

*These amounts do not include the appropriation for the Pharmacology/Toxicology Program of the National Institute of General Medical Sciences.

Source: Report of the OST-CEQ Ad Hoc Committee on Environmental Health Research, quoted in *Chemicals and Health*, p. 89.

Table 3.7
Distribution of Federal Support for Environmental Health Research

Agency	In-house	Percent Contract	Grant
AEC	0	100	0
EPA	59	37	4
DOD	74	25	1
HEW			
NIOSH	60	30	10
NCI	0	51	49
NIEHS	27	3	70
FDA	53	47	0
Average	39	42	19

Source: Report of the OST-CEQ Ad Hoc Committee on Environmental Health Research, qouted in *Chemicals and Health*, p. 89.

this research amounts to $11.9 million, and 80% of this research is supported by NIH (64% of it within the National Cancer Institute).

[Table 3.7] is an estimate of the distribution of these research funds according to their route of expenditure (in-house, contract, or grant).

The NIOSH budget devoted to general occupational health problems is actually smaller than the figures above indicate because about 20% of the research is performed in order to fulfill the coal mining research responsibilities under the Coal Mine Health and Safety Act of 1969. An internal NIOSH document circulated in 1973 states:

The operation of NIOSH is contingent upon the commitment of dollar and personnel resources by HEW/OMB to implement responsibilities under the Occupational Safety and Health Act and the Coal Mine Health and Safety Act. There is a continuing disparity of allocation of resources with counterpart activities in both the Department of Labor and the Department of the Interior.

Under the Federal Coal Mine Health and Safety Act of 1969, the allocation of dollars to the Bureau of Mines for "health research" in coal mining exceeds by several fold the allocation to NIOSH. The irony of this is that health research authority was directed specifically to the Secretary of HEW under this Act. The Department of the Interior has responsibility only for safety research. A key executive decision would be needed to re-allocate the federal coal mine health research dollar into the appropriate agency.[107]

The reader is referred to Sections 6.3.2 and 6.4.2 for further analysis of NIOSH's research.

The *Chemicals and Health* report thus indicates that the priorities in environmental health research are neither balanced nor coordinated. While much environmental health research is relevant to the occupational health problems, in fact, the interaction between NIOSH and the other governmental agencies is minimal. The research is of a basic nature and not immediately useful for attacking the critical occupational health needs.

3.5.2 Analysis of Research Needs

Three useful studies have contributed to the delineation of the research needs in environmental health sciences: *Chemicals and Health* (1973), *Man's Health and the Environment–Some Research Needs* (1970), and *Man, Materials and Environment* (1973).

The *Chemicals and Health* report indicates that aside from the problems of smoking, alcohol, and illicit drugs, a substantial proportion of the deaths related to chemical factors arise from unknown chemical factors in the environment which initiate or promote cancer.[108] The report states that exposure of workers on their jobs is of much less significance than general environmental exposure.* If cancer is, in fact, contributing significantly to the deaths of workers and if one counts cancer in the figures for occupational exposure as well as respiratory disease, the job-related causes of death may be much more significant than the general environmental causes. The committee that prepared the report placed cancer in the general environmental category. Placing it with the causes in the workplace environment may be more justified, especially when one realizes that the standards for workplace exposure are one to two orders of magnitude less stringent than those for the general environment[111] –if they are enforced at all (see Table 3.8).

A much more valuable report entitled *Man's Health and the Environment–Some Research Needs* was completed by the HEW Task Force on Research Planning in Environmental Health prior to the passing of the OSHAct. The report contains an extensive discussion of the research needs regarding industrial exposures. Among the findings of the report were the following:

- Advances in technology are rapidly changing the industrial environment and in many instances are creating new health problems. (p.12)
- Skill and experience in certain industrial accident studies emerge as preventive factors in reducing accident rates. This possibility should be explored and correlated with investigations of the role of improved product design in the acquisition of skill. (p.13)

*The Panel on Chemicals and Health of the President's Science Advisory Committee displayed considerable naiveté when it concluded, "Another example of voluntary acceptance of high risk is the field of occupational exposures. Some occupations can represent very high probability of death and injury when compared to other involuntary activities. Of course, it can be argued that voluntarism does not completely describe occupational choices."[109] Volunteerism does *not* characterize exposure to toxic materials whose effects are undetermined or unkown. (See Sections 7.4 and 7.5 for a discussion of risk analysis.) In another part of the report, the Committee says, "Fortunately, many industrial chemicals do not reach the general public."[110]

• The biochemical, pharmacologic and immunologic mechanisms which lead to lung damage by environmental agents are insufficiently known. These gaps in our knowledge are to a large extent the consequence of: one, the non-existence of an effective, systematic, coordinated and basic-science-oriented national program on the causation, prevention, and treatment of lung disease; two, the non-existence of an effective, systematic, coordinated and basic-science-oriented national program in occupational health. (p.79)
• There is abundant evidence that the great majority of malignant neoplasms–probably over 90% of the total–are induced, maintained, or promoted by specific environmental factors. Many of the known environmental causes of cancer are physical and chemical agents that directly concern the environmental health professions. *Carcinogenesis must therefore be regarded as one of the most significant potential consequences of environmental contamination.* (p.147)
• Rather than depending on hope, luck and hindsight (as has often been the case up to now) more careful surveillance of the health of the population and more systematized testing of the safety of individual chemicals now in use, or proposed for future use, must be employed. (p.15)
• Among the potential effects from chemicals released into the environment, three are of particular concern. These are carcinogenicity, mutagenicity, and teratogenicity. Contributing to the complexity of scientific investigations on these effects are the following shared features: their insidious nature; the relatively long time lag between exposure and overt effect; the irreversible nature of the resulting disease; the relatively great susceptibility of immature or developing tissue; the ability of etiologic factors, other than the one of interest, to produce the same endpoint; and the aggravation of effects from synergistic interactions. As a result, it is difficult to develop simple and reliable tests for evaluation of the products of new technologies and, in fact, *new products are being introduced into our society faster than their safety can be ascertained.* (p.15; emphasis added)
• A separate report is made for the problems of carcinogenesis, mutagenesis and teratogenesis since the problems inherent in assessing carcinogenic, mutagenic, or teratogenic potential of any physical, chemical, or other toxic influences are not simple extensions of the general difficulties which are dealt with in estimating the pharmacologic or toxicologic properties of such agents. *The problems of assessing carcinogenesis, etc., can differ both qualitatively and quantitatively from those encountered in routine toxicity evaluations.* (p.167; emphasis added)
• In spite of the large investment of time and effort in research in toxicology, there are several areas that are relatively unstudied, particularly with respect to the needs of environmental health science. These include: (a) the problems of extrapolating results from animals to man; (b) the detection of early or relatively low levels of injury; and (c) approaches to the study of interactions of toxic agents to produce synergism or antagonism. (p.16)
• *Social and psychological* processes help to create environmental *hazards that contribute as much to morbidity and mortality as do many of the hazards that act through purely physical and chemical mechanisms.* In addition, there are innumerable instances in which the technology is available for the prevention or elimination of environmental hazards, but it is not employed for personal, social, or economic reasons. Recognition of the importance of these factors is exemplified by the fact that consideration of their implications may be found in nearly every chapter of the Task Force report. (p.16; emphasis added)

Table 3.8
Recognized Threats Arrayed Both by Their Size (% of Deaths in 1967) and by the Character of the Actions Involved

Deaths Linked to By-Products of Health-Favoring Actions	Deaths Linked to One's Own Actions or Choices	Deaths Linked to Unknown Sources	Deaths Linked to Collective Actions of Others	Deaths Linked to Individual Actions of Others[1]
	Cigarette smoking (17%)			
Adverse reactions to medication (4%)	Dietary composition (0-20%)	Unknown promoters and initiators of cancer[3] (3-8%)		
	Alcohol abuse (1.5%)			Alcohol abuse (1.5%)
			Air pollution (0.5%)	Airborne particles[4] (0.5%)
	Narcotic and addicting drugs (0.3%)			Narcotic and addicting drugs (0.3%)
	Suicide (0.25%)			
	Coffee (0.15%)			
	Accidents involving chemicals (0.05%)			Accidents involving chemicals[4] (0.5%)
Oral contraceptives (0.01%)				

Table 3.8 (continued)

Deaths Linked to By-Products of Health-Favoring Actions	Deaths Linked to One's Own Actions or Choices	Deaths Linked to Unknown Sources	Deaths Linked to Collective Actions of Others	Deaths Linked to Individual Actions of Others[1]
	Swordfish (0-0.0001%)			
	DDT[2] (0-0.00001%)			

1. Individual firms and organizations as well as individual persons.
2. The effects of accumulated body burden of DDT are unknown; the range given is for recognized acute poisoning, which accounts for at most a fraction of a death per year. No deaths traceable to the intended uses of DDT have been recorded.
3. Probably occupationally related to an unknown but significant extent (author's note).
4. Occupationally related (author's note).

Source: *Chemicals and Health*, p. 36.

Another blue-ribbon report by the National Academies of Science and Engineering entitled *Man, Materials and Environment* addresses itself to broad materials policy issues concerning the manufacturing, use, and disposal of material resources. While the report does not deal specifically with occupational health issues, its findings are relevant to these problems:

- It is in the national interest that policies and practices aimed at satisfying the nation's need for materials essential to social well-being should reflect and accommodate at all stages, from extraction to waste disposal, *considerations of environmental cost to human health*, quality of habitat, and stability of ecosystems.
- This principle should be included in all pertinent policies and legislation stipulating *the right of each citizen to a healthful environment* in accordance with Principle I of the Declaration of the United Nations Conference on the Human Environment.
- The resources of all countries should be regarded as part of an interdependent habitat rather than merely as possible sources of supply; and *our national policy should therefore conform to the principles of conduct adopted by the community of nations* in a common effort to protect the human habitat and its resources.
- While we can today identify and deal with environmental problems relevant to a materials policy, and while we are prepared to propose appropriate remedies, the incomplete state of our knowledge required urgent, *systematic expansion of research and of the institutional arrangements needed to widen the data base.* (Emphasis added)

Most U.S. safety research has concerned itself with automobile safety, transportation safety, fire prevention, and, to a lesser extent, accident prevention at home. Concern for product safety and consumer protection is a relatively new phenomenon. Increased interest in occupational safety is even more recent.

The National Institute for Occupational Safety and Health (NIOSH) is responsible for research in occupational safety as well as occupational health. Due primarily to its health orientation and lack of expertise in safety areas, NIOSH does very little occupational safety research itself. The priorities for safety research have been determined under an outside contract.[112] The objectives and conclusions of the NIOSH-contracted study of needed *safety* research were summarized in a policy analysis by the Government Research Corporation as follows:

The objectives of this study . . . were to assist NIOSH in the development of an occupational safety research program by:

1. summarizing the present status of occupational safety research;
2. assessing the need for developing additional research;
3. developing a priority rating system for occupational safety research projects; and
4. preparing a priority list of specific research projects for NIOSH safety research.

The study focused on the development of criteria and standards, and psychological, behavioral, motivational, engineering, equipment, and educational research.

In line with its objectives, the results of the study were fourfold:

1. the study concluded that safety research in the nation as a whole was limited and inadequate;
2. the study demonstrated the need for a balanced program of safety research in three principal areas: (1) accident causative factors, accident investigations, and statistical studies; (2) general occupational safety research in psychology, motivation, education, training, and management safety practices; and (3) physical safety, particularly the development of criteria and standards;
3. the study developed a priority rating system to rank occupational safety research projects; and
4. the study prepared a priority list of specific occupational research projects.[113]

As mentioned earlier, for the most part, the relevant occupational safety research is done by private organizations (such as the National Safety Council, National Fire Protection Association, and American Society for the Testing of Materials), by insurance organizations, or by a few scattered but important schools for safety and industrial engineering (for example, the New York School for Safety, Texas A&M University, and Auburn University). The private organizations devote a very small proportion of their efforts to industrial safety, while more emphasis is given by some insurance organizations and the universities. The University of Wisconsin School for Workers in Madison also performs safety "research," but of a different kind. Aside from analysis of workmen's compensation records to clarify the causes of accidents, the university develops effective methods for the training and education of workers in occupational health and safety (among other things).

In sum, the three blue-ribbon studies focusing on environmental health science and NIOSH's contract work in occupational safety all indicate that we are seriously deficient in our knowledge of the causation and prevention of occupational (and environmental) disease and industrial injury. Both the lack of knowledge and reluctance to discard old ideas in the light of new evidence have contributed to intense scientific-medical disputes, which exacerbate labor-management conflict, and to the reluctance of government to take any major steps to improve job health and safety. A description of these scientific-medical controversies follows.

3.6 CAUSATION AND PREVENTION OF OCCUPATIONAL INJURY AND DISEASE

It can be said that occupational and environmental injury and disease is the result of man's inability to adapt socially or biologically to rapidly changing technology. Further, technology (and medicine/science) is seen as the instrument by which we may save ourselves from ourselves. Some, feeling futility in the latter statement, would advocate an abandonment of at least some technology. Solving occupational health and safety problems will necessitate both the development of new technology and some abandonment of the old. The form that the solutions take will depend critically on how we view causation, and hence prevention, in specific cases.

3.6.1 Accident Proneness

An excellent short history of scientific management as it applies to accident prevention and of the origin of the concept of accident proneness was given by George Hagglund in an unpublished Ph.D. thesis in 1966.[114] The following abbreviated excerpt is reproduced here with permission of the author.

Early workers in the occupational safety movement relied upon the system of investigating and developing solutions to safety problems that had been originated by Frederick W. Taylor and his associates to increase overall industrial efficiency.[115] A well-known member of the safety movement, H. W. Heinrich, adopted a considerable portion of Taylor's system of scientific management, promulgating what he called "principles of accident prevention," as follows: (1) *Executive Interest and Support*—necessary because accident prevention in the typical firm was not closely enough interwoven with regular business routine, thus making the fixing of responsibility for accidents difficult; (2) *Detailed Cause Analysis*—accident prevention included the need for diagnosis and analysis of every problem before a solution could be decided upon, and solutions varied from one problem to the next because of the distinctive characteristics of individual manufacturing plants; (3) *Selection and Application of Remedy*—since accident causes were related to other causes of faulty production and general industrial inefficiency, an understanding of appropriate remedies in individual cases became clear when probable accident causes were well understood; (4) *Enforcement of Corrective Practices*—methods had to be developed, as for other areas of the industrial enterprise, to see that safe practices were carried out, interest in the problem cultivated, responsibility fixed, and progress recorded.[116]

Heinrich's system of accident prevention met with widespread acceptance with employers in the years following passage of workmen's compensation laws in the United States. His system included careful investigation of accidents when they occurred, but Heinrich stressed avoidance of slavish usage of a limited number of what he called "superficial" solutions to safety problems. Much of his efforts were devoted to the development of a careful system of gathering information and calculating costs of accidents, and he developed systems of machine guarding and safety devices that are still in use today. Heinrich's system was based upon accident prevention by careful collection of information and development of improved working and safety conditions in the industrial workplace.

The National Safety Council subsequently adopted most of Heinrich's approach to accident prevention, and they have since carefully emphasized the need for greater understanding of the nature of potential accident hazards in the work environment. They continue to disseminate information emphasizing the "science" of accident prevention through application of engineering principles.[117]

The term "accident proneness" was coined by Eric Farmer[118] following publication of a study by Greenwood in 1919.[119] Their findings resulted in a change in direction of a considerable number of scholars away from the dimension of the work environment to a study of the individual and his personal contribution to industrial accidents. The quality of "accident proneness" in certain individuals was believed to increase their relative likelihood of involvement in an accident over that of others in society. Some persons appeared to be more likely to suffer minor injuries such as cuts, scratches, or bruises from contact with inanimate objects than their fellow workers; it was generally believed that the accident liability of these persons was due to characteristics associated with carelessness or light-heartedness, and that accidents could be avoided by exercise of proper care and thoughtfulness.[120]

Following earlier findings, a large number of individual researchers, particularly those in the field of industrial psychology, became interested in the problem of accident proneness. They tried by various means, with varying degrees of success, to explain why some individuals appeared to be more liable to accident and injury than the general population. Hundreds of explanations were forthcoming, some supported by voluminous research data and others almost completely speculative in nature. General explanations of accident proneness can be grouped into categories of (1) psychological explanations which traced higher individual accident rates to mental and psychological characteristics; and (2) physical explanations which attempted to demonstrate that certain physiological characteristics led to higher or lower accident frequencies. Still other variables were loosely grouped under the heading "accident susceptibility," which attempted to explain variations in accident frequency in terms of environmental factors not directly associated with the individual. A classic compilation of early studies in these areas was done by Viteles,[121] who also did some of the early research on accident proneness. He concluded that factors appearing to be related to differences in accident rate included (1) age and experience; (2) physical condition; (3) nationality; (4) fatigue; (5) rate of production; (6) atmospheric conditions; and (7) illumination. In a later study Viteles added the factor of sex.[122]

Boyd Fisher, an early writer in the field of industrial safety, compiled the following list of psychological factors believed to be associated with accident proneness: (1) lack of awareness of potential accident hazards; (2) excitability of temperament; (3) a sluggish temperament; (4) deficiency of intellect; (5) inattention; (6) psychomotor retardation; and, (7) a deficiency of English.[123] Unfortunately, many of the preceding were not directly measurable, so Fisher had to content himself with speculating about their association with accident proneness.

In 1949 Alexander Mintz and Milton Blum published an article that challenged the significance of the contribution of accident proneness to the total accident frequency rate.[124] Reviewing the literature on the theoretical function of accident proneness, they observed that the proportion of accidents resulting from suspected personal factors appeared to be evident in only a relatively small proportion of the total number of cases—about 21.2%. They estimated that 60-80% of all accidents were probably traceable to "unpredictable factors." They noted that the commonly used method of comparing the percentage of persons having many or no accidents proved nothing concerning differences in accident proneness. They pointed out that a number of studies related to accident distribution suggested that accident susceptibility varied considerably from one situation to the next. They concluded that accidents are probably due in large measure to specific chance factors which vary enormously from one situation to the next, and, hence, it is almost impossible to determine their contribution to total accident liability.

The criticism of Mintz and Blum evoked a considerable flurry of protest from psychologists who sprang to the defense of the concept of accident proneness. Typical of these was the response of T. A. Ryan and P. C. Smith who contended there was strong evidence for the existence of accident proneness.[125] They recommended that individuals demonstrating higher than expected accident rates should be transferred to less dangerous occupations. They charged that Mintz and Blum drew an extreme conclusion from their data, that low but positive correlations were nevertheless useful indices of accident-prone individuals, that only a negligible number of "safe" individuals would be harmed by a vigorous policy of transferring high-accident individuals to other jobs, and that an individual with a high record of accidents of any kind should not be placed in any hazardous occupation because there was the definite possibility that he would be accident-prone in that occupation also.

A considerable number of excellent reviews of the literature on accident susceptibility and accident proneness have been published in recent years. One of the most complete was published by Leon Brody in 1962.[126] He commented on a large number of studies analyzing a wide variety of factors believed to be related to industrial accidents, and he noted that in recent years there had been a marked increase in the number of investigations into physiological and environmental determinants of accidents.

Hagglund's research sought to clarify the effects of aging and experience on industrial injury. He identifies experimental and physiological factors, rather than psychological, cultural, or social characteristics, as significant determinants of accidents. He also states that the effects of environmental and individual factors are difficult to measure given the present research techniques and data available to the accident research specialist. The literature indicates that age and experience are difficult to study independently of each other. The effects of experience (or lack thereof) are believed to be most evident during a worker's early career, while the deleterious effects of aging are believed to be most evident in later years. Thus, if aging and experience are significant factors in the study of industrial injury, workers in the middle years should show the lowest injury frequency rates.

Hagglund's findings are summarized below:

• A considerable proportion of industrial injuries appears to be associated with youth and inexperience. Youths appear to be more often injured in accidents involving manipulation or operation of equipment and machines.

- Another considerable proportion of injuries sustained by workers appear to be associated with the aging process. Older workers are more often injured in accidents resulting in falls or by being struck by objects.
- When the injury pattern for the population of Wisconsin was computed, older workers were found to incur the largest proportion of industrial injuries. A bimodal pattern was evident, however, with a peak injury rate also evident among youthful workers.
- Injury rates among females appeared to increase steadily with age. Overall injury rates indicated that women are less likely than men to be exposed to hazardous environmental conditions.
- Males suffered the largest number of injuries between the ages of 20 and 25, although adverse effects of both inexperience and age were evidenced by the bimodal injury distribution.
- Youthful workers appeared to suffer from a larger number of temporary injuries, while older workers suffered a larger number of serious injuries.
- Injury patterns differed markedly by age for different occupations. Overall injury rates showed an inverse relationship with occupational level. Effects of age and inexperience were evident at all levels.
- Differential effects of aging and experience were evident when the data were grouped by accident and injury type. For most accident types, however, injury patterns showed a bimodal distribution by age.
- Workers under age 18 and over 65 appeared to be less exposed to hazardous environmental conditions, while workers between ages 30 and 50 are least likely to show adverse effects of aging and inexperience.
- The California data appeared to show that youthful workers suffer the largest number of injuries resulting in short-term disability, and that injury-age patterns differ from one industry to the next.
- Contradictions in findings made from earlier studies may be due in part to dissimilar injury severities analyzed in different analyses.[127]

Hagglund's conclusions in no way preclude the possibility that hazards in the work environment contribute to industrial injuries. The identification of aging and inexperience as significant contributory causes for accidents does not mean that the "worker is to blame" or that work conditions are safe. Accidents are the result of man-environment interactions, and in that sense most accidents are the result of an "unsafe act." If work conditions were made safer and accounted for normal human deficiencies, including aging or inexperience, the unsafe act might not have a chance to occur.

A fashionable modern-day version of accident proneness states that *unsafe acts* rather than *unsafe conditions* are responsible for 90% of industrial injuries. (When I questioned Howard Pyle, then president of the National Safety Council, if he could really say that these "unsafe acts" were in fact the worker's fault, he replied that the state of the art was not sophisticated enough to tell.)[128] Yet management often blames the worker for most accidents.

W. T. Singleton, in a monograph for the World Health Organization entitled *Introduction to Ergonomics*, makes a strong case against the concept of accident proneness:

The general problems of safety and accidents are not usually approached in the traditional way by the ergonomist. For example, he will blame the situation rather than the operator even when the cause is clearly "human error." Human errors are an important source of information about equipment design faults. Thus the attitude is often taken that, if the task has been properly designed from an ergonomics point of view, the solution of problems of errors and safety will have been arrived at almost as a by-product. The study of accidents as accidents is not usually fruitful for a number of reasons. Accidents are rare, and very large samples are needed for statistical significance to be established. By the time this has been done, there are usually so many variables that cause/effect relationships cannot be established. . . .

. . . Accidents are usually literally accidental in their effects if not in their causes. Thus the same human error repeated on many occasions may have had no ill effects, and then, committed once more, it may result in a fatality. Furthermore, statistics can only be about reported accidents, and the tendency to have accidents may be unrelated to the tendency to report accidents. The effect on an individual's accident record on our opinion of his skill is also often unfair, in that exposure to risk is so varied.

In general, then, the ergonomist prefers to approach the problem from a "people-centered," rather than an "incident-centered," point of view. Various concepts of accident proneness have been developed, but some evidence suggests that the concept may not be as simple as it looks at first sight. The established statistical differences between individual accident rates with the same exposure may be due, in some situations at least, to the fact that *each individual has spells when he might be more likely to have an accident, rather than that some individuals are consistently more likely to have accidents.* This would account for the difficulties found in attempting to devise selection tests for the accident prone. This would seem superficially to be one obvious way of reducing accidents, but more practical possibilities are to reduce either the real risk or the apparent risk. Reduction of real risks by better equipment and task design is obviously valuable, but it may not change accident rates. For example, better acceleration and braking characteristics for motorcars improve the design aspects, but unfortunately the driver often adjusts his tolerable margins of error accordingly so that there is no change in the accident rate. This is one illustration of the general point that accidents are more often related to attitude than to skill or equipment design. Most individuals have an unfortunate tendency to regard accident risk as applying to other people rather than to themselves. This is one reason why it is difficult to ensure that workers use the safety devices provided. Although this attitude may be susceptible to slight change by indoctrination and propaganda, *the only real solution is to design in such a way that safety devices and procedures are an inherent and necessary part of machines and tasks, as well as being emphasized in instructions and training schemes*—for example, a fuse box so designed that it is physically impossible to open it without switching off the power is obviously safer than one that merely has a notice on it saying "switch off power before opening box."[129]

Ralph Vernon, head of the Hygiene and Safety Engineering Division, Department of Industrial Engineering, Texas A&M University, has expressed similar views:

I suppose if we wish to be completely honest we could demonstrate that all *acute* exposures result from human error as well as many chronic exposures. However, the human error should not be charged to the injured party. Assume that a man has a finger amputated at the point-of-operation on a punch press. Usually, the human error does not rest with the injured employee; rather, it rests with the human failure of management to assure themselves that the die designer, the set-up man, the foreman, or others did not fail in their responsibilities. However, it has been usual for the "human failure" to be charged to the injured party.

Now, it stands to reason that this *acute* exposure of the unguarded point-of-operation is a *physical exposure*, and any assignment of responsibility should rest there.

I have not accepted the 85-15 [85% unsafe act, 15% unsafe condition] syndrome as factual. It is seldom possible to subscribe any accident to strictly a personal or physical cause. To do this is sheer folly and serves no important function. In fact, Heinrich suggests that one should carefully consider the advisability of assigning accidents to the personal-physical categories.

I think it has become increasingly evident that it is important to consider the seriousness of an accident in classifying "causes." I am more convinced each day that the ratio of physical causes goes up as the seriousness of the accident increases. I believe this to be true of personal injury and property damage accidents. Indeed, *if* the 85-15 ratio is true for all accidents, then the ratio probably approaches 40-60 when the severity of the injury is considered as an important parameter.

. . . We also have the chronic exposure that often results in an occupational disease. Even here, I think some of the practitioners would be willing to charge the events to personal causes even though the agent was clearly known. Such a decision could be made because "the employee was told about the exposure and he exposed himself" or some other comparable excuse.

. . . [Only] very limited data is available on accident causation. Furthermore, when employee injury is involved, it is usual for any fact-finding to be "injury-prevention" rather than "accident-prevention." It is past time when our preventive efforts should be directed to preventing the event that preceded the injury.

How could this be accomplished? I believe consideration should be given to epidemiological studies. Insofar as communicable diseases are concerned, they were rampant until epidemiological studies were utilized in determining causes in population groups. Surely, some of these techniques are as useful in the prevention of occupational trauma as they have been in occupational disease.[130]

Recent unpublished research, in fact, indicates that the 90-10 ratio of unsafe acts to unsafe conditions is a myth. An analysis of accident reports in the state of Wisconsin based on a 10% sample of the reported injuries investigated for causation by the Wisconsin Safety Specialists yielded the data displayed in Table 3.9. A follow-up study of *all* deaths investigated by the Wisconsin Safety and Buildings Division (constituting 19% of the reported job deaths) yielded the data of Table 3.10.

Table 3.9
An Analysis of a Sample of Accident Reports from the State of Wisconsin

Causation	No. of Cases	Percentage
Unsafe conditions:		
Wisconsin Safety Code violation	77	30
Other unsafe working conditions	62	24
Unsafe act or worker's mistake	91	35
Neither (no clear indication on report)	17	7
Both unsafe act and unsafe conditions	10	4
Total	257	100

Source: George Hagglund, School for Workers, University of Wisconsin, Madison, Wisconsin, letter to author, March 11, 1974.

Table 3.10
An Analysis of Death Reports Investigated by the Wisconsin Safety and Buildings Division

Causation	No. of Cases	Percentage
Unsafe conditions:		
Wisconsin Safety Code violations	37	39
Other unsafe working conditions	18	19
Unsafe act or worker's mistake	24	26
Neither (no clear indication on report)*	12	13
Both unsafe act and unsafe conditions	2	2
Total	93	100

*Heart attacks included in this category.

Source: Same as Table 3.9.

While a national extrapolation from one state's data is not possible, these Wisconsin results indicate that a more reasonable estimate is a third or less of accidental injuries or deaths traceable to unsafe acts, not 90%. Even this third may not be the worker's fault. Of course, efforts to caution or train the worker to be vigilant *might* reduce accidents due to unsafe conditions, but this approach in itself has not been very successful in the past, as was illustrated earlier by Singleton. George Guenther, former Assistant Secretary of Labor for OSHA, stated that accident proneness (and its successor, the "unsafe act") is a myth which has no place in a modern technological society.[131] The OSHAct puts the responsibility on the employer to provide a workplace free from hazards. The mandates of the OSHAct, if implemented fully, could rid the industrial workplace of this excuse for the employer not providing a "safe and healthful workplace."

3.6.2 Exposure to Toxic Materials and Carcinogens

The extent to which a particular health hazard should be controlled depends both on its severity and the *dose-response relationship.* In simple terms, the dose-response relationship indicates how a biological organism's response to a toxic substance changes as its exposure to the substance increases. For example, a small dose of carbon monoxide induces drowsiness, but a larger dose can cause death.

In the determination of a dose-response relationship, it is often assumed that a *threshold exposure* exists below which no harmful effect occurs. This relationship is shown graphically in curve A of Figure 3.2. However, there is considerable doubt that the threshold concept holds for radiation damage and carcinogenesis.[132] To put it another way, radiation damage and the initiation of cancer may exhibit a *zero threshold*, as indicated in curve B of Figure 3.2, which means there is *no* dosage that can be considered safe.

Herbert Stokinger has defined toxicity to be the net result of two competing reactions, as shown in Figure 3.3. One reaction (shown on the left) represents the effects of the toxic agent on the body. The other reaction (shown on the right) is the body's adaptive or homeostatic reaction which counteracts the toxin's effect and attempts to return the body to equilibrium. The body's natural cleansing actions and the production of antibodies are good examples. Not until a certain level of toxin is present does the body's adaptive mechanisms fail to counteract the toxin; hence, a threshold is said to exist. It should be recognized that the threshold will usually be higher for *acute* effects than for slower acting and longer-term *chronic* effects; i.e., the amount of toxin needed to produce chronic effects is likely to be very much less than the minimum amount required to demonstrate an immediate, acute toxic effect. (This may hold for the total amount over time as well as the unit dosage at a given time.) The mechanisms for the acute and chronic damage may be very different and may manifest themselves in different parts of the body.

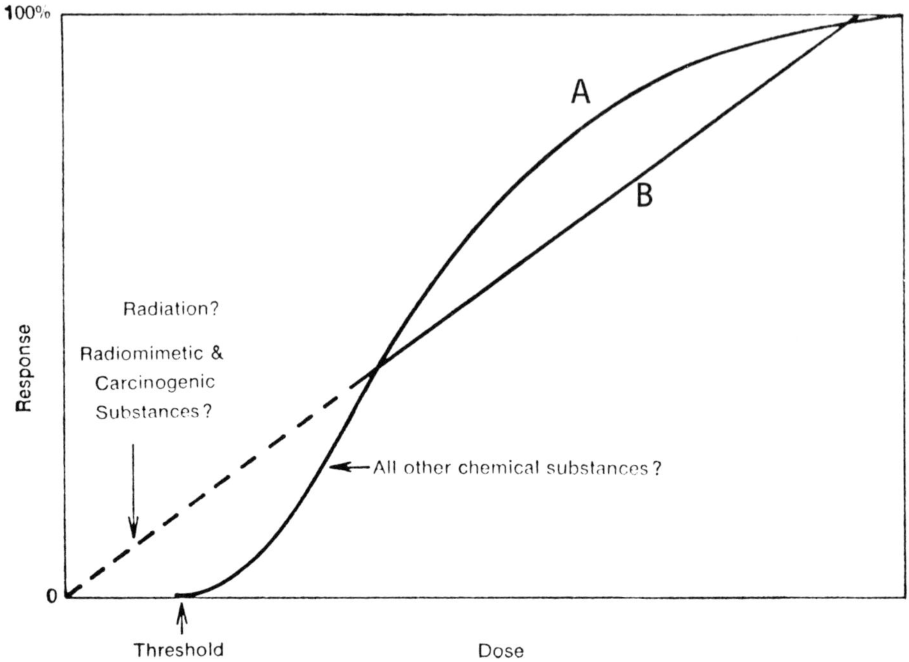

Figure 3.2 Dose-response relationship for a noncarcinogenic (A) and a carcinogenic (B) toxic substance. From Herbert E. Stokinger, "Concepts of Thresholds in Standards Setting," *Archives of Environmental Health*, Vol. 25 (September 1972), p. 155.

Predicting chronic effects from acute reactions is a very difficult task, laden with uncertainties.

The task of setting standards for safe levels of exposure is likewise difficult. It is not possible to determine all the points on a dose-response curve: the lower the dose, the lower the response—and we are limited by inherent difficulties in measuring both near zero. Thus, even for acute effects, we must extrapolate the curve to find the threshold; and this threshold for acute effects may bear no relation to the threshold for chronic effects. (Indeed, not all persons who exhibit chronic effects will display acute symptoms.) Furthermore, if we were to test a large number of people on any particular day, we might find a distribution of thresholds among them (see Figure 3.4).

The distribution can be caused by many factors: (1) natural variance, (2) genetic differences, and (3) other interacting toxins and stressors. The distribution is conveniently viewed as being made up of three parts: the average, the sensitive, and the resilient populations. The concept behind this distribution needs to be stressed because it is often misunderstood and abused in attempts to explain why some people get disease and why some do not.

Because of natural variations in a person's physiology from day to day, he may be sensitive one day but resilient the next. Repeated measurements could determine the extent of this natural variance for acute effects, but the effect of natural vari-

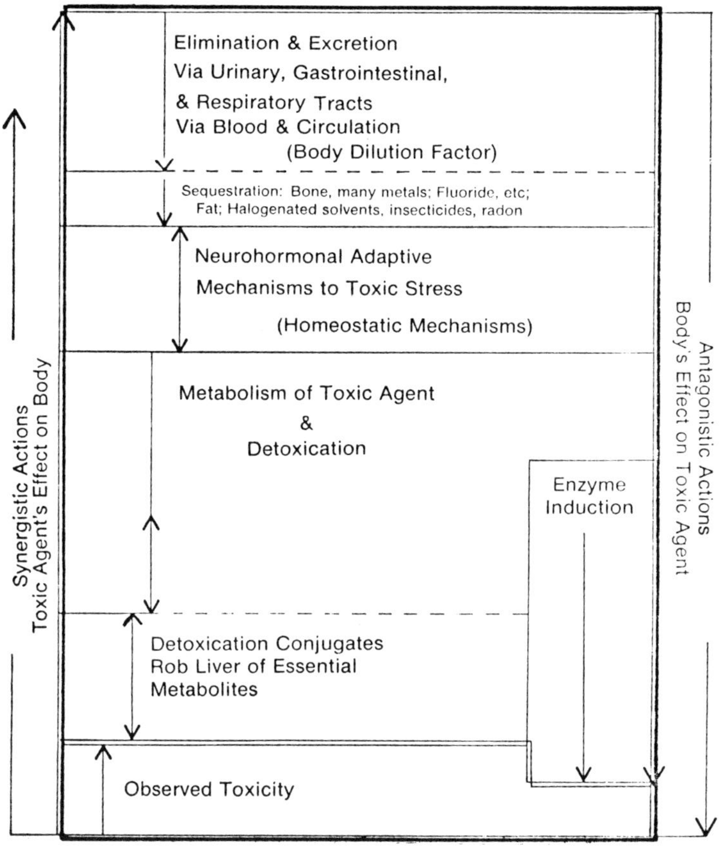

Figure 3.3 Analysis of the effects of a toxic substance. From Stokinger, "Concepts of Thresholds."

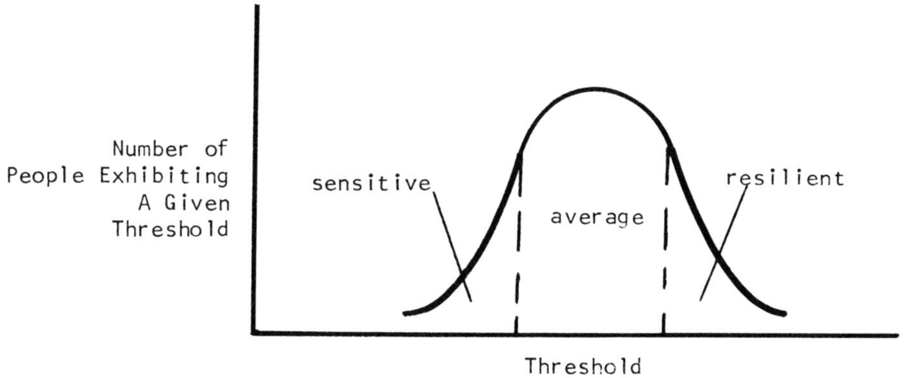

Figure 3.4 Normal "bell-shaped" distribution of thresholds for toxic effects.

ance in the distribution for chronic effects is impossible to determine since an organism only lives once.

The fact that there are genetic differences which can explain the predominance of disease in some people has given rise to the unfortunate concept of *hypersusceptibility* to describe those persons found in the "sensitive" part of the distribution.[133] The term is abused in precisely the same way as the term accident proneness. By pursuing a policy of detecting the hypersusceptible worker for his own good, an industrial physician may perpetuate a "blame the worker" attitude on the part of management. Unfortunately, to many industrial physicians, "preventive medicine" constitutes identifying the hypersusceptible worker and recommending him for a job transfer—or, in the case of a preemployment examination, for no job at all. The removal of health hazards, rather than hypersusceptible persons, should be the goal of preventive medicine.

Of the 92 human disorders for which a genetically determined specific enzyme deficiency has been identified, 5 are reported to meet the "prerequisites for industrial application for better job assignment, and improving coverage of the industrial air limits, and hence, reducing risk to worker health."[134] It may be true that 2% of the population may be allergic (hypersusceptible) to tomatoes, but the 34% of those 500,000 workers alive who will die from asbestosis or asbestos-caused lung and pleural cancer can hardly be called hypersusceptible. A more reasonable explanation for differing susceptibilities to chronic disease might be variations in diet and in other cocausative factors, including stress. It should be recalled, for example, that genetic factors do not explain the bulk of heart disease.[135]

This assertion is not contradicted by the fact that workers in hot jobs are a highly select population. It is obviously true that people can differ in their tolerances to chemical and physical hazards. Workers who feel that they cannot cope with the prevailing, acute heat stress change their job for a less acutely demanding one.[136] The key word here is "acute." A worker who "adapts" easily to a hot environment on a day-to-day basis may nonetheless develop longer-term effects such as heart disease. Health hazards whose most serious effects are chronic often do not have early warning signs that would result in only the "resilient" workers remaining on the job.

In sum, the pursuit of hypersusceptible workers is likely to be about as fruitful as the identification of those alleged to be accident-prone. Job selection makes sense to a point—i.e., persons with heart disease should not be given hard physical labor—and perhaps industry may be justified in hiring only nonsmokers in some jobs, but the bulk of chronic poisoning, respiratory disease, and heart disease should be attacked by cleaning up the work (and general) environment. Detecting hypersusceptible workers is probably not only misguided but impractical, given the proliferation of health hazards in the work environment.

In spite of all the difficulties, exposure guidelines called "threshold limit values"

(TLVs) are being developed by the American Conference of Governmental Industrial Hygienists (ACGIH) and adopted as standards by the government. TLVs have been developed for 450 chemical and physical agents used in industry. The ACGIH is careful to state that the TLVs do not denote a precise line of demarcation between safe and unsafe exposures. They are intended to be used only as guidelines in the control of occupational exposures. The TLVs are not meant to protect all workers or people, even though a "safety factor" is used subjectively to calculate the TLV so that it usually lies below the smallest level thought to cause any toxic effect. While some TLVs are determined from human data, many are calculated from data collected from animal experiments (whose extrapolation to humans is often of questionable validity).

There is strong evidence that the mechanisms for radiation damage and carcinogenesis are quite different than those for ordinary toxic effects. The effects of exposure to harmful radiation are cumulative and irreversible. Each exposure causes permanent damage which adds to the damage caused by previous radiation exposures. Because every exposure, however small, can do permanent damage, the threshold below which no damage occurs is said to be zero.

Cancer is thought to proceed through a two-step mechanism: *initiation* and *promotion.*[137] The irreversible initiation process may result from the insult of a single molecule of a carcinogen on a single susceptible cell.[138] Because exposure of a carcinogen, however small, can initiate a cancer, the threshold below which no "damage" occurs is said to be zero. Even if initiation were the rate-determining step in the cause of cancer, every exposure to a carcinogen would not result in cancer. The carcinogen-cell interaction would have to occur "correctly"—just as a full score in a bowling game requires a "correct hit." Thus, probability considerations can partly explain why all people exposed to carcinogens do not have cancer.

Promotion of the cancer to the status of a tumor (or detectible change in the blood or lymphatic system) may follow at different speeds in different species. The body's adaptive or immunological mechanisms may delay or avoid the manifestation of the cancer into recognizable form, but the potential for promotion to occur is there if initiation has occurred. The observation that skin cancer can be speeded up by the addition of a promoter (e.g., croton oil)[139] after exposure and removal of the cancer initiator indicates that a nonzero threshold may exist for the *promotion* step, but much controversy exists as to the general correctness of this statement for all cancers.[140]

Decisions about what standards to adopt for carcinogens have to be made with the recognition of the good possibility that a zero threshold exists for these substances. This may or may not require the adoption of a zero standard (i.e., zero within the limitations of measurement). For example, if the standard for radiation is set at one-hundredth of the level to which human beings are exposed from naturally occurring uranium in the general environment, a nonzero (but small) standard may be justifiable. On the other hand, since no safe level has been determined for the liver carcinogen vinyl chloride, and humans are not exposed to vinyl chloride from natural sources, the recommendation for reducing workplace exposure "below measurable limits" seems sensible.*[141]

For chronic effects, the extrapolation from animals to humans is difficult, partly because of biological differences and partly because of differences in life span. Whatever problems exist for predicting safe levels for ordinary toxic materials from animal studies, the problems are further complicated in the case of cancer. Many carcinogenic materials are also toxic (e.g., arsenic and vinyl chloride). The fact that a particular cancer is easily produced by certain agents in test animals is useful in identifying potentially powerful human carcinogens. The usefulness of animal data in the prediction of human cancer is diminished when (1) the cancer is a rare occurrence, (2) a long latency period is required (possibly longer than the animal's life span), or (3) different test species react differently to a carcinogen. Furthermore, the *relative* effects of a chemical's carcinogenic and toxic potential differ among species. Animal data indicated that vinyl chloride did not cause cancer in the majority of species tested. Yet it has turned out to be a powerful human carcinogen. Where animal data indicates cancer, materials should be treated as suspect. However, where no indication of cancer is found, the suspect agent cannot be assumed safe.

The Food and Drug Administration Advisory Committee on Protocols for Safety Evaluation addressed itself to the problems of using test animals for the prediction of human cancer:

> Although a positive answer to the question as posed can be given in some particular instances, no unqualified negative answer is ever possible. Thus, if none out of n test animals exposed to the agent develop cancer, and n is a large number, the elevation, if any, is certainly small, but no matter how large the value of n, it is not logically possible to conclude on the basis of such evidence that no elevation has occurred. The agent might in fact induce one cancer in every $2n$ animals, and the experiment, being too small to detect it, would be incapable of providing cogent evidence for the absence of any elevation whatsoever. This is true no matter how large is the value of n.
>
> Even with as many as 1000 test animals and using only 90% confidence limits, the upper limit yielded by a negative experiment is 2.3 cancers per 1000 test animals. No one would wish to introduce an agent into a human population for which no more could be said than that it would probably produce no more than 2 tumors per

*It is critical to understand the importance of impurities. Most chemicals contain impurities, many of which would be harmful in large doses. It is generally expensive to purify chemicals. Thus, very often, we tolerate impurities of the order of 0.1-1.0%. With carcinogens, where the likelihood that a zero threshold exists, a 0.1% impurity could cause cancer. Congressman William Steiger's remark, at the National Symposium on Occupational Safety and Health of June 4-6, 1973, that it was ridiculous for OSHA to ban alpha-naphthylamine with less than 1% impurity of the carcinogen beta-naphthylamine, indicated a lack of recognition of this point.

1000. To reduce the upper limit of risk to 2 tumors per one million with confidence coefficient 0.999 would require a negative result in somewhat more than *three million test animals.*

The impracticably large numbers required to detect carcinogenic effects of possible public health importance, together with the uncertainties of downward extrapolation, set limits to the amount of protection that can be achieved with a routine testing program.[142]

The Food and Drug Administration has thus indicated that very large numbers of animals are required to give meaningful results. However, it is very difficult to breed scientifically comparable animals in large numbers of the order needed to control the necessary experimental variables. Thus, the National Center for Toxicological Research at Pine Bluff, Arkansas, is unlikely to yield useful results on their "megamouse experiments" done with huge numbers of mice—except for very acutely toxic materials.

The OSHAct requires that the Department of Labor promulgate and enforce standards to assure "that no employee will suffer material impairment of health or functional capacity."[143] Clearly, the TLVs were not designed to satisfy this mandate. Medical-scientific dispute continues to rage as to whether or not there exists a finite threshold for cancer. On one side of the argument are the toxicologists and chemists who insist that a finite safe level exists upon which a TLV or standard can be based. On the other side are the cancer specialists who believe the zero threshold is likely enough to be true to warrant making decisions as if it were true—i.e., no exposure should be allowed.* In support of this position, a NIOSH criteria document recommending a standard for asbestos states:

Any carcinogen (initiator) must be assumed, until otherwise proven, to have discrete, dose-dependent, irreversible, and additive effects to cells that are transmissible to the cell progeny. Thus, initiation of malignancy following single small exposures to asbestos is possible, but of a low probability. With frequent or chronic exposure and a low dose rate, the probability of initiation of malignancy is increased. Yet, even under optimal conditions of cell proliferation (in the presence of promotors), these malignant transformations do not lead to instantaneous cancer, but remain insidious for a number of years (latent).

In protracted exposure, some of the total accumulated exposure is "wasted" (or irrelevant) as far as the initiator of cancer is concerned. Exposures in excess of the minimal initiation dose conceivably may shorten the latent period to some extent by substituting for other contributing factors that would have eventually been effectual in converting the latent tumor into a frank malignancy. Analytic methods used in the epidemiology of asbestos-induced cancers are unable to discriminate between the initiating dose and subsequent (wasted) exposure.

Consideration must also be given to the concept that an inverse relationship exists

*It is somewhat inconsistent that Stokinger recognizes the difficulty in establishing a finite threshold for chemical free radicals or free-radical formers while claiming a finite threshold exists for carcinogens, which are thought to proceed through free-radical mechanisms, at least in some cases.[144]

between dose rate and the latent period. As the dose rate becomes progressively lower, the latent period may approach or exceed the life span of exposed individuals.

Adherence to these concepts would argue toward reducing asbestos exposure substantially below those levels currently demonstrated to be associated with the disease. Such a course of action is consistent with the Surgeon General's Ad Hoc Committee on Evaluation of Low Levels of Environmental Chemical Carcinogens statement that, "*for carcinogenic agents, a safe level for man cannot be established by application of our present knowledge.*"[145]

Putting aside for the moment the issue of a threshold for carcinogens, it is not clear that an organism's adaptive mechanisms can be relied upon to combat *continuous* exposures. Although exercising a muscle creates a stressful condition, that exercise (when not continuous) is viewed as desirable. However, hanging a hundred-pound weight from one's arm continuously cannot be said to be healthful. Toxic materials also initiate stresses which are measurable by physiological changes in the blood and in the nervous system. The whole body chemistry is affected by both the toxic material and the body's reactions as it tries to return to equilibrium. A common example of short-term adaptation is the body's development of a fever to counteract an infection. However, calling on the body's adjustment mechanisms continuously cannot be said to be harmless. Constant exposure of the body to toxic materials demands a continuous adjustment of the body chemistry which does not represent the normal state of body equilibrium even though the body system may be stable. On a cellular level, the body is asked to adjust to a stress not unlike hanging a weight from a muscle continuously.

The greater stringency of the Soviet Union's limits for exposure to both carcinogens and ordinary toxic substances is based on a philosophy that "regardless of the value set, the optimal value and goal (the norm) to be sought is zero concentration."[146] Under present Soviet philosophy, air quality standards are based on the premise that Soviet workers should not be subjected to the additional stress of chemical toxicity and irritation above and beyond the normal stresses of everyday living.[147]

The discussion of exposure to toxic materials thus far has focused on the establishment of a safe level for an isolated substance. This make-believe world of isolated hazards does not exist in either the workplace or in the general environment. Most potential human exposure to chemicals is exposure to mixtures.[148] Toxic materials exist in combination with one another, and their effects may be further complicated by stress caused by noise, abnormal temperatures, ergonomic factors, and psychosocial factors. Two toxic substances in combination can have effects on the body which are either (1) *additive* in an expected way, (2) *synergistic* (i.e., combining in such a way that the resultant effect is greater than the sum of the individual effects), or (3) *antagonistic* (i.e., combining in such a way that the resultant effect is less than the sum of the individual effects). Examples of synergism have been given in previous discussions. The question remaining is how commonplace and how predictable the synergistic effect is.

Research on joint (acute) toxic action indicates that about 5% of the combinations investigated in one study exhibited synergistic (or antagonistic) behavior.[149] The remaining 95% combined in a predictable way, i.e., additively. It has been pointed out, however, that the research data indicated that in many cases the substances tested were *not* the most highly toxic of industrial substances, and it is possible that the 5% figure for other-than-additive effects would have been enhanced had pairs of highly toxic substances been used in the investigations.[150] It is not possible to "predict" which 5% of the combinations of *all* substances will behave synergistically.

Most current evidence of toxic interactions is based on acute responses,[151] and there have been relatively few studies to evaluate possible interactions of chronic exposures to mixtures of chemicals.[152] Stokinger believes that the combined effects of combinations at lower doses would be "physiologically inconsequential."[153] The *acute* effects of combinations at low dosages may be inconsequential, but it is not clear what the *chronic* effects will be.

A closer look at what the 5% figure means should convince the reader that the problems of synergism are very great indeed. There are 12,000 toxic materials in commercial use today.[154] For n substances, the number of possible pairs is $n(n-1)/2$. Of the 12,000 x 11,999/2 pairs, 5% or 3.6 million pairs may behave synergistically. To this number is being added 500 new chemicals annually, which increases the number of possible combinations further. Even if we were to confine our investigation to combinations of toxic materials for which TLVs are already individually determined—i.e., for 450 substances—the number of combinations expected to give unpredictable synergistic effects would be (5%) x 450 x 449/2, or approximately 5000. One would have to investigate almost the entire set of 100,000 combinations to find the 5000 pairs! The picture is further complicated by the fact that hazards often exist in combinations larger than pairs—and then the numbers are really staggering.

The problems are even worse than indicated above. Some stressors, such as heat and noise, may enhance the toxicity of many toxic materials. Indeed, the NIOSH criteria document for hot environments states:

> These data [on the role of heat stress as a potentiator or mediator of response to other physical and toxic agents] indicate the complexity of the interactions of multiple stresses. They raise fundamental questions as to the validity of those TLVs based on single stress experiments for toxic substances in the presence of heat.[155]

Toxic materials *as stressors* may promote cancer. The extent to which the synergistic effects of stressors may be important in both toxicity and in the promotion of cancer has not been determined. Research is seriously deficient in these areas.

Thus, serious consideration should be given to the question of whether we shall ever be able to generate the data base necessary to set safe levels for toxic materials and carcinogens. An alternative and more sensible approach may be to redesign our industrial processes, to automate where necessary, and to reduce unnecessary proliferation of new chemicals.

3.6.3 Stress

Research on stress as a precursor of both mental and physical disease—and as a co-causitive factor of disease in combination with toxic or carcinogenic agents—is in its infancy. Hence, the importance of stress continues to be a source of scientific-medical dispute. In the HEW Report *Work in America*, stress was identified as having important consequences for job dissatisfaction and for mental and physical occupational disease. The high use of alcohol and drugs by some groups of workers and its relationship to job stress and mental health is of increasing concern.[156] The difficulties in identifying the elements of job satisfaction or dissatisfaction from an investigation of "role conflict or ambiguity," "job content," or the "organization of work" may be complicated by the presence of physical stressors such as noise, or chemical stressors such as mercury, which affect the nervous system and hence a worker's psychological state. In reaction to the long-time neglect of psychosocial factors in both mental and physical disease, behavioral scientists are now emphasizing their importance to such an extent that they may be inadvertently neglecting physical and chemical stressors.

NIOSH is mandated by the OSHAct to carry out research in psychosocial factors. Alexander Cohen and Bruce Margolis of NIOSH have indicated the way in which NIOSH views its responsibilities in this area:

> Sections 20(a)(1) and 20(a)(4) direct NIOSH to include psychological, behavioral and motivational factors in researching problems of worker safety and health, and to consider these factors in developing new methods, techniques, or approaches in handling such problems. In Section 20(a)(3) there is the requirement to determine exposure limits to physical conditions at the workplace which are not only safe for health, but which pose no threat to one's *functional capacity*. These functional capacities could be taken to mean behavioral or performance *capabilities* as well as neurophysiological functions. Section 20(a)(7), in referencing the need to undertake industry-wide studies of certain worker groups, authorizes the evaluation of job stresses on the potential for illness, disease, or loss of functional capacity in aging adults. Job stresses here could be interpreted to include those of a psychological nature. That is, they may consist of undue tasks or unnatural work regimens, which apart from or combined with exposures to physical and chemical hazards, may be degrading to the worker's physical or mental health.[157]

The five types of psychological research projects being supported by NIOSH are given below.

- Determination of changes in human behavioral capacities and neurophysiological functions as they may be correlated with body burden measures and/or environmental indices of exposure to neurotoxic chemicals or physical stressors at or below recommended limits of health protection.

- Development, evaluation, and application of behavioral test methodologies (sensory, psychomotor, and cognitive task measures) and neurophysiological indicators to scan for irritant and toxic properties of different chemicals and to provide early indications of potentially toxic or unsafe conditions at the workplace.
- Elaboration of human individual differences, both physical (age, sex, physical fitness) and psychological (work attitudes, perception of risk, emotional state) and specific situational variables which may account for differential tendencies toward accident or health problems found among those engaged in hazardous or stressful jobs.
- Determination of relationships between specifiable job stress factors, apart from physical and chemical hazards, and physical and mental disorders found in worker groups, and suggestion of approaches toward stress alleviation.
- Elaboration of factors underlying safety and health motivation in different occupations or industries and shaping of strategies for accident reduction and disease control based on behavioral science principles and techniques.[158]

The intent of these types of studies is to determine the adequacy of existing or newly proposed TLVs for chemical agents to protect against the *behavioral disturbances* that could lead to a mishap.[159] These studies have misplaced emphasis on the importance of differences in individual factors to explain hypersusceptibility and to aid in job selection.[160] In addition, they are seriously deficient in that they do not emphasize the effects of *combined stressors*, such as noise *and* toxic materials, on both psychological and physical health.

The investigation of *single* stressors should not be relied upon for an accurate assessment of the importance of the stressor. Dr. Lennart Levi, Director of the Karolinska Institute's Laboratory for Clinical Stress Research in Sweden, believes that two of the most important causes of stress occurring in working life are *noise* and *the pressure of piecework.* His findings are that "within reasonable limits," the effects of noise as a stressor depends "upon the *attitude* of the person who hears the noise, not on the physical volume of noise itself. [A] positive attitude towards the noise went a long way towards modifying both the psychological and the endocrine stress-reactions normally produced by noise."[161] The reader is cautioned, however, that long-term chronic effects may still be serious, especially where noise occurs in combination with other stressors. Dr. Levi confirmed in a letter to the author the fact that not enough is known about chronic exposure to combinations of stressors.[162]

It is beyond the purpose of this report to dwell on the complex problems of job stress and occupational mental health. I would, however, like to refer the reader to several publications which contain extensive discussion of stress and occupational mental health problems. NIOSH sponsored the Cornell Conference on Job Stress in 1972, held at the Cornell University Medical Center, which concentrated on both psychological and physiological consequences of stress and on identification of critical research needs.[163] The Cornell University Center for Occupational Mental Health, a unit of the Department of Psychiatry at the Medical Center, publishes a valuable journal, *Occupational Mental Health*, in which some papers given at the conference were printed. Also of considerable value is a collection of articles entitled *Mental Health and Work Organizations*, edited by Alan McLean. These publications indicate that some of the topics of concern to the researchers in the field of stress are—

- the effects of noise
- the relationship between occupational stress and coronary heart disease
- information overload
- role ambiguity and stress
- irregular work and company dependency as determinants of a high rate of psychiatric disorders in coal miners[164]
- executive stress.[165]

In sum, the importance of job stress for both mental and physical disease is being increasingly recognized, although important research is seriously lacking. The general confusion and the conflicting results from studies of job satisfaction and job enrichment unfortunately are creating a climate in which the concern for job stress may diminish. However, a demonstration of the interaction between stress and chemical or physical hazards may serve to renew interest in the subject.

3.7 SUMMARY AND FUTURE IMPLICATIONS

The reader by this time should be convinced of the enormity and complexity of the scientific-medical problems in occupational health and safety. Occupational health and safety problems had somewhat different historical origins, and the two are distinguished by the health and safety professionals, the workmen's compensation system, and employer activities. Nonetheless, it is rather futile to separate occupational health from occupational safety. Unsafe conditions may create stress which can be a precursor of mental and physical disease; health hazards can lead to accidents. Job health and safety are both intimately connected to workers' well-being and job satisfaction.

Attempts to separate occupational from general health, or mental from physical health, are just as unproductive. Unidisciplinary approaches to solving just part of the set of problems lead to strange results. The goal of removing DDT from the general environment has led to the use of a substitute pesticide that is much more dangerous for the farmworkers who handle it. The EPA's air quality standards for the general environment do not seem to bear any relationship to the OSHA standards for the workplace. For example, as Table 3.11 shows, the EPA's standards for sulfur dioxide and particulates are much more stringent than those of OSHA, al-

Table 3.11
Air Quality Standards for the General Environment and for the Workplace

Pollutant	Environmental Standard (EPA)*	Occupational Standard (OSHA)
SO_2 (sulfur dioxide)	Annual arithmetic mean: 0.03 ppm Max. 24 hr once/year: 0.14 ppm Max. 3 hr once/year: 0.5 ppm	5 ppm
CO (carbon monoxide)	Max. 8 hr once/year: 9 ppm Max. 1 hr once/year: 35 ppm	50 ppm
NO_2 (nitric oxide)	Annual arithmetic mean: 0.5 ppm	5 ppm
Particulates	Annual geometric mean: 0.075 mg/m^3 Max. 24 hr once/year: 0.26 mg/m^3	Respirable fraction: 5 mg/m^3 Total dust: 15 mg/m^3
Lead	0.7 μg/m^3 (USSR)	10 μg/m^3 (USSR)

*Ambient air standards are given several values for short-term concentrations; e.g., "maximum *n*-hour concentration not to be exceeded more than once per year, etc." All in-plant exposures are presumably for an eight-hour day.

Source: Barry Commoner, "Labor's Stake in the Environment/The Environment's Stake in Labor," paper presented at the Conference on Jobs and the Environment–Whose Jobs? Whose Environment?, sponsored by the Institute of Industrial Relations, University of California, Berkeley (San Francisco, November 28, 1972).

though exposure is particularly severe in the workplace. A coordinated policy is critically needed for safeguarding man's total health both on and off the job. Research needs, manpower development, and regulatory activities are seriously out of phase.

It hardly needs repeating that rapidly expanding technology is greatly responsible for the current state of affairs in environmental and occupational health and safety. This is true for two reasons: (1) the technology creates new problems, and (2) there is little coordinated effort to pass on the knowledge about safety and health hazards attending the new technology. Ignorance abounds as to what substances actually exist in the environment and as to the importance of taking protective or preventive measures.

High-speed steels and equipment development have increased the noise in the workplace. Dry drilling techniques and other new technologies have increased the dust problem. Both the practice of running equipment at full capacity and the economic incentives given employees to work quickly contribute to accidents. The rate of production of hydrocarbons from oil and natural gas has been growing at the rate of 15% per year compounded.[166] This is equivalent to a quadrupling every ten years. (It has been noted that emphysema and chronic bronchitis have increased at precisely the same rate.)[167] Table 3.12 illustrates the consumption of raw materials for the manufacture of synthetic organic chemicals and the rate of production of various classes of consumer products from these chemicals between 1949 and 1969.

However serious the exposure to chemicals is in the general environment, the situation is more critical for the work environment. In support of this view, the report *Chemicals and Health* states that

> Certain chemical exposures occur only in occupational environments; e.g., chemical intermediates which neither exist in consumer products nor are released intentionally or inadvertently to the environment.
>
> Most occupational exposures involve substances with which the public has some contact. The occupational exposure will generally involve higher concentrations, more frequent and extended contact periods, and, very often, different routes of exposure (e.g., through skin contact or inhalation rather than the diet) than the consumer experiences. In many cases, partly because of these differences, the industrial exposure may be better monitored and controlled, and the industrial worker more trained to exercise appropriate precautions. Unfortunately, the appropriate safeguards are often not available or are disregarded.[168]

Barry Commoner has sharply criticized the proliferation of chemicals into the general and workplace environment—particularly the chlorinated and other synthetic organic chemicals which are "noxious, toxic, nonbiodegradable, or otherwise

Table 3.12
Production of Synthetic Organic Chemicals in the United States between 1949 and 1969

	1949 (lb)	1969 (lb)	% Increase
Raw materials and intermediates*	8×10^9	1×10^{11}	1150
Consumer products (Total)	1.6×10^{10}	1.1×10^{11}	581
Pesticides and related products	1.4×10^8	1.1×10^9	686
Medicinal chemicals	4.2×10^7	2×10^8	376
Flavors and perfumes	2.4×10^7	1.2×10^8	400
Plastics and resins	1.5×10^9	1.9×10^{10}	1167
Elastomers	9.5×10^8	4.5×10^9	374
Surfactants	4.3×10^8	3.9×10^9	807
Plasticizers	1.7×10^8	1.4×10^9	724
Rubber chemicals	8×10^7	3×10^8	275
Dyes	1.4×10^8	2.4×10^8	71
Organic pigments	3.7×10^7	6.1×10^7	65
Miscellaneous	1.2×10^{10}	7.6×10^{10}	533

*Includes crude products from petroleum and natural gas, and intermediates derived therefrom.

Source: Data from the U.S. Tariff Commission as quoted in *Chemicals and Health*, p. 78.

deleterious." Commoner suggests that "petrochemical plants must be designed as wholly contained and closed systems to keep their toxic and otherwise harmful wastes out of the environment."[169] It would seem that serious consideration should be given to curtailing the rapid proliferation of chemical products. The commercial life of products in the chemical industry is estimated to be about five years.[170] A study of the pharmaceutical industry indicates that increased regulation will not curtail the economic growth of the industry, but in fact may contribute to its growth by increasing research and development as a result of more stringent testing requirements.[171] The number of new chemical products may decrease but their commercial life will probably increase as a result of more stringent regulation.

The patent system encourages monopolistic competition and is partly responsible for the soaring number of new chemical products. The President's Science Advisory Committee concludes that there is no need to reduce the diversity of chemicals to which humans and the environment are exposed and that "under our present schemes of regulation for safety, it is reasonable to anticipate that the new will be at least as safe as the old."[172] This viewpoint ignores the fact that rapid product changeover does not permit us to catch up with even the "old" problems. The "old" chemicals may eventually be tested—or their effects may show up epidemiologically. The introduction of new materials—at a doubling rate every five years—makes determination of safety virtually impossible for these materials. *Chemicals and Health* states that "traditionally, members of the scientific fraternity tend to be conservative in drawing inferences from the raw data of their experiments."[173] The standard of proof demanded to establish *scientific* facts is not the standard required for *legal* facts to be applied in regulatory decisions in the administrative law process.

The systematic researching of and the establishment of "safe levels" for every new chemical product to which workers will be exposed is not achievable in the practical sense. Furthermore, the rate of increase of new substances for which standards must be set is much greater than the rate at which we are able to improve our standard-setting or enforcement ability. Thus, consideration of new process design—possibly requiring closed systems—should be a first-priority item for long-term solutions to the problem of exposure to chemicals. Many scientists have not been very interested in this "engineering" solution, and as long as OSHA moves slowly on developing health standards for toxic materials, there will be little incentive for new process design, for the development of safer substitute materials, or for sufficient in-house testing by the chemical industry.

As noted by Alan McLean, technology has also had its effect on the nature of work and mental health:

> Technology over the years has created several demands. One demand is for conformity, which means that any time the individual resists, his actions are apt to be

called heresy, treason, protest, or, in the last few decades, mental illness. The organization demands a consolidation of the majority; this demand for consensus creates anxiety because man does not always conform willingly, and, when he does conform, he must be assured that this is a normal process. It may not be a totally conscious process. . . .

. . . Technology's principal gift is time—time to be happy or unhappy; time to be introspective; time for anxiety; time to learn skills useful in or outside of a person's job; and time to obtain understanding so that one can be the person his ego-ideal says he might be and wants to be. . . .

The benefits of science and technology are very real, but to get these benefits we must be willing to give up much. We haven't yet learned to live with technology, and this causes both anxiety and tension within us—conflicts between order and freedom, between individualism and conformity, between the need to belong and the need to be alone.[174]

Finally, it must be recognized that mere recognition of the environmental problems is not enough. The report *Man's Health and the Environment: Some Research Needs* states that

the acquisition of new knowledge and understanding provided by research is only a first step, a step which will remain unrewarded unless that new knowledge and understanding are transmuted into action. In consequence, your efforts, however fundamental their scope, cannot be considered independent of their significance and relevance to the problems of standard-setting, monitoring, regulation, and control.

The issue of *control* is crucial to the whole discussion of job health and safety. The social control of technology in the workplace is being attempted through the OSHAct. As the reader will see, the use of economic incentives and the regulatory process* to improve occupational health and safety are not likely to be successful in the absence of the worker's interest, responsibility, and control of some aspects of his work environment. Occupational health and safety is becoming a political issue and contains elements of conflict which are pervasive throughout labor-management relations and the debates concerning environmental control, the development of industrial democracy, and the design of the health-care system. How we approach the present and future problems in occupational health and safety will be a reflection of our sense of equity and justice as it applies to the *individual* in our

*The economist's classic suggestion for pollution control has been the levying of an effluent tax on the producer.[175] This mechanism could never succeed for the work environment. The number of serious hazards prevalent within the industrial workplace are vast; the number of standards are few; inspection and measurement capability is grossly inadequate; and the job of assessing and collecting a tax for every one of the 12,000 toxic materials that could be taxed is enormous. If a single firm emitted a certain toxic material which could be effectively dispersed so that it was present at a low level in the air or water, no effluent tax would be levied for the general environment. However, it would still have to be levied for the environment within the firm, since exposure there could be serious. Furthermore, harmful dusts settle around the source producing it in the factory. Unannounced remote monitoring could not be as effective as with smokestack emissions.

society. The American worker should not be subjected to "environmental blackmail" whereby *he* must choose between demanding safer and healthier working conditions and losing his job because the employer cannot afford corrective measures and threatens to close down.[176]

The use of market mechanisms and the regulatory process to control health and safety in the workplace are the focus of the remainder of this report.

REFERENCES

1. For example, D. Hunter, *The Diseases of Occupation* (5th ed., Little, Brown, Boston, 1974); F.A. Patty, *Industrial Hygiene and Toxicology* (2 volumes, 2d ed., Wiley, New York, 1958 and 1963); Alice Hamilton and Harriet L. Hardy, *Industrial Toxicology* (3d ed., Publishing Sciences Group, Acton, Mass., 1974); Ethel Browning, *Toxicity of Industrial Solvents and Plasticizers* (Am. Elsevier, New York, 1965); *Toxicity of Industrial Metals* (2d ed., Butterworth, Reading, Mass., 1969); International Labour Organization, *Encyclopedia of Occupational Health and Safety* (2 volumes, McGraw-Hill, New York, 1972); J. M. Stellman and S. M. Daum, *Work Is Dangerous to Your Health* (Vintage Books, New York, 1973).

2. Jerome B. Gordon, Allan Akman, and Michael L. Brooks, *Industrial Safety Statistics: A Re-examination* (Praeger, New York, 1971). [Hereinafter cited as *Gordon Report*]

3. Dr. R. van der Vlist, Netherlands Institute for Preventive Medicine, interview with author, Leiden, Netherlands, September 13, 1972.

4. Raytheon Service Co. (Burlington, Mass.), *Industrial Noise and Worker Medical, Absence and Accident Records*, G.P.O. Document No. 1972-759-570/1311 (June 1972).

5. Phyllis Lehmann, "Vibration," *Job Safety and Health*, Vol. 2, No. 1 (January 1974), p. 9.

6. Van der Vlist, interview.

7. *Occupational Exposure to Hot Environments*, National Institute for Occupational Safety and Health, HEW (1972), p. III-10.

8. *The Target Industries: Profiles of Five Hazardous Occupations*, Occupational Safety and Health Administration, U.S. Department of Labor (1970), inside cover.

9. World Health Organization, WHO Technical Report Series, No. 66 (1963), p. 4.

10. Louis Beliczky, Director of Industrial Hygiene, United Rubberworkers of America, interview with author, Washington, D.C., June 5, 1973.

11. Dr. John Peters, Harvard University School of Public Health, telephone interview with author, January 3, 1974.

12. Joseph R. Anticaglia and Alexander Cohen, "Extra-Auditory Effects of Noise as a Health Hazard," *American Industrial Hygiene Association Journal*, Vol. 31 (May-June 1970), p. 277.

13. Lennart Levi, ed., *Society, Stress and Disease* (Oxford University Press, London, 1971).

14. *Occupational Exposure to Noise*, National Institute for Occupational Safety and Health, HEW (1972), p. II-2; emphasis added.

15. Dr. Alexander Cohen, NIOSH, interview with author, February 14, 1973.

16. *The Target Health Hazards*, Occupational Safety and Health Administration, U.S. Department of Labor, Bulletin No. 2051 (1972).

17. *Occupational Exposure to Asbestos*, NIOSH, HEW (1972).

18. *Occupational Disease . . . The Silent Enemy*, Public Health Service, HEW (n.d.).

19. Phyllis Lehmann, "Job Health Hazard: Beryllium," *Job Safety and Health*, Vol. 2, No. 1 (January 1974), p. 27.

20. *Ibid.*, p. 25.

21. *The Target Health Hazards.*

22. *The Toxic Substances List, 1973 Edition*, NIOSH, HEW (June 1973).

23. *Slow Progress Likely in Development of Standards for Toxic Substances and Harmful Physical Agents Found in Workplaces*, Report to the Committee on Labor and Public Welfare, U.S. Senate, by the Comptroller General of the United States (September 28, 1973).

24. "Occupational Health Programmes," *WHO Chronicle*, Vol. 26, No. 12 (December 1972).

25. *Ibid.*

26. *NIOSH Recommended Standard for Occupational Exposure to Inorganic Arsenic*, transmitted to OSHA on January 21, 1974.

27. *NIOSH Recommended Standard for Occupational Exposure to Sulfur Dioxide*, undated.

28. Phyllis Lehmann, "Early Clues to Job Health Hazards," *Job Safety and Health*, Vol. 1, No. 11 (October 1973), p. 16.

29. *Man's Health and the Environment–Some Research Needs*, Report of the Task Force on Research Planning in Environmental Health Science, National Institute of Environmental Health Sciences, HEW (March 1970).

30. C. E. Searle, "Chemical Carcinogens and Their Significance for Chemists," *Chemistry in Britain*, Vol. 6, No. 1 (January 1970), p. 7.

31. "Asbestos Also Increases Risk of Colon, Rectal Cancer Deaths," *Internal Medicine News* (November 15, 1973), p. 14.

32. George Clack, "Controlling Coke Oven Emissions," *Job Safety and Health*, Vol. 2, No. 3 (March 1974), p. 5.

33. Marcus M. Key, "1973 Priority List for Criteria Development of Toxic Substances and Physical Agents," mimeograph of speech (n.d.).

34. *Occupational Exposure to Hot Environments*, p. III-11.

35. Lehmann, "Vibration," pp. 11-12.

36. *Ibid.*, p. 12.

37. *Work in America*, Report of a Special Task Force to the Secretary of Health, Education, and Welfare (MIT Press, Cambridge, Mass., 1973), p. 79.

38. "Heart Attack Among Japanese Executives," *Occupational Mental Health*, Vol. 2, No. 4 (winter 1973).

39. National Safety Council, *Accident Facts, 1972.*

40. Maurice F. Bresnahan, "Availability of Occupational Injury and Illness Data and Future Plans," speech presented before the National Symposium on Occupational Safety and Health, Washington, D.C., June 4, 1973.

41. *Occupational Safety and Health Reporter* (the Bureau of National Affairs, Inc.), Vol. 3, No. 34 (January 24, 1974), p. 1067. [Hereinafter cited as *OSHR*, BNA]

42. Hearings Before the Senate Subcommittee on Labor on the Occupational Safety and Health Act of 1970, Part I (1970), p. 77.

43. Hearings Before the Subcommittee on Environmental Problems Affecting Small Business of the H.R. Select Committee on Small Business, 92d Congress, 2d Session, on Small Business and the Occupational Safety and Health Act of 1970, June 22, 1972, p. 299: Testimony of George C. Guenther. [Hereinafter cited as Small Business Hearings]

44. Bresnahan, "Availability of Data."

45. *OSHR*, BNA, Vol. 3, No. 34 (January 24, 1974), p. 1067.

46. *Gordon Report*, p. 3.

47. "Occupational Health Programmes," p. 537.

48. National Safety Council, *Accident Facts, 1972.*

49. Quoted in *Man's Health and the Environment*, p. 73. The rate of six days per year refers to 1969.

50. *Protecting the Health of Eighty Million Americans: A National Goal for Occupational Health*, Special Report to the Surgeon General of the United States Public Health Service, HEW (November 19, 1965).

51. *The President's Report on Occupational Safety and Health*, G.P.O. Document No.

4000-00288 (December 1973), p. 138, and Dr. Joe Wagoner, NIOSH, private communication.

52. *President's Report*, 1973, p. 140, and Wagoner, private communication.

53. *President's Report*, 1973, p. 144.

54. John R. French and Robert D. Caplan, "Organizational Stress and Individual Strain," unpublished paper, Institute for Social Research, University of Michigan (1972).

55. Quoted in Ray Davidson, *Peril on the Job* (Public Affairs Press, Washington, D.C., 1970), p. 5.

56. *Occupational Health Survey of the Chicago Metropolitan Area*, HEW (1970), Table 56(a), p. 50.

57. Joe Collier, "Inspection and Enforcement in the Workplace," *Monthly Labor Review* (August 1973), p. 35.

58. Stellman and Daum, *Work Is Dangerous*, p. 35.

59. *Occupational Disease . . . The Silent Enemy*, p. 3.

60. *Ibid.*

61. Clack, "Controlling Coke Oven Emissions."

62. Dr. Joe Wagoner, NIOSH, private communication.

63. Bresnahan, "Availability of Data," p. 1.

64. John T. Inanza, "The New Survey of Occupational Injuries and Illnesses," *Monthly Labor Review* (August 1973), p. 55. See also the forthcoming work by Ken Rosen on the social cost of accidents in the construction industry.

65. *OSHR*, BNA, Vol. 3, No. 34 (January 24, 1974), p. 1067.

66. *Ibid.*

67. *Ibid.*

68. Statement of Assistant Secretary of Labor George C. Guenther Before the House Subcommittee on Agricultural Labor Regarding Occupational Safety and Health, September 25, 1972, p. 1.

69. Small Business Hearings, p. 97.

70. Key, "1973 Priority List."

71. National Commission on State Workmen's Compensation Laws, *Compendium on Workmen's Compensation*, G.P.O. Document No. 496-632 (1973), p. 6.

72. R. S. Ralston, Manager, Technical Programs, American Pulpwood Association, quoted in Daniel M. Berman, *Death on the Job: The Politics of Occupational Health in the United States* (unpublished manuscript, 1974), p. 37.

73. Fred M. Foulkes, "Learning to Live with the OSHAct," *Harvard Business Review* (November-December 1973), p. 67.

74. Harry Chapman, letter to Daniel Berman, November 9, 1970; emphasis added. Dr. Chapman is the author of *A Textbook of Clinical Psychiatry* (Lippincott, Philadelphia, 1967) and numerous other medical and popular books and articles.

75. *The Report of the National Commission on State Workmen's Compensation Laws*, G.P.O. Document No. 477-053 (July 1972), Table B.1, p. 143.

76. Quinn et al., *Monthly Labor Review* (November 1973), p. 35.

77. *Ibid.*

78. Harry A. Donoian and Irving Brotslaw, " 'I Am a Man—Not a Machine': A Survey of Worker Dissatisfaction," unpublished paper (1973).

79. *Ibid.*, Table 29, p. 36.

80. *Ibid.*, p. 37.

81. *Gordon Report*, p. 10. (It is interesting to note that Howard Pyle, Director of the National Safety Council, said in an interview that there are really no better figures than in the *Gordon Report.*)

82. See Davidson, *Peril on the Job*, for a trade unionist's view of employers' occupational injury and illness reporting practices.

83. *Occupational Disease . . . The Silent Enemy*, p. 3.

84. U.S. Department of Labor, Bureau of Labor Statistics, "Occupational Injuries and Illnesses by Industry, July 1-December 31, 1971," BLS Bulletin No. 1798, Table 4c, p. 114.

85. P.C. Sexton and B. Sexton, *Blue Collars and Hard Hats: The Working Class and the Future of American Politics* (Random House, New York, 1972), quoted in *Work in America*, p. 26.

86. *The President's Report on Occupational Safety and Health*, G.P.O. Document No. 2915-0011 (May 1972), p. 111.

87. Richard Thaler and Sherwin Rosen, "The Value of Saving a Life: Evidence from the Labor Market," unpublished discussion paper, Department of Economics, University of Rochester (November 1973), p. 29.

88. *Chemicals and Health*, Report of the Panel on Chemicals and Health of the President's Science Advisory Committee, Science and Technology Policy Office, National Science Foundation (September 1973), p. 152.

89. *Work in America*, p. 79.

90. Statement of Marvin A. Kastenbaum, *Congressional Record*, Vol. 118, No. 108 (June 30, 1972), p. 1.

91. Searle, "Chemical Carcinogens," p. 5.

92. *Man's Health and the Environment*, p. 147.

93. Searle, "Chemical Carcinogens," p. 7.

94. Frederick P. Li et al. "Cancer Mortality Among Chemists," *Journal of the National Cancer Institute*, Vol. 43, No. 5 (November 1969), p. 1160.

95. *Statistical Bulletin* (Metropolitan Life), V-154 (July 1973), p. 6.

96. *Man's Health and the Environment*, p. 75.

97. *Statistical Bulletin*, p. 6.

98. *Man's Health and the Environment*, p. 75.

99. *Statistical Bulletin*, p. 9; emphasis added.

100. *Respiratory Diseases*, National Heart and Lung Institute Task Force Report, DHEW Pub. No. (NIH) 73-432 (October 1972), p. 5.

101. *National Tuberculosis and Respiratory Disease Association Bulletin* (October 1972), p. 2.

102. Agnes Fahy, "Enemies in the Dust: Occupational Respiratory Diseases," *National Tuberculosis and Respiratory Disease Association Bulletin* (June 1972), p. 15.

103. *Chemicals and Health*, p. 157.

104. See, for example, Arthur Kornhauser, *Mental Health of the Industrial Worker: A Detroit Study* (Wiley, New York, 1965). Though Kornhauser's specific findings are in dispute, the underlying notion that there is likely to be a relationship between occupational factors and the incidence of mental health problems remains important.

105. D. P. Discher et al., *Pilot Study for Development of an Occupational Disease Surveillance Method*, HEW Publication No. (NIOSH) 75-162 (1975).

106. *Chemicals and Health*, p. 88.

107. *The Plans and Programs of NIOSH 1973*, NIOSH, HEW (n.d.).

108. *Chemicals and Health*, p. 27.

109. *Ibid.*, p. 136.

110. *Ibid.*, p. 27.

111. Barry Commoner, "Labor's Stake in the Environment/The Environment's Stake in Labor," paper presented at the Conference on Jobs and the Environment–Whose Jobs? Whose Environment?, sponsored by Institute of Industrial Relations, University of California, Berkeley (San Francisco, November 28, 1972).

112. Arthur D. Little, Inc., *The Present Status and Requirements for Occupational Safety Research* (1972).

113. Government Research Corporation, "Occupational Safety and Health: A Policy Analysis" (1973), p. 6.

114. George Sigvard Hagglund, "Some Factors Contributing to Wisconsin Occupational Injuries," unpublished Ph.D. dissertation, Department of Industrial Relations, University of Wisconsin (1966).

115. Frederick Winslow Taylor, *Scientific Management* (Harper Brothers, New York, 1947).

116. H. W. Heinrich, *Industrial Accident Prevention: A Scientific Approach* (McGraw-Hill, New York, 1931).

117. National Safety Council, *Accident Prevention Manual for Industrial Operations* (5th ed., National Safety Council, Chicago, 1964).

118. Eric Farmer and E. G. Chambers, *Report No. 38* (Industrial Fatigue Research Board, London, 1926), p. 3.

119. M. Greenwood and H. M. Woods, *Report No. 4* (Industrial Fatigue Research Board, London, 1919); cf. M. Greenwood and U. Yule, *Journal of the Royal Statistical Society*, Vol. 83 (1927).

120. H. M. Vernon, *Accidents and their Prevention* (Macmillan, New York, 1937), pp. 249-250.

121. Morris S. Viteles, *Industrial Psychology* (W. W. Norton, New York, 1932).

122. Morris Viteles and H. M. Gardner, *Personnel Journal*, Vol. 7 (1929), p. 349.

123. Boyd Fisher, *Mental Causes of Accidents* (Houghton Mifflin, New York, 1922).

124. Alexander Mintz and Milton L. Blum, "A Re-examination of the Accident Proneness Concept," *Journal of Applied Psychology*, Vol. 33 (1949), pp. 195-211.

125. T. A. Ryan and P. C. Smith, *Principles of Industrial Psychology* (The Ronald Press, New York, 1954), pp. 480-486.

126. Leon Brody, *Human Factors Research in Occupational Accident Prevention* (American Society of Safety Engineers and the Center for Safety Education, New York, 1962).

127. Summary provided by Daniel B. Johnson in his foreword to Hagglund, "Some Factors."

128. Howard Pyle, President, National Safety Council, interview with author, February 5, 1973.

129. W. T. Singleton, *Introduction to Ergonomics* (World Health Organization, 1972); emphasis added.

130. Ralph J. Vernon, Texas A&M University, letter to author, August 26, 1973.

131. George Guenther, speech at Harvard University, Kennedy School of Government, April 26, 1973.

132. See, for example, Herbert E. Stokinger, "Concepts of Thresholds in Standards Setting," *Archives of Environmental Health*, Vol. 25 (September 1972), p. 154; Eula Bingham, "Thresholds in Cancer Induction," *Archives of Environmental Health*, Vol. 22 (June 1971), pp. 692-695.

133. Herbert E. Stokinger, "Hypersusceptibility and Genetic Problems in Occupational Medicine," an official report to the Seventh International Congress on Occupational Health (September 17-23, 1972).

134. *Ibid.*

135. *Work in America*, p. 79.

136. *Occupational Exposure to Hot Environments*, p. III-1.

137. I. Hieger, *Carcinogenesis* (Academic Press, New York, 1961); "Food and Drug Administration Advisory Committee on Protocols for Safety Evaluation: Panel on Carcinogenesis Report on Cancer Testing in the Safety Evaluation of Food Additives and Pesticides," *Toxicology and Applied Pharmacology*, Vol. 20 (1971), pp. 419-438.

138. Bingham, "Thresholds in Cancer Induction," p. 1.

139. "Food and Drug Administration Advisory Committee . . .," p. 423.

140. Bingham, "Thresholds in Cancer Induction," p. 1.

141. *Wall Street Journal*, March 25, 1974, p. 19.

142. "Food and Drug Administration Advisory Committee . . .," pp. 430-433.

143. OSHAct, Section 6(b)(5); see Appendix.

144. Herbert E. Stokinger, "Philosophy of Air Limits," *American Industrial Hygiene Association Journal*, Vol. 25 (November-December 1964), p. 591.

145. *Occupational Exposure to Asbestos*; emphasis added.

146. H. I. Magnuson et al., "Industrial Toxicology in the Soviet Union—Theoretical and Applied," *American Industrial Hygiene Association Journal*, Vol. 25 (1964), p. 185, as quoted in Stokinger, "Concepts of Thresholds."

147. *Ibid.*

148. H. F. Smyth, Jr., "Joint Action of 40 Representative Industrially Important Chemicals in Over 600 Combinations by the Peroral Route to Rats," Mellon Institute of Industrial

Research Special Report (unpublished).

149. H. F. Smyth, Jr., C.S. Weil, J. S. West, and C. P. Carpenter, "An Exploration of Joint Toxic Action: Twenty-seven Industrial Chemicals Intubated in Rats in All Possible Pairs," *Toxicology and Applied Pharmacology*, Vol. 14 (1969), p. 340.

150. Herbert Stokinger, letter to author, February 22, 1973.

151. *Man's Health and Environment*, p. 181.

152. Sheldon D. Murphy, "Interactions," Report to the Subcommittee on Toxicology of the Task Force on Research Planning in Environmental Health Sciences (1969), p. 6.

153. Stokinger, letter to author, February 22, 1973.

154. *Toxic Substances List, 1973 Edition.*

155. *Occupational Exposure to Hot Environments*, p. III-11.

156. H. M. Trice, "Alcohol, Illegal Drugs and the Work World: The Need for a Comparative Perspective," Report to Management on Drug Abuse in Industry (unpublished). See also *Work in America*, pp. 85-88.

157. *Chemicals and Health*, p. 2; emphasis added.

158. Alexander Cohen, letter to author, March 1973.

159. *Chemicals and Health*, p. 6.

160. *Ibid.*, p. 12.

161. *Occupational Mental Health*, Vol. 2, No. 2, (summer 1972), p. 7.

162. Lennart Levi, letter to author, February 16, 1973.

163. *Chemicals and Health*, p. 15; *Occupational Mental Health*, Vol. 2, No. 4 (winter 1973), p. 2.

164. L. W. Field, R. T. Ewing, and D. W. Mayne, "Observations on the Relation of Psycho-Social Factors to Psychiatric Illness Among Coal Miners," *The International Journal of Social Psychiatry*, Vol. 3 (autumn 1957), pp. 133-145.

165. Harry Levinson, *Executive Stress* (Harper & Row, New York, 1970).

166. H. H. Landsberg, L. L. Fischman, and J. L. Fischer, *Resources in America's Future* (The Johns Hopkins Press, Baltimore, 1963), quoted in *Chemicals and Health*, p. 78.

167. *Man's Health and the Environment*, p. 75.

168. *Chemicals and Health*, p. 78.

169. "Barry Commoner Assails Petrochemicals," *Chemical Engineering News* (November 26, 1973), p. 10. See also Commoner, "Labor's Stake in the Environment."

170. *Chemicals and Health*, p. 108.

171. *Chemicals and Health*, pp. 106-108.

172. *Ibid.* p. 122.

173. *Ibid.*, p. 131.

174. Alan McLean, ed., *Mental Health and Work Organizations* (Rand McNally, Chicago, 1970), p. 6.

175. Robert M. Solow, "The Economics of Pollution Control," paper presented at the Conference on Jobs and the Environment—Whose Jobs? Whose Environment?, sponsored by Institute of Industrial Relations, University of California, Berkeley (San Francisco, November 28, 1972).

176. Leonard Woodcock, "Labor and the Economic Impact of Environmental Control Requirements," paper presented at the Conference on Jobs and the Environment—Whose Jobs? Whose Environment?, sponsored by Institute of Industrial Relations, University of California, Berkeley (San Francisco, November 28, 1972).

4 The OSHAct of 1970. I. The Law and the Federal Framework

4.1 INTRODUCTION

Government action relevant to occupational health and safety occurs in four areas of decision-making: (1) congressional action (both legislative and budgetary); (2) administrative law, including the issuance of regulations and the promulgation of standards; (3) judicial review; and (4) informal administrative policy and program formulation. While each decision-making area is related to the others, each functions within a different framework and time span, represents a different kind of power, has a different impact on occupational safety and health, displays a varying degree of visibility to the public, and is subject to varying degrees of influence by interest groups or other parts of the government.

Congressional action led to the passage of the Occupational Safety and Health Act of 1970.* The Act set forth as national policy the attainment of safe and healthy workplaces for American workers. Congress further specified within the Act a combination of general policy and procedural guidelines, and the overall legal and administrative framework within which the federal government must act. The Act is a comprehensive piece of legislation which has been called the most important labor legislation since the Wagner Act.[1] However, Congress only makes laws; it does not implement them. Congress functions slowly, sets necessarily general national policy, and has little direct control over the conditions of any specific workplace. However, Congress does influence the general level of effectiveness of the Act through its budgetary control, and in one instance, it almost succeeded in preventing the Act from being applied to 30% of the workforce.[2]

The Occupational Safety and Health Administration (OSHA) in the Department of Labor determines more specifically how the law is to be applied to different health and safety situations. OSHA implements the Act by promulgating rules and regulations, setting health and safety standards, and overseeing enforcement. It does this within a framework created by the Occupational Safety and Health Act, the Administrative Procedure Act (which describes the general procedures a federal agency must follow), and the precedents of administrative law that have evolved over the years. The rules and regulations established by OSHA transform the intent of Congress and the will of the executive branch into action affecting employers and employees under OSHA's jurisdiction. In this process, OSHA modifies, limits, and expands the scope of the original legislation. The implementation of the OSHAct to a great extent hinges upon administratively derived regulations. Furthermore, OSHA is the federal agency directly responsible for administering the possible transfer of responsibility for enforcing health and safety regulations back to the states.

The philosophy behind OSHA's administrative rules and regulations is influenced

*The text of the OSHAct can be found in the Appendix.

by the action or nonaction of two other bodies: the Occupational Safety and Health Review Commission and the Office of the Solicitor in the Department of Labor. The Review Commission functions as an independent agency, and its actions are more visible than those of the Office of the Solicitor. The Review Commission modifies the law through its interpretation of the statutes and procedures of OSHA as applied to specific cases. Its decisions are subject to judicial review by the federal courts. The Office of the Solicitor acts in an informal way, and its legal interpretations of the Act are challenged indirectly through challenges to actions of OSHA or the Review Commission. Over a period of time, the sum total of these statutory and other interpretations helps determine the policies and effectiveness of the Act's implementation and enforcement.

The fourth level of decision-making, embodied in administrative policy and program directives, is less formal and the least visible to the public at large. The administrative policies and programs of OSHA directly affect the number of workplaces inspected, the quality of the inspection and inspection officers, the number and quality of standards promulgated, and a number of other activities which, taken together, critically influence the effectiveness of the Act in improving workplace conditions.

If we are to understand and evaluate the implementation of the OSHAct, we must do so within the framework of the interactions between the various governmental institutions and their decision-making processes. This chapter deals with the provisions of the Act itself and the federal framework in which the Act is to be implemented. Chapter 5 examines the role of the states in occupational health and safety; and Chapter 6 presents a description and evaluation of the federal role in administering the law.

4.2 A BRIEF DESCRIPTION OF THE OCCUPATIONAL SAFETY AND HEALTH ACT OF 1970*

4.2.1 Purpose

The Occupational Safety and Health Act of 1970 was the first *comprehensive* attempt by the federal government to assure safe and healthful working conditions for working men and women. Prior to 1970, the federal government's involvement in industrial health and safety regulation for nongovernment employees was limited to certain industries such as maritime, construction, and mining, and certain businesses with federal contracts;[3] the primary regulation of industry was at the state level. Increasing awareness of the extent of industrial injuries and health problems

*No attempt is made here to undertake a comprehensive analysis of the OSHAct. The reader is referred to more complete analyses in the bibliography. Reference to sections of the Act related to state participation is to be found in Chapter 5.

and of the ineffectiveness of state enforcement led the 91st Congress to declare its intention to regulate safety and health conditions in the workplace. Congress found that "personal injuries and illnesses arising out of work situations impose a substantial burden upon, and are a hindrance to, interstate commerce in terms of lost production, wage loss, medical expenses, and disability compensation payments."[4]

Under its powers to regulate interstate commerce, Congress enacted the Occupational Safety and Health Act to encourage "employers and employees in their efforts to reduce the number of occupational safety and health hazards at their places of employment, and to stimulate employers and employees to institute new and to perfect existing programs for providing safe and healthful working conditions."[5] A primary purpose of the Act is to ensure *uniformly applied* occupational safety and health standards.[6]

4.2.2 Coverage

The Act applies to any employer who is engaged in a business affecting commerce, including agriculture, in the United States or any territory administered by the United States. In addition to providing coverage to many employees not previously covered by any legislation, the OSHAct continued in effect regulatory provisions of several existing federal laws such as the Longshoremen's & Harbor Workers' Compensation Act, the Walsh-Healey Act, the Construction Safety Act, and others. In 1970 the Bureau of Labor Statistics estimated that the OSHAct applied to approximately 4.1 million businesses employing some 57 million employees.[7]

Places of employment already regulated under provisions of the Atomic Energy Act of 1954 or other specific federal safety and health statutes[8] are preempted from OSHA coverage *only* to the extent that the other regulating agencies have exercised their statutory authority over specific working conditions (Section 4(b)(1)). This is a very complex and contested area, and the Review Commission has held that *no* industry is totally exempted from statutory occupational health and safety coverage. (The matter is now before the Court of Appeals.) Employees of the federal government are covered under a separate program established by executive order. In Section 19(a), the OSHAct places responsibility on the head of each federal agency to establish and maintain an occupational safety and health program which is consistent with the standards set forth by the Secretary of Labor. While the Act does not cover state and local government employees, it does provide that they be covered under plans submitted by each state, to the "extent permitted by its law."[9]

4.2.3 Duties of Employers and Employees

Each employer has the specific duty to comply with occupational safety and health standards promulgated under the OSHAct. In all cases not covered by specific standards, the employer has a general duty "to furnish to each of his employees employment and a place of employment which are free from recognized hazards that are causing or are likely to cause death or serious physical harm to his employees."[10] The latter requirement, commonly referred to as the "general duty clause," has been the basis of many citations issued against employers,[11] and has been the subject of much legal analysis.[12]

The duties of employees are specified in Section 5(b) of the act: "Each employee shall comply with occupational safety and health standards and all rules, regulations, and orders issued pursuant to this Act which are applicable to his own actions and conduct."

Since the purpose of the Act is to provide for safe workplaces, the duties imposed under the Act in no way affect any workmen's compensation law. Furthermore, the Act does not diminish or affect in any other manner the common-law or statutory rights, *duties*, or liabilities of employers and employees under any law with respect to injuries, diseases, or death of employees arising from the course of employment.[13]

The Act specifically prohibits discrimination against employees for the filing of OSHA complaints or similar exercise of rights under the Act (Section 11(c)). The district courts have jurisdiction over suits alleging such discrimination.

4.2.4 Administration

Three agencies have been set up within the federal government to administer and enforce the OSHAct. The agency responsible for administering the provisions of the Act is the Occupational Safety and Health Administration (OSHA), located within the Department of Labor. OSHA is required to set occupational safety and health standards and to conduct inspections of workplaces covered under the Act for the purpose of assuring compliance with both standards and the general duty obligation. The agency has the power to issue citations against employers and to assess penalties for violations.

The Occupational Safety and Health Review Commission was established under the OSHAct as an independent quasi-judicial review board; it consists of three members appointed to six-year terms by the President. The Commission rules upon all challenged enforcement actions of the Occupational Safety and Health Administration. If an employer who has been issued a citation or penalty by OSHA fails to lodge a challenge within 15 days, then that failure automatically constitutes a decision of the Review Commission to uphold the citation or penalty. Such a decision by default is not subject to review by any court or agency. If, on the other hand, an employer challenges the agency's decision, the Commission appoints a hearing examiner who hears the case and reports to the Commission. The report of the hearing examiner becomes the final order of the Review Commission unless one of the Commission members calls for the report to be reviewed by the full Commission. Having gone through this procedure, the employer may then seek review of the Commission's decision in the courts. OSHA may also seek review of the Commission's decisions. Employees have limited right of Commission review on the "reasonableness" of the abatement period.

To assist the Secretary of Labor in enforcing the OSHAct, Congress raised to national institute status the former Bureau of Occupational Safety and Health. The National Institute for Occupational Safety and Health (NIOSH), a research body located within the Department of Health, Education, and Welfare, is responsible for developing and recommending occupational safety and health standards. The agency is specifically required to publish a list of all known toxic substances and the concentrations at which these substances exhibit toxic effects. In order to accomplish these tasks, NIOSH is authorized to conduct research and experimental programs in occupational health and safety, promote training of an adequate supply of manpower, and conduct hazard evaluations.

In addition to these three agencies, Section 7 of the OSHAct establishes a National Advisory Committee on Occupational Safety and Health (NACOSH). NACOSH consists of "representatives of management, labor, occupational safety and occupational health professions, and of the public." NACOSH is required to advise, consult with, and make recommendations to the Secretary of Labor and the Secretary of Health, Education, and Welfare on matters relating to the administration of the Act.[14] The Secretary of Labor may also appoint other advisory committees to aid in the setting of standards.[15] (There was also continued in existence a previously established federal advisory committee dealing primarily with safety in federal employment.)[16]

4.2.5 Occupational Safety and Health Standards

The primary function of OSHA is to promulgate and enforce standards governing the work environment in businesses under its jurisdiction. Three types of standards are mandated by the OSHAct: (1) interim (initial) standards consisting of federal standards from other Acts and national consensus standards existing at the time of implementation of the OSHAct; (2) permanent standards to replace or augment the interim standards; and (3) temporary emergency standards which may be issued immediately upon the finding of serious danger to employees.

Under Section 6(a) of the OSHAct, interim standards could be promulgated by the Secretary of Labor for a two-year period ending April 28, 1973. Congress was of the opinion that many basic safety and health standards had already been subject to public debate and could be promulgated immediately to protect employees. Accordingly, the Act required the Secretary to adopt as occupational safety and health standards any appropriate standards which had been promulgated by other federal agencies. Such standards were not subject to rule-making procedures required by the Administrative Procedure Act and could be made effective immediately upon publication in the *Federal Register.* In addition, the Secretary could adopt and promulgate national consensus standards produced by such recognized organizations as the National Fire Protection Association or the American National Standards Institute. (For the definition of "national consensus standard," see Section 3(9).) The Act required the Secretary to promulgate as interim standards only those national consensus standards for which there was substantial agreement among affected parties, thus eliminating standards which unduly favored either employers or employees. The Secretary announced an initial package of standards covering most industries soon after the Act became effective, and a 90-day grace period was afforded employers to allow them time to comply voluntarily.[17]

Under Section 6(b) of the OSHAct, the Secretary of Labor may—upon the basis of information submitted by the Secretary of Health, Education, and Welfare, NIOSH, advisory committees, or other groups—revise, modify, or revoke existing standards and promulgate new permanent ones.[18] Whenever the Secretary concludes that a permanent standard is appropriate, he may either publish the proposed standard directly in the *Federal Register* and ask for comments, or he may first request the recommendations of an advisory committee which may be appointed.[19] Once a proposed standard has been published in the *Federal Register*, interested parties are afforded at least 30 days in which to comment. Once a standard has been adopted, any person adversely affected by the standard may challenge its validity by petitioning the U.S. Court of Appeals within 60 days.[20]

Under Section 6(c) of the OSHAct, if the Secretary finds that employees are exposed to grave danger from a newly determined hazard, a temporary emergency standard may be issued. Such a standard is not subject to the hearing requirements of the Administrative Procedure Act and is effective immediately upon publication in the *Federal Register.* However, the emergency standard promulgated must be necessary to protect employees from such danger.[21] A temporary emergency standard may remain in effect for six months; thereafter, the Secretary must adopt a permanent standard or the protection of the temporary standard will cease.

Individual variances from occupational safety and health standards may be issued to employers by the Secretary for two reasons. An employer may apply for a temporary variance for the purpose of obtaining additional time in the event that he cannot come into compliance for such reasons as unavailability of equipment or personnel.[22] Permanent variances may be granted if the Secretary finds that the safety measures used by an employer are as safe as those specifically required by a standard.

The Act also provides for experimental variances (Section 6(b)(6)(c)) and national defense variances (Section 16).

4.2.6 Enforcement

The OSHAct provides for inspectors to enter at reasonable times, without delay, and without prior notification, any establishment covered by the Act for the purpose of inspecting the premises and all pertinent conditions, structures, machines, and apparatus within.[23] Inspectors may also question privately any employer, owner, operator, agent, or employee.[24]

The inspector must present proper credentials to the employer or employer representative in charge of the workplace. Procedures are provided in Section 15 to assure the confidentiality of trade secrets obtained by OSHA compliance officers. Employees and employers have the right to accompany the inspector during the inspection (the walkaround right). Where no "authorized employee representative" exists, the inspector must consult with a reasonable number of employees.

After completing the inspection, the inspection officer (called the "compliance officer" in OSHA regulations) reports any violation found to the Area Director, who decides whether to issue a citation. In some cases, the inspector may issue it directly.[25] A citation must describe the specific nature of the violation and establish a reasonable time for abatement of the condition in violation. A copy of the citation must be prominently posted at or near the place of violation.[26] Notices, in lieu of citations, may be issued for *de minimis* violations, i.e., ones which have no direct or immediate relationship to safety or health. A citation must be issued with reasonable promptness but in any case within six months following the inspection.

If the Secretary of Labor finds any conditions or practices in a place of employment which could reasonably be expected to cause death or serious physical harm before normal enforcement procedures could eliminate the hazard, he may seek a restraining order in the U.S. District Court.[27]* If the Secretary arbitrarily or capriciously fails to seek relief from an imminent danger, a writ of mandamus may be sought by any affected party to compel the Secretary to carry out his duty.[28]

Within a reasonable time after issuance of a citation, the Secretary must notify the employer by certified mail of any proposed penalties. The employer has fifteen working days from the receipt of the notice in which to contest the proposed penalty or citation. If the employer fails to notify the Secretary within this allotted time, the citation and the assessed penalty are deemed the final order of the Review Commission, barring employee objections, and are not subject to review by any court or agency.[29]

If the employer contests the citation or the penalty, the case goes before an administrative law judge (hearing officer) of the Review Commission. The judge's decision is considered final unless, within 30 days, one of the three members of the

*In state occupational health and safety programs (see Chapter 5), there may be quicker relief. In California, for example, a state safety inspector can stop operations on the spot.

Review Commission exercises his statutory discretionary right of directing review.[30] If a review is directed, the case must be decided by the full three-member Review Commission. Any aggrieved party (not just the employer) may seek judicial review in the U.S. Court of Appeals within 60 days of the administrative law judge's decision or, if a review is directed, the Review Commission's final disposition.[31]

4.2.7 The Abatement Period

A reasonable period for abatement of a hazard must be prescribed in the citation issued by OSHA. The abatement period may begin immediately or up to 15 days after issuance of a citation (some have begun even later) if the employer does not contest the citation. If the employer files a notice of contest in good faith, and not solely for delay or avoidance of penalties, the abatement period begins after the final administrative disposition by the Review Commission. In general, commencement of appeal proceedings in the U.S. Court of Appeals does not operate as a stay of the order of the Review Commission, unless so ordered by the court.[32]

Any employer who fails to correct a violation for which a citation has been issued within the period permitted for abatement may be assessed a civil penalty of not more than $1000 a day for each day the violation continues.[33] The Secretary of Labor must notify the employer in writing of this proposed penalty. This notice and assessment is final unless the employer contests the same by notice to the Secretary within 15 days. The Secretary must afford an opportunity for a hearing if the abatement has not been completed because of factors beyond the employer's control and if the employer shows he has made a good-faith effort to comply with the abatement requirements of a citation.[34]

4.2.8 Penalties for Violations

The OSHAct provides for a range of monetary penalties for violation of the Act or rules promulgated under it. Any employer who willfully or repeatedly violates the general duty clause or any standard, rule, or order may be assessed a civil penalty of up to $10,000 for each violation.[35] If an employer receives a citation for a serious violation, he must be assessed a civil penalty not to exceed $1000 per violation.[36] A serious violation exists where there is a substantial probability that death or serious physical harm could result. Even if a violation is specifically determined not to be of a serious nature, an employer may receive a civil penalty of up to $1000 per violation.

An employer who commits a willful violation which results in the death of an employee is liable upon conviction for a penalty of $10,000 and/or up to six months imprisonment. Subsequent violations can increase the criminal penalty to $20,000 and/or one year imprisonment.[37]

Penalties can also be assessed for interference with enforcement activities. A person giving unauthorized advance notification of an inspection is liable to a $1000 fine and/or imprisonment for six months.[38] Anyone convicted of knowingly making false statements can be assessed a fine of up to $10,000 and/or six months imprisonment.[39] Failure to post official notification required under the Act may result in a civil penalty of up to $1000 per violation.[40] (See Section 6.3.1 for a discussion of the actual penalties assessed.)

4.2.9 Record-Keeping Requirements

Section 2(b)(12) of the OSHAct requires each employer to maintain a record-keeping system for the purpose of obtaining accurate and uniform information about workplace injuries and illnesses. The system provides inspectors with on-site records of occupational accidents and illnesses. The maintenance of records is also expected to heighten employers' awareness of the problems of occupational safety and health.

An injury is deemed recordable if it involves medical treatment, loss of consciousness, restriction of work or motion, or transfer to another job.[41] Employers must also maintain accurate records of employee exposure to toxic or physically harmful substances for which specific standards have been promulgated. The Secretary of Labor, working with the Secretary of HEW, is required to promulgate regulations to govern the monitoring of these toxic substances and to provide employees or their representatives with access to monitoring records and records of their own exposure to these substances.[42] (These requirements are now embodied in the individual standards.)

4.2.10 Statistics

The record-keeping system is the basic source of data for the statistical program required by the OSHAct. The Secretary of Labor, in conjunction with the Secretary of HEW, must compile accurate statistics on work injuries and illnesses, including all disabling, serious, or significant injuries and illnesses. To carry out these requirements, grants to or contracts with private organizations, states, or local political subdivisions may be used. Employers can be required to file reports of work injuries and illnesses for this statistical program.[43]

4.2.11 Research

The National Institute for Occupational Safety and Health is required to carry out the research and educational functions assigned to the Secretary of HEW under the OSHAct.[44] The Institute is authorized to recommend criteria for occupational safety and health standards development and to conduct research and experimental programs determined by the Institute Director to be necessary for developing or improving job safety and health standards.[45]

A prime function of the Institute is to generate and circulate information in the fields of occupational health and safety. For this purpose, the Institute may conduct programs of medical monitoring of employees and monitoring of the workplace environment to measure potentially toxic substances or harmful physical agents. The Institute must publish annually a list of all known toxic substances and the concentrations at which toxicity is known to occur. In addition, industrywide studies on chronic or low-level exposure to a broad variety of industrial materials and processes are mandated.

4.2.12 Education and Training Programs

The Secretary of HEW, in conjunction with the Secretary of Labor, is required to conduct educational programs to provide an adequate supply of qualified persons to enforce the OSHAct. Programs are also authorized for instruction in the proper use of safety and health equipment. Under Section 21, the Secretary of Labor has prime responsibility in the following areas: short-term training of personnel enforcing the Act; general programs for employers and employees on the recognition, avoidance, and prevention of unsafe or unhealthful working conditions; and programs on effective means of preventing occupational injuries or illnesses.

4.2.13 Assistance from the Small Business Administration

Section 28 of the OSHAct amends sections of the Small Business Act to provide for financial assistance to small businesses for the purpose of altering equipment, methods of operation, or facilities to comply with standards and duties required by the Act. Loans are available to a small business if OSHA determines that compliance with the Act is likely to cause substantial economic injury.

4.2.14 Employer and Employee Rights

Since the enforcement activities of OSHA are directed at employers, specific procedural safeguards and provisions for judicial review are specifically mentioned in the Act and need no further comment here. In addition, the Administrative Procedure Act and the Federal Advisory Committee Act apply to the decisions of the three agencies and the advisory committees designated to administer the OSHAct.

The Act does contain provisions which allow employees to become involved in the administrative and enforcement activities of the Act, even though no *enforceable* duties are imposed on employees.

Perhaps the most important employee right is that of requesting an OSHA inspection. Any employee or employee representative who believes that a violation of a safety or health standard, or of the general duty obligation, has occurred which threatens physical harm or which is imminently dangerous, may request in writing an inspection by the Occupational Safety and Health Administration. A signed letter must be sent to OSHA detailing specific information about the alleged violation. OSHA gives a copy to the employer prior to the inspection, and the employee need not notify the employer of his request to OSHA. Moreover, the employee's name must be withheld by the Secretary if the employee so requests. A special investigation must be made if the Secretary determines that there are reasonable grounds to believe that a violation or danger exists. However, if the Secretary finds that there are no reasonable grounds to believe that a violation exists, notification of this finding must be given to the complainant.

Senator Harrison Williams, coauthor of the OSHAct, has referred to certain provisions of the Act as a "worker's environmental bill of rights."[46] Among these rights are the following:

1. The right to observe the monitoring of harmful substances and to have access to records of monitoring (Section 8(c)(3)).
2. The right to request OSHA to perform a special inspection (Section 8(f)(1)).
3. The right to have a representative accompany an OSHA inspector (Section 8(e)).
4. The right to have all citations posted so that employees will know of any violations found by an inspector (Section 9(b)).
5. The right to obtain review within Labor Department if an inspector fails to issue a citation after employees have provided written statement of alleged violations (Section 8(f)(2)).
6. The right to appeal to the Review Commission if the time allowed for abatement of a violation seems unreasonably long (Section 10(c)).
7. The right to oppose any appeal taken by an employer to the Review Commission (Section 10(c)).*
8. The right to participate in the standard-setting process by offering evidence and comments on proposed standards (Section 10(c)).
9. The right to appeal to the U.S. Court of Appeals on the grounds that the Secretary's action in setting a standard is not based on substantial evidence (Section 6(b)).
10. The right to obtain a hearing on an employer's request for a variance to a standard (Sections 6(d) and 6(f)).
11. The right to request the Department of Health, Education, and Welfare to determine whether substances found in the workplace are toxic (Section 20(a)(6)).

*The Act does not, however, allow an employee to appeal to the Review Commission if OSHA refuses to inspect or to issue a citation in response to the employee's complaint. Further, an employee cannot appeal the size (or lack) of a penalty imposed by OSHA.

12. The right to have the Secretary of Labor seek redress for discrimination resulting from the exercise of rights under the Act (Section 11(c)).
13. The right to sue in a mandamus action to require the Secretary of Labor to abate an imminent danger (Section 13(d)).

It is interesting to compare a similar list of employee rights compiled by the United Steelworkers of America:[47]

1. The right to petition the Secretary to commence the procedure for promulgating standards (Section 6(b)(1)).
2. The right to standards which most adequately and *feasibly* assure that no employees will suffer any impairment of health or functional capacity, or diminished life expectancy, even if such employee has regular exposure to toxic or harmful materials (Section 6(b)(5)).
3. The right to standards which prescribe, where necessary, the labeling of hazardous substances, protective equipment, and monitoring (Section 6(b)(7)).
4. The right to medical examinations so as to determine whether exposure is adversely affecting health (Section 6(b)(7)).
5. The right to have the results of his medical examination transmitted to an employee's physician (Section 6(b)(7)).
6. The right to notification of an employer's request for variance and to participate in the hearing called to evaluate the request (Section 6(d)).
7. The right to representation on the standard-setting advisory committees and on NACOSH.
8. The right to have employers keep them informed of their rights under the Act (Section 8(c)(1)).
9. The right to observe the monitoring equipment and have access to the records thereof (Section 8(c)(3)).
10. The right to be notified when concentration levels are being exceeded, and to be informed of measures being taken to correct such a situation (Section 8(c)(3)).
11. The right to accompany the OSHA inspector (Section 8(e)).
12. The right to request a special inspection; to be able to inform the inspector of any alleged violation; and to receive written explanation of failure by the inspector to issue a citation of violation and to receive an informal review thereof. (Section 8(f)(1-2)).
13. The right to demand a hearing on unreasonable (long) time set for the abatement of a hazardous situation by the Secretary (Section 10(c)).
14. The right to participate in hearings to determine whether an employer should receive a modification of the original abatement order (Section 10(c)).
15. The right to protection against disciplinary action for exercising rights under the Act (Section 11(c)(1-3)).
16. The right to request HEW determination of the toxicity of any substance normally found in thc workplace (Section 20(a)(6)).

4.3 INTERPRETATION OF THE ACT AND IMPORTANT DECISIONS AFFECTING ITS IMPLEMENTATION*

4.3.1 Rules and Regulations

In accordance with its congressional mandate, OSHA has promulgated a number of regulations which clarify procedure and set policy for implementation of the OSHAct. These regulations cover a wide variety of topics, including (1) OSHA administrative responsibilities in such areas as standard setting, the granting of variances, and the use of advisory committees, and (2) inspections, citations, and proposed penalties. The Review Commission issues its own rules of procedure.

The reader is referred elsewhere for the details of these important regulations.[49] In the discussions of selected problem areas which follow, reference is made to the regulations where instructive.

Standard Setting

The OSHAct does not contain specific standards; rather, it sets forth the procedures and guidelines by which standards are to be adopted. Section 4(b)(2) provided for existing Federal standards under other Acts—e.g., the Walsh-Healey Service Contract and National Foundation on the Arts and Humanities Acts—to become occupational safety and health standards automatically on the effective date of the law (April 28, 1971).

On May 29, 1971, the Department of Labor promulgated additional standards as provided under Section 6(a) of the OSHAct. Section 6(a) permitted the Secretary, for a period of two years after the date of enactment, to promulgate as standards, without hearings, national consensus standards and established federal standards. The May 29th package contained over four hundred pages of standards, including the consensus standards of the American National Standards Institute and the National Fire Protection Association that were deemed by the Secretary of Labor to be related to job health and safety. (These numerous standards contained many petty regulations not directly aimed at important worker safety problems such as a requirement that toilet seats be split and not round.)

Although the standards promulgated under Section 6(a) were published in the *Federal Register* on May 29, 1971, the Department of Labor allowed a "familiarization period" until August 27, 1971, and in some cases until February 15, 1972, in order "to allow employers and employees to acquaint themselves with the standards, to encourage voluntary compliance, and to provide the states with additional opportunity to further consider and develop their plans to assume program respon-

*The constitutionality of the OSHAct was challenged and upheld in the U.S. Courts of Appeals for the Third Circuit and the Eighth Circuit.[48] As of this writing the Third Circuit case is scheduled for rehearing, and the Eighth Circuit has twice upheld the Act's constitutionality.

sibilities under the Act."[50] However, on August 13, 1971, the Department of Labor made effective immediately those standards whose effective date had been delayed until February 15, 1972.[51] (These included standards for noise, personal protective equipment, hazardous materials, and the threshold limit values (TLVs), which are suggested limits for general worker exposure to chemical and physical agents over an eight-hour period.)

The Special Problems Encountered in Setting Carcinogen Standards

An examination of the difficulties encountered by OSHA in setting permanent carcinogen standards will help to clarify the problems involved in the standard-setting process.

In 1968 the American Conference of Governmental Industrial Hygienists (ACGIH), a private standard-setting organization, adopted a list of some 400 threshold limit values as "guides for the control of health hazards." Several of the substances investigated had been found to be carcinogenic and their recommended exposure limits (zero) were listed in a special appendix. In 1969 the Labor Department published regulations incorporating the ACGIH list into the Walsh-Healey Act.[52] Under the OSHAct, however, the Labor Department has claimed that the *appendix* to the list was not included within Walsh-Healey, even though the 1969 regulations had made no such exception.[53] Hence, the zero tolerances for these carcinogens were not considered federally established standards and were therefore not part of the initial OSHA package. A full hearing was thus required to establish permanent standards for these carcinogens under the Section 6(b) procedure.

On December 29, 1972, the Oil, Chemical, and Atomic Workers International Union (OCAW) and the Health Research Group (HRG) submitted a petition to OSHA, requesting that temporary emergency standards be adopted which set a zero tolerance level for ten carcinogens.[54] These emergency standards were sought under the provisions of Section 6(c) of the OSHAct, which allows OSHA to issue such standards without going through the full rule-making process, and thus to avoid delay.

OSHA, however, chose to proceed cautiously and invited comments from interested parties from February 9, 1973, to March 11, 1973. OSHA sought specific information on (1) the level of exposure that would pose a grave danger to employees, (2) the necessity of an emergency standard, (3) the quantity of the substances used and the number of employees exposed, (4) the types of uses of the substances and the types of processes in which any of the substances were formed or released as an end product or by-product, (5) the extent to which substitutes were technologically and economically feasible, and (6) the feasibility of zero tolerance standards.[55]

In April 1973 OCAW and HRG filed a complaint in the U.S. District Court for the District of Columbia, seeking an order directing the Department of Labor to issue temporary emergency standards. The complaint alleged that OSHA had unlawfully withheld action on the petition originally filed by OCAW and HRG on December 29, 1972.

Temporary emergency standards regulating work practices for employees exposed to 14 carcinogens were adopted by OSHA on May 3, 1973.* OSHA found (1) that the 14 carcinogens are toxic and physically harmful, (2) that exposure to any of the 14 substances poses grave danger to employees, (3) that employees were being exposed to the substances, and (4) that a temporary emergency standard was necessary to protect employees from such exposure.[56]

On October 4, 1973, the U.S. Court of Appeals for the Third Circuit, in a suit brought by a chemical manufacturers' association, struck down two of the temporary standards.[57] The court found that the Department of Labor did not sustain the burden of proof by failing to give sufficient reasons for its action in determining that the two substances were sufficiently dangerous to humans to warrant the issuance of an *emergency* standard.

OSHA, however, proposed *permanent* standards for the 14 carcinogens on July 16, 1973, based on the temporary emergency standards issued on May 3. The proposed standards did not ban the use or manufacture of any of the 14 carcinogens; rather, they set requirements designed to protect employees whose duties involved exposure to any of the substances. An important aspect of the proposed standards was their reliance on a use-permit system.[58]

The proposed permit system would have required employers wishing to continue using carcinogens after October 31, 1975, to apply for a permit before October 1, 1975. The procedure would have required the employer to inform his employees that such an application had been made, and to describe in detail in the application how the carcinogens would be used and the conditions, practices, means, methods, operations, and processes that would be instituted to protect employees. Employers

*The 14 carcinogens are the following:

2-Acetylaminofluorene
4-Aminodiphenyl
Benzidene and its salts
3,3'-Dichlorobenzidene and its salts
4-Dimethylaminoazobenzene
alpha-Naphthylamine
beta-Naphthylamine
4-Nitrobiphenyl
N-Nitrosodimethylamine
beta-Propiolactone
bis-Chloromethyl ether
Methyl Chloromethyl ether
4,4'-Methylene (bis)-2-chloroaniline
Ethyleneimine

would have to provide at no cost to employees an opportunity for at least one medical examination per year; to make the results of the examinations available to employers and NIOSH; to report any conditions or incidents resulting in direct exposure of employees to carcinogens; to maintain accurate records of employee exposure; to provide employees access to exposure records; and to test carcinogen containers for susceptibility to rupture or leaks.[59] Public hearings on the proposed permanent standards were concluded on September 14, 1973.

Strong support for adoption of performance standards (as opposed to specification standards) and the use-permit system has been expressed by the Oil, Chemical, and Atomic Workers Union, the Health Research Group, and the American Chemical Society. HRG in particular has dealt with the legal issues involved in the use-permit proposal: it contends that there is no express limitation in the OSHAct prohibiting the establishment of a use-permit system and, further, that any legal reasoning limiting the use of such a system would have to be implied from the statute.[60] HRG disputed testimony given in the public hearings which implied that the system would allow the Secretary of Labor to close a plant without using the statutory injunctive remedy in a U.S. District Court available in the case of an imminent danger. The system *would* allow the Secretary of Labor to suspend or revoke a use permit under certain circumstances; if the employer continued to operate without a use permit, he would then be violating a health and safety rule and would be subject to a citation or imminent danger action. The HRG has favored provisions for the suspension of permits without a hearing, noting that "the Administrative Procedure Act permits this by specifically relinquishing the notice and opportunity to demonstrate compliance when matters of public health or willfulness are involved."[61] This statutory exemption suggests permissible relaxation of the due process hearing requirement when emergency action is necessary to protect public health.

On October 2, 1973, OSHA submitted an environmental impact statement to the Council on Environmental Quality as required by the National Environmental Policy Act. The statement concluded that regulations would necessitate capital outlays by firms. Firms would be more likely to obtain substitutes than to shut down; in fact, a secondary benefit of the proposed standards would likely be a heightened search for substitute chemicals or processes not involving carcinogens.[62]

On November 29, 1973, the OCAW-HRG group filed suit in the U.S. District Court for the District of Columbia requesting the Labor Department to issue a permanent standard for the 14 carcinogens. (Recall that, under the OSHAct, permanent standards should have been issued on November 3, six months after OSHA's issuance of temporary emergency standards on May 3.) OCAW and HRG contended that OSHA had unlawfully failed to issue a permanent standard as required under Section 6(c)(3). In their brief, HRG and OCAW stated: "Notwithstanding this explicit statutory mandate, no permanent standard has been issued by the defendants and none will be issued until defendants are ordered to do so by the court."[63] Assistant Secretary of Labor for OSHA John Stender contended that OSHA did not have the authority to require a use-permit system. Further, OSHA stated that it preferred to require compliance via the imminent danger provisions rather than the use-permit system proposed by the Standards Advisory Committee.[64] (No doubt, difficulties entailed in administering a use-permit system in both the federal and state systems played a large part in this decision; see below.)

In early December OSHA reported that the final carcinogen statement was being "delayed" by the Council on Environmental Quality, which was reviewing the environmental impact statement.[65] (The CEQ regarded OSHA's statement to be inadequate because it did not discuss the alternative of a use-permit system and subsequent monitoring and therefore did not meet the requirements of the National Environmental Policy Act.) On December 3, 1973, OSHA filed a supplement to the environmental impact statement. In this supplement, the Labor Department expressed serious doubts about the feasibility of a use-permit system and also about its legal authority to establish such a system. OSHA stated that it had no legal authority to issue use permits, that nothing in the legislative history suggested such a system, and that the Act only allowed promulgation of general standards in accordance with specific procedures. The administrative feasibility was also questioned in terms of the enforcement effort and budget which would be required to operate such a system.[66]

The final standards for 14 carcinogens were published in the *Federal Register* on January 29, 1974. In promulgating these standards, OSHA indicated that it was relying extensively but not exclusively on the hazard criteria documents prepared by NIOSH. In addressing the question of carcinogenicity in animals versus humans, OSHA found it improper to afford less protection to workers exposed to substances found to be carcinogenic only in experimental animals. However, OSHA excluded from coverage mixtures containing less than 1% by weight or volume of eight of the carcinogens and mixtures with less than 0.1% for the remaining six. *These exclusions would permit biologically significant exposures.* Furthermore, using a less stringent percentage (1%) for the first group *is* treating the carcinogens differently.

In response to OCAW and HRG's petition which sought to establish zero tolerance levels, OSHA stated that a policy of "no possible exposure to the carcinogens under any circumstance could only be guaranteed by a total ban on the manufacture, use (even for cancer research), and transportation of the substances." OSHA expressed its agreement with the findings of the Ad Hoc Committee on the Evaluation of Low Levels of Environmental Chemical Carcinogens "that a safe level of human exposure to any of the 14 carcinogens cannot be established by application of present knowledge." But OSHA was "not prepared to draw from this state of knowledge the conclusion that such levels do not exist." After consideration of the two points above, OSHA concluded that the intent of the standards is to reduce exposure of workers to any of the listed substances to the maximum extent practicable consistent with continued use.[67]

After considering the administrative and legal aspects of a use-permit system, as opposed to the current enforcement tools provided in the Act for general standard enforcement, OSHA decided not to adopt a use-permit system. In support of its decision, OSHA referred to testimony in the public hearings regarding (1) the use-permit system's provision for administrative action to close down a facility, in contravention of the statutory requirement that such closings should be left to the courts; (2) the need to commit substantial time and resources in order to implement a use-permit system, and (3) the need to tailor each permit to the specific situation for which it is requested. OSHA concluded that "the requirement in the adopted standards for employers to report the uses of carcinogens and the nature of operations involved, together with incidents of release of carcinogens, will permit significant administrative surveillance."[68]

Fourteen *separate* standards were adopted due to the different chemical, physical, and biological properties of each carcinogen. OSHA favors a control strategy of isolating carcinogens in a "closed system," and thus placed a ban on open-vessel operations. When protection is afforded by means of personal protective equipment, the standards require total body coverage. Employers are also responsible for providing training to their employees on the nature of the carcinogenic hazards in the workplace. While the standards do not include requirements for monitoring activities, NIOSH has been requested to develop techniques and instruments to facilitate such monitoring. Medical surveillance by a physician is required at yearly intervals and whenever exposure to a carcinogen may have occurred.

HRG and OCAW sought judicial review of the inadequacies of the permanent standard, including inadequate medical surveillance, the absence of any monitoring requirement, the denial of the use-permit system, and the failure to prescribe a performance standard of no detectable exposure (*Oil, Chemical, and Atomic Workers Union v. Brennan* No. 74-1149). Review was also sought by two industry petitioners, the Synthetic Organic Chemical Manufacturers Association and the Polyurethane Manufacturers Association. The Third Circuit Court of Appeals upheld the carcinogen standard except that it vacated the standard for 4,4'-Methylene (bis)-2-chloroaniline on procedural grounds and it vacated the special provisions governing laboratories in all the standards.[69]

OSHA has now adopted only three new health standards packages: asbestos, the 14 carcinogens, and vinyl chloride. In each case, temporary emergency standards were issued in response to petitions by unions, and in each case, the permanent standards were challenged in the courts—by unions for asbestos and the 14 carcinogens, and by industry for the 14 carcinogens and vinyl chloride. The court ruled against the unions in the asbestos case in agreeing that economic feasibility can be taken into account in the setting of standards.[70] The court ruled for the unions in the vinyl chloride case in upholding the use of animal data as sufficient for supporting a standard for protection of humans against cancer.[71]

The legal and political problems encountered by OSHA in attempting to issue permanent carcinogen standards and then, later, pesticide standards (see Sections 4.4.2 and 12.3) led OSHA in subsequent standards activity—and especially in the case of vinyl chloride—to proceed with considerable care in collecting adequate technical data to support the standards requirements and in setting forth the basis for the standard in the published preamble.

Exemption of Smaller Firms from Record-Keeping Requirements

Under the Act, OSHA also issues rules and regulations concerned with nontechnical and procedural aspects relevant to administering the Act.

On July 25, 1973, George C. Guenther, Assistant Secretary of Labor for Occupational Safety and Health, proposed an administrative rule to exempt from the record-keeping requirements of the Act those employers with seven or fewer employees. He noted that the proposal was "in furtherance of Section 8(d) of the Williams-Steiger Occupational Safety and Health Act of 1970 providing, among other things, that information be obtained with a minimum burden upon employers, especially those operating small businesses."[72]

Under this rule, small employers would meet their obligations of record-keeping by reporting fatalities or multiple hospitalization accidents within 48 hours. The usual log of occupational injuries and illnesses, supplementary record of individual job injuries or illnesses, and the annual injury and illness summary would not be required (except for a small random statistical sampling of such firms which would maintain the log).

OSHA adopted this proposal on October 19, 1973, after publication in the *Federal Register* on October 4, 1973. Thus, OSHA, in effect, substantially modified the original legislation by exercising its rule-making authority. (This change was ratified by Congress in late 1974 when it passed the fiscal 1975 Labor-HEW appropriations bill (H.R. 15580) exempting firms with ten or fewer employees from record-keeping except on fatal and "serious" accidents.)

4.3.2 Interpretations and Important Decisions of the Office of the Solicitor, Review Commission, and Federal Courts

General Duty

The legislative history of the OSHAct does not reveal a clear congressional intent regarding the general duty provision. The original bills, which contained broad general duty requirements and no penalties for violations, were redrafted and amended until in the final Act specific requirements and penalty provisions were adopted.[73] The House-Senate Conference Committee found that the concept of a general duty is firmly rooted in the common law. The Senate Committee on Labor and Public Welfare noted that "under principles of common law individuals are obliged to refrain from actions which cause harm to others. Courts often refer to this as a general duty of care to others. . . . The committee believes that employers are equally bound by this general and common duty to bring no adverse effects to the life and health of their employees. . . . [The Act] merely restates that each employer shall furnish this degree of care."[74] Further, there is existing precedent for the general duty concept in federal and state laws which compel employers to provide a safe and healthful place of employment. Three federal statutes and the laws of over 36 states recognize this concept.[75] The reader is referred elsewhere for extensive legal discussion of general duty obligation.[76]

Early Review Commission cases established the principle that employer liability under the general duty provision does not require the occurrence of an accident or employee injury.[77] Liability is conditioned only on whether the employer has exercised reasonable diligence in providing his employees with a place of employment free from recognized hazards. The question of liability has turned upon the definition of the concept of "recognized hazards" in any particular case. Congressman William Steiger, in discussion of the Conference Report, stated that the "recognized hazards" language of the general duty clause applies to hazards which are of "the type that can be detected on the basis of the human senses. Hazards which require technical or testing devices to detect them are not intended to be within the scope of the general duty requirement."[78] On the other hand, OSHA's position in its *Compliance Operations Manual* is that "A hazard is 'recognized' if it is a condition that is (a) of common knowledge or general recognition in the particular industry in which it occurs, and (b) detectable (1) by means of the senses (sight, smell, touch and hearing) or (2) is of such wide general recognition as a hazard in the industry that *even if it is not detectable by means of the senses, there are generally known and accepted tests for its existence which should make its presence known to the employer*" (p. viii-2).

The Review Commission has rejected the Steiger position in favor of OSHA's interpretation on the basis of other relevant legislative history, including remarks by Congressman Dominick Daniels. In *American Smelting and Refining Co.*, the Commission found that "serious hazards that are not obvious and can be detected by instrumentation are within the scope of Section 5(a)(1) of the Occupational Safety and Health Act, which is not limited to 'recognized hazards' of types that are detectable by basic human senses."[79]

The Eighth Circuit Court of Appeals specifically upheld the Commission's interpretation of "recognized hazard" in its review of *American Smelting.*[80] The court stated that a narrow construction of the term would endanger the health and safety of workers: "Where hazards are recognized but not detectable by the senses, common sense and prudence demand that instrumentation be utilized."[81] In an earlier case, *Vy Lactos Laboratories, Inc.*, the same circuit court had held that "recognized hazards" include an employer's actual knowledge as well as the generally recognized knowledge of the industry.[82]

The Review Commission has found that when a specific standard exists, the citation under the general duty clause is precluded.[83] However, the Review Commission continues to maintain that in the absence of a standard, protection should be afforded employees through the general duty clause.[84] At least one member of the Commission has even suggested that if the Secretary were not sure whether to cite a respondent under a specific standard or the general duty clause, he should issue citations for violation of the specific standard and also the general duty clause and let the Commission decide which section of the OSHAct, if any, was violated.[85]

Given that the employer is under an obligation to provide a workplace free of "recognized hazards," the question remains whether the employer is liable for conditions created by his employees when they are not acting in accordance with his instructions.[86] While the OSHAct specifically states that each employee must comply with occupational safety and health standards and all rules, regulations, and orders issued pursuant to the Act, no mechanism to force compliance by employees exists under the Act. Under Section 5(b) ultimate responsibility lies with the employer. The report of the Senate Committee makes this point clear: "The Committee does not intend the employee-duty provided in Section 5(b) to diminish in any way the employer's compliance responsibility or his responsibility to assure compliance by his own employees. Final responsibility for compliance with the requirements of this Act remains with the employer."[87]

A number of Review Commission cases have concluded that an employer is *not* liable for a violation of the general duty clause if the condition in question results from an unauthorized or disobedient action which could not have been prevented by the exercise of reasonable employee supervision.[88] This position has its basis in the common-law tort concepts which hold that the employer cannot be held negligent in such a situation. In *Richmond Block, Inc.*, the employer had instructed all employees to "lock out" the flow of electric power to a cement mixer prior to entering the mixer for cleaning purposes. The employee responsible for cleaning the mixer had been so instructed twice in the ten days preceding the accident, but on the day in question, he failed to follow instructions and was killed when a coworker accidently started the mixer. The Review Commission affirmed the trial judge's vacating of the citation because the fatality was caused by "the employee's disregard of a simple safety rule and procedure, clearly and recently communicated to the employee, which he apparently understood."[89]

A similar result was reached by the Seventh Circuit in *Republic Creosoting Co., Division of Reilly Tar and Chemical Corp.* In this case, an employee disregarded specific instructions and cut the steel band encasing a package of railroad ties. As a result, five ties fell on the employee, fatally injuring him. The Seventh Circuit affirmed the Review Commission's vacating of the citation, stating that "a reasonably diligent employer would not have foreseen that the employee would injure himself."[90]

In several cases where there has been found evidence of inadequate supervision, a violation of the general duty clause has been cited. In *John B. Kelly, Inc.*, an employee assigned to remove masonry from a freestanding wall was told to brace the wall before starting the work. The employee was observed by his supervisor on at least two occasions working without any bracing and was instructed to erect the bracing. The employee never complied, and the wall later collapsed, killing the employee. The Review Commission affirmed the trial judge's finding of a violation because the supervisor "should have remained on the scene in order to insure that the employee, who had twice contravened his instructions, properly carried them out."[91]

In *REA Express, Inc.*, the Second Circuit Court of Appeals affirmed the Review Commission's finding of a violation of the general duty clause where the employer's supervisory personnel permitted untrained employees to attempt repairs of high-voltage equipment on a wet concrete floor without using protective equipment.[92]

Perhaps the most important case involving the general duty clause is *National Realty and Construction Co.* In this case, a citation for serious violation of the general duty clause was issued to the employer for permitting a foreman on the running board of a moving front-end loader, by which action the foreman was killed. The employer asserted that he had not permitted the foreman to ride on the equipment and that, moreover, the foreman had violated an unwritten rule against such action. In finding the employer in violation of his general duty obligation, the Review Commission concluded that unwritten rules are not a reasonable form of employee supervision: "In any case, the employer in whom 'final responsibility for compliance with the requirements of this Act' reposes cannot shrive itself of responsibility . . . by asserting an unwritten rule." Further, the Commission stated that Congress did not intend the defenses available at common law—contributory negligence, the fellow servant doctrine, and assumption of risk—to be available to employers.[93]

The applicability of the general duty clause may have been severely restricted from a practical point of view under a ruling by the U.S. Court of Appeals for the District of Columbia on the appeal of the *National Realty* case. The Court of Appeals found that the Secretary of Labor had failed to sustain his burden of proof in showing what the employer should have done to improve his safety policy (so that the accident would not have occurred). Congress intended to require elimination only of preventable hazards, and did not intend unpreventable hazards to be considered "recognized" under the general duty clause, thereby imposing strict liability on the employer. The court held that in order to prove a violation of the general duty clause, the Secretary must prove (1) that the employer failed to render his workplace "free" of a hazard which was (2) "recognized" and (3) "causing or likely to cause death or serious physical harm."[94] The burden of proof now required of the Secretary is likely to discourage the use of general duty citations.

The question of employer and employee duties has perhaps been best summarized by an OSHA hearing judge who stated that "despite the mutuality of responsibility which the Act imposes on employer and employee, it is clearly apparent that because the employer controls the work environment, the standard of care he owes to his employees under the general duty clause of the Act is high, but not so high . . . that the employer becomes the virtual insurer of the conduct of his employees, and thus absolutely liable for all their acts of commission and omission."[95]

On its face, the potential strength of the general duty clause lies in the imposition of a general responsibility upon the employer to take affirmative action in approving workplace conditions. The applicability of the employer's general duty even in the absence of injury could, in principle, encourage general preventive responses to injury and illness.

Finally, it should be noted that although violation of the general duty clause cannot be charged in a situation which is covered by a specific standard, the requirements for proving a general duty violation are becoming more stringent. It is not clear what indications of harmful effects need to be supplied in order that the hazard be "recognized" by industry or the employer and hence a violation of general duty in the absence of a specific standard. Nonetheless, since only a relatively small number of health standards exist, it remains the only viable enforcement mechanism for protection from the majority of health hazards. One of the reasons that more health standards have not been issued is that the industrial hygiene data are not conclusive. Therefore, more stringent requirements for proving violation of the general duty clause serve to limit further employees' protection from potential health hazards.

Employee Rights

The OSHAct grants employees several important legal rights (see Section 4.2.14). In general, these rights were designed to facilitate employee action in eliminating safety and health hazards in the workplace. While employees are just beginning to exercise their rights, the scope of their action has already been influenced by a number of judicial and administrative decisions.

Walkaround One of the most developed areas of employee rights is that of employee participation in the "walkaround."[96] In *Chicago Bridge and Iron Co.* and *Wright-Schuchart-Harbor Contractors*, the Review Commission held that the right to accompany the compliance officer on the walkaround was not mandatory under all circumstances.[97] While these cases were concerned with the failure to afford the *employer* the opportunity to participate in the walkaround, the principle has also been applied in cases involving failure to afford employees participation. In *Drake-*

Willamette Joint Ventures, the Review Commission ruled that failure to afford employee representatives the opportunity to accompany the Labor Department representative on his inspection does not require vacation of a citation if it is not shown that the Secretary's action prejudiced the employer. Commission Chairman Moran notes that "this decision says directly what two earlier decisions (*Chicago* and *Wright*) implied: that the employee's walkaround rights, like those of the employer, are not mandatory."[98]

Compensation for the Walkaround Related to the walkaround right is the question of compensation of employees for time spent on the walkaround. Although Section 8(e) makes no mention of compensation for the walkaround, it has been argued that an employer's failure to pay employees constitutes discrimination against employees exercising their rights under the Act, a practice forbidden under Section 11(c). This question first arose in an incident involving the Mobil Oil Company and OCAW, after which OCAW filed a formal charge with the Labor Department alleging a violation of the discrimination provisions of Section 11(c). After a formal investigation of the facts, the Solicitor of Labor ruled that refusal to pay employee representatives for participation in a walkaround is not per se a violation of Section 11(c). The Solicitor, while noting the refusal to pay for the walkaround might tend to discourage employees from participating, held that Section 11(c) does not prohibit employer action which might have the effect of discouraging employees from exercising rights; rather, it makes it unlawful for an employer to take action *because* employees exercise rights. As a guideline, the Solicitor suggested that "an employer's past practice respecting payment for certain *other* safety and health activities on regular working time may well require the conclusion that the failure to compensate for the walkaround is discriminatory."[99] The Solicitor's decision that the walkaround is not a working test and therefore does not require pay under the Fair Labor Standards Act was upheld by the U.S. District Court in *Frank P. Leone v. Mobil Oil Corp.*[100] *

The issue of discrimination against employees by denial of pay for the walkaround was expanded when Stauffer Chemical Co. refused to pay OCAW officials for walkaround time. OCAW filed a complaint with the Solicitor of Labor alleging discrimination in light of the employer's practice of paying for participation in plant safety meetings. Considering this policy of the employer, OCAW contended that Stauffer's refusal to pay was "because of" the employees' exercise of their rights under the Act. The Solicitor found that although the employer had paid employees for participating in plant safety meetings, the plant safety committee had never made on-site investigations of unsafe practices, and thus the employer had

*In a private interview on June 30, 1972, George Guenther stated that *Mobil Oil* would not be a precedent for other cases diminishing the walkaround right. In a second interview on the same day, an attorney in OSHA expressed disagreement with the decision, saying that he would find it difficult to defend on legal grounds and that it would act as a harmful precedent.

never paid employees for activities similar to the walkaround. In short, the employer did not violate Section 11(c) by refusing to pay employees to participate in an OSHA walkaround because he had never paid employees for participation in a non-OSHA walkaround.[101] The Solicitor's findings imply that only if employees have been paid for attending plant safety committee meetings involving on-site investigation will the failure to pay employees for participation in the OSHA walkaround constitute discrimination under Section 11(c).*

Employee Notice Section 10(c) of the OSHAct requires that representatives of affected employees be afforded an opportunity to participate as parties in hearings. The Review Commission has held that notices of hearings and of proposed settlement agreements must be served either personally or by mail on authorized employee representatives.[102] In addition, a motion by the Secretary[103] or an employer[104] to withdraw a contested portion of a citation must be "served" on the authorized employee representative, if there is one, or posted at the worksite.

The importance of the employee notice requirements was recently illustrated in *American Airlines, Inc.* In this case, the Review Commission reversed the hearing judge's order accepting the terms of a settlement agreement because the proposed agreement was not served on the union prior to its approval by the hearing judge. The union objected to the settlement because under its terms only the battery service areas would be provided with facilities for the quick flushing of eyes. The union noted that employees working in other areas, which were listed in the original citation, were also exposed to harmful fluids.[105]

On-Site Consultations with Employers

Section 21(c)(2) requires the Secretary of Labor to "consult with and advise employers and employees, and organizations representing employers and employees, as to effective means of preventing occupational injuries and illnesses." Until recently, OSHA was fulfilling this requirement by providing for consultations (with employers only) at the OSHA Area Offices because the Solicitor of Labor had ruled that OSHA may not conduct *on-site* consultations with employers without issuing appropriate citations.

The Solicitor's ruling stated that whenever a Department of Labor employee goes onto an employer's premises for any purpose under the Act (except for inspection in connection with the issuance of a Section 6(d) variance), he is required to note any violations, and citations must be issued as appropriate. Assistant Secretary of

*See *Secretary of Labor v. True Temper Corp.* in *OSHA*, BNA (July 11, 1974), p. 147, where the Department of Labor has filed suit in behalf of two employees who were denied payment for the walkaround after having been paid for three hours safety work per month.

In an interview on May 8, 1973, a high OSHA official responded to my questions concerning the diminishing of the walkaround right by these two cases by saying that "the inspectors could never learn anything from workers anyway." While this view cannot, of course, be attributed to the entire agency, it is most disconcerting.

Labor Guenther explained the Department's position in testimony on June 22, 1972, before the House Subcommittee on Environmental Problems Affecting Small Businesses:

Since Section 8 . . . makes it clear that any entering upon the employer's premises is regarded as an inspection provided for in Section 9, we have been legally precluded from providing any consultation visits on an employer's premises. We would support an amendment to provide for workplace visits, upon request, for specific consultative purposes by Department of Labor advisors. Voluntary compliance, which we consider essential to the success of the whole occupational safety and health program, would be greatly enhanced by such an amendment.[106]

On May 20, 1975, as a result of a $5 million appropriation for state-conducted on-site consultation, OSHA promulgated a regulation providing for grants of 50% federal funding for on-site consultation programs to states without approved state plans. Under the authority of Sections 7(c)(1) and 21(c) of the OSHAct, states without plans can enter into agreements with OSHA whereby trained state personnel can provide on-site consultative services separate from the federal inspection force.[107] As of the June 30, 1975, deadline, however, only 50% of the eligible states had applied for funding.

While NIOSH is not specifically mandated by the OSHAct to provide consultation to employers, Section 20(a)(6) requires NIOSH to determine, in response to written requests by an employer or employee, whether any substance normally found in the place of employment has potentially toxic effects. Under Section 20(b), NIOSH is authorized to make inspections and to question employers and employees in the same manner as is provided in Section 8 for the purposes of conducting such a health hazard survey. However, NIOSH has no authority to issue citations. Thus, if an employer requests a health hazard survey from NIOSH, he will for all intents and purposes be receiving an on-site consultation in the area of health hazards.

The 1973 *President's Report on Occupational Safety and Health* further emphasizes the consultative activities of NIOSH. It states, for example, that in the area of industrial hygiene,

consultation, technical assistance and other services related to industrial hygiene and engineering aspects of occupational health are provided to Federal, State and local agencies, and to others upon request. The principal purpose of technical assistance is to assure the translation of research knowledge into control practice. Procedures in industrial hygiene related to collection of samples, measurements at the workplace, analysis of samples, and evaluating results are provided to employers and employees to assist them in recognizing, evaluating, and controlling real or potential health hazards at the workplace. (p. 125)

NIOSH has also implemented consultative services in the area of occupational safety. The program, which began in May of 1972, provides technical and consultative services to government and industry on various approaches to prevent or reduce the severity of injuries associated with machines, equipment, devices, and appliances used in the occupational environment.[108]

The probability of a follow-up by OSHA inspectors in part depends on the type of consultation an employer requests. OSHA has stated that an employer who in good faith requests an off-site consultation will not be inspected as a result of that request (see Section 4.3.3). Similarly, it would appear that information that NIOSH might give OSHA indicating the existence of a health problem found during a health hazard survey would not be likely to result in an OSHA inspection and citation of a good-faith employer, except in cases of clear and imminent danger. Under 7(c)(1) agreements for on-site consultation, a consultant is required to seek elimination of any hazard that could reasonably be expected to cause serious physical harm or death; if the hazard is not abated, the consultant is then expected to contact OSHA's enforcement arm and the affected employees.

Constitutional Challenges to the OSHAct, Its Standards, and Its Penalties

Employers have brought several unsuccessful suits to have portions of the OSHAct declared unconstitutional. The inspection process has survived its attack as an unlawful search and seizure under the Fourth Amendment.[109] The fines imposed as penalties have been ruled civil and not criminal, thus denying employers the constitutional protections guaranteed to accused criminals under the Fifth, Sixth, and Seventh Amendments (e.g., right to remain silent, right to confront witnesses, right to a jury trial, due process, etc.).[110]

Several cases before the Review Commission have sought rulings by the Commission on the constitutionality of standards. While the dissenting commissioner in *Divesco Roofing and Insulation Co.* wrote that the Commission can pass upon the constitutionality of a standard, on the theory that this action is indistinguishable from passing upon the legislation itself, the majority have held that no such power exists.[111] In his concurring opinion in *Divesco*, Commissioner Cleary stated that, with regard to standards, "the Commission *does* have the power to decide all questions short of invalidating the standards on constitutional grounds. It can determine the applicability of a standard to specific facts; it can decide the weight to be given to the Secretary's interpretation of a standard; and it can construe a standard to resolve vagueness and ambiguity."[112]

Vagueness of Standards

Most of the challenges to standards have centered around the vagueness and generality of some standards promulgated under the OSHAct. A key issue involved is whether the defendant/employer is deprived of the protection of due process when he is cited for violating a vague standard. In a 1966 case which did not involve the OSHAct, the Supreme Court stated: "It is established that a law fails to meet the requirement of the Due Process Clause if it is so vague and standardless that it leaves the public uncertain as to the conduct it prohibits or leaves judges and jurors free to decide, without any legally fixed standard, what is prohibited and what is not in each particular case."[113]

A majority of the Review Commission has held that the Commission has the authority to rule on whether standards are vague.[114] Although the courts of appeals have not specifically ruled on the Review Commission's authority in this area, the silence of the courts has been viewed as implied approval of the Commission's policy.

Even if the Review Commission does have the authority to rule on questions of vagueness, however, the courts of appeals have been less receptive to the vagueness argument than the Review Commission. In *Ryder Truck Lines, Inc.*,[115] and *McLean Trucking Co.*,[116] the Fifth and Fourth Circuit Courts of Appeals rejected the contentions of the companies that a standard requiring the use of "personal protective equipment" was vague. In *Ryder*, the court stated that "so long as the mandate affords a reasonable warning of the proscribed conduct in light of common understanding and practices, it will pass constitutional muster."[117]

In *Santa Fe Trail Transport Co.*, the Tenth Circuit Court of Appeals reversed the Commission and held that an OSHA standard requiring that, in the absence of a medical facility in "near proximity," a person trained in first aid be present in the workplace, is not unconstitutionally vague. The court emphasized the remedial nature of the OSHAct and the notion that some standards must necessarily be broad.[118]

Another related constitutional question is whether the Review Commission has the authority to rule whether the procedures used by the Secretary in promulgating a standard have violated due process. Although a majority of the Review Commission has ruled that the Commission does have this authority,[119] the question has not been completely resolved.

Equal Protection in Standard Setting

A different kind of constitutional question may be raised in the future under Title VII of the Civil Rights Act and the Fourteenth Amendment. Standards as they are presently set generally protect only the "average" worker. However, exposure to toxic materials may be more serious in the case of pregnant women, for example, or of clearly defined groups that manifest special genetic susceptibilities. Air quality standards for the general environment take the more susceptible parts of the population into consideration. Is the failure to set OSHA standards in this manner a violation of civil rights or of the "equal protection" guarantee of the Fourteenth Amendment? The issue has not yet been tested.

Employer Duties under Section 5(a)(2)

A number of cases have arisen in which employers have been found to be responsible for the acts and omissions of employees with regard to specific OSHA standards. In *Mobile Coatings, Inc.*, and other cases, the Review Commission has held that an employer must establish an effective policy to ensure that employees comply with specific standards regarding the wearing of protective equipment.[120] Employers must not only provide equipment, they must also enforce the use of such equipment through explicitly announced policy and monitoring of programs.[121] However, in *United Auto Workers v. OSHRC and the Budd Co.*, the Third Circuit Court of Appeals affirmed the Review Commission's holding that the employer is *not* obligated to pay for required safety equipment and that the issue of who pays should be left to collective bargaining.[122]

In cases where violative conditions are directly caused by employee disobedience, the Commission's decisions have been similar to those rendered under the general duty clause. That is, so long as the employer has an adequate safety program and close supervision, idiosyncratic misconduct that results in the creation of a hazard will not be found to be the responsibility of the employer under the OSHAct.

In *Standard Glass Co., Inc.*, the Commission ruled that when an employer has made a concerted effort to enforce rules regarding employee protective equipment, isolated employee failure to use equipment does not constitute a violation of the standard by the employer. In such an instance, the employer cannot be expected to guarantee that all employees will observe good safety practices at all times.[123] In *Cape and Vineyard Division, New Bedford Gas and Edison Light Co.*, the First Circuit Court of Appeals reversed the Commission's finding that the employer had violated Section 5(a)(2) when a wireman working on a utility pole contacted a live wire and was electrocuted. The court stated that "an employer cannot be held to guard against hazards created by employee conduct which is not reasonably forseeable."[124]

On the other hand, where there is evidence of inadequate supervision, the Commission has found a violation.[125] Similarly, the Commission has found violations where there is widespread employee misconduct, indicating insufficient enforcement of company safety policy.

The defense that a violation was caused by employee disobedience is an affirmative defense and must be raised by the employer.[126] The Ninth Circuit Court of Appeals has recently ruled in *Raymond Hendrix d.b.a. Alsea Lumber Co.* that once an employer proves that a violation was caused by employee disobedience, the Secretary must show *actual knowledge* by the employer to establish a violation.[127]

A difficult issue concerning employer duties under Section 5(a)(2) is the problem of the multiemployer worksite, a situation which arises frequently in the construction industry. In *Gilles and Cotting, Inc.*, the Fourth Circuit Court of Appeals deferred to the Review Commission's determination that a prime contractor is not liable for the violations of its subcontractors, but the court remanded to the Commission the question of whether the Secretary must prove that the employees of a cited employer had *access* to the area of hazard or that they were actually *present* in the zone of danger.[128] However, in *Underhill Construction Corp.*, the Second Circuit Court of Appeals specifically held that the Secretary has only to prove access to the zone of danger. The court also held that an employer is liable for the creation of hazards, even though none of his own employees are exposed, so long as the employees of some employer on a common worksite have access to the hazard.[129] Finally, in *Anning-Johnson Co.*, the Seventh Circuit Court of Appeals held that a subcontractor who does not create or have control over a hazard, but whose employees are exposed to nonserious violations, cannot be found in violation of Section 5(a)(2).[130]

Reasonable Promptness

One of the most controversial procedural issues before the Review Commission involves the interpretation of Section 9(a) of the OSHAct which requires that citations be issued with "reasonable promptness." A majority of the Review Commission in *Chicago Bridge and Iron Co.* interpreted this language as requiring that a citation must be issued within three working days from the time that the Area Director determines that a violation exists.[131] This position was expressly rejected by the Seventh Circuit Court of Appeals.[132] The case is at this writing on remand to the Commission.[133]

Determination of the Nature of the Violation

Another set of legal issues surrounding standards deals with how they are to be cited. Section 17 of the OSHAct limits the penalties which may be assessed for serious and nonserious violations of standards. The question of whether a particular set of facts constitutes a "serious" violation has arisen in several cases. Section 17(k) of the Act deems a serious violation exists "if there is a substantial probability that death or serious physical harm could result from a condition which exists, or from one or more practices, means, methods, operations or processes . . . unless the employer did not, and could not with the exercise of reasonable diligence, know of the presence of the violation."

Different interpretations of Section 17(k) have been set forth by the hearing judges. In *Crescent Wharf and Warehouse Co.*, the employer claimed that a serious violation can be established only on proof which shows (1) that it is substantially probable that an accident could occur as the result of the condition, and (2) that it is substantially probable that death or serious physical injury would be incurred as the result of the accident. OSHA contended that only the latter condition need be established by evidence. The hearing judge found for the employer, noting that "substantial probability of serious injury then means a situation from which a foreseeable danger of serious injury exists, with great enough likelihood to make a requirement of the specified precaution reasonable." The Review Commission overturned the hearing judge's decision, finding that Section 17(k) does not require the Secretary of Labor to prove a substantial probability that an accident will occur as the result of the conditions arising from the violation. To require such proof, the Commission held, was not intended by Congress, which specifically provided for the abatement of imminently dangerous situations in Section 13.[134]

The Commission's interpretation of Section 17(k) was specifically upheld by the Ninth Circuit Court of Appeals in *California Stevedoring and Ballast Co.* The court stated that "where violation of a regulation renders an accident resulting in death or serious injury possible, however, even if not probable, Congress could not have intended to encourage employers to guess at the probability of an accident in deciding whether to obey the regulation."[135]

Another related legal question is the standard of proof to which the Secretary of Labor should be held in the determination of violations. Three positions have been taken by hearing judges: (1) the degree of proof is one of the "preponderance of the evidence," as is the customary procedure in civil actions, (2) proof should be established by a "clear and convincing test," and (3) the degree of proof should be one of "substantial evidence."* In *Armor Elevator Co., Inc.*, the Review Commission noted that "the preponderance of evidence is the usual standard applied to the evidentiary record in agency proceedings. . . . [Since] the preponderance rule is appropriate to our proceedings, we hereby adopt it."[138]

Penalties

The Review Commission has made numerous adjustments to penalties assessed against employers by OSHA. The Secretary of Labor's formula for proposing penalties begins at the statutory maximum of $1000 and adjusts downward on the basis of good faith, gravity of violation, size of business, and employer history of previous violations (see Section 5.1). In *Nacirema Operating Co., Inc.*, the Review Commission ruled that the precise formula used by OSHA could not always be automatically applied without consideration of the circumstances of the particular case. Finding that the gravity of the violation was the principal factor in *Nacirema*, the Commission reduced the penalty from $700 to $350.[139] In *Hidden Valley Corp. of Virginia*, the Commission stated:

> While we find no reason to disagree with the Secretary's use of guidelines to assist his agents in determining the amount of a penalty to be proposed, we believe that the gravity of the violation charged in this case is of signal importance. The careless disregard of the extremity of the hazard created by the trenching operations of the employer when added to the lack of good faith override any consideration that

*"Preponderance of the evidence" means that, based upon all of the evidence, the facts asserted by the party having the burden of proof are more probably true than false.[136] "Substantial evidence" means such relevant evidence as a reasonable mind might accept as adequate to support a conclusion.[137]

might be given to other criteria such as history of previous violations or size of business.[140]

The Commission accordingly raised the penalty from \$650 to \$1000. The Commission's findings in *Nacirema* and *Hidden Valley* have been cited by many hearing judges in support of their modification of the Secretary's proposed penalties.

It was not until *Dreher Pickle Co.* that the Commission explicitly stated that the Secretary's proposed penalties were only advisory and the Commission alone had the statutory power to assess penalties. In overturning a hearing judge who simply accepted a figure proposed by one of the parties (the Secretary of Labor), the Commission found that even if the Secretary had given due and proper consideration to all the factors, the mandates of the Act would not have been fulfilled: "Subsection 17(j) vests the sole authority to assess civil penalties in the Commission. . . . There is no escaping the clear congressional mandate that the Commission's consideration of penalties when they are contested are to be made independently of the Secretary's computations."[141]

The Review Commission's independence in penalty assessment has been supported by the courts. In *Interstate Glass Co.*, the Eighth Circuit Court of Appeals stated "the congressional intent is . . . that the Commission shall be the final arbiter of penalties if the Secretary's proposals are contested and that, in such case, the Secretary's proposals merely become advisory."[142] In *California Stevedoring and Ballast Co.*, the Ninth Circuit Court of Appeals held that in contested cases, the Commission has the authority to assess penalties in excess of the ones proposed by the Secretary.[143]

In cases involving small proposed penalties for minor violations of job safety and sanitation standards, the Commission has viewed penalties in terms of their effectiveness in encouraging the development of safer workplaces. In *J. E. Chilton Millwork and Lumber Co., Inc.*, and *General Meat Co., Inc.*, the Commission ruled that the purposes of the Act would not be furthered by imposing small monetary penalties for violations of extremely low gravity that would be better characterized as *de minimis.*[144] In *Colonial Craft Reproductions*, the Commission vacated the penalties assessed for nonserious violations because they would have imposed a severe hardship on a very small business that was in extreme financial difficulty.[145]

Reinspections

In *Kesler and Sons Construction Co.*, a majority of the Commission held that the Secretary of Labor could not order abatement of a violation within 15 working days of the issuance of a citation. The Commission also held that the Secretary could not reinspect within 15 working days. The basis for these rulings was that reinspection within 15 working days would infringe upon the employer's allotted time for deciding whether to contest the citation.[146]

The Commission's holding was specifically reversed by the Tenth Circuit Court of

Appeals. The Court's opinion stated that the language of the OSHAct did not support the Commission's interpretation and that the setting of 15 working days as the minimum abatement time might result in the unnecessary exposure of employees to hazardous conditions. As the court stated, "To us, the fact that an employer such as Kesler has 15 working days to *contest* a citation does not in anywise mean or imply that an employer also has a 15-day grace period to *correct* a condition required to be immediately abated."[147]

Civil Actions and the OSHAct

Section 4(b)(4) states that the OSHAct does not enlarge or diminish or affect in any other manner the common-law or statutory rights, duties, or liabilities of employers and employees under any law with respect to injuries, diseases, or deaths of employees arising out of, or in the course of, employment. (Cf. Chapter 8 on workmen's compensation legislation, in which common-law tort rights have been partially "traded off" for a "guaranteed" compensation system.) Interpretations of this section by the Solicitor of Labor and the district courts have precluded many forms of civil action by employees as the result of OSHA enforcement. In *Skidmore v. Travelers Insurance Co.*, the district court affirmed that the OSHAct does not create a private cause of action for damages suffered by an employee as the result of an employer's violation of the Act. (In other words, the Act provides for action *by OSHA* against a recalcitrant employer, but does not create an additional private cause of action for damages by the employee against the employer.)[148]

In his decision on *Skidmore*, the judge referred to the existing precedent in other areas[149] to find that in suits for damages based on violation of federal statutes which lack any express authorization of a damage remedy, relief is granted only if damages are necessary to effectuate the congressional policy underpinning the substantive provisions of the statute. Civil action based on the OSHAct is thus not warranted because (1) congressional policy was to provide for safe workplaces, not compensatory relief, and (2) no precedent exists whereby an imposition of a duty upon the employer automatically creates a liability for performance of that duty. The decision was upheld by the Fifth Circuit Court of Appeals.

While the language of the OSHAct may be interpreted to support the *Skidmore* finding, the issue of whether an employee can use an OSHA determination of the conditions in a workplace as evidence in a civil suit is not as clear. On September 27, 1973, the Solicitor of Labor issued regulations placing restrictions on OSHA personnel called as witnesses in litigation in which neither the federal government nor the Secretary of Labor is involved. OSHA claimed the new regulations were necessary "to conserve the time of employees for conducting official business" and "to minimize the possibility of involving OSHA in controversial issues not related to its mission."[150] (Other agencies have similar regulations for the same purpose.)

The regulations were intended to prevent possible misconstruction by the public of the distinction between the personal opinion of a witness who is a government employee and OSHA policy. Specifically, the regulations require that if an OSHA employee is called on to produce records, the Labor Department will attempt to have the subpoena withdrawn, quashed, or modified.* A government attorney will be provided (1) to explain to a court the regulation that prohibits production of information and (2) to represent the employee if a contempt citation is issued. In *Wodzenski v. Moore-McCormack Lines*, the district court ruled that the Labor Department is not required to disclose, in a civil suit in which it is not a party, a compliance officer's conclusion as to the cause of an accident.[152] The court found that the compliance officer's conclusion constituted a single step in the administrative-law process leading to the ultimate agency decision in the case and, as such, was not subject to disclosure.[153]** In *Pizzo v. National Shipping Corporation*, the U.S. District Court held that a subpoena cannot be issued to compel the Labor Department compliance officers to testify in private litigation as to findings made under the OSHAct:

> To require official investigators of the United States Department of Labor to testify under the compulsion of a subpoena in a private litigation . . . would readily interfere with the functioning of the Department of Labor and the performance of official duties by the investigators; and may have the effect of hampering necessary investigations and the effective enforcement of safety laws. . . . [The Act] was not intended to serve the interests of individual litigants.[155]

A recent ruling (*Arthur v. Flota Mercante, Gran Centro Americana, S.A.*) by the Fifth Circuit Court of Appeals may open the area of civil action by employees.[156] The appellate court affirmed a district court's instructions to the jury, in a civil case brought by an employee, that violation of safety and health regulations constituted negligence per se because the employee had proved that (1) violation of the regulations occurred, (2) the regulations were designed to protect the employee, and (3) the regulations were designed to protect against the type of risk created. While these are the three criteria necessary to establish negligence as a matter of law, it is not clear to what extent this precedent will be applicable to other employer violations of occupational safety and health regulations. The telling point in *Arthur* was

*Note, however, that the recently adopted *Federal Rules of Evidence for United States Courts and Magistrates* may affect the success of this strategy. Rule 803 on Hearsay Exceptions provides: "The following are not excluded by the hearsay rule, even though the declarant is available as a witness: . . . (8) *Public records and reports.* Records, reports, statements, or data compilations, in any form, of public offices or agencies, setting forth (A) the activities of the office or agency, or (B) matters observed pursuant to duty imposed by law, or (C) *in civil cases and against the government in criminal cases, factual findings resulting from an investigation made pursuant to authority granted by law, unless the sources of information of other circumstances indicate lack of trustworthiness.*"[151]

**This case was reversed in December 1974 by the Fourth Circuit Court of Appeals and remanded. The court noted that the compliance officer's remarks had already been made public in a memorandum and were factual in nature.[154]

that the employee had suffered an injury and had successfully used the employer's violation of safety and health regulations as proof of negligence on the part of the employer.

The existence of the OSHAct has provided a general stimulus for the filing of civil suits involving occupational health and safety issues. For example, in January 1974 a civil suit was filed in federal court against PPG Industries and others alleging negligence by the Pittsburgh Corning Corporation in exposing employees at a Tyler, Texas, asbestos plant to "asbestos fibers in extremely dangerous concentrations" (*Yandle v. PPG Industries, Inc.*). The seven plaintiffs in the $100 million class-action suit charge that Pittsburgh Corning knew of the dangerous conditions at the Tyler plant but failed to correct the deficiencies or warn the employees of the dangers of exposure to excessive concentrations of asbestos. Two codefendants, the Industrial Health Foundation, Inc., and the Asbestos Textile Institute, are charged with impeding the flow of information concerning the health hazards associated with asbestos manufacturing, thus contributing to the lack of information available to plaintiff employees. Firms supplying raw amosite-asbestos are also charged with failure to warn either employees of Pittsburgh Corning or the company itself of the health hazards inherent in handling these materials.[157]

The finding in *Borel v. Fibre-board Paper Products Corp.* may be relevant for the pending *PPG Industries* class-action suit. In *Borel*, which was the first case applying strict tort liability principles to a suit involving an occupational disease, defendant asbestos product manufacturing firms were charged with failure to provide warning of the dangers involved in handling asbestos products. In finding for the plaintiff employee, the Fifth Circuit Court of Appeals noted that it was common knowledge that asbestos was dangerous and that asbestos products should thus be deemed "unavoidably unsafe" under Texas tort law. Under this category of torts, the manufacturer has a duty to inform purchasers and users of asbestos of its inherent dangers. The manufacturer should be held to the standard of an expert and must, as a minimum, "keep abreast of scientific knowledge, discoveries, and advances."[158]

Occupational health and safety professionals working for management may themselves be sued for failing to act in a responsible manner toward employees. In a 1973 case, a worker with asbestosis who was not advised of the condition by a physican who examined him periodically for his employer was awarded $351,000 in damages by a California jury.[159]

4.3.3 Informal Policy and Program Directives

The effectiveness of the OSHAct has been influenced not only by the formal rules and regulations issued by OSHA and statutory interpretations, but also by the informal policies that have been adopted. These informal decisions, which are usually quite difficult for interested parties to contest, often involve the establishment of priorities and decisions not to exercise authority. Several important directives and informal policy statements are discussed in this section.

The Compliance Manual and Enforcement Policies of OSHA

On April 28, 1971, OSHA issued a *Compliance Operations Manual* for use by occupational safety and health compliance officers but also available to the public. (A revised version, retitled *Field Operations Manual*, was issued in July 1974.) The manual is designed to ensure uniform application of the requirements of the OSHAct by setting forth basic policy with regard to operational and administrative activities. In the manual, OSHA lists its priorities for inspections as follows: investigation of catastrophes and fatalities; response to employee complaints; inspection of target industries and industries with target health hazards; and finally a random cross-section of all employers.[134] The five following target industries employing some 1.5 million employees and having high injury rates were chosen for initial inspection emphasis: longshoring, roof and sheet metal, meat and meat products, miscellaneous transportation equipment (mobile home manufacturing), and lumber and wood products. The Target Health Hazards Program involves asbestos, cotton dust, silica, inorganic lead, and carbon monoxide.*

The manual describes the workload distribution policy of OSHA with regard to inspections (but cf. Chapter 6 for the actual utilization of enforcement effort):

The primary consideration in conducting compliance operations is the attainment of maximum effective inspection coverage. General guidelines to be used in manhour scheduling for compliance operations should approximate the following:

Catastrophe and/or fatality investigation	10%
Complaint (serious and "other") investigation	30%
Target Industry Program	35%
General Inspection and Related Activities	25%[160]

Also described in great detail are policies governing advance notice; expert assistance; entry of a workplace; contingency plans for an employer's refusal to permit inspections and/or an employer's forcible interference with the conduct of an investigation; selection of employer and employee representatives; the general duty requirement; interviews with employees; inspection techniques; closing conferences; issuance of citations; and follow-up procedures, including penalties.

Informal Formulas for Penalties

Section 17(j) of the OSHAct sets forth the four factors to be considered in assessing penalties: "the size of business of the employer being charged, the gravity of the violation, the good faith of the employer, and the history of previous violations."

*In the Spring of 1975, OSHA adopted regionally determined special emphasis programs as an alternative to the target programs.

(The Review Commission, the final assessor of penalties, has ruled that these factors are not always to be given equal weight; see Section 4.3.2.)

For nonserious violations which are not *de minimis*, initial determination is made to establish the gravity of the violation, taking into account (1) the probability or likelihood of injury or disease occurring as a result of the alleged violation; (2) the severity of injury or disease most likely to result; and (3) the extent to which the applicable provision of the Act was violated.[161] Each of the three gravity factors is given the code A, B, C, or X (A being the least severe), and the three ratings are averaged to obtain a final letter rating. "Unadjusted penalties" are imposed according to the final rating as follows: A, $0; B, $100-$200; C, $201-$550; X, $501-$1000. The unadjusted penalty may be reduced up to 50% on the following basis: a reduction of up to 20% for good-faith efforts to comply with the Act; up to 10% on the basis of size measured by number of employees; and up to 20% on the basis of the company's history of previous violations of the OSHAct or other safety acts. The adjusted penalty is then reduced by up to 50% (abatement credit) if the employer corrects the violation within the abatement period specified in the citation.[162] (As noted above, these formulas cannot be applied mechanically.)

For serious violations, an employer is subject to a penalty of up to $1000.[163] This amount may again be adjusted downward by up to 50% by a penalty adjustment factor based upon the employer's good faith, size of business, and history of previous violations.

For failure to abate a nonserious violation, the employer may be assessed a fine equal to the unadjusted penalty plus the abatement credit. If the employer does not abate a serious violation, he may be assessed a fine equal to the amount of the original unadjusted penalty for each day of noncompliance. If no original penalty was imposed, the daily penalty is $200.[164] (For a discussion of the actual penalties imposed, see Section 6.3.1.)

These procedures for penalties are examples of important informal policy decisions which are difficult, but not impossible, to contest through the administrative-law process.

Policy Concerning On-Site Citations

On September 7, 1973, the Labor Department issued Program Directive 200-20 which instructed certain compliance officers to issue citations for serious violations at the closing conference that is held at the workplace after an inspection. Exceptions were made for cases where laboratory analysis results might be needed to determine a violation or where the compliance officer might find it necessary to consult other specialists, either legal or industrial. In addition, officers would issue citations for nonserious violations if necessary data were available. The directive specified that the closing conference be held in two parts. During the first part, the officer would discuss the hazards noticed during the inspection and listen to the employer's suggestion for abatement. During the second part, the compliance officer would prepare and issue the citation to the employer before leaving the site. Citations would be reviewed by the Area Director and would be subject to his amendment.

Industry spokesmen immediately sought revocation of the policy. John J. Johnson, Jr., Vice-President of the Industrial Relations Department, National Association of Manufacturers, called the program directive "contrary to the intent of Congress. . . . [The] program directive includes a substantive change in procedure and therefore should be exposed to comment via public hearings and/or statements." Industry felt that the new directive would be counterproductive, would solidify the adversary positions of industry and government, and would not provide the "cooling-off period" necessary for the inspector to make an objective judgment.[165]

On November 21, 1973, the OSHA Office of Federal and State Operations suspended the on-site citation authority, apparently in order to avoid court challenges on the policy. (Attorneys for the U.S. Chamber of Commerce were arguing that the legislative history of the OSHAct precluded the issuance of on-site citations.[166] The directive authorizing on-site citations was reissued in April 1974.[167]

Policy Concerning the Applicability of Consensus Standards

The former head of the OSHA Office of Program Operations, Robert Gidel, has pointed out that OSHA compliance officers cite employers for being in violation of consensus standards when the situation under inspection does not fall within the range of consensus of the original standard development committee. Gidel is of the opinion that the practice of issuing such citations is illegal.

> OSHA is directly or unknowingly attempting to bypass Section 6(b) of the Act which requires the Secretary of Labor to provide public notice and an opportunity for a public hearing whenever he wishes to promulgate, revoke, or modify any occupational safety and health standard. . . . [Whenever] an OSHA inspector or approved state inspector tries to apply a consensus standard beyond the scope, purpose, and application stated in the consensus standard, his proposed action is improper and should be vigorously contested.[168]

The then Assistant Secretary of Labor for Occupational Health and Safety, George Guenther, in his testimony before the House Subcommittee on Environmental Problems Affecting Small Business, also addressed this issue:

> Legislation has been introduced which presumably would require the promulgation of separate sets of standards not only for each type of industry, but for various subcategories within each industry. To attempt to categorize the standards in such a way would be a mammoth administrative task involving a massive commitment of resources and the development of a far greater bureaucracy than this Department cares to contemplate. Rather than follow such an approach, we have set our policy from the beginning toward the development of general standards which are applicable equally to all establishments in all industries. . . . [The] Act recognizes that employers may do things in many different ways and be as effective in terms of safety and health as would be compliance with the standards issued. An employer

may either petition for a change in the standard or else apply for a variance from its requirements.[169]

Policy Concerning the Relationship between Consultation and Inspection

As stated before, the Solicitor of Labor has ruled that OSHA may not consult with employers at their establishment without triggering an inspection. However, in accordance with its overall policy of encouraging voluntary compliance, OSHA has stated that it will *not* make inspections of employers when they seek OSHA help at a place outside their establishment, although inevitably a few are inspected as part of a random sampling effort or as the result of an employee complaint.[170]

Performance vs. Specification Standards

A question being considered by OSHA is whether to adopt performance standards in place of current specification standards. Many of the current standards specify the type of equipment and facilities that must be used by employers. For example, a simple specification standard might state that in a certain situation the railings must be 42 inches in height and made of 3-inch steel tubing. The intent would be to ensure that the railing is strong enough to give adequate support and high enough to prevent falls. Such a standard would certainly not be the only possible way to provide adequate protection, yet a plant with an equally adequate, but different safety system would still have to install the specified railings.

Until recently, under the existing law and regulations, employers seeking to avoid major and expensive changes in their plants were required to apply for variances from the existing specification standard. Obtaining a variance often took considerable time, and the procedure was often initiated only after a citation was issued. A recent Review Commission decision in *Industrial Steel Erectors, Inc.*, touched on one aspect of specification standards in finding that when strict adherence to a safety standard would result in diminished, rather than enhanced, safety, the standard need not be completely obeyed.[171]

One solution to such problems would be the adoption of performance standards. A performance standard might simply require a guardrail to withstand a force of 200 pounds per square inch. Such a standard would allow a variety of *means* for compliance and might avoid expensive alteration in facilities. Further, a performance standard could result in better protection for workers in that they would allow each employer to choose the means that are most economically feasible for his business, thereby encouraging compliance.

However, while allowing employers greater freedom in complying, performance standards would make it more difficult for OSHA to prove the existence of a violation. In the guardrail example, a compliance officer would obviously find it more difficult to test whether a railing met a performance standard than to measure the height and composition of the railing. While it is possible to argue that performance standards encourage more voluntary compliance, then, it is also true that performance standards are in some cases likely to result in fewer citations and more work for the already understaffed and underbudgeted compliance division.

On the other hand, performance standards in the form of work-practice guides can contribute to greater worker protection in certain instances, especially in the occupational health area. For example, NIOSH has been slow in developing criteria documents for carcinogens and other toxic chemicals. If a performance-standard approach were adopted, it would be possible simply to categorize these substances into groups on the basis of toxicity. Performance standards could then be adopted specifying the speed at which ventilating systems should operate in order to lower the overall concentration of substances. This approach is supported by labor groups because such ventilation performance standards could be issued more quickly than detailed specification standards and could cover more substances (see Section 6.3.2). OSHA has allowed the states to adopt the performance-standards approach (see Chapter 5).

4.3.4 Amendments to the Act

In 1972 both the House and the Senate passed bills preventing the use of federal funds to inspect employers having a small number of employees. On June 15, 1972, the House voted 312-154 to exempt businesses with 25 or fewer employees from compliance for a period of one year.[172] It was estimated that this provision would affect 90% of all workplaces and 30% of all employees covered by the OSHAct (including 98% of all farmworkers).[173] On June 27, 1972, the Senate, while voting 41-44 to reject the House version, approved 45-41 an exemption for businesses with 15 or fewer employees (which would have affected 85% of the workplaces and 20% of the workers covered by the Act).[174] Both bills were attached to the fiscal 1973 Labor-HEW appropriations bill (H.R. 15417). The conference bill included the Senate's version of the exemption, but it never became law because the entire appropriations bill was vetoed by President Nixon.

Since the promulgation by OSHA of the regulation exempting businesses with seven or fewer employees from reporting requirements, the pressure for the enactment of small-business exemptions has diminished only slightly. A 25-employee exemption was again added to the fiscal 1975 appropriations bill by the House, while the Senate version did not include an exemption. The Conference Committee rejected the House exemption, but the trade-off was an extension of the exemption from record-keeping requirements to employers with ten or fewer employees.

4.4 UNRESOLVED LEGAL ISSUES

4.4.1 Proposed Amendments

In addition to the small-business exemption, over 100 bills aimed at modifying the

OSHAct were introduced in the 92d Congress. Since that time, a steady stream of bills has been introduced, including 30 bills in the opening days of the 94th Congress. Among these bills, many are similar or identical to those introduced in previous sessions. The following House bills introduced in January 1975 are representative:[175]

- H.R. 1050, to provide coverage under OSHA for state and local employees.
- H.R. 109, 130, 445, 1119, and 1129, to amend the law to require the Secretary of Labor to recognize the differences in hazards between light residential and heavy construction.
- H.R. 609, 919, 930, and 1019, to provide additional assistance to small employers.
- H.R. 444, 1120, and 1128, to provide that where violations are corrected within the prescribed abatement period, no penalty will be assessed.
- H.R. 347 to extend the protection of OSHA to high-school and college athletic contests.
- H.R. 131 and 1086, to repeal the OSHAct.

The bills that have received the most support have been those to provide federal on-site consultation assistance to small businesses. Congressman William Steiger (R-Wisconsin), cosponsor of the OSHAct, filed a bill in August 1973 to allow such consultations for employers with 25 or fewer employees, for the purpose of discussing interpretation of the Act, application of standards, or possible alternative means of compliance. The bill would have established two classes of OSHA officers: inspection officers and consultation officers. No new funds were to be provided: existing OSHA training and education funds were to be used. The bill did not provide for a worker walkaround right during consultative visits or for consultation with employees.[176] Although this bill failed, efforts continue in Congress to amend the OSHAct to provide for federal on-site consultation; Congressman Dominick Daniels (D-New Jersey) has proposed such a bill in the 94th Congress.[177]

Management and labor groups have reacted strongly to on-site consultation programs. Management groups have traditionally supported on-site consultation and the separation of consultation from enforcement. The National Association of Manufacturers, however, has withdrawn its support for the May 1975 OSHA regulation on on-site consultation because of the requirement that consultants notify enforcement personnel if they find unabated hazards which can be expected to cause serious injury or death.[178] On the other hand, many unions believe that on-site consultation seriously weakens the impact of first-instance sanctions.

In June 1974 Congress held oversight hearings on the implementation of the OSHAct in order to study the following issues:

1. An analysis of the OSHA compliance manual for content and policy.
2. Whether the walkaround provision of the Act is mandatory or discretionary.
3. The relation between state and local authorities in imminent-danger situations.
4. Failure to inspect farm workplaces, and agriculture as a possible target industry.

5. Pressures from management exerted upon the Secretary and Assistant Secretary of Labor.
6. The role of the Office of the Solicitor of Labor in delaying or watering down citations.
7. The role of OSHA area and regional offices before citations are issued in "hot" cases.
8. Who runs OSHA, the Assistant Secretary or the Deputy Assistant Secretary, and what are the lines of communication from the Assistant Secretary or Under Secretary of Labor?
9. Whether legislation is needed for on-site consultation, or whether the Act conveys authority.
10. The gray areas of coverage, such as possible overlap between the OSHAct and the Railway Safety Act, the Federal Aeronautics Act, the Gas Pipeline Safety Act, the Federal Metal and Non-metallic Mine Safety Act, and the Atomic Energy Act.
11. Analysis of state plans in relation to the Act.
12. The policy of unloading federal responsibility upon the states.
13. The variation among approved state plans.
14. The role of OSHA in pushing state plan approval.
15. The need for total coverage of employees in the public as in the private sector.
16. The quality of OSHRC judges.
17. Whether the Review Commission is necessary, or whether it has been the big enforcement bottleneck, water-downer, and legal-precedent-creator in a way Congress did not expect.
18. The placing of the National Institute for Occupational Safety and Health in the Department of Health, Education, and Welfare hierarchy.
19. NIOSH morale and long-term needs.
20. Training of personnel, existing resources, and the overreliance on private medical and industrial training.
21. OSHA's failure to train workers.
22. Assistance to small businesses.

4.4.2 Pesticide Jurisdiction*

In order to avoid unnecessary duplication and to achieve coordination between the OSHAct and other federal laws, OSHA has entered into negotiations with other federal agencies. The Department of Labor has ruled that OSHA standards apply to all working conditions unless another agency has issued an enforceable standard

*This issue has now been resolved, as noted in the text.

that will reduce or eliminate the particular hazardous condition. Where another federal agency has failed to exercise its statutory authority to correct a hazardous condition, OSHA is bound by its responsibilities under the Act to apply its standards to make certain that the affected workers are protected.[179]

A major jurisdictional battle arose between EPA and OSHA concerning the promulgation and enforcement of pesticide regulations. The difficulties stemmed in part from the fact that a farm is both a workplace and a part of the general environment.

On July 24, 1973, representatives of the Office of Management and Budget, EPA, OSHA, the Department of Agriculture, and the Domestic Council met to resolve the jurisdictional question involved. EPA was designated to draft a memorandum of agreement which was circulated in August. On the basis of the Federal Insecticide, Fungicide, and Rodenticide Act and the fact that in the past EPA had exercised authority in setting down field reentry standards on pesticide labels, EPA concluded that the consensus of the July 24 meeting was that EPA should review, set, and issue pesticide standards, after completion of current hearings and consultation with the Department of Labor and other federal agencies.[180]

The Department of Labor drafted a countermemorandum asserting that the agencies had agreed on July 24 to consult each other before issuing standards for farmworkers "in order to assure that the respective standards . . . will be compatible with one another."[181] The Department of Labor further recalled that EPA and DOL had agreed to coordinate their enforcement of their respective pesticide standards.

The jurisdictional question was first raised when OSHA issued a temporary emergency standard for 21 pesticides on May 1, 1973. On June 29, 1973, OSHA revised the emergency standard to exclude 9 of the 21 pesticides. The standards were never enforced because the Fifth U.S. Circuit Court of Appeals ordered a stay of the effective date of the standard on July 10 in a case which was not related to the EPA jurisdictional problem.

On January 9, 1974, the Fifth U.S. Court of Appeals vacated the OSHA temporary emergency standard for pesticides. In *Florida Peach Growers Association, Inc. v. Secretary of Labor*, the appellate court found no substantial evidence indicating the existence of "grave danger" as required for emergency standards under Section 6(c) of the Act. In the absence of more serious effects than headache, fatigue, and vertigo, the court held that the presence of pesticide residues in treated fields does not represent a grave danger.[182]

Meanwhile, EPA was in the process of holding public meetings. On May 10, 1974, EPA issued its pesticide standard. Prior to that time, on January 15, 1974, the Organized Migrants in Community Action, Inc., The RAZA Association of Spanish Surnamed Americans, and a farmworker from Florida filed suit against OSHA seeking issuance of a permanent pesticide standard. The case, *Organized Migrants in*

Community Action, Inc. v. Brennan, was decided in favor of the Department of Labor, and supports EPA's preemption of standard-setting authority for pesticides.[183] (For further discussion of the problems involved in setting pesticide standards, see Section 12.3.)

4.4.3 Federal Preemption and State Plans

Another unresolved legal issue facing the Occupational Safety and Health Administration is the extent of federal control over state occupational safety and health activities. Under Section 18 of the OSHAct, federal enforcement of standards promulgated by OSHA preempt state enforcement until an effective state plan is approved. States may retain jurisdiction only over existing state standards for which OSHA does not promulgate federal standards.

Section 18 provides guidelines for states to develop plans which are "at least as effective" as the OSHAct "in providing safe and healthful employment and places of employment." Many crucial administrative-law and policy issues turn on legal and operational interpretations of the phrase "at least as effective." Also, problems exist as to what legal avenues employees may pursue if they are unsatisfied with state enforcement and how much authority OSHA is required to relinquish to the states.

The issues of concurrent federal-state jurisdiction, phasing out of federal inspections, and interpretation of state effectiveness present significant problems which are currently the subject of much labor-government dispute. These issues are further discussed in Chapter 5. Suffice it to say that there will be many unanswered legal questions which will have to be resolved at the state and federal level through administrative-law decisions, court tests, and legislative action.

4.4.4 Antitrust Action

In attempts to minimize the total financial costs of compliance with the Act, some industry groups have sought to develop standardized practices for the entire industry. For example, the Owens-Illinois Glass Container Division, in attempts to reduce excessive noise exposure, sought "to locate and evaluate commercially available means for reducing sound levels to meet the standards required by OSHA without reducing the quality of the products and while maintaining production quantities which will make each plant an economically viable unit."[184] Toward this goal, Owens-Illinois works with other glass container producers (who together constitute the Glass Container Manufacturers Institute) to develop techniques to reduce sound levels, in the belief that joint efforts will "increase the rate of development of controls and provide the entire industry with the controls best suited for the benefit of employees, consumers and the manufacturers."[185]

A major legal problem in such efforts concerns the considerable amount of proprietary information involved. Owens-Illinois, in a letter to OSHA, claimed that even if the individual companies were not reluctant to disclose such proprietary information, the Justice Department's Antitrust Division has noted that there could be a pooling of proprietary information in such a joint development program that would result in a violation of antitrust laws. In a similar situation involving the match manufacturing industry, the Justice Department had given the following advice:

> Since no specific joint research program has been agreed upon, we, of course, cannot make any determination as to whether such joint research would raise serious antitrust problems. . . . For your guidance, however, we can [only] make suggestions. Potential antitrust concerns can be lessened by limiting the joint efforts to what might be considered 'basic' as distinguished from 'applied' research.[186]

The Justice Department feels that it is impossible to determine whether joint ventures among firms to eliminate workplace health hazards would violate antitrust laws before such ventures occur. Factors to be considered are size of the firms, the nature of the program, and the effect on finished products. For example, it appears that a number of small firms that do not possess individual research ability could undertake joint ventures for safety and health research with little fear of antitrust action, provided the effects on the finished product involved do not affect competition.[187]

Since the Antitrust Division does not issue advisory opinions, there is little incentive for cooperation on health and safety improvements between firms. The issue has yet to be tested.

I have covered only a few unresolved issues in this section: proposed amendments, pesticide jurisdiction, federal preemption and state plans, and antitrust action. Many other troublesome areas are discussed throughout this chapter and the next two chapters. In addition, as the next section notes, the relationship of the OSHAct to other labor law is in constant flux.

4.5 THE RELATIONSHIP BETWEEN THE OSHAct, EXISTING LABOR LAW, ARBITRATION, AND COLLECTIVE BARGAINING

The purpose of this section is to examine the relationship between the OSHAct and preexisting labor relations practice. It appears likely that the OSHAct will have a number of important though difficult to anticipate consequences for our system of labor relations. Conversely, preexisting labor relations practice has many implications for the interpretation, enforcement, application, and impact of the OSHAct. This mutual interaction between the OSHAct and our labor relations system will be examined in a preliminary way in this section. Among the issues which will be raised, though certainly not settled, are the following:

- What are the implications of the fact that there are likely to be more safety- and health-related grievances than previously, partly as a result of the OSHAct?
- How will the OSHAct influence the way in which arbitrators rule on such grievances?
- How will the OSHAct affect employee self-help remedies such as the right to refuse hazardous work without fear of reprisal?
- Does the OSHAct affect the availability of injunctive relief to labor unions wishing to forestall employer actions which subject workers to hazardous conditions?
- To what extent can and will the grievance and arbitration procedures be used as a method of implementing the spirit of the OSHAct?
- To what extent will the OSHAct serve as a collective-bargaining lever for labor unions in negotiating a less hazardous workplace?
- In general, what are likely to be the primary areas of overlap and interaction between the OSHAct, preexisting labor legislation, and the practice of arbitration and collective bargaining?

4.5.1 The "Common Law" of Arbitration and the Right to Refuse Hazardous Work

Over the years a "common law" of arbitration has developed: where there is a grievance machinery established for labor-management relations, employees ordered to perform a particular task which they think is improper should follow orders and then file a grievance—*except* when carrying out the order would be illegal or would involve *dangerous hazards.* These exceptions were summarized in *FMC Corp.* (1965):

> The Company seems to rely on the general principle that an order, even a wrongful one, should be obeyed, and a grievance filed afterward. But that principle has always been held to be subject to certain exceptions. One of these is that a man is not required to obey an order that he reasonably believes will endanger his health or safety.[188]

In *Hegler Zinc Co.*, employees refused to work outdoors on a cold winter day with only the light-weight work clothes they wore for their usual furnace-room duties. The arbitrator ruled in favor of the employees, noting as follows:

> Resort to the grievance procedure is not an antidote for pneumonia nor can it be said that where there is an actual hazard to life and limb, it is the intent of the parties to the collective bargaining agreement that no matter what the circumstances, the order must be obeyed. The collective bargaining agreement is concerned with human beings. In the normal employment relationship, it is true there are emergencies which require prompt and disciplined obedience, but discipline essential in an army, that of an unqualified duty to obey, has no place in the ordinary industrial management relationship.[189]

In principle, this "well-accepted doctrine" might protect employees from penalties arising out of refusal to obey an order which, if followed, would subject them to a physical hazard. In practice, relatively little protection has been afforded employees by this means. Arbitrators' rulings have varied widely in the degree of protection afforded employees. In one particularly striking case—*International Harvester and the Auto Workers*—an arbitrator held that a temporary suspension was justified when an employee refused to crawl into a large die for the purpose of cleaning it:

In brief, [the employee] was simply frightened at the prospect of crawling into a hydraulic press and working under a five-ton die which was suspended by bolts and U-clamps. He had been instructed during his apprenticeship never to do this, and had the dramatic experience of having to clean out a die in which a die maker's arm had been crushed. He had also seen a man lose a finger in such an operation.[190]

However, other die makers had performed this task, and the job description was held to encompass such work. The arbitrator, finding that more effective safety procedures could have been utilized, nevertheless stated that "obviously, grievant [the employee] was not completely rational in his fear." The arbitrator ordered the employee reinstated in his job with a deduction of "five days' pay for the time he might properly have been suspended."[191] In another decision, an arbitrator reduced a worker's suspension from five and one-half days to a half-day for refusing to climb a utility pole. The arbitrator commented that safety and fear were "inextricably entwined" in the case and the grievant felt "an intense anxiety" that probably precluded completion of the walk without a mishap.[192] These cases serve to demonstrate that under case law, workers have little guarantee that refusal to perform hazardous work will not result in penalties.

4.5.2 Legislative Embodiment of the Right to Refuse Hazardous Work without Incurring Penalty

In 1935 Congress enacted the National Labor Relations Act (perhaps better known as the Wagner Act) to provide a legal framework for labor-management relations as a substitute for industrial strife. The NLRA was amended in 1947 by the Labor Management Relations Act (Taft-Hartley). The intent of Congress is stated in the preamble of the LMRA:

Industrial strife which interferes with the normal flow of commerce and with the full production of articles and commodities for commerce, can be avoided or substantially minimized if employers, employees, and labor organizations each recognize under law one another's legitimate rights in their relations with each other, and above all recognize under law that neither party has any rights in its relations with any other to engage in acts or practices which jeopardize the public health, safety, or interest.[193]

The Taft-Hartley Act protects the rights of employees to organize and to bargain collectively with employers over terms and conditions of employment through representatives of the workers' choosing. Further, Taft-Hartley defines certain practices of employers and unions as "unfair labor practices" (Section 8), in order to protect the rights of employees and employers, and to prevent labor disputes that could seriously affect the welfare of the nation. The National Labor Relations Board (NLRB), established by the Wagner Act and continued under Taft-Hartley, is empowered to prevent any person from engaging in an unfair labor practice, such as employer reprisals against employees engaged in "protected concerted activities" covered by Section 7. The important question discussed below is whether refusing to perform hazard work is a protected concerted activity.

Is Refusal of Hazardous Work a "Protected Concerted Activity"?

In *NLRB v. C and I Air Conditioning, Inc.*, the Ninth Circuit Court of Appeals reversed a decision of the NLRB:

NLRB held not warranted in finding that employer violated Section 8(a)(1) of LMRA by discharging employee because *he had voiced complaints concerning safety on job site*, there being no substantial evidence that the discharged employee had engaged in concerted activity protected by Section 7 of the Act.

Discharged employee, under the circumstances not only was acting by himself, but also was acting "for" himself; *there is no evidence* that purpose of his conduct was for mutual aid and protection of other employees, or that it was an attempt to enforce provisions of collective bargaining contract.[194]

In reversing the NLRB, the court distinguished those cases in other circuits applying the *Interboro Contractors* rationale in finding protected concerted activity, although the discharged employee was acting by himself.[195] The court did not disagree in principle with the *Interboro Contractors* rationale that: "while interest on the part of fellow employees would indicate a concerted purpose, activities involving attempts to enforce the provisions of a collective bargaining agreement may be deemed to be for concerted purposes even in the absence of such interest by fellow employees." The court went on to say, however:

We recognize the emergence of that [*Interboro*] rule but fail to see any evidence showing that the purpose of [the employee's] conduct was for the "mutual aid and protection" of other employees, or that it was an attempt to enforce the provisions of a mutual bargaining agreement. In our opinion, [the employee's] complaint was an individual protest, "on behalf of the discharged employee himself."

Thus, in *NLRB v. C and I Air Conditioning, Inc.*, the court did not rule on the question of whether an employee who refused to perform hazardous work and was subsequently discharged could win an unfair labor practice suit against his employer by giving evidence that he did so in the mutual interest of his fellow employees, or in an attempt to enforce a provision of a collective-bargaining agreement.

The case appears somewhat anomalous, because no grievance was filed by the employee and no mention of the OSHAct was made in either the NLRB or court decisions. (California safety regulations had been incorporated by reference in the collective-bargaining agreement, but were not material to the decision.) As was mentioned earlier, the OSHAct does not give an additional source of relief against employer discrimination under its Section 11(c) unless the discrimination was directed against the employee for exercising specific rights listed therein, such as the filing of an OSHA complaint.

Does the "Right to Refuse Hazardous Work" Have Special Status, Apart from Protection Afforded "Protected Concerted Activities"?

In addition to the concerted activities protected under Section 8, Section 502 of the Taft-Hartley Act explicitly protects employees who refuse work because of abnormally dangerous working conditions: "nor shall the quitting of labor by an employee or employees in good faith because of abnormally dangerous conditions for work at the place of employment be deemed a strike under this Act." It has been thought by some students of labor relations that the purpose of this section was to extend "protected activity" status to refusals of hazardous work in situations governed by collective-bargaining contracts containing a no-strike provision.

The National Labor Relations Board must make the determination of whether the condition involved is abnormally dangerous, based upon the evidence presented in complaints brought before it. In an early case involving Section 502–*Knight-Morley* (1957)–the NLRB and the Sixth Circuit Court of Appeals found that employees of Knight-Morley who quit work due to an abnormally dangerous condition did not violate a no-strike clause of their collective-bargaining contract. The NLRB stated:

> The board found that the buffers quit work in good faith because of abnormally dangerous working conditions and that respondent [the employer] violated Section 8.a (1) and (3) of the Act in discharging them [i.e., an unfair labor practice] . . . in spite of the no-strike provision of the bargaining contract, the cessation of work . . . was protected under Section 502.[196]

The Nature of the Evidence Required to Prove the Existence of Abnormally Hazardous Conditions

While *Knight-Morley* did provide some protection for workers, since that decision the NLRB and the courts have tended to interpret Section 502 very strictly on the basis of the evidence presented, thereby seriously limiting its usefulness and applicability. The opinion of the court in *Knight-Morley* was that "Section 502 expressly limits the right of management to require continuance of work under what the employees in *good faith* believe to be 'abnormally dangerous' conditions." This language appeared to indicate that the existence of a *subjective, good-faith belief* on the part of employees that abnormally dangerous conditions prevailed was the controlling issue concerning the availability of Section 502 protection. A subsequent decision, however, restricted the applicability of Section 502 by imposing the requirement of *objective* proof. In *Redwing Carriers*, the NLRB states:

> We are of the opinion that the term (abnormally dangerous conditions) contemplates . . . an *objective*, as opposed to a subjective, test. What controls is not the state of mind of the employees or employers concerned, but whether the actual working conditions shown to exist by competent evidence might in the circumstances be considered "abnormally dangerous."[197]

A major problem arising out of *Redwing* and subsequent cases was the nature of the evidence required to establish "objective" proof of an "abnormally dangerous condition." The requirement of objective proof was not expressly included in Section 502, which only mentions "good-faith belief." Such a requirement creates an extremely difficult situation for workers who have to decide on the spot whether to perform hazardous work, without knowing whether they will be protected from employer reprisal if they refuse (see also Section 4.5.4). While lay testimony has been admitted as evidence in establishing the existence of dangerous conditions, such evidence is only one facet of a well-documented case, and typically must be bolstered by expert testimony and/or proof of the employer's explicit violation of the safety laws.

Other court decisions have had the effect of further limiting the protection afforded employees by Section 502. For example, working conditions which are merely "highly unpleasant" have been held not to merit Section 502 protection, nor do dangerous working conditions which are no more than "normally" so. In general, there appears to have been substantial erosion over time of Section 502 protection, since the findings favorable to employees in early cases such as *Knight-Morley*.[198]

While the results of arbitration over safety and health issues have been unpredictable, employees have in general fared better under arbitration than in the courts. In balancing the prerogative of management to direct the labor of its employees against the right of employees to working conditions which are free from abnormal hazards, arbitrators have tended to be relatively more solicitous of the latter than the courts. For example, at least some arbitrators still adhere to the subjective "good-faith" standard of proof regarding the existence of abnormal hazard which was explicitly rejected by the NLRB in *Redwing Carriers*.[199] An arbitrator is often in a better position than the courts to understand and weigh the concerns of both labor and management. The relative informality of the arbitration proceeding and the arbitrator's better familiarity with the workplace have generally provided employees with a more favorable forum for the protection of their rights and interests than the courts. In addition, an examination of NLRB cases, reviewed on appeal by the courts, indicates that the NLRB has also been more protective than the courts of workers' rights under Section 502.

4.5.3 Implications of the OSHAct for Arbitration and Section 502

The question arises as to whether an employer's violation of occupational health and safety standards provides objective evidence for Section 502 protection under Taft-Hartley or in arbitration cases in support of employees disciplined for refusal of hazardous work. Chairman Moran of the Occupational Safety and Health Review Commission has questioned whether Section 502 and violations of OSHA standards are to be construed together.[200] OSHA "imminent-danger" or "serious-violation" citations *may* help to establish the fact of "abnormally dangerous conditions" for the purposes of Section 502. In this case, employees would be allowed to walk off the job without penalty if the conditions were such as to produce a "substantial probability" or a "reasonable expectation" of "death or serious physical harm." The issue is less clear for OSHA citations involving either "nonserious," "willful," or "repeated" violations.* However, none of these questions has yet been fully tested in the courts.

As mentioned before in regard to civil actions, introduction of convincing evidence proving violation of OSHA standards can be used to demonstrate negligence per se on the part of the employer, although a plaintiff may find it impossible to subpoena OSHA records or employees to establish relevant and material facts. It is not certain that the fact that a Review Commission decision went against an employer would suffice to establish liability, especially since Review Commission proceedings are less formal, the rules of evidence are relaxed, and the proceedings may be conducted without an attorney. Similar problems in introducing evidence are likely to arise in attempts to use violation of OSHA standards to demonstrate the existence of "abnormally dangerous conditions" under Section 502 of the LMRA. (See the note on the new Federal Rules of Evidence for U.S. Courts and Magistrates on p. 54.)

Thus, it has not yet been demonstrated that the OSHAct and Section 502 may be used together successfully. From a practical point of view, it is certainly possible that employees may have an expanded right of immediate protection against hazardous conditions currently not provided separately under either the OSHAct or the Taft-Hartley Act. For example, worker rights may be greatly expanded where OSHA requires monitoring equipment at the workplace which provides employees with an immediate "objective" measure of unsafe work conditions. The basic flaw

*It is interesting to note from the legislative history of the OSHAct that one important bill—introduced by Congressman Dominick Daniels—actually envisaged employee self-help methods such as refusal of hazardous work as functioning in tandem with the OSHAct to facilitate its enforcement. Daniels's bill even provided that companies would have to continue paying employees engaging in a work stoppage in protest of OSHAct violations at their place of employment.[201]

in Section 502 protection, namely the difficulty facing workers in making immediate, defensible decisions about stopping work under "abnormally dangerous" conditions, still remains. Workers are still likely to be subjected "to the horns of a dilemma which forces them to risk either their jobs or their health."[202] However, the OSHAct may have the effect of making the horns of this dilemma somewhat less sharp in that in some cases an employee's refusal to perform work likely to cause death or serious physical harm would be a protected activity for the purposes of Section 11(c) of the OSHAct.

4.5.4 Collective Bargaining and the Right to Strike because of Unsafe Conditions

One important question is whether a labor union can be enjoined from engaging in a work stoppage over health or safety issues during the life of a contract if its collective-bargaining agreement contains a no-strike provision and/or a broad compulsory-arbitration clause. This has been a particularly crucial question due to the large number of unions that have general no-strike provisions and broad arbitration clauses in their collective-bargaining agreements.*

One must examine the recent history of labor injunctions and public policy regarding arbitration to understand the potentially far-reaching implications of the findings of the Supreme Court in the recent *Gateway Coal Co. v. United Mine Workers* case. The court held that safety disputes are presumed to be subject matter for arbitration unless the collective-bargaining contract explicitly exempts safety and/or health disputes from arbitration. This finding may require employees to continue to work in unsafe or unhealthful conditions until an arbitrator decides the issue, except in those limited cases when Section 502 affords protection for the work stoppage.

History and Recent Developments–The Labor Injunction

The labor injunction was widely used in the years prior to 1932 to forestall the spread of unionization and to restrain labor organizations and their members from engaging economic self-help activity, even when nonviolent. In keeping with increasing public sentiment that such activity on the part of labor organizations was legitimate and should not ordinarily be subject to judicial interference, Congress enacted the Norris-LaGuardia Act in 1932. This Act greatly reduced the power of the federal judiciary to issue injunctions in labor disputes. Section 4 of the Act states that "no court of the United States shall have jurisdiction to issue any restraining order or temporary or permanent injunction in any case involving or grow-

*Among major unions, to my knowledge, only the United Auto Workers has obtained specific contractual provisions exempting disputes over safety and/or health issues from binding arbitration. Similar contract language was sought by the United Mine Workers during the 1974 coal industry negotiations, but the union did not prevail on this point.

ing out of any labor dispute to prohibit any person or persons participating or interested in such dispute."[203]

The Norris-LaGuardia Act was effective in achieving its purpose. Prior to enactment, over 500 labor injunctions had been issued by federal courts; in the following 18 years only 66 such injunctions were issued.[204]

The National Labor Relations Act did permit district courts to issue temporary restraining orders for prescribed purposes, such as enforcement of NLRB rulings. Moreover, Section 301(a) of the Taft-Hartley Act gave federal courts jurisdiction for civil damage suits involving violations of collective-bargaining contracts.

After passage of the Taft-Hartley Act, legal debate arose over the relationship between Section 4 of Norris-LaGuardia and Section 301(a) of Taft-Hartley. The key question was whether Section 4 precluded suits for injunctive relief under Section 301(a) against work stoppages in violation of collective-bargaining contracts. If Section 4 were controlling, then only suits for money damages could be brought under Section 301(a).

After many years of controversy over the meaning of Section 301(a), the Supreme Court in *Sinclair Refining Co. v. Atkinson* (1962) held that despite the fact that a work stoppage or strike may constitute a breach of a labor agreement, it is still within the meaning of a "labor dispute" under the Norris-LaGuardia Act.[205] Therefore, the federal courts have no jurisdiction to issue an injunction unless Section 4 of the Norris-LaGuardia Act is overridden by Section 301(a). As noted below, *Sinclair* was overruled in 1970.

Public Policy Favoring Arbitration

The 1947 Labor Management Relations Act declares in Section 203(a) that it is national policy to encourage arbitration of labor disputes: "Final adjustment by a method agreed upon by the parties is declared to be the desirable method for settlement of grievance disputes arising over the application or interpretation of an existing collective bargaining agreement."

Section 203(a) in conjunction with Section 301(a) formed the basis for Justice Douglas's majority opinion in *Textile Workers Union v. Lincoln Mills* (1957).[206] The Supreme Court ruled that Section 301(a) provided for orders of specific performance of labor-contract provisions requiring arbitration of labor disputes. Section 301(a) was interpreted as an expression of federal policy that industrial peace is best served through judicial enforcement of agreements to arbitrate disputes which occur during the life of a collective-bargaining contract.

In the 1960 "*Steelworkers Trilogy*" cases, the Supreme Court further extended the *Lincoln Mills* doctrine that public policy favors resolution of labor disputes through the arbitration process.[207] Under this doctrine, the court's role is not to consider the merits of the case, but must, rather, refer consideration of the matter to the arbitrator, if the following very minimal conditions are met: (1) a labor contract exists; (2) the contract contains a binding-arbitration clause; and (3) there is an allegation that a provision of the contract has been violated. In the language of the court: "An order to arbitrate the particular grievance should not be denied unless it may be said with positive assurance that the arbitration clause is not susceptible to interpretation that covers the asserted dispute. Doubts should be resolved in favor of coverage."[208] Arbitration awards are to be upheld by the courts unless it can be shown that the arbitrator clearly overstepped the bounds of the contract in reaching his decision.

The Reemergence of the Labor Injunction as an Outgrowth of Public Policy Favoring Arbitration

Not surprisingly, the court's interpretation in *Sinclair* of Section 301(a) in light of the Norris-LaGuardia Act was soon found to constitute a serious impediment to the evolving thrust of national labor policy in the United States favoring arbitration of labor disputes. Accordingly, *Sinclair* was reversed in a highly significant 1970 Supreme Court decision, *Boys Market v. Retail Clerks Union, Local 170* (1970).[209] In *Boys Market*, the Supreme Court reasoned that a no-strike obligation on the part of the union, expressed *or implied*, is quid pro quo for the employer's agreement to submit disputes arising during the life of the contract to binding arbitration. The incentive for the employer to enter such an agreement would be greatly diminished if the union could not be forced (via injunction, if necessary) to live up to its part of the bargain (i.e., not to strike). To this end, injunctive relief under Section 301(a) was held to be necessary for carrying out the national labor policy favoring arbitration as a means of promoting industrial peace. In allowing an injunction to issue under Section 301(a), the court in fact found that suits for money damages under 301(a) did not always adequately protect employers from the imminent and irreparable harm of a work stoppage. In rejecting Justice Black's minority opinion that 301(a) was not intended "to clear the way for judicial obliteration of that [Norris-LaGuardia] Act under the soft euphemism of 'accommodation,' " the majority of the court held that Section 4 of Norris-LaGuardia did not preclude issuance of injunctions against strikes in breach of a collective-bargaining agreement. The court reasoned that although the central objective of Norris-LaGuardia was to prevent judicial interference with the growth and legitimate economic self-help activities of labor organizations, congressional intent in subsequent legislation (LMRA) was to promote the peaceful settlement of industrial disputes through arbitration. Congress, by not enacting amendments to Norris-LaGuardia, was held to have left it to the courts to accommodate the old law to the new policy.

The Labor Injunction in Safety Disputes: *Gateway Coal Co. v. United Mine Workers*

In *Gateway Coal Co. v. United Mine Workers*, an injunction was granted against the union in a case involving a safety dispute. In so ruling, the Supreme Court extended the presumption of arbitrability enunciated in the *Steelworkers Trilogy* to cover safety issues. On April 18, 1973, workers of the United Mine Workers Local No. 6330 in Greene County, Pennsylvania, voted not to work unless Gateway Coal Co. suspended three foremen who had allegedly falsified records of air flows in the mine. The company acquiesced to this demand and the miners returned to work. Subsequently, criminal prosecutions were instigated against the three foremen and an investigation was undertaken by the Pennsylvania Department of Environmental Resources pursuant to possible decertification proceedings.

While criminal charges were still pending, the Pennsylvania Department of Environmental Resources notified Gateway Coal on May 29, 1973, that the foremen could be reinstated. On June 1, 1973, the foremen resumed work and miners on all shifts struck to protest the alleged safety hazard created by the foremen's presence in the mine. On June 8, 1973, the union refused a company offer to arbitrate the matter. Meanwhile, the foremen pleaded *nolo contendere* to criminal charges for falsification of records and paid fines of $200 each.

The union's refusal either to return to work or to submit the dispute to arbitration prompted Gateway Coal (a subsidiary of U.S. Steel) to seek an injunction against the strike in district court under the provisions of Section 301(a) of the Labor Management Relations Act. The district court issued a preliminary injunction based upon the finding that the broad arbitration clause of the collective-bargaining contract did not exclude arbitration of safety disputes; in other words, despite the absence of an explicit no-strike clause, the court recognized an *implied* no-strike contract. Significantly, the court made the grant of injunctive relief contingent upon the company's suspension of the foremen pending arbitration of the dispute.

The Third Circuit Court of Appeals then overturned the injunction, reversing the district court's decision. The court relied upon Section 502 of the LMRA in finding that the usual federal policy favoring arbitration of labor disputes need not extend to disputes over hazardous conditions. In the words of the appellate court:

> If employees believe that correctable circumstances are unnecessarily adding to the normal dangers of their hazardous employment, there is no sound reason for requiring them to subordinate their judgement to that of an arbitrator, however impartial he may be. The arbitrator is not staking his life on his impartial decision. It should not be the policy of the law to force the employees to stake theirs on his judgement.[210]

On this basis, the appellate court concluded that the union had no contractual duty to submit the controversy to arbitration, and hence no injunction could issue even under the traditional considerations of equity set forth in *Boys Market.*

The company appealed this ruling, and the *Gateway* case found its way to the Supreme Court. In granting *certiorari*, the Supreme Court addressed three questions. First, did the collective-bargaining agreement impose upon the parties a compulsory duty to submit safety disputes to arbitration? If so, did that duty to arbitrate imply a no-strike obligation on the union supporting issuance of a *Boys Market* injunction? Finally, did the circumstances of the case satisfy the traditional equity criteria concerning the availability of injunctive relief?

Addressing the first issue, the court found that the facts of the case and the union's contractual obligation to arbitrate "any local trouble of any kind [arising] at the mine" on its face required submission of the safety dispute to arbitration. Further, the Supreme Court disagreed with the appellate court's argument that public policy favoring arbitration did not extend to safety disputes. Relying on legislative history, the Supreme Court noted:

> On its face, [Section 502 of the LMRA] appears to bear more directly on the scope of the no-strike obligation rather than on the arbitrability of safety disputes. Indeed, there is nothing in the legislative history to suggest that Section 502 was intended as a limit on arbitration.[211]

Accordingly, the Supreme Court held that the dispute in *Gateway* was covered by the arbitration clause in the parties' collective-bargaining agreement, thereby extending to safety disputes the "presumption of arbitrability" set forth in *Steelworkers Trilogy*.

As to whether a *Boys Market* injunction could issue, the court found that "the aggreement to arbitrate and the duty not to strike should be construed as having coterminus application." While the Supreme Court agreed with the finding of the appellate court that "a refusal to work because of good faith apprehension of physical danger is protected activity and not enjoinable even where the employees have subscribed to a comprehensive no-strike clause in their contract," it concluded that the Court of Appeals erred in applying the principle to *Gateway*.[212] The Supreme Court relied upon the finding in *Redwing Carriers* that employees' good-faith belief that an "abnormally dangerous condition" exists is not proof enough. In any event, the Supreme Court noted that the district court had resolved this issue by expressly conditioning injunctive relief on the suspension of the three foremen pending the outcome of arbitration, thereby removing the alleged "abnormally dangerous condition."

To the question of whether the circumstances of the case satisfy traditional equity considerations, the Supreme Court responded:

> It is also evident that injunctive relief was appropriate in the present case under the equitable principles set forth in *Boys Market*. . . . The District Court found that [the union's] continued breach of its no-strike obligation would cause irreparable harm to the petitioner. It eliminated any safety issue by suspending the two foremen, pending a final arbitral decision. In these circumstances, we cannot say that the District Court abused its discretion.[213]

However, in the future, *Gateway* may be distinguished on its facts and not set a precedent beyond the special situation found in the case.

4.5.5 Implications of the *Gateway* and *Boys Market* Decisions

A major question arising out of *Gateway* and *Boys Market* is whether unions can also obtain an injunction to force employers to cease dangerous practices pending arbitration.* It is significant in this regard that in *Gateway* the district court's grant to the company of injunctive relief was conditioned upon the suspension of the foremen. At least one legal scholar has speculated that the *Boys Market* doctrine may encourage unions to seek injunctive relief pending arbitration to forestall employer actions which could cause "irreparable harm."[214] If so, the *Boys Market* doctrine ironically, under certain conditions and without fear of employer reprisal, may provide employees with injunctive relief from being compelled to perform hazardous work. In view of the fact mentioned earlier that arbitrators have typically been more solicitous of employee rights than the courts, it is likely that arbitrators would subsequently rule in employees' favor in many instances in which the courts had granted prior injunctive relief.

Of course, the union's claim for injunctive relief would have to be based upon the usual equity criterion of "irreparable damage," and the harm the union would suffer if the injunction were denied would have to be judged to exceed that which the employer would suffer if the injunction were granted. Nevertheless, it is precisely in cases involving disputes over hazardous conditions that unions may stand the best chance of obtaining injunctive relief on equity grounds. In such cases, a union's claim that the employer was in violation of OSHAct standards may prove to be a significant factor in the court's decision to grant injunctive relief, particularly in cases where such OSHAct standards have been explicitly incorporated into the collective-bargaining agreement.

Alternatively, some unions may still prefer the self-help route when faced with employer action (or inaction) resulting in hazardous working conditions. If so, it is possible that employer violations of OSHA standards can be used to prove the existence of "abnormally dangerous conditions" in support of a Section 502 defense to forestall the employer's access to injunctive relief. In *Gateway*, a *Boys Market* injunction could issue, in part, because the protection of Section 502 was eliminated when the union could give *objective* proof of the existence of an "abnormally dangerous condition." If the employer had been cited for violation of an OSHA (or Coal Mine Health and Safety Act) standard, the union might have been successful in invoking Section 502 protection for engaging in a work stoppage.

Finally, the presumption of arbitrability set forth in *Gateway* explicitly states that safety disputes are arbitrable unless the collective-bargaining agreement specifi-

*As discussed previously, Section 13(d) of the OSHAct can be used by an employee to compel the Secretary of Labor to take action to correct an "imminent danger." The relationship between what constitutes an "imminent danger" under the OSHAct and what constitutes "irreparable harm" for general injunctive relief has never been clarified.

cally exempts safety issues from arbitration. One possible outcome of this decision is that some unions will now seek modification of contract language to exempt safety disputes from the arbitration procedure (as the United Auto Workers has done). Unions may also seek a contractual "502" right to refuse hazardous work. In hazardous industries such as coal mining, employees are particularly unlikely to be willing to "subordinate their judgement to that of an arbitrator" who "is not staking his life on his impartial decision." Not surprisingly, the United Mine Workers sought (but did not achieve) new contract language in their 1974 negotiations that would have given miners more latitude to strike in cases of imminent danger. In this industry, at least, it seems likely that recent court decisions such as *Gateway* may have the effect of limiting the extent to which labor unions are willing to rely on the arbitration process for the resolution of safety disputes.*

4.5.6 Deferral to Arbitration under the OSHAct

In view of the broad public policy favoring arbitration of labor disputes, it seems likely that the Occupational Safety and Health Administration may begin to defer to arbitrators' rulings in certain instances. If so, arbitration could become an important method of implementing statutory rights under the OSHAct. (OSHA has, in fact, issued guideline regulations regarding agency deferral.)[215]

Legal scholars have already begun to speculate that the policy of prearbitral agency deferral, set forth by the NLRB in *Collyer Insulated Wire,*[216] may be actively pursued by the Solicitor of Labor in safety matters. The likelihood of this development grows as the spirit—and in some cases the letter—of the OSHAct is increasingly embodied directly in collective-bargaining contract language.

Deferral to arbitration may arise in several types of situations. Attempts by employees to force employers to pay for time spent in exercising the walkaround right may well find their way to an arbitrator. If so, it is possible that arbitrators will be more inclined than the Solicitor of Labor to rule in favor of employees on this matter.

Alleged discriminatory action by management against employees exercising other OSHAct rights may also become a subject for arbitration, if this forum is available, instead of the employees filing an 11(c) complaint under the OSHAct. However, at least one commentator has warned that the arbitration of safety disputes can have disastrous consequences for organized workers' rights under the OSHAct for a number of reasons:

*Students of labor relations may be struck by the irony of the notion that unions may begin to rely less strongly upon the arbitral forum as a result of the *Gateway* decision. The status accorded the arbitral forum in the 1960 *Steelworkers Trilogy* decision was a source of consternation to many employers who had always viewed the courts as a relatively more favorable forum than arbitration. Fourteen years later, employers have learned to live with arbitration and have even adapted to it quite successfully. Meanwhile, at least with regard to safety matters, *unions* may be becoming increasingly reluctant to rely upon the arbitration process.

1. Health and safety matters require an ever greater degree of expertise, perhaps going beyond what may be reasonable to expect from arbitrators.
2. Arbitrators only need to look at the collective-bargaining contract in arriving at a decision, and not to OSHA regulations and Review Commission decisions.
3. Arbitration determinations are almost never overturned,* while OSHA compliance officer actions are reviewable by the Review Commission and the Review Commission decisions by the courts. Thus, there is "institutionally greater assurance of an equitable ruling" under OSHA procedure.
4. Filing complaints under 11(c) of the OSHAct is "free," while the arbitration process is not, and therefore unions may not be willing to pursue the resolution of the issue in the arbitration forum.
5. Furthermore, there may not be harmony between employees and their unions (a necessary prerequisite for fair disposition of a complaint).
6. Finally, it can be argued that arbitration is inappropriate because it forces workers to pay for what should be theirs as of right under the OSHAct to begin with.[218]

4.5.7 The Implications of the OSHAct for Collective Bargaining

In spite of the difficulties mentioned in the previous section, collective bargaining may provide unionized employees with more protection from health and safety dangers than underbudgeted OSHA efforts in some instances. The OSHAct has been important in providing a basis for union demands that basic safety and health standards be included in collective-bargaining contracts. Prior to the OSHAct, collective bargaining did not adequately focus on or protect employee safety and health. Safety and health issues had low priority on the bargaining agenda and were often bartered for more pressing objectives.** Recent developments indicate that this unfortunate state of affairs may now be changing for the better, due in large measure to the stimulus provided by the OSHAct.

One result may be the emergence of health and safety committees with stronger roles. An interesting question arises as to whether employer participation in the committees, which is encouraged by the OSHAct, constitutes an illegal activity un-

*But see *Banyard v. NLRB*, in which the court refused to enforce an arbitrator's decision which would have forced truck drivers to violate safety laws.[217]

**Indeed, it was not until 1966 that the NLRB ruled definitively that job safety was a mandatory subject of collective bargaining, rather than an exclusive prerogative of management. In *Gulf Power Co.*, the union and Gulf Power were in the process of contract renegotiation when Gulf Power refused to meet with the union (IBEW) to negotiate issues of safety on the grounds that safety was the exclusive responsibility of management and therefore not a bargainable issue. The National Labor Relations Board found "that safety provisions constitute an essential part of the employer's terms and conditions of employment, and, as such, are a mandatory subject of bargaining." The NLRB referred to Department of Labor statistics "which indicate that joint participation in the formulation and administration of safety regulations by management and labor organizations has been recognized by management and labor organizations as a proper subject of collective bargaining."[219]

der Section 8(a)(2) of the NLRA. It is likely, however, that overall policy favoring management and labor cooperation in the health and safety area would override objections except in flagrant instances of employer domination.

A unique and positive feature of the recent United Steelworkers-U.S. Steel agreement is the provision of pay incentives for workers who help reduce toxic emissions (as well as the augmentation of work crews in certain areas so that the exposure per worker is decreased).[220] This cooperative and nonadversarial provision may represent the beginning of a new era in management-labor relations in the health and safety area.

The rapid growth of union activity in this area may well mean that collective bargaining will prove to be of great importance in the implementation of the OSHAct. The United Steelworkers of America, for example, are attempting to incorporate OSHA standards by reference in negotiations with the steel, aluminum, can, and nonferrous metal industries. In addition, the Steelworkers have obtained the right to expedited arbitration when an employee files a grievance involving a change in work practice that has led to the creation of an unsafe condition.

The Oil, Chemical, and Atomic Workers Union (OCAW) has won a four-point safety and health clause in recent collective bargaining contracts with several major companies. The OCAW contract provides for (1) the establishment of joint labor-management health and safety committees, (2) periodic health surveys of plants by independent industrial health consultants, approved by labor and management, to determine whether workers are being exposed to "toxic" or injurious substances; (3) medical examinations of workers at company expense, the scope to be determined by the joint labor-management committee; (4) annual employee morbidity and mortality data provided by the company to the union.

Finally, collective-bargaining agreements have the potential of going beyond the provisions of the OSHAct. For example, it could be agreed that the safety committee has the right to stop an operation if it determines that "imminent" danger exists. It deserves repeating, however, that such agreements, in going beyond what the law provides, may be reached only by trading off wages or other employee benefits.

4.5.8 Summary

The OSHAct could have significant implications for the right of employees to refuse hazardous work without penalty, both under arbitration and in the NLRB and the courts. Explicit employer violations of OSHAct standards may provide the kind of "objective proof" regarding the existence of "abnormally dangerous conditions" required by the courts under the *Redwing* doctrine as a condition for granting Section 502 protection to employees who refuse hazardous work. Similarly, the OSHAct may shift the balance in arbitration awards in favor of the right of employees to working conditions that are free from abnormal hazards and away from the traditional prerogative of management to direct the labor of its workforce. As the volume of safety-related grievance activity increases under the impetus of the OSHAct, the possibility that arbitrators may become more solicitous of employees' rights and interests in these matters takes on added significance, even though arbitration in the health and safety area has many pitfalls.

The OSHAct may also have several significant implications in regard to the labor injunction. It is possible that injunctive relief under the *Boys Market* doctrine will become available to labor unions seeking to forestall harmful employer actions pending arbitration of health and safety disputes. If so, the existence of employer violation of OSHAct standards may assist unions to meet the usual equity criteria for injunctive relief. Similarly, employer violations of OSHAct standards may legitimate employee self-help methods of OSHAct enforcement by forming the basis for a union's Section 502 defense against the grant of a *Boys Market* injunction to employers in safety disputes.

Finally, and perhaps most importantly, the OSHAct is already beginning to serve as a bargaining lever for labor unions in negotiating a less hazardous workplace. The OSHAct is providing an important stimulus to labor union collective-bargaining activities in regard to occupational health and safety. Conversely, the growth of such activity on the part of labor unions may prove to be of great importance in the implementation of the OSHAct, particularly in view of the highly inadequate level of resources budgeted by Congress for the Act's enforcement. Contractual embodiment of OSHAct provisions in collective-bargaining agreements and their implementation through revitalized national and local union safety committees of greatly expanded scope and function are likely to be of considerable importance in implementing the OSHAct in the years ahead.*

REFERENCES

1. W. L. White, Jr., and J. T. Carey, "OSHA Comes of Age: The Law of Work Place Environment," *The Business Lawyer* (July 1973).

2. "The Small Business Exemption," *Congressional Quarterly* (July 27, 1972), p. 1829.

3. Atomic Energy Act, 42 U.S.C. 2021 (1954), Section 274; Walsh-Healey Act, 41 U.S.C. 35 et seq.; Service Contract Act, 41 U.S.C. 351 et seq. (1965); PL 91-54, 40 U.S.C. 333 (August 9, 1969); PL 85-742, 33 U.S.C. 941 (August 23, 1958) (Longshoremen's & Harbor Workers' Compensation Act); National Foundation on Arts and Humanities Act, 20 U.S.C. 951 et seq.

4. OSHAct, Section 2(a).

5. OSHAct, Section 2(b)(1).

6. Legislative History, OSHA, 3 U.S. Code Cong. and Admin. News (1970), p. 5177.

*Three law review articles bearing on the topics discussed in this section have recently been published: Gregory G. Tobin, "OSHA, Section 301 and the NLRB: Conflicts of Jurisdiction and Rights," *American University Law Review*, Vol. 23 (1974), p. 837; Leslie R. Stellmen, "Occupational Safety: A New Dimension in Labor-Management Relations," *Syracuse Law Review*, Vol. 25 (1974), p. 838; and Lawrence F. Doppelt, "OSHA: Impact on the NLRA and Arbitration," *Wayne Law Review*, Vol. 20 (1974), p. 1015.

7. *The Job Safety and Health Act of 1970* (the Bureau of National Affairs, Inc., Washington, D.C., 1971), p. 1.

8. E.g., Federal Coal Mine Safety Act, 30 U.S.C.A. 801 et seq.; Federal Metal and Non-Metallic Mine Safety Act, 30 U.S.C.A. 721 et seq.; Safety Appliance Acts, 45 U.S.C.A. 1-16; OSHAct, Section 653 (b)(1)(Supp. 1972).

9. OSHAct, Section 18(c)(6).

10. OSHAct, Section 5(a).

11. Robert D. Moran, *Occupational Safety and Health Reporter* (the Bureau of National Affairs, Inc.), Vol. 1, No. 21 (October 1971), p. 729. [Hereinafter cited as *OSHR*, BNA]

12. See Richard S. Morey, "The General Duty Clause of the Occupational Safety and Health Act of 1970," *Harvard Law Review* Vol. 86 (1973), p. 988.

13. OSHAct, Section 4(b)(4).

14. OSHAct, Section 7(a)(1).

15. OSHAct, Section 7(b).

16. Executive Order No. 11807, September 28, 1974.

17. For initial standards, see 36 *Fed. Reg.* 10466-714 (1971).

18. See also *A Handy Reference Guide to the Williams-Steiger Occupational Safety and Health Act of 1970*, U.S. Department of Labor.

19. OSHAct, Section 6(b)(1). See OSHAct, Section 7(b)(1) for the composition of such advisory committees.

20. OSHAct, Section 6(f).

21. OSHAct, Section 6(c)(1).

22. OSHAct, Section 6(b)(6) provides that an employer may obtain a temporary variance if he establishes that: (1) he is unable to comply with a standard by its effective date because of unavailability of professional or technical personnel or of materials and equipment needed to come into compliance with the standard, or because necessary construction or alteration of facilities cannot be completed on the effective date; (2) he is taking all available steps to safeguard his employees against the hazards covered by the standard; and (3) he has an effective program for coming into compliance with the standard as quickly as practicable.

23. OSHAct, Section 8(a)(1).

24. OSHAct, Section 8(a)(2).

25. OSHA Program Directive 200-28 (April 22, 1974).

26. OSHAct, Section 9(b).

27. OSHAct, Section 13(a).

28. OSHAct, Section 13(d).

29. OSHAct, Section 10(a).

30. OSHAct, Section 12(j).

31. OSHAct, Section 11(a). See also R. L. Moore, "The Federal Occupational Safety and Health Act of 1970," *Wisconsin Bar Bulletin* (February 1972), p. 46.

32. OSHAct, Section 11(a).

33. OSHAct, Section 17(d).

34. OSHAct, Section 10(c).

35. OSHAct, Section 17(a).

36. OSHAct, Section 17(b).

37. OSHAct, Section 17(e).

38. OSHAct, Section 17(f).

39. OSHAct, Section 17(g).

40. OSHAct, Section 17(i).

41. See *Recordkeeping Requirements under the Williams-Steiger Occupational Safety and Health Act of 1970*, OSHA, p. 7.

42. OSHAct, Section 20(a)(5).

43. OSHAct, Section 24.

44. OSHAct, Section 22(a).

45. OSHAct, Sections 22(c) and 22(d).

46. Harrison Williams, Jr., "Be Prepared," *Trial*, Vol. 9, No. 4 (July-August 1973), p. 14.

47. "Occupational Safety and Health Act of 1970, Law and Explanation" (Commerce Clearing House, Chicago), pp. 3, 4.

48. *Frank Irey, Jr., Inc. v. OSHRC and Brennan*, 2 OSHC 1283 (3d Cir. 1974) and *Beall Construction Co. v. OSHRC*, 507 F.2d 1041 (8th Cir. 1974).

49. See *Code of Federal Regulations*; also *OSHR*, BNA, Reference Files 1 and 2.

50. U.S. Department of Labor, Office of Information, News Release, April 28, 1971, p. 3.

51. *Federal Register*, Vol. 36, No. 157 (August 13, 1971), p. 15102.

52. *Federal Register*, Vol. 34, No. 96 (May 20, 1961), Part II.

53. Joseph Page and Mary-Win O'Brien, *Bitter Wages* (Grossman, New York, 1973), p. 203.

54. *OSHR*, BNA, Vol. 2, No. 30 (December 1972), p. 868.

55. *OSHR*, BNA, Vol. 2, No. 37 (February 15, 1973), p. 1051.

56. *Federal Register*, Vol. 37 (May 3, 1972), p. 10929.

57. *Dry Color Manufacturers Association, Inc.*, 486 F.2d 98 (3d Cir. 1973).

58. *OSHR*, BNA, Vol. 3, No. 7 (July 19, 1973), p. 189.

59. *Ibid.*, p. 190.

60. *OSHR*, BNA, Vol. 3, No. 18 (October 4, 1973), p. 590.

61. *Ibid.*, p. 591.

62. *OSHR*, BNA, Vol. 3, No. 20 (October 18, 1973), p. 645.

63. *OSHR*, BNA, Vol. 3, No. 27 (December 6, 1973), p. 861.

64. *Ibid.*, p. 859.

65. *OSHR*, BNA, Vol. 3, No. 31 (January 3, 1974), p. 969.

66. *OSHR*, BNA, Vol. 3, No. 32 (January 10, 1974), p. 998.

67. *Federal Register*, Vol. 39, No. 20 (January 29, 1974), p. 3758.

68. *Ibid.*

69. *Synthetic Organic Chemical Manufacturers Association v. Brennan*, 506 F.2d 385 (3d Cir. 1974), cert. den. 2 OSHC 1654 (May 1975).

70. *Industrial Union Department, AFL-CIO, v. Hodgson*, 499 F.2d 467 (D.C. Cir. 1974).

71. *Synthetic Organic Chemical Manufacturers Association v. Brennan.*

72. U.S. Department of Labor, News Release, July 25, 1972.

73. Marjorie E. Gross, *Loyola University Law Journal*, Vol. 3 (1972), p. 246; see also H. Rep., No. 91-1291, 91st Cong., 2d Session (1970), p. 21.

74. S. Rep., No. 1282, 91st Cong., 2d Session (1970), p. 9.

75. "Federal Common Law Remedies Under the Occupational Safety and Health Act of 1970," *Washington Law Review*, Vol. 47 (1972), p. 629.

76. Richard S. Morey, "The General Duty Clause of the OSHAct of 1970," *Harvard Law Review*, Vol. 6 (1973), p. 988.

77. *OSHR*, BNA, Vol. 1 (1972), p. 1029. See also White and Carey, "OSHA Comes of Age."

78. *Congressional Record*, Vol. 116 (December 17, 1970), p. 11899.

79. *American Smelting and Refining Co.*, Occupational Safety and Health Review Commission (hereinafter OSHRC) Docket No. 10 (August 17, 1973), as reported by the Bureau of National Affairs, in Vol. 1 of *Occupational Safety and Health Cases* (OSHC), p. 1256.

80. 501 F.2d 504 (8th Cir. 1974).

81. *Ibid.*, p. 511.

82. 494 F.2d 460 (8th Cir. 1974).

83. *Brisk Waterproofing Co., Inc.*, OSHRC Docket No. 1046 (July 27, 1973), 1 OSHC 1263.

84. *American Smelting and Refining Co.*, p. 107.

85. *Sun Shipbuilding and Refining Co.*, OSHRC Docket No. 161 (October 3, 1973, Cleary, Commissioner, concurring), 1 OSHC 1381.

86. *Robbins Painting Service*, OSHRC Docket No. 789 (December 14, 1973), 1 OSHC 1482.

87. S. Rep., No. 1282, p. 91.

88. White and Carey, "OSHA Comes of Age," p. 1320.

89. *OSHR*, BNA, Vol. 2 (1972), p. 119; *Richmond Block, Inc.*, OSHRC Docket No. 82 (January 11, 1974), 1 OSHC 1505. See also *Arnold Hansen d.b.a. Hansen Brothers Logging*, OSHRC Docket No. 141 (October 13, 1972), 1 OSHC 1060.

90. 501 F.2d 1200 (7th Cir. 1974).

91. *John B. Kelly, Inc.*, OSHRC Docket No. 154 (August 3, 1973), 1 OSHC 1301.

92. 495 F.2d 822 (2d Cir. 1974).

93. *National Realty and Construction Co., Inc.*, OSHRC Docket No. 85 (September 6, 1972), 1 OSHC 1049; see also 1973 *President's Report on Occupational Health and Safety*, p. 47.

94. 489 F.2d 1265 (D.C. Cir. 1973).

95. *Richmond Block, Inc.*

96. See "OSHA: Unresolved Issues and Potential Problems," *George Washington Law Review*, Vol. 41 (1972), p. 304.

97. *Chicago Bridge and Iron Co.*, OSHRC Docket No. 224, appeal docketed, No. 75-1163, 3 OSHC 1056 (7th Cir. 1975). *Wright-Schuchart-Harbor Contractors*, OSHRC Docket No. 559 (February 15, 1973), 1 OSHC 1138.

98. *OSHR*, BNA, Vol. 1 (1972), p. 1183; see also *Drake-Willamette Joint Ventures*, OSHRC Docket No. 117 (April 18, 1973), 1 OSHC 1181.

99. 377 F. Supp. 1302 (D.C.D.C. 1974) as discussed in *OSHR*, BNA, Vol. 2 (1972), pp. 962-964.

100. *OSHR*, BNA, Vol. 4, No. 7 (July 18, 1974), p. 172.

101. White and Carey, "OSHA Comes of Age," p. 1325.

102. *Sinaico Brothers Co.*, OSHRC Docket No. 4958 (August 28, 1974), 2 OSHC.

103. *Pacific Intermountain Express Co.*, OSHRC Docket No. 4303 (September 24, 1974), 2 OSHC 1228.

104. *Newspaper Agency Corp.*, OSHRC Docket No. 1902 (May 6, 1974), 1 OSHC 1711.

105. *American Airlines, Inc.*, OSHRC Docket No. 6087 (December 4, 1974), 2 OSHC 1391.

106. Testimony of Assistant Secretary of Labor Guenther on June 22, 1972, before the Subcommittee on Environmental Problems Affecting Small Businesses.

107. *OSHR*, BNA, Vol. 4, No. 34 (January 23, 1975), p. 1028; Vol. 4, No. 51 (May 22, 1975), p. 1714.

108. *President's Report on Occupational Safety and Health* (1973), p. 126.

109. *Accu-Namics, Inc.*, OSHRC Docket No. 477 (May 30, 1974), 1 OSHC 1751, petition for review docketed, No. 74-2979, 5th Cir., July 26, 1974.

110. E.g., *Frank Irey, Jr., Inc. v. OSHRC and Brennan*; *Beall Construction Co. v. OSHRC.*

111. *Divesco Roofing and Insulation Co.*, OSHRC Docket No. 345 (August 13, 1973), 1 OSHC 1279.

112. *OSHR*, BNA, Vol. 3, No. 41 (March 14, 1974), p. 1280.

113. *Giaccio v. Pennsylvania*, cited in *OSHR*, BNA, Vol. 3, No. 119 (October 11, 1973), p. 614.

114. *Santa Fe Trail Transportation Co.*, OSHRC Docket No. 331 (December 20, 1973), 1 OSHC 1457.

115. 497 F.2d 230 (5th Cir. 1974).

116. 503 F.2d 8 (4th Cir. 1974).

117. 497 F.2d 233.

118. 505 F.2d 869 (10th Cir. 1974). Cf. Commissioner Cleary's dissent at 1 OSHC 1460.

119. *United States Steel Corp.*, OSHRC Docket Nos. 2975 and 4349 (November 14, 1974), 2 OSHC 1343.

120. *Mobile Coatings, Inc.*, OSHRC Docket No. 226 (May 1, 1972), 1 OSHC 3007.

121. *General Electric Co.*, OSHRC Docket No. 2739 (April 21, 1975), 2 OSHC.

122. Nos. 74-1256, 74-1781 (3d Cir., March 24, 1975).

123. *Standard Glass Co., Inc.*, OSHRC Docket No. 259 (July 27, 1972), 1 OSHC 1045.

124. 512 F.2d 1148 (1st Cir. 1975); quote is at 1152.

125. *Lebanon Lumber Co.*, OSHRC Docket No. 184 (March 19, 1973), 1 OSHC 1165.

126. *Mississippi Valley Erection Co.*, OSHRC Docket No. 524 (December 26, 1973), 1 OSHC 1527.

127. 511 F.2d 1139 (9th Cir. 1975).

128. 504 F.2d 1255 (4th Cir. 1974).

129. 513 F.2d 1032 (2d Cir. 1975).

130. Nos. 74-1381, 74-1382 (7th Cir., May 27, 1975).

131. *Chicago Bridge and Iron Co.*, OSHRC Docket No. 744 (January 24, 1974), 1 OSHC 1485.

132. No. 74-1214 (7th Cir., April 22, 1975).

133. Cf. Commissioner Cleary's proposal in his dissent at 1 OSHC 1487.

134. *Crescent Wharf and Warehouse Co.*, OSHRC Docket No. 1 (April 27, 1973), 1 OSHC 1219, 1223.

135. No. 73-3103 (9th Cir., May 21, 1975).

136. See *Burch v. Reading Co.*, 240 F.2d 574, 579 (3rd Cir. 1957), cert. denied 353 U.S. 965 (1957).

137. See *Consolidated Edison Co. of New York v. National Labor Relations Board*, 305 U.S. 197, 229 (1938).

138. *Armor Elevator Co., Inc.*, OSHRC Docket Nos. 425 and 426 (November 20, 1973), 1 OSHC 1409; see also *OSHR*, BNA, Vol. 1 (1972), p. 1411.

139. *Nacirema Operating Co., Inc.*, OSHRC Docket No. 4 (February 7, 1972), 1 OSHC 1001.

140. *Hidden Valley Corp. of Virginia*, OSHRC Docket No. 11 (February 8, 1972), 1 OSHC 1005.

141. *Dreher Pickle Co.*, OSHRC Docket No. 48 (February 13, 1973), 1 OSHC 1132.

142. 487 F.2d 442 (8th Cir. 1973).

143. *California Stevedoring and Ballast Co.*, OSHRC Docket No. 72 (June 16, 1972), 1 OSHC 1035.

144. *J. E. Chilton Millwork and Lumber Co.*, OSHRC Docket No. 123 (May 15, 1972), 1 OSHC 1034; General Meat Co., Inc. OSHRC Docket No. 250 (June 20, 1972), 1 OSHC 1032.

145. *Colonial Craft Reproductions*, OSHRC Docket No. 881 (October 27, 1972), 1 OSHC 1063.

146. *Kesler and Sons Construction Co.*, OSHRC Docket No. 306 (July 8, 1974), 2 OSHC 1096.

147. 513 F.2d 557 (10th Cir. 1975).

148. U.S. District Court of the Eastern District of Louisiana, Docket No. 72-1981 (March 9, 1973), as reported in 1 OSHC 1173. For a useful discussion of these issues predating *Skidmore*, see "Federal Common Law Remedies under the OSHAct," p. 629.

149. E.g., *Bivins v. Six Unknown Agents of the Federal Bureau of Narcotics*, 403 U.S. 388, 402 (1971).

150. U.S. Department of Labor, News Release, September 27, 1973.

151. 93d Cong., 2d Session, H.R. 5463 (October 11, 1974); emphasis added.

152. U.S. District Court for the District of Maryland, Civil No. 71-522-HM.

153. *OSHR*, BNA, Vol. 3, No. 1 (June 7, 1973), p. 12.

154. *Moore-McCormack Lines v. ITO Corp.* (4th Cir., December 27, 1974).

155. U.S. District Court for the Southern District of New York, Civil No. 70-5011 (February 19, 1973).

156. No. 75-2597 (5th Cir.).

157. *OSHR*, BNA, Vol. 3, No. 32 (January 10, 1974), p. 995.

158. *OSHR*, BNA, Vol. 3, No. 16 (September 20, 1973), p. 522.

159. *Vela v. Wise*, Docket No. 116083 (California Superior Court, 1973).

160. *Compliance Operations Manual*, OSHA (hereinafter cited as *COM*), p. IV-1.

161. *COM* p. XI-2.

162. *COM*, p. XI-3.

163. OSHAct, Section 17(b).

164. *COM*, p. XI-8.

165. *OSHR*, BNA, Vol. 3, No. 24 (November 15, 1973), p. 776.

166. *Ibid.*

167. OSHA Program Directive No. 200-28 (April 22, 1974), superseding Program Directives Nos. 200-20 (September 7, 1973) and 200-20A (September 28, 1973).

168. *OSHR*, BNA, Vol. 3, No. 26 (November 29, 1973), p. 839.

169. Testimony of Assistant Secretary of Labor Guenther.

170. *Ibid.*

171. *Industrial Steel Erectors, Inc.*, OSHRC Docket No. 703 (January 10, 1974), 1 OSHC 1497.

172. *Congressional Quarterly* (July 27, 1972), p. 1829.

173. *Ibid.*, p. 1561.

174. *Ibid.*, p. 1829.

175. *OSHR*, BNA, Vol. 4, No. 34 (January 23, 1975), p. 1004.

176. H.R. 5996, filed in August 1973.

177. Ray Randlette, Department of Labor, private communication, July 21, 1975.

178. *OSHR* BNA, Vol. 5, No. 2 (June 12, 1975), p. 31.

179. *OSHR*, BNA, Vol. 3, No. 25 (November 22, 1973), p. 805.

180. *OSHR*, BNA, Vol. 3, No. 24 (November 15, 1973), p. 771.

181. *Ibid.*

182. *OSHR*, BNA, Vol. 3, No. 33 (January 17, 1974), p. 1019.

183. U.S. District Court for the District of Columbia, No. 74-54 (October 1974).

184. *OSHR*, BNA, Vol. 3, No. 24 (November 15, 1973), p. 777.

185. *Ibid.*, p. 778.

186. Thomas E. Kauper, Assistant U.S. Attorney General, letter to the Bureau of Product Safety, quoted in *OSHR*, BNA, Vol. 3, No. 24 (November 15, 1973), p. 778.

187. *Ibid.*

188. *FMC Corp.*, McCoy, 45 LA 293 (1965), as referenced in John J. Slote, "Employee Protection Against Abnormally Dangerous Working Conditions," *Industrial and Labor Relations Forum*, Vol. 6, No. 1 (March 1969).

189. *Hegler Zinc Co.*, Elson, 8 LA 826 (1947), as quoted in Slote, "Employee Protection."

190. 6 ALAA par. 69.601.

191. *Ibid.*

192. *Daily Labor Report*, No. 232 (December 2, 1974), p. 2.

193. 49 Stat. 449 (1947), Section 1(b).

194. 84 LRRM 2625 (9th Cir. 1973); emphasis added.

195. 388 F.2d 495, 500, 67 LRRM 2083, 2087 (2d Cir. 1967).

196. *Labor Law Reports* (Commerce Clearing House, Chicago, 1967), Court Decisions, p. 95-543.

197. *Redwing Carriers, Inc. v. International Brotherhood of Teamsters,* in *Labor Law Reports*, (Commerce Clearing House, Chicago, 1961), p. 15202; emphasis added.

198. Sheldon Friedman, "The Right to Refuse Abnormally Dangerous Work Under Section 502 of the Taft-Hartley Act," unpublished paper, Massachusetts Institute of Technology (1970).

199. Leslie Boden and Sheldon Friedman, "The Right to Refuse Abnormally Dangerous Work: Managerial Prerogatives vs. Workers' Health," paper in progress, Massachusetts Institute of Technology (1975).

200. *OSHR*, BNA, Vol. 2 (1972), p. 597.

201. Roger B. Jacobs, "Industrial Safety in the Seventies: The Occupational Safety and Health Act of 1970," *Industrial and Labor Relations Forum*, Vol. 9, No. 2 (May 1973), pp. 5-6.

202. Friedman, "The Right to Refuse Work Under Section 502."

203. 47 Stat. 70 (1932).

204. Ellen Chapnick, "The Labor Injunction," unpublished paper, p. 4.

205. 370 U.S. 195, overruled by *Boys Market v. Retail Clerks Union*, 398 U.S. 235 (1970).

206. 353 U.S. 448 (1957).

207. 363 U.S. 564 (1960); 363 U.S. 574 (1960); 363 U.S. 593 (1960).

208. *Ibid.*

209. 398 U.S. 235 (1970).

210. *Gateway Coal Co. v. United Mine Workers*, 414 U.S. 368 (Sup. Ct. 1974).

211. *Ibid.*

212. *Ibid.*

213. *Ibid.*

214. Professor Stanley Jacks, comments at an Industrial Relations Seminar, Massachusetts Institute of Technology, 1973.

215. 29 CFR Sections 1972.1-1972.23 (1973).

216. 192 NLRB No. 150, 77 LRRM 1931 (1971).

217. No. 73-1609 (D.C. Cir., August 14, 1974).

218. Anne Trebilcock, "OSHA and Collective Bargaining: An Overview," unpublished paper (1974).

219. *Gulf Power Co. and Local Unions 1055 and 624 of the International Brotherhood of Electrical Workers (AFL-CIO)*, NLRB Case No. 15-CA-2556 (January 5, 1966).

220. *Daily Labor Report,* No. 221 (November 15, 1974) p. A-12.

5 The OSHAct of 1970. II. State Participation

5.1 INTRODUCTION

5.1.1 A Brief History of State Activities and the Passing of the OSHAct

Early state efforts in occupational health and safety were concentrated in two areas of legislative action and subsequent judicial development. The first was the passage of state factory laws which attempted to remove children from factories and to protect women in the plant. The second was the passage of state workmen's compensation laws which sought to broaden the scope of an employer's responsibility for maintaining a hazard-free workplace and to provide an incentive for reducing injury in the workplace.

These early efforts were moderately, but by no means entirely, successful. In agriculture, women and children are still working under substandard conditions, and the workmen's compensation system came under sharp criticism in the 1972 report by the National Commission on State Workmen's Compensation Laws set up by the OSHAct.[1]

A third and later development in state activities, arising naturally in response to the rapid industrialization of the United States, was a shift from a concern for hiring practices and compensation to a concern for the nature of the workplace itself and the elimination of hazards. Hence, the field of industrial safety, with the development of safety standards and practices to be used with discretion, gave rise to state inspection and the enforcement of those standards. A parallel and often separate development in industrial health resulted in state public health activities centered around the control of tuberculosis and the prevention of other contagious diseases that would present an especially severe problem in a highly concentrated and unsanitary workplace.

For a period prior to 1961, the combination of new health and safety laws and increased professional activity in industrial health and safety resulted in a decrease in the industrial accident rate. However, these efforts were not sufficient to deal with the problems produced by rapidly expanding technologies and the increased use of toxic materials. The industrial accident rate increased nearly 29% from 1961 to 1970.[2] After the Farmington, West Virginia, mine explosion in 1968, Congress responded by passing the Coal Mine Health and Safety Act of 1969. Then, in response to increasing pressure from an awakened and alarmed organized labor, Congress passed the far-reaching Occupational Safety and Health Act of 1970.

A more detailed and anecdotal account of the recent state activities and their failings is given by Joseph Page and Mary-Win O'Brien in *Bitter Wages* and by Franklin Wallick in *The American Worker: An Endangered Species.*[3] Suffice it to say that, in general, prior to the passing of the OSHAct, state enforcement of safety and health standards (1) involved too few and insufficiently trained personnel, (2) was encumbered by a slow-moving state bureaucracy, (3) was not infrequently corrupted by the abuse of discretionary authority which accompanied the vague and confusing state standards, (4) did not include first-instance citations, but depended on warnings, and (5) was heavily safety-biased, to the detriment of acute health problems and the virtual exclusion of chronic occupational disease. It became clear to Congress that "voluntary compliance" with the then established state safety and health standards and practices was not sufficient.

Mandatory standards, developed through a consensus method or adopted by administrative fiat on a firm data base, an informed and strong enforcement force, and a greater emphasis on occupational disease were seen to be necessary components of a sincere federal effort to reduce industrial accidents and disease. The embodiment of this effort was the Occupational Safety and Health Act of 1970.

5.1.2 The OSHAct and State Participation

The language of the OSHAct makes it clear that complete and permanent federal preemption was not envisioned where competent and responsive state action could be effected. In Section 2(b) it is stated:

> The Congress declares it to be its purpose and policy . . . to assure every working man and woman in the Nation safe and healthful working conditions . . . by *encouraging the States to assume the fullest responsibility* for the administration and enforcement of their occupational safety and health laws . . . [and] to develop plans in accordance with the provisions of this Act. (Emphasis added)

And further in Section 18:

> Nothing in this Act shall prevent any State agency or court from asserting jurisdiction under State Law over any occupational safety and health issue with respect to which no standard is in effect under section 6 [Federal standards].
>
> (b) Any State which, at any time, desires to assume responsibility for development and enforcement therein of occupational safety and health standards relating to any occupational safety and health issue with respect to which a Federal standard has been promulgated under section 6 shall submit a State plan for development of such standards and their enforcement.

It is not clear, however, from the congressional hearings on the Act whether Section 18 was meant merely to be a "valve" through which the few good existing state programs could be salvaged, or whether Section 18 was envisioned to be the mechanism through which all states would regain authority over occupational health and safety—once the federal government showed them "how to do it." The latter view has been competently defended by Sally Seymour in a well-written article for the *Monthly Labor Review.*[4]

Little discussion of the issue occurred during the congressional hearings, and organized labor was told by friendly congressmen that the Act could not have been passed without a state section.[5] What is clear is that neither the Occupational Safety and Health Administration nor organized labor initially expected the overwhelming activity of the states directed toward regaining local control.

As of July 1975, only five states had not submitted plans for consideration, nine states had withdrawn their plans, and twenty-two had been approved. Widespread submission of state plans was most likely stimulated by adverse business reaction to the new Act and by President Nixon's "New Federalism." In the 1972 *President's Report on Occupational Safety and Health*, it is stated: "Above all else, if we are to be successful, the full collaboration of private industry, the States and the employees must be enlisted." And further into the report we find that "the purpose of the Act is not to eliminate State safety and health programs but to include the States as major participants in the implementation of the Act as clearly stated in the 'Congressional Findings and Purposes.' "

The Act provides under Sections 23(a), (b), and (f) that the federal government may pay up to 90% of the cost for developing state plans, while under Section 23(g) it may pay up to 50% of the cost of administering the plans. In practice, the federal government has provided the maximum amount of funds for development and administration and has thus encouraged state participation to the fullest degree.

The Nixon administration turned what appeared to be a rather obscure issue into what former Assistant Secretary of Labor George Guenther called one of the four top priorities of the Act.[6] The matter of state participation continues to be an area of the most intense dispute between organized labor and the government. The national unions are strongly resisting the return of control to the states, while local unions sometimes support the effort.

In interviews, several staff members within the Department of Labor expressed their feeling that the policy of New Federalism for occupational health and safety had been determined by the Office of Management and Budget from the beginning and that the framework for the system had been surrepticiously constructed in Section 18 of the OSHAct. In any event, the new state involvement has raised both old and new issues within a different framework and has created a new uncertainty for the future of occupational health and safety in the United States.

5.2 STATE JURISDICTION AND THE IMPLEMENTATION OF STATE PLANS

5.2.1 The Provisions of the Act with Regard to State Plans

In the following discussion, various personalities, offices, and functions may be described with different titles because of organizational changes. Figure 5.1 shows in a schematic way how the office overseeing state programs has fit into the overall structure of OSHA through three organizational stages.

If no federal standard is in effect on a particular occupational safety or health issue, the states can assert jurisdiction, whether or not there exists a federally ap-

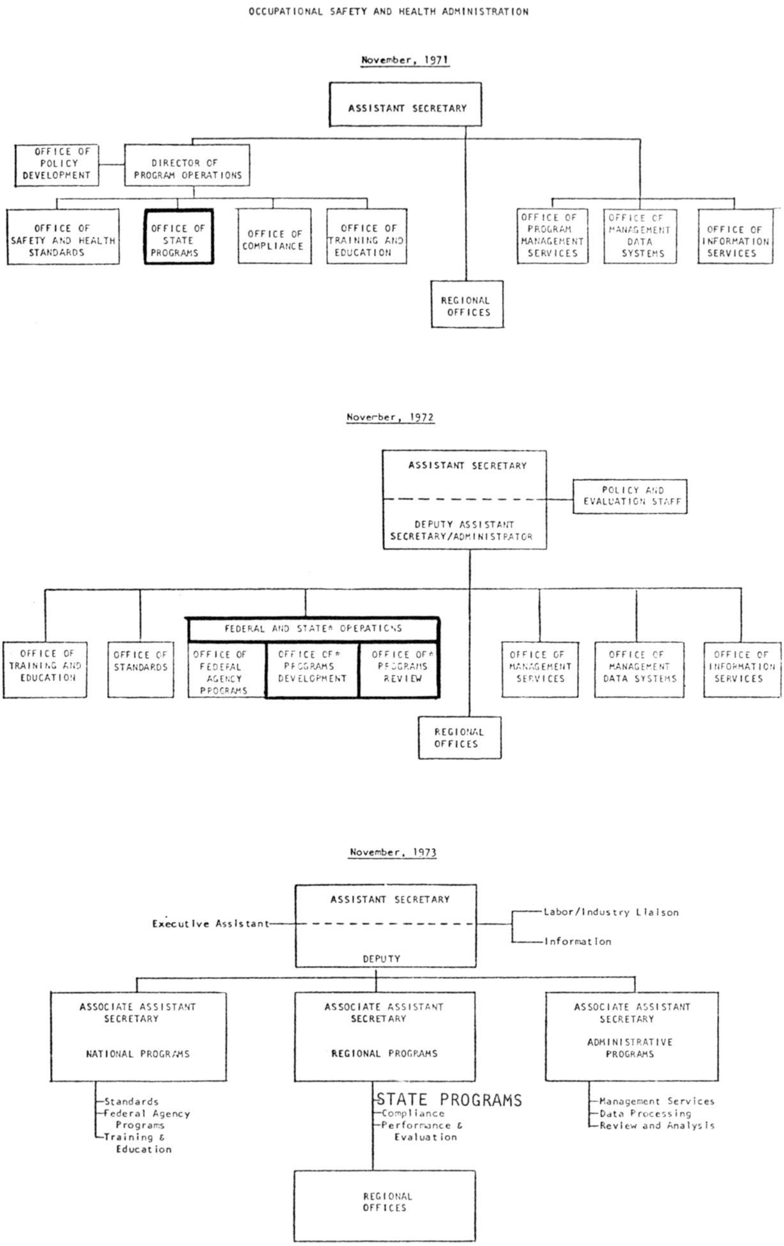

Figure 5.1 The status of the office overseeing state programs in the administrative hierarchy of OSHA, 1971-1973.

proved state plan (see Section 18(a) of the Act).

According to Section 18(c), an acceptable state plan must meet the following specifications:

1. It must designate a responsible state agency or agencies to administer the plan.
2. It must provide for state standards that are or will be at least as effective as federal standards, and it must ensure that the standards, when applicable to products distributed or used in interstate commerce, are required by compelling local conditions and do not unduly burden interstate commerce.
3. It must provide for a right of entry and inspection at least as effective as the federal procedure, and it must include a prohibition of advance notices of inspection.
4. It must contain satisfactory assurances that the designated agency has legal authority and qualified personnel, and that the state will devote adequate funds to administration and enforcement.
5. It must contain satisfactory assurances that public employees will be protected to the extent permitted by state law.
6. It must require employers to make reports to OSHA in the same manner as if the plan were not in effect.
7. It must require the state agency to supply any information required by OSHA.

After a state plan is approved, the Secretary of Labor may exercise his authority until the plan is determined to meet all of these criteria. (Such a determination must be made on the basis of the operation of the plan and may not be made until at least three years after the plan's approval.) Once the plan has been judged to be in compliance, the applicable federal provisions are preempted by those of the state plan (Section 18(e)).

The Secretary of Labor must make a continuing evaluation of the state plan, retaining the right to withdraw approval for cause after affording a state adequate notice and an opportunity for a hearing (Section 18(f)). Judicial review is available for decisions of the Secretary of Labor pertaining to rejection or withdrawal of a state plan (Section 18(g)).

Finally, under special agreement, state inspectors may be employed for a short term by the federal government to act as federal inspectors (Section 7(c)(1)).

5.2.2 The Mechanics of Adoption of the State Plan

The purpose of this section is to provide the reader with a chronological sequence of the important steps in the adoption of a state plan without unnecessary reference to the pertinent sections of the OSHAct.

The governor of an interested state must first designate an appropriate agency to administer occupational safety and health activities within the state. Section 23(a) of the Act then allows the federal government to fund up to 90% of the cost of developing the state plan. This support permits states to identify their special needs and responsibilities in occupational safety and health, to develop their proposed plans, to establish information systems, to increase the expertise and enforcement capabilities of their personnel, or otherwise to improve the administration of their program, laws, and standards. (Such activities culminated in all but five states submitting plans by December 1973.)

The designated state agency develops the plan, the governor supports it, and then either the state legislature enacts new legislation or improves existing legislation to authorize the carrying out of its provisions, or the designated agency provides OSHA with a draft of legislation along with a letter of support from the governor and a legal opinion as to the likelihood of the legislation's being passed.

The OSHA regional office is the first to receive the plan officially on behalf of the Secretary. Under the concept of decentralized operations, the plan is subjected to an intensive review by people who both know the requirements of the federal program and understand the problems of the particular state. As quickly as possible, the plan is referred to the national office, which requests public comments. It is then reviewed again by all the offices concerned with the various aspects of plan approval. A summation of both internal and public comments and the state's responses or promises to incorporate recommended changes is prepared by the project officer along with his office's final recommendations as to approval of the plan. This information is presented to the Assistant Secretary, and if acceptable (with the concurrence of the Secretary), the plan is approved.

If the plan or any assurance submitted with it is considered in some way deficient, a formal letter explaining the failings and the recommended improvements is transmitted through the regional office to the state. After discussing the issues involved with federal officials at both regional and national levels, the state decides how far it will go toward meeting the recommended changes. If the state's revised plan is not considered satisfactory, and the Assistant Secretary has determined to reject the proposed plan, he must provide the state due notice and an opportunity for a hearing. His decision is subject to further review in the court of appeals.

Once a plan has been approved, the federal government may provide up to 50% of the funds necessary to run the state's program through an operational grant (Section 23(g)). OSHA's professed policy is always to provide the full 50%.

All plans submitted to date have been developmental. All developmental steps agreed to as part of the plan, including legislation, must be accomplished within three years after the effective date of the plan.

Throughout the period after the initial approval of its plan, a state may submit new material in the form of changes, modifications, or corrections to the Secretary, for approval and addition to the plan. Evaluation of the plan is conducted through such means as written reports, periodic on-site inspections, and case file review, and through the administrative mechanisms of CASPA (Complaints About State Program Administration).

Once the state has completed the developmental steps, the (Assistant) Secretary will observe operations under the plan for at least one year and then make his determination as to final approval based on effectiveness of the plan relative to the federal effort.

Even after final approval, the Secretary of Labor is required to make a continuing evaluation of the state's performance, and to step back in and withdraw approval (after notice and hearing) for substantial failure to comply. This action is also subject to review in the court of appeals. The only indications of the eventual effectiveness of the federal watchdog role are to be found in the monitoring activities currently underway and prior to final approval.

5.2.3 OSHA Rules and Regulations, Program Directives, and Policy

To assist in carrying out the provisions ofthe Act with regard to state plans, the Secretary of Labor has promulgated regulations covering the following specific areas of activity: state plan approval (29 CFR 1902); state plan operational grants (1951); approved plans (1952); changes to state plans (1953); and evaluation and monitoring of approved state plans (1954).

Other regulations covering the federal program, which are necessary guidelines for use in the state programs if they are to be at least as effective as the federal program, deal with the following areas: compliance programs (1903); reporting procedures (1904); variance procedures (1905); and standards promulgation (1911).

Various sections of these regulations are quoted from or referred to in the following discussions of important problem areas.

The Meaning and Implications of "At Least as Effective as"

This concept is the key to the formulation and evaluation of state plans and the hardest to define. The federal program has been described correctly as a "moving target."[7] The ultimate test is that a state's enforcement program, standards, legal provisions, and reporting system must go just as far as do the federal ones toward reducing injuries and deaths in the workplace. Currently, the Secretary accepts the state's promises on paper that it will be "at least as effective as" the federal effort, after a point-by-point review of the plan by his staff. These promises are what give a plan its "developmental" nature. The review is necessarily subjective. The issues which follow are all important examples of instances in which the (Assistant) Secretary has tried to establish guidelines for the states.

Walkaround Right Section 8(e) of the Act provides the basis for the "walkaround" right:

> (e) Subject to regulations issued by the Secretary a representative of the employer and a representative authorized by his employees *shall* be given an opportunity to accompany the Secretary or his authorized representative during the physical in-

spection of any workplace under subsection (a) for the purpose of aiding such inspection. Where there is no authorized employee representative, the Secretary or his authorized representative shall consult with a reasonable number of employees concerning matters of health and safety in the workplace. (Emphasis added)

The appropriate regulation applicable to state plans is 1902.4(c)(2)(ii):

(2) The Assistant Secretary will determine whether the State plan:
(ii) Provides an opportunity for employees and their representatives before, during and after inspections to bring possible violations to the attention of the State agency with enforcement responsibility in order to aid inspections *by such means as* affording a representative, the employer and a representative authorized by employees an opportunity to accompany the State representative during the physical inspection of the workplace, or where there is no authorized representative, by providing for consultation by the State representative with a reasonable number of employees. (Emphasis added)

Sheldon Samuels of the Industrial Union Department, AFL-CIO, has pointed out in some detail that the intent of Section 8(e) was to *mandate the walkaround right.* He contends that the language of the regulation—"by such means as"—has diluted the effectiveness of the Act by allowing the state to choose an *alternative* to the walkaround.[8]

It would appear from the discussion of legal developments in the 1973 *President's Report on Occupational Safety and Health* that the federal right has been whittled down as well.[9] The Review Commission has taken up one case concerning the *employer's* right, and has decided that the right of the Secretary to make the inspection is a primary and public right, while the right of accompaniment is secondary and private.[10] In that case, because of the geographical dispersion of the workplace, a walkaround by employer representatives would have been unacceptably cumbersome in the Commission's view; a compliance officer's failure to offer the right to the employer was upheld. In a second case concerning the same right for *employee* representatives, the Review Commission ruled that the failure to afford an employee representative the opportunity to accompany the inspector does not require vacation of a citation if it is not shown that the Secretary's action was prejudicial against the employer.[11] The mandatory nature of the walkaround right therefore no longer remains inviolate.* In practice, however, OSHA has not had occasion to pass judgment on specific alternatives to the walkaround in the state plans, and the issue will probably be tested on a case-by-case basis. In August 1973, in an interview with an OSHA employee working with state plans in the Solicitor of Labor's Office, it was explained that, in practice, the wording "by such means as" had not (yet) proved to be any problem with regard to the walkaround right.[12] In fact, all approved state plans contain the walkaround right provision; some plans even re-

*One would interpret this case narrowly, of course, as merely denying the employer the right to raise this defense on procedural grounds.

quire payment for time spent during the walkaround or consider refusal to pay grounds for a discrimination complaint.

Qualifications of the State Enforcement Personnel Section 18(c) of the Act demands "satisfactory assurances [of] qualified personnel necessary for the enforcement of such standards." Regulation 1902.3(h), on Personnel, requires:

The State plan shall provide assurance that the designated agency or agencies have, or will have, a *sufficient* number of *adequately trained* and *qualified* personnel necessary for the enforcement of the standards. For this purpose *qualified* personnel means persons employed on a merit basis, including all persons engaged in the development of standards and the administration of the State plan. *Conformity with the Standards for a Merit System* of Personnel Administration, 45 CFR Part 70, issued by the Secretary of Labor, including any amendments thereto, and any standards prescribed by the U.S. Civil Service Commission pursuant to section 208 of the Intergovernmental Personnel Act of 1970 (Public Law 91-648; 84 Stat. 1915) modifying or superseding such standards, *will be deemed to meet this requirement.* (Emphasis added)

A continuing controversy surrounds the question of whether state inspectors—subject only to a state merit system, usually paid less than their federal counterparts, and subject to more pressure from local businesses, either directly or through their own agency—can be really as effective as federal inspectors in carrying out the aims of the Act.

OSHA Program Directive 73-11, issued November 29, 1973, mandated the following program aimed at monitoring the quality of state inspection personnel:

When an approved 18(b) plan becomes operational, the OSHA assistant regional director may work with the state designee to establish an on-the-job evaluation program for certain state enforcement personnel. It is to involve observation by experienced OSHA compliance officers. All state enforcement officers who are participating in the OSHA-funded state program may be evaluated on the job.

The OSHA assistant regional director will determine the number and type of OSHA compliance personnel required. The scheduling of specific on-the-job evaluations will be worked out between the assistant regional director and the state designee.

The OSHA personnel responsible for on-the-job evaluation will work with the assistant regional director and the state designee to determine the number of evaluation inspections. The assistant regional director shall keep records on state personnel so evaluated, including the number of visits on which they were accompanied, and the name of the accompanying federal representative.

The OSHA compliance officer will not interfere in the conduct of the inspection itself. However, the employer will be advised of the fact that OSHA is monitoring the state inspection.

At appropriate intervals, as established with the state designee, the OSHA assistant regional director will discuss with the state the results of the monitoring of on-the-job evaluation activity, advising the designee of OSHA's findings regarding problem areas in the state's enforcement activity evidenced during the evaluation.[13]

The program directive seems to be indicating that, although the 1902 regulation merely calls for conformity with an approved state merit system, OSHA will try to ensure that state enforcement personnel will in fact become, through quantity and/ or quality, "at least as effective as" the federal inspectors. (However, since the quality of the federal inspections leaves much to be desired, the attainment of this standard is not particularly impressive.)

Standards Section 18(c) of the Act deals with the question of state standards:

(2) [The State plans must provide] for the development and enforcement of safety and health standards relating to one or more safety or health issues *which standards* (and the enforcement of which standards) are or *will be at least as effective* in providing safe and healthful employment and places of employment as the standards promulgated under section 6 which relate to the same issues, and which standards, when applicable to products which are distributed or used in interstate commerce, are required by compelling local conditions and do not unduly burden interstate commerce. (Emphasis added)

Regulation 1902.4(b)(2)(ii) requires that the plan provide "an adequate method to assure that its standards *will continue to be at least as effective* as Federal standards, including Federal standards relating to issues covered by the plan which become effective *subsequent to any approval of the plan*" (Emphasis added).

To satisfy this criterion, states with approved plans have had to promise, subject to their own administrative procedures, that they would adopt any *federal* standards coming out in the future; or that they would propose to OSHA their own standards which would be at least as effective as the (new) federal ones. The only states retaining state standards so far are California, Washington, Oregon, Michigan, and Hawaii.[14] Unless a state already has an extensive body of standards and an enforcement program, it would be prohibitively difficult and time-consuming to attempt to create a body of state standards different from and at least as effective as the federal ones. Another problem related to the effectiveness of standards has to do with performance vs. specification standards. (See Section 4.3.3; California's standards are based on performance rather than specifications.) As long as OSHA can find their *approach* at least as effective as its own *approach*, the state standards will be acceptable.[15] While performance standards may allow more flexibility for the states (and in some cases more protection for workers), their relative "effectiveness" is harder to evaluate and can therefore result in less protection than the federal specification standards—especially if the total federal evaluative competence is not strengthened.

State Standards and the General Duty Clause Another very interesting issue connected to the problem of standards is what happens when there is neither a state nor a federal standard for a given situation. Existing regulations do not require that the states institute a general duty clause. If the state does not have the clause, then in theory federal enforcement authority still exists, although given its general "hands off" policy toward the states, it is doubtful that OSHA would cite general duty violations in states with approved plans. Even if a state does have a general duty clause, however, it is not clear under what conditions it can or will be used. The criteria for compliance with the federal general duty requirement are hardly developed, and the state criteria are even less so. It is difficult to see how OSHA can determine if the state's enforcement under its general duty clause is as effective as federal performance. The problem is particularly acute because of the slow development of federal health standards: three new health standards and about 400 TLVs have been adopted, but this has hardly scratched the surface of the more than 12,000 toxic materials in commercial use. The general duty clause has been used in the past, and it should be used more extensively while the health standards are being developed. However, wide usage under the state plans is doubtful since citing a violation of general duty requires the exercise of expert and discretionary judgment.

First-Instance Sanctions Another issue that has raised much controversy has been whether or not states must impose first-instance sanctions (penalties) for *serious* violations. In the case of the Washington state plan, approval was not granted until this provision was included. However, Barry White, Associate Assistant Secretary for Regional Programs at OSHA, has stated that "it could be difficult to prove in court that a state plan without first-instance sanctions is not as effective as the Federal law."[16] Since many states have not instituted mandatory penalties for serious violations, and since no clear policy exists, a case-by-case determination will probably continue to be made.[17] Meanwhile, penalties for nonserious violations are discretionary at the federal level and hence can be discretionary in the states.

Review Mechanism Regulation 1902.4(e)(2)(xii), under Enforcement, states that "an employer [shall] have the right of review of violations alleged by the State, abatement periods, and proposed penalties and for employees or their representatives [shall] have an opportunity to participate in review proceedings, *by such means as providing for administrative or judicial review*, with an opportunity for a full hearing on the issues" (emphasis added).

With regard to some state plans, critics have been concerned about the extent of employer and employee rights to participate in the review of alleged violations, abatement periods, and proposed penalties. In Illinois, for example, there are two designated agencies responsible for occupational health and safety.* The Industrial Commission sets standards and also exercises the review function, while the State Department of Labor does the enforcing. Critics indicate that since the Department of Labor is in reality (though not in law) the agent of the Industrial Commission for enforcement purposes, the Commission is in effect reviewing itself if it hears cases contested against the State Department of Labor. The state AFL-CIO and the Chicago Area Committee on Occupational Safety and Health, a public-interest group, believe this procedure is not as effective as the separate and independent Review

*The Illinois state plan has now been withdrawn.

Commission set up under the federal OSHAct, even though the final disposition of the decision of the Industrial Commission is reviewable in the state court.

Consultation Neither the OSHAct nor the federal regulations allow federal OSHA *inspectors* to visit a workplace solely for the purpose of consulting with or advising the employer. Many businessmen claim that this would be a more realistic approach, and one that they would sincerely appreciate and solicit. Labor, on the other hand, fears that this would put the situation right back where it was before OSHA, when an employer did not have much incentive to correct conditions until after a visit from the inspector. Therefore, the provision that state plans may include on-site consultation has been controversial. Regulation 1902.4(3)(2)(xiii) calls for encouraging voluntary compliance by such means as training and consultation with employers and employees. OSHA Program Directive 72-27 carefully pointed out that although consultation is allowed, personnel engaged in on-site activities must be divorced from inspecting, and their existence must not detract from the state's ability to be "as effective as" the federal enforcement effort. A rule of thumb attributed to Thomas Brown, former Director of Federal and State Operations at OSHA, was that no state could be funded for an on-site consultation staff larger than 15% of the enforcement staff.[18]

Oregon has incorporated a voluntary compliance unit into its plan's activities.[19] Personnel assigned to the unit will have no enforcement or inspection responsibilities but will be required to ask for compliance or immediate termination of work in an imminent-danger situation. If their request is not honored, they must call on enforcement personnel. If serious violations are observed, the employer must be advised and a reasonable abatement period established after which a follow-up inspection will be made by enforcement personnel.

It is too early to make a judgment on the effectiveness of this approach—with regard to both its salutory effects in encouraging voluntary compliance and educating employers and its possibly adverse effects on a state's enforcement efforts.

Completed vs. Developmental Plans and the Absence or Inadequacy of Enabling Legislation

OSHA's ability to approve plans that are developmental has been challenged on many occasions, by the AFL-CIO, NACOSH, and others. Many critics feel that if a plan has many operational areas covered only by promises on paper, it cannot be as effective as an on-going federal program. An especially loud cry in this regard has been raised when a state does not yet have legislation enacted to enable it to carry out its promises. Some legislatures meet only every two years, and bills in any legislature are affected by many competing interests.

The Assistant Secretary of Labor for Occupational Safety and Health sent out a memo to regional offices in July 1973 in which he noted that failure to complete planned legislative action will affect the review of plans under consideration. A rather complete explanation of why legislation might not pass, when it would be proposed again, and what other actions could be taken in the meantime were items to be requested from the states. In an interview with an employee in the Solicitor of Labor's Office, it was emphasized that the Department of Labor is now asking for more assurances that the necessary legislation will be passed and that the state will have adequate enforcement authority at the time of plan approval.[20]

The Health Research Group has challenged OSHA's right even to "approve" a state plan in its developmental form, saying that in fact this is a form of administrative rather than statutory approval, and thus not within the original intent of the OSHAct. It is further claimed that OSHA does not have the right to withdraw federal enforcement activity during the three-year period prior to final approval under Section 18(e) on the basis solely of this administrative "approval" of a developmental plan.[21] The withdrawal of federal enforcement authority during a period when a state does not yet have a fully operational plan would certainly not encourage vigorous state activity. This is particularly true in the health area, in which the states have always been particularly deficient.

Concurrent Federal and State Jurisdiction

The issue of concurrent federal and state jurisdiction has caused considerable conflict. It is useful to consider separately the problems of concurrent jurisdiction in the periods prior to and after the "approval" of a (developmental) state plan. Under Section 18(h) of the OSHAct, until December 28, 1972, a state that had indicated an intent to develop a state plan could enter into an agreement with OSHA to continue enforcing some of its own standards until the plan was approved. An attempt by OSHA to extend this date by six months was successfully challenged in the courts, thus ending all *state* enforcement of *state* standards in those states with unapproved plans.[22] In an interview, an OSHA spokesman indicated that in these states, the state inspectors continued to inspect; however, they no longer had the authority to enforce.[23] Section 7(c)(1) agreements, by which state inspectors can be used to enforce certain *federal* standards, still remain in effect, though. Prior to "approval" of the state plan, full federal enforcement is *mandated* and is exercised by OSHA.

In the period *after* "approval" of the (developmental) state plan, the problems are different. As pointed out earlier, the Act provides that after the Secretary "approves" a plan, he *must* continue to monitor the state's performance for three years, and he *may* continue to enforce certain provisions at the same time that the state is active. Many businessmen and state governments oppose the idea of concurrent enforcement activity. The businessmen object to living under two systems, however similar; the state governments are not pleased with the implication that although their program sounds good, they are not quite to be trusted. It is instructive to follow the development of policy regarding concurrent jurisdiction.

While joint inspections were being conducted, Richard Wilson, Deputy to the Director of the OSHA Office of State Programs, indicated that the federal inspectors would try to defer to state inspectors whenever possible in the case disagreement on whether to cite for a violation.[24] *Current policy was to try to phase out federal activity within six months of approval*, according to Assistant Secretary of Labor for OSHA John H. Stender.[25] He said that it was not OSHA's intention to duplicate state efforts or to issue citations in states with approved plans *and* enabling legislation.

Associate Assistant Secretary for Regional Programs Barry J. White said in December 1973 that although there would be *no more joint inspections* (really meaning no more joint enforcement), OSHA would continue to be a *presence* in those states whose programs need assistance.[26] Benjamin Mintz, Associate Solicitor of Labor, pointed out that the level of federal enforcement in a state with an approved plan is up to the discretion of the Secretary of Labor, although the Secretary is subject to court action for abuse of authority.[27]

Barry White explained more explicitly in a presentation to NACOSH that one situation OSHA is trying to avoid is concurrent inspections with states meeting approved operational plan criteria. Employers should not be placed in the position of having to comply with, in some cases, two different sets of standards when dual inspections occur. Under the new *monitoring* guidelines, White said, "we don't plan to be in a workplace and cite employers concurrent with state inspectors but we will go to the state later and point out effectiveness or ineffectiveness of state safety inspectors."[28] As we shall see next, OSHA has now codified its policy shift from enforcement to monitoring.

Evaluation and Monitoring of Approved Plans

OSHA Regulations On January 15 and 20, 1974, OSHA adopted Regulations 1954 and 1954.3 for the evaluation and monitoring of approved state plans.[29] These regulations were long overdue, as the first plan was approved in November 1972. Among the critics of state takeover, this delay raises a serious question as to just how thoroughly OSHA really intended to scrutinize plans once they had received initial approval.

These new regulations were effective immediately, although interested parties had until February 14, 1974, to submit written data, views, and arguments. By the regulations, the Assistant Secretary established a State Program Performance Monitoring System. Evaluation under this monitoring system covers the period both before and after an 18(e) (i.e., final) determination. The evaluation system is divided into three phases in order to ensure that the developmental steps are completed, that the plan meets the "at least as effective as" test, and that the implementation of the plan will be reviewed on a continuing basis. Phase I lasts from initial approval to 18(e) determination. Phase II is concerned with the actual 18(e) determination. Phase III is the continuing evaluation thereafter. The system provides for, but is not limited to, (1) quarterly and annual reports of state program activity; (2) visits to state agencies; (3) on-the-job evaluation of state compliance officers; and (4) a program for investigation of complaints about state program administration (known as CASPA).

This last provision could be an effective check on the state program, but it could also become an administrative bottleneck resulting in an essential breakdown of the monitoring function.* CASPA provides that any interested party may complain about the operation or administration of any aspect of a state plan. The Assistant Regional Director for Occupational Safety and Health has sole discretion to determine whether or not to act upon the complaint, to offer an informal hearing if he does not plan to act, and to affirm or modify his own decision. His decision on whether or not the complaint warrants action is to be based on such factors as (1) the extent to which the complaint affects any substantial number of persons; (2) the number of complaints on the same or similar issues and their relation to safety and health conditions at a particular establishment; (3) whether the complainant has exhausted applicable state remedies; and (4) the extent to which the subject matter of the complaint is pertinent to the effectuation of federal policy. States with approved plans were required to inform OSHA by July 1, 1974, of the procedure they would use to notify employees, employers, and the public of their right to complain to OSHA under these regulations.[31]

The above descriptions of the workings of CASPA found in the regulations really say nothing about what the philosophy behind the monitoring will be or what the federal role will be in states with approved plans prior to (final) 18(e) determination. Most decisions are within the discretionary powers of the Assistant Regional Director and, as such, cannot be effectively challenged through the administrative-law or judicial processes. However, an examination of activity and statements made prior to the issuance of the regulations is instructive as an indicator of what OSHA policy is likely to be.

Policy behind the Regulations There has been preparatory activity in each region. The Assistant Regional Directors for OSHA have been developing systems for monitoring the effectiveness of approved state plans under the Act. It is interesting to note that in November 1973 none of the Assistant Regional Directors interviewed by the *Occupational Safety and Health Reporter* (Regions 1, 2, 3, 5, and 6) reported any intention of reducing federal inspection forces in the foreseeable future in states with approved plans. Among the reasons given were the need for extensive monitoring of the state programs, maritime responsibilities (which the states do not assume), and the fact that some states do not assume full responsibility for job safe-

*In an interview in August 1973, an employee in the Office of the Solicitor of Labor who works with state programs expressed concern that no real understanding or potentially effective procedure had been developed for CASPA.[30]

ty and health enforcement under their approved plans.[32]

In December 1973 Assistant Secretary John Stender, while discussing federal activities in states with approved plans, said that federal compliance officers would be used increasingly to inspect for maritime standards violations *and to monitor state plan operations.* He indicated that they were also being used to help train state inspectors and to evaluate inspections conducted by state compliance officers.[33]

Soon thereafter, however, as was noted in an earlier section of this chapter, Stender said that federal enforcement activity would be phased out in states with approved plans, generally within six months of approval. (This does not necessarily mean removal of federal personnel, but rather a shifting of their activities from enforcement to monitoring.) He stated that it is not OSHA's intention to duplicate state efforts or to issue citations in states with approved plans *and* enabling legislation. States must meet strictly defined criteria to provide plans "as effective as" OSHA, and Stender believes they will do a good job. If OSHA *monitoring* indicates states are deficient, OSHA will *probably* withdraw approval. Stender also emphasized that OSHA needs state participation. It cannot possibly hire enough inspectors to ensure that the law will be enforced vigorously, and it therefore needs state people.[34]

On December 6, 1973, at the NACOSH meeting, Barry White, Associate Assistant Secretary for Regional Programs, said that no OSHA activity was more important than the *monitoring* of state programs. Many critics feel that states will fail, he said, and OSHA will conduct daily monitoring to ensure that this does not happen. In his speech, he went on to make the following points:

> *All complaints about state performance will be investigated.* . . . OSHA's goal is not to harass the states, but to assist the states in developing "as effective as" programs. . . .
>
> *The OSHA state plan monitoring system will enable the agency to determine the level of Federal enforcement needed*, the level of technical assistance needed, and whether the state should be allowed to continue enforcement. . . .
>
> OSHA is conducting inspections in all the states. . . . Many states have limited coverage and OSHA fills the gap. Employee complaints are referred to the states and OSHA monitors the follow-up. *There will be no more OSHA-state joint inspections, though OSHA will conduct on-the-job evaluations of state inspections.* . . .
>
> OSHA begins monitoring a state plan as soon as enabling legislation is passed. . . . *If states do not have sufficient personnel, OSHA will continue to enforce.* . . .
>
> The average of penalties proposed by the states should equate to those proposed by OSHA. . . .
>
> If a state like California has a full-blown program, obviously full Federal enforcement is not necessary. . . . On the other hand, if a state program is "sketchy," a full Federal force is necessary. . . .
>
> A state that feels no obligation to provide qualified industrial hygienists would not be approved for industrial hygiene inspections.[35]

On December 21, 1973, NACOSH's Subcommittee on State Plans concluded that "Federal enforcement in states with approved plans should continue until an initial evaluation of *actual* operation is made."[36] Six weeks later, Barry White was again the spokesman for OSHA, but his policy statements clearly contradicted those of December 6:

"Existing" Federal policy . . . is to *shift to monitoring* of state plans rather than continuing Federal enforcement activities upon commencement of enforcement by states with operational approved plans. . . .

The new monitoring guidelines will supersede the scheduled six-month evaluations of states with approved plans, he said, in response to a question on why [sic] "overdue" six-month Federal evaluations on nine states with approved plans have not been conducted by DOL yet. . . .

Under existing policy, Federal enforcement authority would *not* be exercised in states with enabling legislation (including operational review procedures),* with adopted standards (whether or not they have been approved at least as effective as Federal ones), and with competent enforcement personnel. . . . *Regardless of whether evaluation has taken place when these criteria are present, Federal personnel would shift emphasis from enforcement to monitoring. . . .*

. . . we don't plan to be in a workplace and cite employers concurrent with state inspectors but we will go to the state later and point out effectiveness or ineffectiveness of state safety inspectors.[38]

Thus, in light of the shifting philosophy espoused prior to the issuance of the regulations and the vagueness of the regulations themselves, both the extent of federal "presence" in states with "approved" plans and the effectiveness of the monitoring scheme remains highly uncertain.

The following remarks of Assistant Secretary Stender about CASPA do not indicate that the Federal government will be effective in influencing the future of state plan operations:

Stender said that the OSHA decision to investigate complaints for the first six months after a state plan is operational was a compromise between state opposition to dual jurisdiction after a state plan has been approved and a National Advisory Committee on Occupational Safety and Health subcommittee unanimous recommendation that concurrent federal-state enforcement continue throughout the three-year monitoring period.

*After the first six month period, CASPAs will be referred back to the states.*** However, if monitoring reveals that states do not respond to CASPAs, federal authority will be exercised again in that area.[39]

5.3 A Brief Description of Present State Participation

5.3.1 Number of State Inspectors

It is difficult to determine the exact number of available state inspectors which will

*Regulation 1954.3 requires review authority to be operational as well.[37]

**There is no indication as of yet that CASPAs are being referred back to the states.

be added as a result of state plan approval. In November 1973 Assistant Labor Secretary Stender estimated that 1600 additional compliance officers would be available as a result of OSHA's approval of 19 state plans. Stender also noted that these 1600 new compliance officers would provide protection for approximately 44% of the private-sector workforce.[40] In December Stender stated that the approved state plans would add 1800 state compliance officers to the overall enforcement effort.[41] As of June 1975, the federal force was made up of 759 inspectors and 174 industrial hygienists.[42] Additional federal manpower is acquired through the use of state personnel to enforce federal standards under Section 7(c)(1) agreements with the states.

Difficulties arise in estimating the actual number of state compliance officers for a variety of reasons: (1) State plans in the approval process are subject to changes in the number of compliance officers. (2) The total number of state plans currently approved or likely to be approved must be considered in estimating the number of compliance officers. (3) In figures referring to total state personnel, administrative staff must be subtracted in order to arrive at a total *available* inspection force. Despite all of these caveats, though, it is possible to assume at this writing that the number of state compliance officers available for occupational safety and health inspections is between 1500 and 1800.

5.3.2 Qualifications of State Inspectors

In June 1973 OSHA undertook a study of the statutory qualification *requirements* for nonsupervisory state compliance officers. This study, which did not attempt to evaluate the state inspectors in the field, grew out of a recommendation of the state plans subcommittee of NACOSH.

From an examination of the personnel requirements contained in submitted state plans, Charles Boyd, a project officer in the Office of Federal and State Operations, estimated that 14 of the 49 submitted plans appeared to have safety inspector qualification requirements which were *less* stringent than those of OSHA. One state appeared to have less stringent requirements for health inspectors. State qualification requirements were similar to OSHA requirements for safety in 13 states, for health in 24 states. Twelve states had safety inspector qualifications more stringent than OSHA, and 14 states had stricter health inspector qualification requirements.[43]

In light of the fact that the present state inspection capability inherited its personnel from the previous inadequate effort, these "paper" comparisons should not be relied upon as a guarantee of well-qualified inspectors manning the state activities. The American Society of Safety Engineers has unsuccessfully attempted to get OSHA to issue definitive guidelines beyond low minimum requirements for qualified safety professionals to implement the state and federal efforts.[44] An additional problem is that the inspection force continues to be heavily safety-biased, to the detriment of occupational health.

In December 1973 guidelines for the review of state plans by regional directors of NIOSH were submitted to NACOSH. The manpower resource guidelines which were suggested for the states called for a minimum ratio of one industrial hygienist to eight safety inspectors, with a preferred ratio of one to four. (The present estimated ratio is one to eight.)[45] The minimum ratio of industrial hygienists to workers was set at 1 to 100,000, with a preferred ratio of 1 to 50,000. The guidelines suggested minimum ratios of one to six for industrial health supervisors to hygienists, with a ratio of one to four preferred by NIOSH.[46]

It has been repeatedly argued by OSHA personnel concerned with state plans that the larger number of inspectors that would be available under such plans would alone ensure a more effective enforcement of occupational health and safety standards. It is difficult to understand the basis of this belief—especially in the occupational health area.

5.3.3 Status of State Participation

As of July 1, 1975, OSHA had approved the occupational safety and health plans of twenty-two states, and seven of these states had received final approval under Section 18(e) (Colorado, Kentucky, North Carolina, Oregon, Tennessee, Utah, and Vermont). Sixteen states had submitted plans which were under OSHA review and had been requested to fulfill additional requirements before plan approval. Two of these states had passed enabling legislation (Rhode Island and Guam). Formal rejection proceedings had been instituted against the New Mexico and Virginia plans, but both states were in process of submitting plan changes. Only five states had chosen not to submit plans (Kansas, Louisiana, Nebraska, South Dakota, and Ohio).

During the first six months of 1975, four states withdrew plans previously underway (New Jersey, New York, Illinois, and Wisconsin).[47] Prior to 1975, five other states had withdrawn their plans (Georgia, New Hampshire, North Dakota, Pennsylvania, and Montana). The reasons for withdrawal vary. According to Barry White, economic conditions have been a major factor in the recent withdrawals: 50% funding is no longer attractive to the states.[48] Other states have withdrawn their plans in response to intense state labor opposition.

5.3.4 Section 7(c)(1) Agreements

By March 1973 a total of eighteen states and territories had interim 7(c)(1) grant agreements with OSHA. As noted elsewhere, Section 7(c)(1) permits the Secretary to contract with the states for inspection services utilizing state personnel to enforce federal standards. The agreements cover all industry categories and utilize existing OSHA standards. State inspectors working under the agreements can be assigned to cover any area of the federal program, including target industries, target health hazards, complaint investigation, and catastrophe or fatality investigation. Funding is on a 50-50 basis, except for 100% OSHA funding of certain training expenses.

The first series of agreements, involving 157 safety inspectors and 9 industrial hygienists with total federal funding at $1,005,561 and total state funding at $873,527, terminated June 30, 1973, with an option to renew.

In addition, OSHA also entered into target health hazard inspection agreements with nine states employing 27 full-time industrial hygienists and 13 back-up hygienists. The agreements specified a total of 1584 primary inspections (not including follow-ups). This first round of agreements for target health hazard inspections terminated on June 30, 1973, and consisted of $1,053,556 in federal funding.[49]

In July 1973 OSHA announced a continuation of funding for 7(c)(1) agreements.[50] Bruce Hillenbrand, Chief of the Division of Planning, Office of Federal and State Plans, OSHA, stated that manpower data indicated that state officers working under 7(c)(1) agreements had more time for compliance activities than did federal officers because the state inspectors did not have to become involved in training, instruction, and other administrative tasks. Under the 7(c)(1) agreements, state inspectors had been attending the OSHA Training Institute in Chicago, but on February 5, 1974, it was announced in the *Wall Street Journal* that OSHA expected to terminate training of state inspectors to enforce federal standards in states whose plans had not been approved. However, after July 1, 1975, approximately fifteen states will enter into 7(c)(1) agreements whereby state personnel will conduct on-site consultation in states without approved plans (see Section 4.3.2).

5.4 ARGUMENTS FOR AND AGAINST STATE TAKEOVER

5.4.1 The Criticism

State takeover is bitterly opposed by organized labor at the national level for three reasons: (1) The national unions will not be able to exercise sufficient influence at the state level, since (a) management organizations are relatively more powerful there, (b) local unions lack the expertise and manpower to be active, and (c) grass-roots worker support for occupational health and safety issues is just beginning to be significant, but the effectiveness of local leadership does not yet compare to that of the more policy-oriented national union leadership. (2) The poor record of the states prior to the OSHAct gives support to the belief that the states will not do a good job. (3) Whatever the arguments are for state takeover with regard to safety, the states do not have the research capability necessary to tackle the more severe and prevalent occupational health problems. In many interviews, I was unable to find a single person who believed that the states would give equivalent protection in the health area. With the exception of a few states, organized labor has cause for great concern.

The Industrial Union Department of the AFL-CIO and the Illinois state AFL-CIO have filed suit to enjoin the Department of Labor from approving state occupational safety and health plans that do not meet statutory requirements. AFL-CIO President George Meany issued a statement on March 12, 1974, which expressed his support for the suit:

> For the past two years, the Department of Labor has been approving state plans which are merely unsecured promissory notes of future performance. Even on paper these plans are inadequate. They fail to provide development and enforcement of standards at least as effective as the federal law. They fail to provide necessary skilled workers to inspect and enforce, or necessary funds to administer, the programs. In short, we believe the Labor Department's approval of these state plans is illegal.[51]

The District Court for the District of Columbia dismissed the suit, upholding the challenged state plan's regulations. The case is now on appeal.

The IUD has been extremely active in its efforts to oppose the return of occupational safety and health programs to the states and the use in state plans of consultation to replace inspections and first-instance sanctions. The IUD has published a document containing recommended language for state enabling legislation for occupational safety and health programs.

The Steelworkers have also opposed the return of programs to state jurisdiction. In a union newsletter, the Steelworkers acknowledged that labor has probably lost ground in its attempt to obtain strict compliance with the "at least as effective as" test for state plans:

> Since we cannot be assured of adequate ground rules for the state plans, our negative reactions must be focused on the plans themselves. . . . [In] this effort all local unions in your district should be urged to contact state legislators to urge rejection of the pending occupational safety and health enabling legislation.

Activist groups in Chicago and Pittsburgh (CACOSH and PACOSH) are also opposed to OSHA's policies regarding the return of control to the states. CACOSH is completely opposed to the *concept* of state plans, which they view as a step backwards for workers' health. PACOSH, on the other hand, has resigned itself to the inevitability of state takeover and has sought to promote those aspects of state plans which will afford maximum protection to workers. Ironically, CACOSH saw the adoption of a state occupational safety and health plan for Illinois (since withdrawn), while PACOSH did not see a plan approved for Pennsylvania.

Ralph Nader's Health Research Group has compiled a comprehensive position paper which criticizes the state plan approach being offered by OSHA. The HRG found wide variations in the protection that state plans afford workers. Specifically, the dollar difference in each plan was analyzed by HRG:

> The ten most effective state programs spend as high as $4.83 and an average of $2.35 per worker, the ten least effective states spend approximately *1/8 of this* or an average of $.30 per worker per year. Echoing past problems with Medicaid, Mississippi, spending 10 cents per year for each worker, had the least effective state program.
>
> It is highly unlikely that this fourth [OSHA] goal of equal protection [of worker health] will be realized if states continue to spend such widely disparate amounts of money on worker health.
>
> A more rational approach would be to equalize expenditures on a per capita basis. This is supported by Department of Labor surveys showing that the accident and illness rates do not differ substantially from state to state.[52]

OSHA's own national advisory committee, NACOSH, has voiced some general criticisms of the OSHA state plan policies in a report issued February 12, 1974. A NACOSH subcommittee recommended that first-instance citations be provided for in state plans, but it did not recommend that first-instance penalties be required. It also suggested that federal enforcement in states with approved plans should continue until an initial evaluation of actual state operations has been made by OSHA. NACOSH is of the opinion that (1) federal evaluations should be made available to the public in order to elicit comment before any OSHA decisions are made on the withdrawal of federal compliance personnel from the states; (2) federal enforcement should increase or decrease in proportion to the evaluation finding and according to the extent to which states have completed developmental steps; (3) monitoring and evaluation procedures should be revised to provide for *federal* enforcement where conditions are not identical at the time of federal and state inspections or where imminent danger or a serious or willful violation exist. NACOSH, expressing concern over safety and health programs for public employees, suggested that OSHA also monitor state efforts in this area. NACOSH also noted that some states are not meeting assurances (especially promises to enact enabling legislation) given OSHA in plan-approval notices.

The states have also expressed dissatisfaction with current OSHA policy and CASPA investigation procedures. State officials at the first annual federal-state workshop on January 28, 1974, told OSHA officials that federal investigations of complaints against state programs would result in dual jurisdiction which would give rise to confusion among state legislators and employers. State officials learned that state plans will not be approved by OSHA unless enabling legislation is in place, and that states will have no input into federal policy decisions affecting all state job safety and health plans.[53]

5.4.2 The Defense

There are two points made in defense of state plans. First, they increase local control. To the extent that local control increases the responsiveness of programs to the specific needs of people in the area, this is a potentially good policy. Second, a significant increase in the number of available compliance personnel would result from state takeover. Assistant Secretary of Labor John Stender has estimated that state programs would increase the compliance force by 1500 to 1800 inspectors.[54]

5.5 CONCLUSIONS

The return of control over occupational safety and health programs to the states poses very real problems, for it was the manifest incompetence of the states in this area that led to the assertion of federal authority in the first place. Decentralization of control may well result in unequal protection and a decreased overall level of protection, a watering down of the concept of the general duty clause, and a diffusion of the efforts of labor unions to play a watchdog role and to lobby for improved working conditions.

These problems have been compounded by the mechanisms adopted by OSHA in its attempts to return control to the states. The monitoring system invests a great deal of discretionary authority in the Assistant Regional Directors. OSHA has allowed significant deviations from the "at least as effective as" criterion in the area of first-instance sanctions, has not dealt with the lack of coverage of state employees, and has permitted plans with nonexistent or inadequate state enabling legislation. Further, OSHA has not ensured that the states have an adequate number of trained industrial hygienists to enforce OSHA health standards and, in its haste to divest itself of control, has withdrawn federal enforcement prematurely in many instances.

The prospect of state takeover has been well summarized by George Meany:

> Unless this regressive process can be stopped, every American worker covered by this Act will witness a return to the situation which caused the Act to be passed in the first place—a fragmentized, weak-kneed and wavering decentralized system, impossible to police, becoming progressively more impotent as big and small business pick it to pieces at the state level. The result will be the unabated continuation of the tragic toll of deaths, injuries and illnesses to the American worker.[55]

In the first six months of 1975, four important states withdrew their plans. This action heralds a rather sharp reversal of the trend toward state takeover and leaves in doubt the future role of the states in the field of occupational health and safety.

REFERENCES

1. *The Report of the National Commission on State Workmen's Compensation Laws*, G.P.O. Document No. 477-053 (July 1972).

2. Testimony of Assistant Secretary of Labor George Guenther at the Hearings Before the Subcommittee on Environmental Problems Affecting Small Business of the H.R. Select Committee on Small Business, 92d Congress, 2d Session, on Small Business and the Occupational Safety and Health Act of 1970, June 22, 1972, p. 299.

3. Joseph A. Page and Mary-Win O'Brien, *Bitter Wages* (Grossman, New York, 1973). Franklin Wallick, *The American Worker: An Endangered Species* (Ballantine Books, New York, 1972).

4. Sally Seymour, "Forging a Partnership with the States," *Monthly Labor Review*, Vol. 96, No. 8 (August 1973), p. 28.

5. George Taylor, AFL-CIO, interview with author, June 28, 1972.

6. George Guenther, lecture at the Kennedy School of Government, Harvard University, April 26, 1973.

7. Seymour, "Forging a Partnership," p. 28.

8. Sheldon Samuels, "State's Rights or Worker's Rights," *Trial*, Vol. 9, No. 4 (July-August 1973).

9. *The President's Report on Occupational Safety and Health*, G.P.O. Document No. 4000 00288 (December 1973), p. 50.

10. *Chicago Bridge and Iron Co.*, OSHRC Docket No. 224, appeal docketed, No. 75-1163 (7th Cir., February 24, 1975). See also *Wright-Schuchart-Harbor Contractors*, OSHRC Docket No. 559, 1 OSHC 1138.

11. *Drake-Willamette Joint Ventures*, OSHRC Docket No. 117 (April 18, 1973).

12. Mary-Win O'Brien, Office of the Solicitor of Labor, U.S. Department of Labor, interview with author, Washington, D.C., August 7, 1973.

13. *Occupational Safety and Health Reporter* (the Bureau of National Affairs, Inc.), Vol. 3, No. 28 (December 13, 1973), p. 899. [Hereinafter cited as *OSHR*, BNA]

14. *OSHR*, BNA, Vol. 3, No. 27 (December 6, 1973), p. 862.

15. *OSHR*, BNA, Vol. 3, No. 9 (August 2, 1973), p. 259.

16. *OSHR*, BNA, Vol. 3, No. 26 (November 29, 1973), p. 828.

17. For the case of Oklahoma, see *OSHR*, BNA, Vol. 3, No. 24 (November 15, 1973), p. 784.

18. *OSHR*, BNA, Vol. 3, No. 1 (June 7, 1973), p. 9.

19. *OSHR*, BNA, Vol. 3, No. 25 (November 22, 1973), p. 813.

20. Mary-Win O'Brien, interview with author.

21. *OSHR*, BNA, Vol. 3, No. 30 (December 27, 1973), p. 933.

22. *OSHR*, BNA, Vol. 2, No. 40 (March 8, 1973), p. 1155.

23. Richard Wilson, Deputy to the Director of the OSHA Office of State Programs, interview with author, Washington, D.C., August 7, 1973.

24. *Ibid.*

25. *OSHR*, BNA, Vol. 3, No. 27 (December 6, 1973), p. 861.

26. *OSHR*, BNA, Vol. 3, No. 28 (December 13, 1973), p. 891.

27. *Ibid.*, p. 892.

28. *OSHR*, BNA, Vol. 3, No. 30 (December 27, 1973), pp. 932-933.

29. *Federal Register*, Vol. 29, No. 10 (January 15, 1974), pp. 1837-1839, 29 CFR 195 2 [sic]. 3, "Concurrent Authority"; Vol. 39, No. 120 (January 20, 1974), pp. 22124-22147.

30. Mary-Win O'Brien, interview with author.

31. *OSHR*, BNA, Vol. 3, No. 26 (November 29, 1973), pp. 831-833.

32. *OSHR*, BNA, Vol. 3, No. 24 (November 15, 1973), pp. 771-772.

33. *OSHR*, BNA, Vol. 3, No. 27 (December 6, 1973), p. 859.

34. *Ibid.*

35. *OSHR*, BNA, Vol. 3, No. 28 (December 13, 1973), p. 891; emphasis added.

36. *OSHR*, BNA, Vol. 3, No. 30 (December 27, 1973), p. 932; emphasis added.

37. *Federal Register*, Vol. 39, No. 120 (January 20, 1974), pp. 22124-22147.

38. *OSHR*, BNA, Vol. 3, No. 30 (December 27, 1973), p. 932; emphasis added.

39. *OSHR*, BNA, Vol. 3, No. 35 (January 21, 1974), p. 1102; emphasis added.

40. *OSHR*, BNA, Vol. 3, No. 22 (November 1, 1973), p. 710.

41. *OSHR*, BNA, Vol. 3, No. 27 (December 6, 1973), p. 859.

42. *OSHR*, BNA, Vol. 5, No. 4 (June 26, 1975), p. 125.

43. *OSHR*, BNA, Vol. 3, No. 1 (June 7, 1973), p. 9.

44. *OSHR*, BNA, Vol. 3, No. 36 (February 7, 1974), p. 1134.

45. Mary-Win O'Brien, interview with author.

46. *OSHR*, BNA, Vol. 3, No. 31 (January 3, 1974), p. 969.

47. *OSHR*, BNA, Vol. 5, No. 6 (July 10, 1975), p. 183.

48. *OSHR*, BNA, Vol. 4, No. 46 (April 17, 1975), p. 1505.

49. *OSHR*, BNA, Vol. 3 (1974), p. 1219.

50. *OSHR*, BNA, Vol. 3 (1973), p. 169.

51. *OSHR*, BNA, Vol. 3 (1974), p. 1275.

52. Sidney M. Wolfe, "Occupational Health and Safety: Present Problems, Future Prospects," speech before the American Public Health Association Annual Meeting, Atlantic City, N.J., November 14, 1972.

53. *OSHR*, BNA, Vol. 3, pp. 932, p. 1165.

54. *OSHR*, BNA, Vol. 3, No. 27 (December 6, 1973), p. 859.

55. *OSHR*, BNA, Vol. 3 (1974), p. 1275.

6 The OSHAct of 1970. III. An Evaluation of the Administrative Mechanisms

The purpose of this chapter is to provide a critical evaluation of the government agencies, actors, policies, and programs involved in implementing the OSHAct.* Sections 6.1-6.3 contain a description of the important agencies and programs. Section 6.4 is an attempt to evaluate these agencies and programs with respect to (1) their stated goals and purposes, and (2) the appropriateness of these goals and achievements for meeting the perceived needs in occupational health and safety.

6.1 RESPONSIBILITIES DESIGNATED BY THE OSHAct

The Occupational Safety and Health Act designated five federal agencies or bodies to provide the administrative, scientific, and legal support necessary to implement the Act. These five are (1) the Occupational Safety and Health Administration (OSHA), in the Department of Labor; (2) the National Institute for Occupational Safety and Health (NIOSH), in the Department of Health, Education, and Welfare; (3) the Occupational Safety and Health Review Commission (OSHRC), an independent quasi-judicial body; (4) the National Advisory Committee on Occupational Safety and Health (NACOSH), appointed by the Secretary of Labor to advise the Secretaries of Labor and HEW on matters relating to the administration of the Act; and (5) the National Commission on State Workmen's Compensation Laws, a Presidential Commission authorized to carry out a one-year study and evaluation of state workmen's compensation laws, reporting in July 1972. Of these five, four were newly established by the Act, while NIOSH drew on existing manpower and programs from the former HEW Bureau of Occupational Safety and Health.

The Act mandated certain responsibilities and functions to each of these agencies. The agencies interact in a rather intricate and interlocking system which has not resulted in a consistent or coordinated set of procedures for issuing and enforcing standards, performing research, or developing the manpower necessary to implement the Act. In an attempt to clarify the interactions of the agencies, the discussion which follows will describe the delegated responsibilities, while Sections 6.2 and 6.3 will discuss in a more detailed way the agencies' organization, manpower, budgets, and activities. An assessment of the agencies' policies and programs will be attempted in Section 6.4.

OSHA has primary responsibility for enforcement, standard setting, and the evaluation and approval of state plans; it is therefore the principal administrative unit for implementing the Act. The enforcement powers provided under the Act give OSHA two critical authorities which had been lacking from most state or federal occupational safety and health programs in the past. The first is the right to enter any workplace covered by the Act, without advance notice, for the purpose of inspecting for violations of the Act. The second is the authority to propose penalties upon discovery of violations of either standards or the "general duty clause." Under

*A similar evaluation of nongovernmental activities is to be found in Chapter 10.

federal safety programs in the past, sanctions could be proposed only for failure to correct a violation. The Act's standard-setting provisions give the Secretary of Labor, and hence OSHA, the ultimate decision-making authority for the promulgation, revocation, and modification of standards.

The Act gives OSHA, through the Secretary of Labor, the discretionary authority to approve state plans which return the responsibility for occupational health and safety to the states. The Secretary is also obligated to maintain a continuous evaluation of state plans as carried out and to withdraw approval if a state plan does not remain "at least as effective as" the federal program.

OSHA is required to issue regulations governing record-keeping of occupational injuries and illnesses by employers. The Bureau of Labor Statistics is required to compile occupational health and safety statistics based on these records.

NIOSH has the primary responsibility for conducting research, preparing criteria documents, recommending standards, and developing occupational safety and health manpower. NIOSH may also provide on-site technical consultation for employers on their health and safety problems. The Act repeatedly requires NIOSH to consult with the Secretary of Labor—in developing plans for research, in preparing criteria documents, and in conducting training and employee education programs. These requirements give NIOSH a role that is more supportive than authoritative. This is particularly true in the standard-setting area, where OSHA can either accept or reject, in whole or in part, the recommendations for standards proposed by NIOSH.

The Review Commission was assigned the single responsibility of carrying out adjudicatory functions under the Act. Three categories of cases may be brought before the Commission. An employer may contest (1) the citation, (2) the abatement period specified in the citation, and (3) the proposed penalty assessed. An employee, or a representative of employees, may only contest the length of the abatement period specified, although he may appear and testify as an interested party in all proceedings. Associated with its adjudicatory responsibilities, the Commission has the sole authority to assess civil penalties when the penalties proposed by OSHA are contested. If the OSHA-proposed penalties are not contested within fifteen working days, they are deemed to be a final order of the Commission and are not subject to review by any court or agency.

The National Advisory Committee on Occupational Safety and Health and the National Commission on State Workmen's Compensation Laws serve strictly in an advisory capacity and lack any administrative responsibilities. NACOSH was established by Section 7(a)(2) to advise the Secretaries of Labor and HEW on the administration of the Act—nothing more specific was stated in the OSHAct as to its responsibilities. Section 27(a)(2) assigned the National Commission of State Workmen's Compensation Laws the more specific task of conducting a study to evaluate state workmen's compensation laws "to determine if such laws provide an adequate, prompt, and equitable system of compensation for injury or death arising out of or in the course of employment." The Commission's task was completed in July 1972, as required by the Act, and the Commission was subsequently dissolved.

The OSHAct also provided that loans be made available to small business for the purpose of altering equipment, methods of operation, or facilities to comply with standards and duties required by the Act. The Small Business Administration (SBA) was designated to administer the loans, yet in practice the SBA must coordinate closely with the Department of Labor to be certain that if an applicant purchases specific equipment, he will be in compliance. The program got off to a very slow start, and due to the problems of coordination, only eight loans were made in the first two years after the Act's passage.[1] Activity increased following congressional hearings on the program in 1972, but the program is still underutilized: only 70 SBA loans had been made by the end of 1973.[2]

6.2 THE ORGANIZATIONAL STRUCTURE OF THE FEDERAL AGENCIES

6.2.1 The Occupational Safety and Health Administration

OSHA was established in the Department of Labor, with a new Assistant Secretary for Occupational Safety and Health appointed as the chief administrative officer. The Assistant Secretary is for all practical purposes the chief decision-maker, reporting to the Under Secretary of Labor. The Solicitor of Labor also has certain important decision-making powers (see, for example, the discussion of the walk-around right in Chapter 4).

OSHA is organized on three levels: the national office, 10 regional offices, and 54 area offices. The national office has been reorganized twice since the Act was implemented in April 1971, and a third reorganization is in process at this writing. The current structure, depicted in Figure 6.1, divides OSHA into three program areas under the direction of three Associate Assistant Secretaries. These positions are intended to focus major management responsibility on three critical areas: national programs, regional programs, and administrative programs. The information and liaison activities have been separated from the programs and are handled out of the Office of the Assistant Secretary. The three program categories and their functions are as follows:

1. National Programs (Alexander Reis, Associate Assistant Secretary):
   - *Federal Agencies:* assists in establishing federal agency plans to assure effective compliance to standards.
   - *Safety and Health Standards:* for both safety and health of workers, develops and promulgates standards and issues regulations.
   - *Training:* provides program for training Department of Labor personnel and

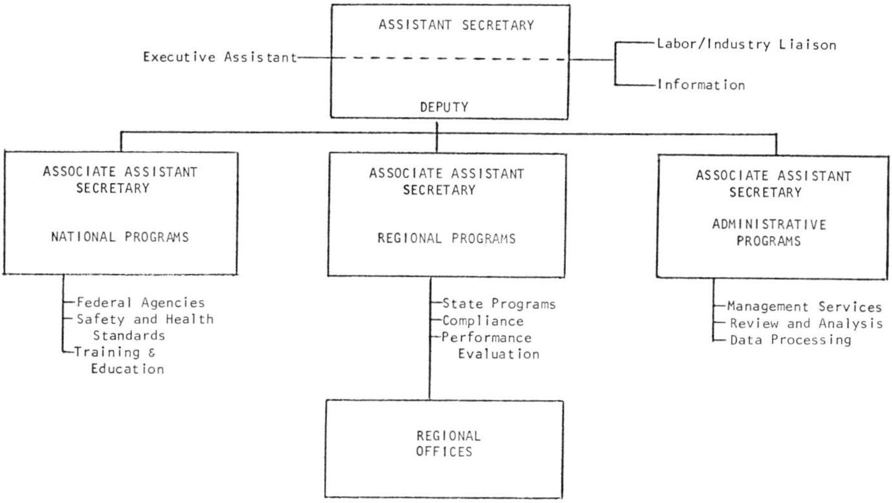

Figure 6.1 The administrative structure of the Occupational Safety and Health Administration after the reorganization of November 1973.

employers and employees to improve unsafe and unhealthful conditions in the workplace.

2. Regional Programs (Barry White, Associate Assistant Secretary):
   - *State Programs:* provides for national support in development of state compliance programs; approves same.
   - *Compliance:* assures the execution of an effective compliance program through inspection and technical assistance.
   - *Performance Evaluation:* reviews and assesses the effectiveness and efficiency of the federal and state compliance activities.
3. Administrative Programs (John Plummer, Associate Assistant Secretary):
   - *Management Services:* provides financial, personnel, and administrative services for OSHA.
   - *Review and Analysis:* performs management reviews and analysis to effect improvements in the delivery of services to OSHA clientele.
   - *ADP:* provides for data collection and the analysis and design of information systems to facilitate the management of OSHA programs.

For administrative purposes, the country is divided geographically into ten regions. Each regional office is headed by a Regional Administrator who is supported by three Assistant Regional Administrators for State Programs, Compliance, and Technical Support. Broad policy and program directives, which serve as the guidelines for the regional operations, are developed in the national office. Overall national guidelines have also been developed for the Area Directors, who are responsible for scheduling and executing investigations of complaints and regular target industry inspections in the 54 OSHA area offices.

The budget and manpower of OSHA are presented for the fiscal years 1972 to 1976 in Tables 6.1 and 6.2. The Act specifically provided that a new Assistant Secretary's position and 25 top-level management positions (GS levels 16, 17, 18) be added in the Department of Labor. These positions ensure OSHA a high-level program position within the Department of Labor. Total manpower is divided between the national, regional, and area offices. The split is approximately one-third in the national office and two-thirds in the regional and area offices. The average grade level of the OSHA staff is above GS 10.

The budget is divided between operating funds and grant monies for the states for the development and operation of state plans. The large increase in budget from 1972 to 1973 was due primarily to the increase in grant monies for the states.

6.2.2 The National Institute for Occupational Safety and Health

NIOSH was reorganized from the Bureau of Safety and Health Services (BOSH) in the Health Services and Mental Health Administration to carry out HEW's responsibilities under the OSHAct. Prior to its reorganization, BOSH was fulfilling the technical and consultative role assigned to HEW under the Public Health Services Act and the more specific mandates of the Federal Coal Mine Health and Safety Act. These functions were relatively new to BOSH, which had been reorganized seven times in the five years preceding the enactment of the OSHAct.[3]

The creation of NIOSH did not mark the end of the program upheaval, as NIOSH itself has been reorganized since April 1971. This reorganization was announced on the second anniversary of the effective date of the OSHAct, as the result of a major

Table 6.1
OSHA Budget History (Thousands of Dollars)

	FY 1971 Budgeted and Actual	FY 1972 Budgeted and Actual	FY 1973 Actual	FY 1974 Budgeted	FY 1975 Actual	FY 1976 Estimate
Standards	1062	2220	2983	4355	4802	5153
Training	455	2294	6517	4491	4874	4838
State programs	5764	7781	27,080	25,080	41,000	48,500
Federal enforcement	6068	14,275	24,161	27,811	42,177	48,050
Statistics	758	3345	4814	5141	5526	5607
Executive direction and administration	1082	5969	3719	3530	3984	3877
Total budget	15,189	35,684	69,274	70,408	102,327	116,025

Source: Gary Steele, Budget Officer, OSHA, letter to author, March 5, 1974; Ray Randlett, OSHA, conversation with author, June 22, 1975.

Table 6.2
Manpower Allocations

	FY 1971	FY 1972	FY 1973	FY 1974	FY 1975	FY 1976 Estimate
Standards	N/A	N/A	130	130	141	141
Training	N/A	N/A	146	146	146	146
State programs	N/A	N/A	85	85	N/A	N/A
Federal enforcement	N/A	N/A	1133	1274	1689	1689
Compliance officers[1]		(400)[2]	(456)[3]	(800)[4]	(933)[5]	
Statistics	N/A	N/A	51	51	51	51
Executive direction and administration	N/A	N/A	154	144	144	144
Total	970	1696	1699	1830	2184	2171

1. In a telephone interview, Gary Steele stated that the correct figures for FY 1973, 1974, and 1975 are 659, 800, and 920. The latter two figures include a projected utilization of 150 and 180 industrial hygienists, respectively. In 1972 the compliance staff numbered 638, of whom 95 were industrial hygienists. (Note that these higher figures are probably arrived at by including state manpower enforcing federal standards under 7 (c) (1) agreements.)
2. *President's Report*, 1972, p. 9.
3. *President's Report*, 1973, p. 1.
4. *OSHR*, BNA, Vol. 3, No. 36 (February 7, 1974), p. 1131.
5. *OSHR*, BNA, Vol. 5, No. 4 (June 26, 1975), p. 125.

Source: Gary Steele, Budget Officer, OSHA, letter to author, March 4, 1974; Ray Randlett, OSHA, conversation with author, June 22, 1975.

reorganization of HEW which dissolved the Health Services and Mental Health Administration. The reorganization transferred NIOSH to the Center for Disease Control, which has its headquarters in Atlanta, Georgia. NIOSH's administrative headquarters, however, remain in Rockville, Maryland.

The organizational structure of NIOSH is set out in Figure 6.2. NIOSH has six operating divisions and eight staff offices. There are two field facilities located in Cincinnati, Ohio, and Salt Lake City, Utah, an Appalachian Laboratory for Occupational Respiratory Diseases in Morgantown, West Virginia, and a national office in Rockville, Maryland. The six operating divisions are organized under two Associate Directors, and the staff offices report to the Director of NIOSH. In the discussion that follows, special attention will be focused on the activities of the starred (*) divisions and offices described below.

The Associate Director for Cincinnati Operations has responsibility for four divisions:

*1. The Division of Laboratories and Criteria Development, responsible for health and safety research.

*2. The Division for Field Studies and Clinical Investigations, responsible for industrywide epidemiological studies.

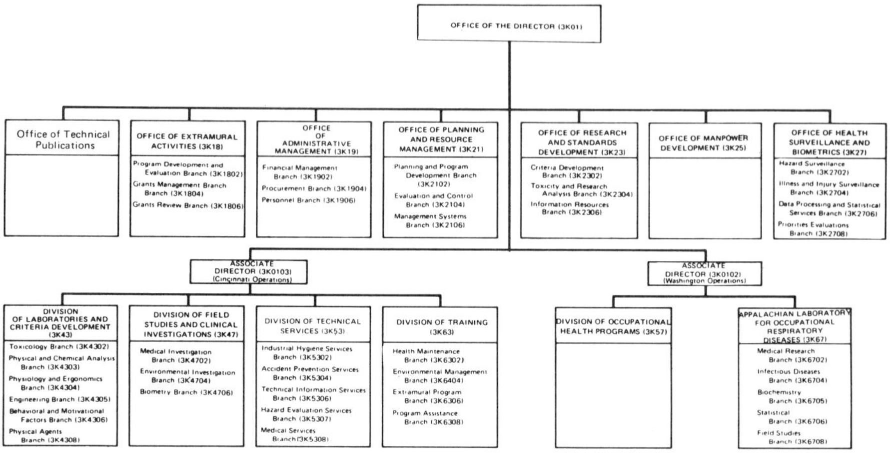

Figure 6.2 The administrative structure of the National Institute for Occupational Safety and Health.

*3. The Division of Technical Services, responsible for technical and clinical services, including back-up resources for hazard evaluations.
*4. The Division of Training, responsible for professional training programs.

The Associate Director for Washington Operations has responsibility for two divisions:
1. The Division of Occupational Health Programs, responsible for the promotion of new occupational health programs at the state and local levels and in industry and agriculture.
2. The Appalachian Laboratory for Occupational Respiratory Diseases, responsible for implementation of HEW's responsibilities under the Coal Mine Health and Safety Act (except for black lung benefits), and for testing and certifying the devices used for evaluation and control of occupational hazards.

The eight offices in the national headquarters are as follows:
*1. Office of the Director, responsible for directing and evaluating the operations of the Institute, and for coordinating its program with the health programs of other federal agencies.
2. Office of Technical Publications, responsible for public relations activities and the development and distribution of various technical reports.
3. Office of Extramural Activities, responsible for the management of research, training, and demonstration grants.
4. Office of Administrative Management, providing general office, personnel, and fiscal management services.

5. Office of Planning and Resource Management, responsible for program planning and evaluation.
*6. Office of Research and Standards Development, responsible for reviewing criteria development and toxicity research.
7. Office of Manpower Development, responsible for policy guidance and evaluation of the Institute's training and education programs.
*8. Office of Health Surveillance and Biometrics, responsible for statistical and data research on occupational injuries and illnesses, and for establishing priorities for research and standard development.

The budget and manpower available to carry out these activities are presented for the years 1972 to 1975 in Tables 6.3-6.5. The general allocation of funds for Coal Mine Health and Safety Act research is a significant portion of the NIOSH budget and accounts for more than 35% of the total research effort. The remainder is utilized for research on hazards in all other industries.

The federal government employment cutbacks instituted in 1973 required NIOSH to reduce its staff by 72 employees before fiscal 1974 to meet the lower authorizations and the employment ceiling set for the Health Services and Mental Health Administration.[4] The quality of the NIOSH staff is also suffering from NIOSH's inability to offer competitive salaries. The average grade level is about GS 7.6, roughly equivalent to a good secretary.[5] This sharply contrasts with the grade level usually given chemists and other professionals, which is between GS 12 and GS 15.

Table 6.3
NIOSH Budget History[1] (Thousands of Dollars)

	FY 1972 Actual	FY 1973 President's Revised[2]	FY 1974 Actual	FY 1975 President's Budget
Total funds authorized	26,267	25,670	31,282	27,888
Grants	4121	2852	2252	2252
Direct operations	21,085	21,548	26,877	23,578
Regional operations			884	742
Program direction	1061	1270	1269	1316
Positions				
Budget authorization	701	708	611	651
End-of-year ceiling		661	594	634

1. Includes funding under the Coal Mine Health and Safety Act of 1969.
2. The President vetoed the congressional appropriations bill, so the President's revised budget was the actual operating budget.

Source: Personal communications from various staff members at NIOSH.

Table 6.4
NIOSH Expenditures by Activity (Thousands of Dollars)

	FY 1972 Allocation	Percent	FY 1973 Allocation	Percent	FY 1974 Allocation	Percent
Surveillance	1128	4	1297	5	1474	6
Criteria documentation	918	4	1737	7	2649	10
Industrywide studies	2659	10	3108	12	3641	14
Laboratory research and services	8490	33	8442	33	9004	33
Federal Coal Mine Health and Safety Act research	6138	24	5317	21	5561	21
Technical assistance	2704	11	2599	10	2671	10
Manpower development	2919	11	1900	7	600	2
Program direction	843	3	1270	5	1204	4
Total	25,709	100	25,670	100	26,804	100

Source: General Accounting Office, Report B-163375 (September 28, 1973).

6.2.3 The Review Commission

The Review Commission is composed of three commissioners and a staff of hearing examiners. The commissioners are appointed by the President, with consent of the Senate, for six-year terms. The commissioners are required to oversee the reports and final dispositions of the proceedings before the hearing examiners. Any commissioner can request, within 30 days, that the report of a hearing examiner be reviewed by the Commission. If such a request is not made within 30 days, the report of the hearing examiner becomes the final order of the Commission. The Commission's decisions are subject to review by the appellate court.

Each case brought before the Commission for review of an OSHA citation, proposed penalty, or specified abatement period is assigned to a hearing examiner by the Chairman of the Commission. The hearing examiner presides over the case and, based on findings of fact, issues a recommended decision to affirm, modify, or vacate OSHA's citation or proposed penalty, or directs other appropriate relief.

The OSHAct initially provided ten positions in GS 16, 17, and 18 for the Review Commission. It was further specified that the hearing examiners were to receive compensation at a rate not less than that prescribed for GS 16.[6] By April 1973 the Commission had expanded to a staff of 41 judges, and Chairman Moran testified at the congressional appropriations hearings that the Review Commission needed an additional 21 judges in 1974 to meet the projected caseload. The 1975 budget request was for $5,720,000, an increase of $979,000 over the fiscal 1974 appropri-

Table 6.5
NIOSH "Professional Personnel"*

	9-1-70	9-1-71	9-6-72
Civil service personnel			
GS 16 and above	2	5	5
GS 15	3	5	8
GS 14	12	24	28
GS 13	16	26	32
GS 12	8	26	32
GS 11	13	22	26
Total	54	108	131
Public Health Service commissioned officers			
0-6 and above	13	22	21
0-5	10	13	17
0-4	19	26	32
0-3	31	38	49
0-2	13	29	56
Total	186	128	175
Total PHS and GS 11 and above	140	236	306
Total NIOSH personnel	277	453	625
PHS and GS 11 and above as percent of total NIOSH	51	52	49
GS 11 and above as percent of total NIOSH	20	24	21
PHS as percent of total NIOSH	31	28	28

*Defined as all commissioned officers and all GS personnel grade 11 and above.

Source: General Accounting Office, Report B-163375 (September 28, 1973).

ation. This increase was more than the $891,000 increase for 1974, which raised the Review Commission authorization from $3,850,000 in fiscal 1973 to $4,741,000 in fiscal 1974.[7] The budget for the first six months of 1971, in contrast, was only $75,000, and contained no provisions for hearing examiners.

6.2.4 The National Advisory Committee on Occupational Safety and Health

NACOSH consists of twelve members appointed by the Secretary of Labor and four members designated by the Secretary of HEW. The members are to be chosen "upon the basis of their experience and competence in the field of occupational safety and health." The Act specifies that the members are to be representatives of management, labor, the occupational safety and occupational health professions, and the public, with a public member serving as chairman.

While the Committee is required by the Act to meet at least twice a year, in its projected budget (filed with its charter) NACOSH "contemplated that no more than six meetings a year will be held." The budget projected $60,000 to $70,000 in annual expenses to cover travel and compensation for the members, office expenses, and staff assistance at the level of two man-years annually.

6.2.5 The National Commission on State Workmen's Compensation Laws

The National Commission on State Workmen's Compensation Laws was composed of fifteen members, "appointed by the President from among members of State workmen's compensation boards, representatives of insurance carriers, business, labor, members of the medical profession having experience in industrial medicine or in workmen's compensation cases, educators having special expertise in the field of workmen's compensation, and representatives of the general public." The Secretaries of Labor, HEW, and Commerce served as ex officio members.

The members were appointed on June 15, 1971, and their report was required by the Act to be ready by July 31, 1972–giving them only a little more than twelve months to complete the study and develop their recommendations. The report was completed on time, and the Commission ceased to exist 90 days after it was submitted.

Compensation to the members was provided for all expenses incurred and for services on a per diem basis, designated by the Act to be at the rates prescribed for GS 18. The budget also included funds for a full-time staff of thirty professionals and for contracts to carry out research and surveys.

6.3 DESCRIPTION OF AGENCY PROGRAMS

Although this section will attempt to describe, without evaluating, the various agencies' programs, the criticisms of others are included. In addition, it is recognized that statements such as "health standards were not enforced," while descriptive, are inherently critical. A more comprehensive evaluation will be presented in Section 6.4.

6.3.1 The Occupational Safety and Health Administration

Many of the responsibilities delegated to OSHA by the OSHAct have already been discussed in detail in Chapters 4 and 5. This section will present a more specific description of the activities related to (1) the setting and enforcement of standards, (2) the training and educational activities of OSHA in general and at the OSHA

Training Center in particular, (3) collection and compilation of health and safety statistics in conjunction with the Bureau of Labor Statistics, and (4) OSHA's role in the health and safety effort for the employees of federal agencies. OSHA's responsibility for effecting the return of authority to the states was discussed in Chapter 5.

OSHA set out its priorities for its third year of operation at a press briefing on May 3, 1973. While no major changes in enforcement were announced, there was increased emphasis on training, development of health standards, and use of enforcement personnel. Chain Robbins, then Deputy Assistant Labor Secretary, announced a new program emphasis on pushing criteria documents into health standards under the Section 6(b) standard-setting provision of the Act. Robbins also said that OSHA was dissatisfied with the rate of inspections and would take action to improve inspection coverage.

The Development of Standards

In commenting on the purpose of the OSHAct, George C. Guenther, former Assistant Secretary of Labor for Occupational Safety and Health, has written that it "is essentially a labor standards law. Being such, its heart is the development and enforcement of safety and health standards."[8]

The allocation of funds for standards development in fiscal 1972-1975 was approximately \$2.22, \$2.98, \$4.36, and \$4.80 million out of a total budget of \$35.7, \$69.3, \$70.4, and \$102.3 million respectively. The proposed fiscal 1976 allocation is \$5.15 million out of a total budget of \$116.0 million. In a recent policy analysis of occupational safety and health, it was reported that

> since fiscal 1972, OSHA has authorized 130 positions for its standards-activity staff. In actuality, however, the office of standards has a staff of between 80 and 90 people, about half of whom are professionals. Most of the increases in the standards budget have been earmarked for an increased number of standards advisory committees.[9]

As will be discussed below, most of the standards adopted by OSHA have been safety standards. A detailed description of the three types of standards—initial (interim) standards, permanent standards, and temporary emergency standards follows.

Standards Promulgated under Section 6(a) On May 29, 1971, OSHA promulgated an initial standards package of 250 pages which incorporated existing federal safety and health standards (see Section 4.2). Also included were national consensus standards previously adopted by the National Fire Protection Association and the American National Standards Institute.

This initial package contained mostly safety standards, but it also included 400 so-called threshold limit values as health standards. Of the 120 ANSI consensus standards contained in the OSHA initial standards package, only 20 can be considered health standards to the extent that they set exposure limits on 20 hazardous substances.[10] As mentioned in Section 4.3.1, the carcinogen standards included under Walsh-Healey were *not* included in the package. OSHA preferred to develop its own standards for carcinogens.

While OSHA did move rapidly to promulgate the initial package within one month after the effective date of the Act, many of the standards involved were obsolete or not related to workers' health or safety. According to then Deputy Assistant Secretary Chain Robbins and then Office of Standards Director Gerald Scannell, one of OSHA's first priorities has been to analyze standards currently on the books. Robbins has stated that "before we rush into any additional consensus standards or any brand-new standards, we will make the necessary clean-up of the existing standards."[11] According to the 1973 *President's Report*, by the end of 1972, "more than 200 new or existing standards had been adopted or revised."[12] A careful examination of the facts reveals that only about 150 changes were actually made in the standards. All but 14 were clarifications, revisions, amendments, or revocations of the initial safety standards.[13]

OSHA has set a goal of eliminating those standards which have, in Scannell's words, "only a tenuous relationship to occupational safety and health. OSHA seeks to retain only those standards which reduce injuries, illnesses and fatalities in the workplace."[14] Some of the initially adopted consensus standards did not meet this criterion, such as the standard prohibiting ice in drinking water (developed in the days when ice was chipped out of polluted rivers).

These initial standards have been criticized not only for being tenuously related to job health and safety, but also for being both incomprehensible and inaccessible.[15] While there is some truth to the latter two criticisms, compliance with the standards need not mean that each firm must purchase $4000-$6000 worth of relevant ANSI standards,[16] or that every responsible person must understand every standard. The OSHAct represents an attempt to establish compliance with standards which will control satisfactorily the technology in the workplace. Access to relevant technical information *and* expertise is a necessary practical requirement for a firm's compliance with the Act—just as is the case with environmental pollution control requirements.

In an analysis (based on an unpublished OSHA "Briefing Book") of the initial standards package by the Government Research Corporation, the following observations were made:

Sub-Part G: Occupational Health and Environmental Control. This sub-part is perhaps the most important in terms of long-range consequences and is by far the least developed. For all practical purposes, these standards will demand the greatest share of NIOSH's medical, scientific and technical research efforts.

The purpose of these standards is "to provide protection from occupational diseases resulting from exposures to chemical contaminants and physical agents. . . ."

The initial package contained standards on:

1. air contaminants (these are the TLVs, covering about 450 substances, which were originally contained in the Walsh-Healey Act);
2. ventilation;
3. occupational noise exposure;
4. ionizing radiation; and
5. nonionizing radiation.

To these standards has been added the asbestos standard promulgated under Section 6(b) on June 7, 1972.

In commenting on these standards (except for asbestos), the briefing book stated that there were no:

- satisfactory or available coverage of requirements for local exhaust and general ventilation systems;
- available standards covering the control and use of coherent light (lasers) or incoherent light;
- existing standards covering heat stress;
- available standards dealing effectively with infra-noise or ultrasonics;
- standards on the use of carcinogenic materials.

The briefing book did not review the adequacy of the TLVs, since its primary concern was an analysis of existing Federal and national consensus standards to be promulgated under 6(a) authority. In any case, *it is highly doubtful that those who produced the book had the scientific or medical capabilities to judge the adequacy of the TLVs.*[17]

Standards Adopted under Section 6(b) Until April 1973 OSHA continued to issue safety standards primarily through the consensus mechanism provided for in Section 6(a). While OSHA has often expressed its intent to give top priority to the issuance of health standards, it has so far promulgated (through 6(b) proceedings) only three permanent health standards, those for asbestos, 14 carcinogens, and vinyl chloride.

OSHA has relied heavily upon standards advisory committees in the 6(b) process. There are two types of advisory committees: ad hoc and standing. Ad hoc committees, such as the asbestos committee, are created to deal with a single problem and are dissolved when a solution to the problem is found. Standing committees, such as the agriculture or construction standards committees, are formed to address problems in particular areas or industries. The 1973 *President's Report*, following the mandates of Section 7(b) of the Act, states that the

> committees generally consist of 15 members and are designed for balance. A typical committee will have an equal number of employer and employee representatives, representatives of State agencies, technical or professional representatives chosen for their specialized knowledge, at least one member appointed by the Secretary of Health, Education, and Welfare, and, when another Federal agency is interested in the standards under development, at least one representative of that agency. Com-

mittees are sometimes broken down into subcommittees to develop recommendations for specific standards.[18]

Gerald Scannell, former director of the OSHA Office of Standards, has expressed the intention of the Department of Labor to utilize consensus organizations in lieu of advisory committees for the development of standards to be promulgated through the procedures set forth in Section 6(b) of the Act.[19] The use of the consensus organizations in an advisory rather than consensus function is said to be motivated by OSHA's desire to respond more rapidly to the needs in the health area.

The designation of the American Society for Testing of Materials as a consensus organization for this purpose has evoked strong criticism from labor groups. The AFL-CIO has stated that–

1. ASTM does not have expertise in the health area;
2. the consensus approach does not lend itself to the development of health standards where scientific precision, not the political consensus of interest groups, is needed;
3. ASTM is business-oriented and has never asked for labor participation.[20]

OSHA officials have estimated that at least four months are required to process under 6(b) a comprehensive recommended standard received from NIOSH. In fact, it takes much longer. The steps "required" include

1. requesting recommendations from an advisory committee;
2. reviewing and evaluating the recommendations received;
3. publishing the proposed comprehensive permanent standard in the *Federal Register*;
4. considering objections to the proposed standard and possibly holding a public hearing on such standards; and
5. issuing the comprehensive permanent standard.[21]

The use of advisory committees in part explains the fact that so few health standards have been adopted under 6(b) proceedings. On May 27, 1972, OSHA stated its intention to establish advisory committees on carbon monoxide, cotton dust, lead, silica, toxic metals, carcinogens, toxic chemicals, halogenated hydrocarbons, heat stress, noise, fibrous glass, hazardous material labeling, and radiation.[22] In January 1973 OSHA announced that advisory committees had been appointed for noise and heat stress. At that time the committees on beryllium and carbon monoxide were still being appointed. Approximately six months elapsed from the time OSHA received NIOSH documents until appointments to advisory committees were made. The advisory committees themselves are taking a very long time to complete their tasks. During the congressional appropriations hearings on the fiscal 1973 budget, it was revealed that the bulk of the $600,000 increase OSHA received for standards development actually went to support the increased number of standards advisory committees.[23]

While the use of advisory committees seem to be partly responsible for the delay in issuing standards under 6(b), they are important in that they provide the only mechanism for equal participation of labor and management in the setting of health standards. Participation in the consensus organizations is difficult for labor because it neither has sufficient expertise nor is prepared to pay experts to "volunteer" in the consensus process. Industry already has its experts on board. In contrast, participation in the OSHA advisory committees is remunerated by OSHA.

No doubt, the difficulties involved in selecting individual committee members are somewhat responsible for the delay in appointing advisory committees after the receipt of a NIOSH recommendation for a standard—and also explain why OSHA would prefer to use the (management-oriented) consensus organizations, which have everything to gain by helping OSHA in setting standards.

The use of advisory committees is not the only reason for the delay in the issuance of permanent health standards. The carcinogen standards were delayed by OSHA itself which opposed the recommendation for a use-permit system made by its advisory committee. The history of the pesticide standard reveals other kinds of delay (see Sections 4.4.2 and 12.3).

It is worth repeating that the three new permanent health standards adopted—asbestos, the 14 carcinogens and vinyl chloride—were all processed through the 6(b) mechanism. Court action played an integral part in determining the final form of these standards.

Temporary Emergency Standards Adopted under Section 6(c) and the Subsequent Adoption of Permanent 6(b) Standards There have been four 6(c) emergency temporary standards issued: for asbestos, the 14 carcinogens, pesticides, and vinyl chloride. These emergency standards were issued only after petitions by labor and public-interest groups (see Sections 4.3.1 and 4.4.2). While Section 6(c)(3) requires OSHA to promulgate a permanent standard "no later than six months after publication of the emergency standard," only two comprehensive standards (asbestos and vinyl chloride) were adopted within approximately six months. In the case of asbestos, NIOSH officials have noted that it would not have been possible to develop a scientifically accurate and comprehensive recommended standard within six months, had not the necessary research on the toxicity of asbestos already been done prior to the OSHAct.[24] For the carcinogens, a period of nine months passed from the time an emergency temporary standard was adopted until a 6(b) permanent standard was issued. In the case of vinyl chloride, NIOSH turned all of its attention to generating a recommended standard (delivered March 11, 1974), and in doing so necessarily delayed other critical tasks.

The General Accounting Office has recommended that the Act be amended to allow "more than the six months now specified in the Act to promulgate a permanent standard after issuing an emergency temporary standard."[25] (Any such amendment should allow a longer period for the temporary standard to remain in effect, because after the six months expires it is likely that no protection exists if OSHA has not issued a permanent standard.)

OSHA has not taken initiative on its own to adopt temporary emergency standards (or permanent health standards). Although it has been estimated by the Director of NIOSH and by academic researchers that there are 1000 to 2000 toxic agents and substances whose potential harmful effects are serious enough to warrant federal standards, and although a priority list of 113 toxic substances and harmful physical agents has been drawn up by NIOSH, OSHA has taken no further action.[26] Of course, it is difficult to proceed without a firm data base, but temporary emergency standards could be utilized by OSHA to stimulate the presentation of industry-held data and to foster discussion, instead of waiting until workers awaken to a danger already known by some scientists or industries. The tragedy of the asbestos workers in Tyler, Texas, presents a vivid example of how such situations should not be handled (see Section 4.3.2 and the series of articles by Paul Brodeur published in *The New Yorker* on October 29-November 26, 1973).

OSHA did not move quickly in adopting either a temporary emergency standard or a permanent standard after learning of the death of workers at Goodrich Rubber from exposure to vinyl chloride (which was in the eleventh most serious category on NIOSH's priority list).[27] Although Marcus Key, then Director of NIOSH, indicated that no safe level could be established for vinyl chloride and recommended that work practices be expeditiously adopted which would reduce exposure below measurable limits, OSHA's temporary emergency standard recommended an exposure level no higher than 50 ppm.[28] Six months later OSHA promulgated a standard of exposure no higher than 1 ppm averaged over an eight-hour day. These delays are explained by the complexity of the standard-setting process, lack of close coordination with NIOSH on health problems, and a desire on the part of OSHA that the standard hold up in court.

The Economic Impact of Standards The OSHAct specifies in Section 6(b)(5) that

> The Secretary, in promulgating standards dealing with *toxic materials or harmful physical agents* under this subsection, shall set the standard which most adequately assures, *to the extent feasible*, on the basis of the best available evidence, that *no* employee will suffer material impairment of health or functional capacity even if such employee has regular exposure to the hazard dealt with by such standard for the period of his working life. (Emphasis added)

The word "feasible" is open to interpretation. It was probably intended to mean *technologically* feasible, but it has been administratively interpreted by OSHA to mean *economically* feasible. Furthermore, the court decision on the asbestos standard gives OSHA authority to consider economic feasibility in setting standards.[29] Also, since OSHA must consider the economic impact of its standards, as required by the National Environmental Policy Act (NEPA), the consideration of economic feasibility was probably inevitable in its formulation of standards.

Part of OSHA's concern over the economic impact of proposed standards is in response to pressure from the Office of Management and Budget.[30] While OSHA's Chain Robbins has stated, "We can't willy-nilly promulgate standards without regard to economic factors," he has also said that he doesn't "like any approach which trades dollars for lives." Robbins further stated that while better standards would, necessarily, dictate increased costs, they would also result in economic benefits to employers, such as lower workmen's compensation costs and less production time lost due to injuries. Robbins concluded that OSHA's standards-setting job was to protect workers and to ensure that the standards are economically feasible. To this end, OSHA has contracted with Arthur D. Little, Inc., and Systems Technology Corporation for economic impact analyses.[31]

According to the 1973 *President's Report*, OSHA's economic impact studies "are not traditional cost effectiveness evaluations. They contain inherent limitations—the difficulty of economic prediction; the paucity of reliable injury and illness data; the understandable inability to objectively quantify pain, suffering, and loss of bodily function and loss of life. Impact studies are, nonetheless, valuable safeguards against adopting standards that might otherwise result in unnecessarily high compliance costs to employers."[32] (See Chapter 7 for a further discussion of the use of cost-benefit studies.)

In response to industry complaints about the difficulty of complying with standards as they have been written, OSHA has stated its intention to review existing standards to determine if any are unreasonably specific or rigid. Robbins has stated that OSHA is committed to the development of performance rather than specification standards and will, if possible, modify existing standards to reflect performance criteria. In further deference to employer complaints, OSHA has issued a set of program directives to help ensure that standards are uniformly interpreted and applied throughout the country.[33]

Applicability of Standards In 1972 then Assistant Secretary Guenther rejected suggestions that standards be developed specially for each industry (see Section 4.3.3), but OSHA has since undertaken a policy of developing guideline standards for 40 or 50 key industries.[34] OSHA has just begun to develop standards applicable to problems in special industries, e.g., agriculture. Standards adopted under 6(a) for the construction and maritime industries had been formulated previously through the consensus method, and federal standards already existed for the mining and atomic energy industries.

The Enforcement of Standards

Due to its limited manpower resources, OSHA has pursued a goal of encouraging "voluntary compliance" on the part of employers. One aspect of OSHA's voluntary compliance policy is to enforce health and safety standards in a selective and effective manner so that the fear of being inspected will act as an incentive for the uninspected employers to improve their worksites. The analogy of the Internal Revenue Service auditing tax returns in order to secure honest reporting of income by wage earners is often cited as a successful example of encouraging voluntary compliance.*

To be successful, "voluntary compliance" requires–

1. understandable and accessible standards and regulations;
2. an inspection capability large enough to present a significant chance of the employer being randomly inspected;
3. significant penalties on being found in violation;
4. well-qualified inspectors; and
5. employees well-informed about workplace hazards and their rights of protection under the law.

In addition, if health hazards are to be eliminated, "voluntary compliance" with the *general duty* obligation is crucial.

Insofar as OSHA cannot police the four million establishments covered by the Act, its enforcement activities must be judged ultimately not only by the inspections it does make, but also by the extent to which its programs foster voluntary compliance.

Inspection Priorities OSHA has established priorities for the inspection of workplaces. The initial inspection priorities were: (1) inspections involving reported accidents and catastrophes; (2) inspections in response to employee complaints; (3) special inspection programs; and (4) random inspections of workplaces covered under the OSHAct.

OSHA Program Directive 200-18, which became effective on July 1, 1973, revised the inspection priorities of the agency. Under these new priorities, OSHA will first inspect in response to employee complaints or other allegations of the existence of *imminent-danger* situations. Second priority is given to catastrophe and fatality investigations. Third priority goes to investigations of "valid" complaints. Fourth on the priority list is investigation of situations covered by special emphasis programs, (Target Industries, Target Health Hazards, and Trenching and Excavation). Random inspections of a cross section of industries now ranks lowest on OSHA's priority list.

The program directive spells out specific procedures for investigating inspections. For example, investigation of fatalities or catastrophes will be scheduled if the occurrence results in five or more fatalities, five or more hospitalizations, or *significant publicity*, or if it indicates a history of such events.

Table 6.6 shows the allocation of inspections carried out under OSHA's original priority list. In a sense, the overall figures are not very meaningful, since enforcement manpower utilization is different in some particularly hazardous industries.

*Note that the term "voluntary compliance" is used slightly differently in Europe. There it refers to the system of the inspector warning the employer to abate a hazard–i.e., allowing him a chance to comply voluntarily–before issuing a citation or penalty (see Chapter 11).

Table 6.6
OSHA's Inspection Priorities

Activity	% of Initial Inspections[1]	OSHA Manpower Hours Allocation Policy[2]	% of All OSHA Violations[3]
Reported accidents	8	10	
Employee complaints	13	30	
Target industries	40	35	25
Target health hazards	3		
Random inspections and others	36	25	

1. Period: July 1971 to January 1973.
2. OSHA's policy regarding the allocation of manpower hours as set forth in the OSHA *Compliance Manual* (see Section 4.3.3).
3. Period: July 1972 to January 1973.

Source: Joe Collier, "Inspection and Enforcement in the Workplace," *Monthly Labor Review* August (1973), pp. 35-42.

For example, based on a (possibly poor) sample of members of the National Safety Council Sections on Rubber and Plastics, Chemicals, Petroleum, and Research and Development, it was noted that 70% of the inspections were triggered by employee complaints (as opposed to 13% overall), and 30% were a result of random inspections (equal to the 30% overall).[35] The sections mentioned represent industries whose injury and illness incidence rates per 100 full-time workers are 19.3, 10.9, and 10.9. (There is no illness incidence rate for Research and Development.) The high number of complaint-triggered inspections probably results from union activity, but I have not been able to confirm this.

As part of its inspection priorities, OSHA has established three special enforcement programs to concentrate on particular safety and health issues: the Target Industries Program, the Target Health Hazards Program, and the Trenching and Excavation Program. The overall inspection is concentrated almost entirely on safety violations as opposed to health violations. (These programs are now being revised.)

Target Industries Program The Target Industry Program (TIP), which went into effect in mid-1971, focuses attention on five industries: marine cargo handling, roof and sheet metal work, meat and meat products, miscellaneous transportation equipment (primarily mobile home manufacturing), and lumber and wood products. These industries had injury rates twice as high as the average rate for manufacturing industries (which, in 1970, was 15.2 injuries per million man-hours worked). Other factors considered by OSHA in selecting industries for enforcement concentration were the number of employees affected, the geographic distribution of the industry across the country, and the existence of standards which, if enforced, could significantly reduce injuries. Relevant data for the TIP industries are shown in Table 6.7.

Included as part of the program were OSHA efforts to stimulate voluntary com-

Table 6.7
The Target Industries Program

Industry	Million Manhours Worked[1]	Number of Employees[1]	Injury Frequency Rate (Injuries per Million Man-hrs Worked)[2]	% of TIP Inspections in FY 1972[1]	% of TIP Inspections in FY 1975[3]
Marine cargo handling	69.9	112,000	69.9	44	22.5
Roof and sheet metal work	43.0	115,000	43.0	8	16.5
Meat and meat products	43.1	356,000	38.5	10	14.9
Misc. transportation	33.3	116,000	37.6	5	6.6
Lumber and wood products	34.4	596,000	36.1	33	39.5
Total	250.7	1,295,000			

1. From Joe Collier, "Inspection and Enforcement in the Workplace."
2. From "The Target Industries–Profiles of Five Hazardous Occupations," U.S. Department of Labor, OSHA, Publication No. 2034.
3. OSHA Report No. 1, Supplement, "Summary of Basic and Target Industry Enforcement Activity," computer output, May 19, 1975, for the period July 1, 1974 through May 1, 1975.

pliance through interaction with trade associations and employee organizations, development of safety and health programs within the industries, and dispersion of information to employers and employees.

In the first six months of inspections under the Target Industries Program, 45% of the workplaces in these industries which were inspected had no violations. In 1972, 14,830 inspections (41% of the total number of OSHA inspections) were conducted under the Target Industries Program. Almost all TIP firms with 20 or more employers were inspected at least once, and 25% of all OSHA citations were issued to the target industries. In 1972 OSHA made 7(c)(1) arrangements with 13 states specifically to inspect TIP industries, resulting in the participation of 79 additional inspectors.[36]

Target Health Hazards Program OSHA has chosen to concentrate its health hazard enforcement efforts on five substances: asbestos, cotton dust, silica, lead, and carbon monoxide. These substances were chosen for inclusion in the Target Health Hazards Program for three reasons: they were deemed to be serious health hazards by NIOSH; they can all be effectively measured and monitored; and large numbers of employees are exposed to them (asbestos, over 200,000 workers; cotton dust, 800,000; silica, 1.1 million; lead, 1.6 million; carbon monoxide, unknown but high).[37] A high probability of death or serious impairment of health exists for these substances. For example, it has been estimated that of the 500,000 workers now or previously exposed to asbestos, approximately 100,000 will die of lung cancer, 35,000 of abdominal or chest cancers (mesotheliomas), and 35,000 of asbestosis (scarring of the lungs).[38]

The five target hazards were included in the TLV list originally adopted under Section 6(a) of the OSHAct. The TLVs specify the maximum allowable exposure of employees to these substances. The concentration levels are eight-hour time-weighted averages based upon samples collected by instruments worn by a few employees. A later standard for asbestos, based on a NIOSH recommendation, is more stringent. Currently, both time-average (5 fibers per cc) and maximum-ceiling (10 fibers per cc) standards exist for asbestos. The maximum-time-average standard will be reduced to 2 fibers per cc in 1976.

Health standards include not only concentration requirements but also requirements concerning work practices. Wherever feasible, engineering controls are to be used to minimize employee contact with health hazards. Isolation of employees from the hazards, enclosure of equipment, exhaust ventilation, and dust collection are acceptable engineering controls. Interim measures, such as rotation of employees in and out of hazard areas and the use of respirators, can be used while permanent protection devices are being installed. Special monitoring devices, protective clothing, and regular employee medical examinations are required for employees exposed to asbestos.

During the first half of fiscal 1973, OSHA made approximately 1400 inspections under the Target Health Hazards Program. Almost 500 of these inspections were conducted by state inspectors under OSHA's 7(c)(1) agreements with states. Total inspections under the Target Health Hazards Program represent a mere 3% of total OSHA inspections,* yet over 50% of all employee complaints received by OSHA prior to the January 1972 commencement of the program were related to occupational health issues. During the second seven months of the program, over 35% of all employee complaints were related to health issues. Since the general inspections performed by OSHA essentially ignore health standards, the Target Health Hazards Program is a necessary corrective measure. Inspections made under the program resulted in the issuance of citations in 65% of the inspections.[39]

Trenching and Excavation Special Program On March 20, 1973, OSHA launched a special program of inspections to reduce the number of fatalities and injuries incurred in trenching and excavation work. Chain Robbins stated that OSHA chose this area for a special program because the deaths and serious injuries incurred by workers in these activities were the "most needless and preventable." While OSHA's standards are adequate to reduce injuries, they are not being followed by contractors, and thus OSHA must do more to ensure compliance. A key part of this pro-

*During the period July 1974 to May 1975, 1831 inspections were made under the program (2.8% of the total number of inspections).

gram involves the training of construction supervisors in the hazards of trenching and excavation operations. Thomas Brown has stated that OSHA intends to investigate every fatality and injury incurred in trenching accidents.[40]

Inspection Profile In the nineteen months from July 1971 to January 1973, OSHA conducted initial inspections of over 45,800 workplaces, representing only about 1% of the total number of establishments covered by the OSHAct. Approximately 8.5 million workers, 14% of the employees covered by the Act, were employed in these workplaces.[41]

From its inception on April 28, 1971, through January 1975, OSHA made a total of 206,163 inspections resulting in 140,467 citations alleging 724,582 violations with proposed penalties totaling $18,186,627.[42]

Data for the first eleven months of fiscal 1975 are presented in Table 6.8. The bulk of OSHA inspections in this period were conducted in workplaces that had never before been inspected; however, these workplaces employed only 40% of the 11.5 million workers covered by the inspections. A comparison of enforcement activities during OSHA's first nineteen months and the first eleven months of fiscal 1975 indicates that the inspection coverage is increasing. In the latter period, OSHA inspected workplaces employing 36% more workers in 42% less time than in the former period.

Table 6.9 presents a profile of OSHA inspections by size of workplace. These data show that the probability of being inspected is relatively greater for larger workplaces. In 1972, 29% of all inspections were conducted in workplaces with fewer than 25 employees; these workplaces employ 30% of the workforce and represent approximately 90% of all workplaces covered by the Act. Data for fiscal 1975 show almost exactly the same profile; moreover, in the third quarter of the year over 75% of all workplaces with over 500 employees were inspected, while only 1.7% of workplaces with fewer than 25 employees were visited.

A breakdown of OSHA enforcement activities by industries is presented in Table 6.10. The relative distribution of activity among the industries has remained relatively constant. (The seeming relative decrease in maritime inspections is the result of a change in OSHA's record-keeping procedures. While previously OSHA counted a visit to each ship as a separate inspection, current practice is to count a visit to a port as a whole as a single inspection.)

In 1974-1975 approximately 75% of OSHA's initial inspections took place in the manufacturing and construction industries. The manufacturing industries alone accounted for 72% of the violations cited by OSHA compliance officers.

The data of Table 6.11 indicate that the rate at which OSHA has been conducting initial inspections has been increasing. Almost twice as many initial inspections per month were conducted in 1974-1975 as compared to 1971-1973. This difference is also reflected in the rate of follow-up inspections to determine the progress being made in the abatement of cited hazards.

Table 6.8
OSHA Inspections in the First Eleven Months of Fiscal 1975

	Number	% of Total
Workplace Inspections		
New-initial*	41,218	63
Repeat-initial	12,021	18
Follow-up	12,568	19
Total	65,807	
Employees in Workplaces Inspected	4,635,734	40
New-initial	4,635,734	40
Repeat-initial	5,079,124	44
Follow-up	1,846,364	16
Total	11,561,222	
Inspections by Type		
Accident	1,592	2
Complaint	5,771	9
General schedule	45,876	70
Follow-up	12,568	19
Total	65,807	

*An "initial inspection" is the first of an inspection sequence. "New-initial inspections" are of workplaces never before inspected, while "repeat-initial inspections" are of workplaces that have been visited at some previous time.

Source: OSHA Compliance Activity Report, May 19, 1975, p. 1.

Table 6.9
Distribution of OSHA Inspections by Size of Workplace

Number of Employees in Workplace[1]	% of Such Workplaces Covered by the OSHAct[2]	% of Total Inspections July-December 1972[2]		% of Workplaces in This Class Inspected during the 3d Quarter of Fiscal 1975[3]
1-9	77	13		
10-24	14	16	47	1.7
25-49	5	18		
50-99	2	19		
100-249	1	18	45	29.7
250-499	0.5	8		
500+	0.5	8		76.9

1. Number of employees at a given worksite, not the size of the overall business (which may have many separate worksites).
2. From the 1973 *President's Report on Occupational Safety and Health.*
3. From the OSHA Compliance Activity Report, May 19, 1975, p. 19.

Violations and Penalties Proposed by OSHA As of January 1973 over 167,300 violations had been alleged by OSHA. The bulk of these violations were found in the manufacturing industry (72% of all violations in the latter half of 1972). Table 6.12 indicates that as of January 1973, approximately 98% of all alleged violations were considered by OSHA to be of a nonserious nature. In 1975 the distribution of alleged violations was almost identical to the distribution in January 1973, but a larger percentage of penalties being proposed were for violations of an imminent-danger, wilful, or repeated nature.

Although it is difficult to measure the quality of inspections from the statistics compiled by OSHA, there is some evidence that the inspectors are becoming more thorough. In 1971-1973 OSHA inspectors found an average of 3.23 violations per inspection; by 1975 their rate has increased to 3.82 violations per inspection. Over this period, however, the number of citations issued per violation found remained almost constant at 0.2 (one citation for every five violations found).[43]

In the period July 1971 to January 1973, the average penalty proposed for serious violations was $625, while that for nonserious violations was $45. Major dollar differences in penalties existed across industries for wilful, repeated, and imminent-danger penalties. As shown in Table 6.10, penalties for such violations averaged $9500 in the construction industry, $2000 in the maritime industry, and $400 in the manufacturing industry.

Enforcement of the General Duty Obligation

The first citation and penalty that OSHA issued for a serious violation occurred on

May 28, 1971, and was based upon the general duty clause. This inspection and citation was the result of an employee's complaint concerning an imminent-danger situation. Significantly, this violation of the general duty clause was for an occupational health hazard involving excessive concentrations of airborne mercury in a Moundsville, West Virginia, chlorine plant. OSHA cited the employer for exceeding the TLV level for airborne mercury, even though the violation occurred during the 90-day grace period before the TLVs were to be adopted and enforced as standards. (It appears that the employer did not appeal the citation; however, a later finding of the Review Commission did concur in a general duty citation based upon a concentration of inorganic lead in excess of the TLV level which was about to become effective as a standard.)[44]

Of the first sixty cases filed before the Review Commission, more than half were based on alleged violations of the general duty clause. While this figure showed some initial promise that the general duty clause would be used to protect workers' occupational health, the trend was curtailed by subsequent Review Commission decisions. The Review Commission first limited the applicability of general duty to those cases where a standard did *not* exist. In a later case it further limited the applicability of general duty clause by imposing a substantial burden of proof upon OSHA to show what the employer could have done to avoid the condition (see Section 4.3.2).

Use of the general duty clause has also been reduced as a result of a change in policy by OSHA. In a law review article, Earl Heath, OSHA Director of Training and Education, has quoted former Assistant Secretary Guenther as saying:

> The general duty clause was used only in serious cases involving recognized hazards where it seemed OSHA's clear duty to issue a citation and where, at that time, there was no effective standard. The use of the general duty clause, prior to the effective date of the standards [August 27, 1971] was a conscious one on our part to meet critical program needs at that time, but under no circumstances should be interpreted as an indication that it is our general intention, now that the standards are effective, to continue issuing citations based on the general duty clause. The general duty clause shall be used only in cases involving alleged serious violations where (1) no standard exists and (2) where the specific and stringent requirements of the general duty definition [involving "recognized" hazards and "death or serious physical harm"] are met.[45]

Some members of OSHA's field staff choose not to cite violations of TLVs, though they are legally enforceable, because they do not regard them as good standards. Furthermore, a violation of the general duty obligation cannot be cited if a standard exists. Thus, control of those 400-odd substances for which TLVs exist is generally inadequate. Where no TLV exists, the burden of proof required to prove a violation of general duty is now more difficult, so serious health hazards will continue to be ignored by OSHA, except in the small Target Health Hazards Program

Table 6.10
OSHA Enforcement Activity by Industry

Industry Group	% of Initial Inspections 1971-1973[1]	1974-1975[3]	% of Total Accident Investigations 1972-1973[2]	1974-1975[3]
Manufacturing	44	42	35	31
Construction	23	32	35	36
Maritime	23	3	8	0.5
Service, Financial, Mercantile, Agricultural, Transportation	10	23	22	28

1. Period: July 1971 to January 1973. From Joe Collier, "Inspection and Enforcement in the Workplace."
2. Period: July 1972 to January 1973. From Joe Collier, "Inspection and Enforcement in the Workplace."
3. Period: July 1974 to May 1975. From the OSHA Compliance Activity Report, May 19, 1975.

effort and hopefully in the cases of asbestos, the 14 carcinogens, and vinyl chloride for which new standards have been developed.

Educational and Informational Activities and the Voluntary Compliance Goal

As previously mentioned, OSHA has pursued a policy of encouraging employers to comply voluntarily with OSHA standards. Informing both employers and employees is a necessary prerequisite for voluntary compliance to work.*

In March 1973 Earl Heath set forth the policies of the OSHA education and information programs. Top priority for OSHA in dealing with employers is to increase their knowledge of and acquaintance with the OSHAct and the Department of Labor's programs. Of secondary importance are efforts to train employers in the means of voluntary compliance with the Act. OSHA also seeks to acquaint employees with the Act and its administration. Second priority with regard to workers involves efforts to help them support the purposes of the Act through programs such as hazard-recognition training.[46]

During 1972 OSHA processed approximately 1.2 million separate requests for information (700,000 telephone inquiries, 400,000 written inquiries, and 100,000 in-person inquiries).[47]* OSHA personnel addressed approximately 700 meetings reaching an estimated 150,000 persons. OSHA estimates that its exhibits at trade

*In an interview on February 5, 1973, one of the OSHA Regional Directors stated that most of the effort OSHA has made with regard to educational and informational activities has been with employers. He added that "labor doesn't request speakers and assistance from OSHA in any real way."

% of Total Complaints Received 1972-1973[2]	1974-1975[3]	% of Violations Cited 1972-1973[2]	Average Penalty for Imminent-danger, Willful, or Repeated Violation[1]
60	60	72	$400
17	14	13	$9500
		5	$2000
13	16		
		10	
10	10		

centers, conventions, and fairs reached about one million persons. In 1972 OSHA released twenty radio and six television public service announcements and issued 176 press releases.[48]

OSHA issued a number of publications during 1972 containing technical and general information. The monthly magazine *Job Safety and Health* reached a circulation of 30,000 during the year. Specific technical publications included a complete revision and update of the general industry, maritime, and construction standards and the initiation, through the Government Printing Office, of a loose-leaf subscription service for standards and revisions to standards. The *General Industry Guide for Applying Safety and Health Standards* was designed to help employers deter-

Table 6.11
OSHA's Rate of Inspections and Follow-ups, 1971-1975

	1971-1973[1]	1974-1975[2]
Total initial inspections	45,800	53,239
Average no. of initial inspections per month	2544	4840
Total follow-up inspections	6000	12,568
Ratio of follow-up to initial inspections	0.12	0.24

1. Period: July 1971 to January 1973. From Joe Collier, "Inspection and Enforcement in the Workplace."
2. Period: July 1974 to May 1975. From the OSHA Compliance Activity Report, May 19, 1975.

Table 6.12
Profile of Violations Found and Penalties Proposed by OSHA

Type of Violation	% of Alleged Violations 1971-1973[2]	1974-1975[3]	% of Proposed Penalties[1] 1971-1973[2]	1974-1975[3]
Nonserious	98	98	60	51
Serious	0.7	1	31	29
Imminent-danger, willful, or repeated	1.3	1	3	15

1. Does not include penalties assessed for failure to abate a violation.
2. Period: July 1971 to January 1973. From Joe Collier, "Inspection and Enforcement in the Workplace."
3. Period: July 1974 to February 1975. From *OSHR*, BNA, Vol. 4, p. 1311.

mine the relevant problems in their workplaces, while the publication *Training Requirements of the Occupational Health and Safety Standards* contains specific information on those standards which require employee training in work practices or the proper use of equipment.

In 1972 OSHA conducted approximately 1200 one-day seminars to inform employers and employees of their rights and duties under the OSHAct and OSHA regulations. A 30-hour audiovisual presentation entitled "Construction and Safety and Health Training" has been developed as a self-instruction course for employers and employees. A more generalized 100-hour correspondence course in occupational safety and health is being developed by OSHA in conjunction with NIOSH. Finally, OSHA has developed a 30-hour programmed instruction course, covering recordkeeping, training, first aid, etc.[49]

The Occupational Safety and Health Training Institute offers three courses for employers and employees. Two of the courses are one-week sessions dealing with the construction industry; these were attended by approximately 730 industry representatives between 1971 and 1973. OSHA estimates that these industry representatives have in turn trained an additional 70,000 construction supervisors who oversee some one million employees. In addition, the Institute has conducted a modified basic construction course for 181 college and university instructors. In 1972 approximately 150 industry representatives attended the third course on voluntary compliance.

OSHA reports that 10,000 union officials took part in Department of Labor training programs from 1971 to 1973. The 1973 *President's Report* does not elabo-

*When I attempted to get xeroxed material from the OSHA Office of Public Information on June 30, 1972, I was told it would cost 30¢ per page. At the lecture at the Kennedy School of Government, George Guenther denied knowing of this charge when asked about it by another person in the audience. (April 26, 1973).

rate on this figure, but describes the various programs of the unions. The figure should thus be treated cautiously since it probably does not represent extensive training by OSHA.

Training Programs

Another critical component of OSHA's voluntary compliance goal is the development and maintenance of a top-quality compliance inspection force. Training of state and federal enforcement personnel is conducted on a short-term basis at the Occupational Safety and Health Training Institute, near Chicago, Illinois. Originally, there were meant to be four or five centers.[50]

The Training Institute conducts a four-week course for federal compliance officers and industrial hygienists. In addition, as part of its training policy, OSHA requires compliance officers to take a minimum of two technical courses at the Institute. OSHA claims that its management information system enables it to identify training deficiencies and thereby develop an individualized training program for each compliance officer (see Section 9.2.2). In selecting compliance officers, OSHA gives special consideration to persons "who have received training under practical conditions, such as working on safety and health committees in industry. . . . New compliance personnel without recent experience are given special training *before* taking the four-week basic course."[51]

After a new compliance officer or industrial hygienist has completed the four-week basic course, he is assigned to an area office where training continues. "The area director determines the type and amount of formal and on-the-job training. On-the-job training consists of assisting an experienced compliance officer on his inspections. Only when the area director is convinced that a new compliance officer/industrial hygienist is ready is he permitted to inspect on his own."[52]

A source within the OSHA Training Institute has indicated that the quality of the people being trained at the Institute is shockingly low. They are either in their twenties or their fifties, and the older trainees often have little experience that might benefit the younger trainees. All are essentially dropouts. Contrast these statements with those made by George Guenther: "Enforcement of the standards is in the hands of the OSHA compliance officers, a select group of safety engineers, industrial hygienists, aerospace experts, instructors, etc., who have been given intensive training to prepare them for their jobs in the field."[53]

Collection and Compilation of Occupational Injury and Illness Statistics

The record-keeping and reporting system required by the OSHAct is administered by the Bureau of Labor Statistics, while OSHA has responsibility for making and enforcing record-keeping regulations. All employers, with the exception of small firms employing ten or fewer persons, are required to maintain at the workplace statistics of work-related deaths, illnesses, and injuries other than those requiring only minor first aid. (A few of the small firms are required to report for sampling purposes.)

In July 1971 the Bureau of Labor Statistics mailed a record-keeping requirements booklet to the more than five million employers. Three record forms were also devised and mailed to employers: OSHA 100, a log in which every recordable case is entered; OSHA 101, a supplementary record of each recordable case; and OSHA 102, an annual summary of recordable occupational illnesses and injuries. Recordable cases are divided into three categories: fatalities, lost workday cases, and nonfatal cases without lost workdays. By late 1971 approximately six and one-half million sets of records were being kept at workplaces around the country.

The BLS conducted a survey in 1971, prior to the establishment of the new record-keeping system, which indicated a continuing increase in the rate of disabling work injuries in manufacturing to a level of 15 per million man-hours worked. The first statistics available under the new reporting requirement indicated an average rate of recordable occupational injuries and illnesses of 13.8 per 100 man-years worked. (In this survey, covering the last six months of 1971, the BLS abandoned the measure of injury used prior to the OSHAct, the "disabling-injury frequency rate," in favor of an "incidence rate" statistic that is defined as the rate of occurrence of recordable occupational injuries and illnesses per 100 man-years of work—equivalent to 200,000 man-hours on the basis of 100 full-time workers working 40 hours per week for 50 weeks per year. Because pre-OSHA and BLS occupational safety and health surveys are based on a million man-hour figure, the results of the two different survey methods are not directly comparable, although, initially, both systems were used to establish a comparative base.) Selected statistics from the survey are shown in Table 6.13.

According to BLS Director Maurice F. Bresnahan, the BLS did not intend its 1971 survey to be a benchmark against which all future surveys would be measured. Bresnahan notes that the survey covered only six months and was therefore subject to seasonal effects. Further, Bresnahan warns that "changes between the first and second surveys may reflect other factors than progress in improvement in occupational safety and health."[54]

The results of the 1972 and 1973 surveys were not released until December 1974.[55] Work-related fatalities decreased 7% between 1972 and 1973—from 5500 to 5100 in the industries surveyed—although the overall number of injuries and illnesses per 100 full-time workers (the incidence rate) was about the same for both years. Most important was the decline in incidence rates for the five target industries, ranging from 2.6 to 5.1%. (These figures should be regarded with cautious optimism since variations can be due to many sources. However, a *decrease* in the number of deaths in contract construction from 1500 to 1000 may reflect the special OSHA attention given this industry. This reduction is responsible for the overall reduction in deaths; in fact, examination of the figures above indicates an

Table 6.13
Injury and Illness Incidence Rates in Private, Nonagricultural Employment

Industry	Total Overall Incidence Rate 1971[1]	1972[2]	1973[2]	Lost-Workday Cases Incidence Rate 1971[1]	1972[2]	1973[2]
Contract construction	22.4	19.0	19.8	6.8	6.0	6.1
Manufacturing	16.7	15.6	15.3	4.3	4.2	4.5
Transportation & public utilities	12.1	10.8	10.5	4.8	4.5	4.6
Wholesale & retail trade	8.7	8.4	8.6	2.9	2.8	2.7
Services	7.3	6.1	6.3	2.7	2.0	2.0
F.I.R.E.	2.9	2.5	2.4	1.0	0.8	0.8
Overall rate	12.1	10.9	11.0	3.7	3.3	3.4

1. From "Occupational Injuries and Illnesses by Industry, July 1-December 31, 1971," U.S. Department of Labor, Bureau of Labor Statistics, Bulletin No. 1798 (1973).
2. From U.S. Department of Labor Press Release 74-687, Friday, December 20, 1974.

increase in deaths outside of this industry!)

Close examination of the results of the first full eighteen months of data collection under the new statistical reporting systems indicate that the BLS has been seriously deficient in carrying out its congressional mandate to institute *accurate* reporting practices with regard to the incidence of occupational disease. According to BLS statistics, occupational *illness* accounted for under 5% of the total incidence of occupational *injury and illness* between July 1, 1971, and December 31, 1972.* Roughly 133,000 cases of occupational illness, a large proportion of which were relatively nonserious (for example, dermatitis accounted for 33.9% of all *recorded* occupational illness), and only 600 deaths from occupational disease were recorded during the last six months of 1971. But other estimates indicate that the true incidence of the more serious types of occupational disease is substantially greater, and that as many as 100,000 deaths may occur each year as a result of occupational disease.[56] This chronic underreporting of health-related statistics has the effect of distorting the perceptions of employers regarding the relative magnitudes of their occupational health versus safety problems.

Health and Safety Effort for Federal Agencies

The federal government, as the nation's largest employer, has a special obligation to provide safe and healthful workplaces for its employees. Section 19 of the OSHAct

*The 1973 survey shows a greater cognizance of occupational illness and disease, yet fewer cases were actually reported.

requires each federal agency head to establish a comprehensive health and safety program consistent with the standards promulgated by the Secretary of Labor. Executive Order 11612, issued July 26, 1971, required the Secretary to coordinate federal programs by issuing guidelines and consulting with a newly appointed Federal Safety Advisory Council consisting of fifteen members appointed by the Secretary and representatives of federal agencies and federal employees.

The federal safety and health program requirements apply to approximately 120 agencies and departments and cover some three million civilian employees in approximately 5000 occupations. In 1971 employees of the federal government suffered about 37,000 disabling injuries and 255 fatalities. Workmen's compensation and medical expenses resulting from these injuries and fatalities cost an estimated $118 million annually, with related property damage of some $250 million.[57]

According to a 1973 study by the General Accounting Office,

> OSHA's enforcement and inspection practices differ significantly between private businesses and Federal agencies. For private businesses OSHA enforces compliance with safety and health standards through inspections and penalties authorized by the act. The act provides that Federal agencies can be inspected *only with the consent of the agencies.*[58]

The GAO inspected four federal agencies in the Washington, D.C., area and found approximately 200 instances of noncompliance with OSHA standards. The GAO inspectors found many instances of agencies failing to meet Labor Department standards for building safety or health. Some fifty of the conditions observed by GAO investigators were "sufficiently severe that, had they been found in private businesses, the businesses would have been subject to monetary penalties" by the government.[59] In the occupational health area, a GAO questionnaire of 49 agencies resulted in 33 agencies admitting that employees were exposed to toxic agents such as carbon monoxide, lead dust, and agents that can cause forms of cancer or liver damage.

The General Accounting Office found a need for an effective leadership role on the part of OSHA. However, most of OSHA's efforts in the federal safety and health area have been limited to matters involving dissemination of information and planning of future efforts. At the time of the GAO study, nearly two years after the President had issued Executive Order 11612, OSHA had not yet established and issued the regulations required by the President to assist the heads of federal departments and agencies in fulfilling their occupational safety and health responsibilities. The GAO found that OSHA had evaluated only a few federal safety and health programs to determine whether federal departments and agencies were complying with the Act and the Executive Order. "Consequently, OSHA does not know whether Federal employees are assured of the required safe and healthful working conditions."[60]

According to the GAO, the Department of Labor generally concurred with its

recommendation that OSHA take a stronger leadership role in the following areas:
1. Preparing and issuing regulations to further assist and guide the agencies in developing their safety and health programs.
2. Developing a more aggressive and expanded evaluation and inspection program to insure that federal agencies are making adequate efforts to provide safe and healthful workplaces.
3. Continuing to work with federal agencies to resolve the problems with definitions and to make its new reporting system more compatible with the agencies' systems.
4. Assisting federal agencies in developing a system to insure that qualified safety engineers and industrial hygienists inspect federal workplaces and in making comprehensive surveys of their workplaces to determine the specific actions and estimated costs needed to bring the agencies in compliance with the Act.

The GAO recommended that the Senate Committee on Labor and Public Welfare consider having the OSHAct amended to bring federal workplaces under the inspection responsibility of OSHA. Such inspections would supplement, and not replace, inspections by the agencies' own personnel.[61]

On September 28, 1974, Executive Order 11807 was issued to supersede Executive Order 11612. The Order requires each federal agency head to appoint an official, with sufficient authority to represent the support of the agency head, who will coordinate the agency's health and safety programs. Each agency must establish inspection procedures and methods for handling hazard abatement and worker complaints. Employee representatives must be consulted prior to the establishment of any occupational health and safety program. The Secretary of Labor is required to consult with the agencies and to evaluate the programs of those agencies with over 1000 employees at least once a year.[62] The effectiveness of the new program will not be apparent until late 1975.

6.3.2 The National Institute for Occupational Safety and Health

NIOSH's activities include conducting research, preparing criteria documents, recommending standards, providing technical assistance, and training some of the technical manpower needed to implement the OSHAct. It also retains similar responsibilities under the Coal Mine Health and Safety Act of 1969.

The following discussion will examine NIOSH's budget as an indication of what has been happening to these activities since the OSHAct became effective; it will focus on a number of NIOSH's policies that influence the Institute's ability to carry out its responsibilities under the Act.

Research and the Development of Criteria for Standards

The portion of NIOSH's total budget going to research and the development of criteria for standards has increased from approximately 75% in fiscal 1972 to 84% in fiscal 1974. In view of these budget allocations, research and criteria documentation must be considered the primary activities of NIOSH. In fact, NIOSH's expenditures for criteria documentation itself almost tripled between fiscal 1972 and fiscal 1974 (see Table 6.4), and the trend is expected to continue for at least the next two to three years.[63] Criteria documentation, however, represents the end of a process that starts with establishing priorities and compiling research data. These preliminary activities have also received budget increases, but the emphasis is on speeding up the delivery of the final product. The General Accounting Office reports that in order to speed progress,

> the Institute reprogrammed about $600,000 in fiscal year 1973 from other important functions—particularly research—to documentation of comprehensive standards. In the long term, however, the Institute believes this lost research time will inevitably decrease the annual rate of recommended standards developed.[64]

This may be serious since the GAO also found that in most cases NIOSH had not researched *new* substances and physical agents suspected of being harmful to employees. The research facilities in Cincinnati are substandard for the important task given to NIOSH by the OSHAct. An internal NIOSH document circulated in 1973 states:

> The research activities have been seriously handicapped and projects delayed because of these location moves and the inability to acquire suitable laboratory space.
>
> Because of the increasingly acute facility problem in Cincinnati, the need to locate an adequate, coordinated research facility for the Institute has become a major problem.[65]

The Establishment of Priorities

In 1972 NIOSH's Washington Office of Health Surveillance and Biometrics (with a fiscal 1972 budget of $1.3 million and a staff of 54) published a "Priority List for Toxic Substances and Physical Agents" (Table 6.14). This list was prepared by calculating a rating for the substances based on the numerical value for five (weighted) variables:

- *Population Index:* relative evaluation of the number of workers exposed.
- *Relative Toxicity Index:* professional opinion of relative toxicity.
- *Incidence Index:* documented incidence of disease from occupational disease records, physician's first reports of work injury, and workmen's compensation records.
- *Quantity Index:* amount produced or used each year.
- *Trend Index:* estimates on increased or decreased usage.

While such a rating system may look impressive, it has several major weaknesses. For example, it is not obvious that the variables included are the most appropriate

Table 6.14
NIOSH's 1972 Priority List for Toxic Substances and Physical Agents

Criteria Developed
Asbestos
Inorganic Lead
Heat Stress
Untraviolet Radiation
Carbon Monoxide
Beryllium
Noise

In Progress
Benzene
Cadmium and Compounds
Chromic Acid Mist[1]
Cotton Dust
Fibrous Glass
Mercury
Parathion
Silica
Trichloroethylene[1]
Coke Oven Emissions[1]
Toluene Diisocyanate[1]
Toluene[1]
Sulfuric Acid
Sulfur Dioxide
Arsenic

Priorities
1. Bis (Chloromethyl) Ether
   2-Naphthylamine
   Radioactive Products of Uranium Mining (Gaseous and Particulate)
2. Benzidine and Its Salts
   Carbon Tetrachloride
   Ozone
   Tin and Compounds
3. Chromium Compounds
   Dichlorobenzidine
   Ozides of Nitrogen
   Sodium Hydroxide
4. Carbaryl
   Chloroform
   4-Dimethylaminoazobenzene
   Nitric Acid
5. Ammonia
   beta-Propiolactone
   Epoxy Resins
   Methylene Chloride
   4-Nitrodiphenyl
6. Asphalt Fumes
   Ethylene Dichloride
   Fluoridc and HF
   Polychlorinated Biphenyls
   Tetrachloroethylene
7. 2-Acetylaminofluorene
   Chlorobenzene
   Methylene Bisphenyl Isocyanate (MBI)
   Phosgene
   Trichloroethane
8. Acetone
   4-Aminodiphenyl
   Dieldrin
   Malathion
   N-Nitrosodimethylamine
9. Aniline
   Copper and Compounds
   Cyanides
   Styrene
   Zinc and Compounds
10. Chlorine
    Formaldehyde
    Manganese and Compounds
    Phenol
    Platinum and Compounds
11. Acrolein
    Aluminum and Compounds
    Carbon Disulfide
    Methyl Ethyl Ketone
    Vinyl Chloride[1]
12. Creosote
    Methyl Chloride
    Nickel and Compounds
    Phosphorus and Compounds
    Tetrachloroethane
13. Acrylonitrile
    2,4-Dinitrophenol
    Magnesium and Compounds
    Methyl Alcohol
    Paraffin
14. Ammonium Nitrate
    Cold Stress
    Dioxane
    Fluorine
    Microwaves
15. Hydrogen Chloride
    Ethyl Benezene
    Nitroglycerine
    Vibration
    Xylene
16. Methyl Butyl Ketone[2]
    Mineral Spirits
    Oil Mists
    Selenium and Compounds
    Turpentine
17. Arsine
    Gasoline
    Kerosene
    Iron and Compounds
    Petroleum Naphtha
18. Barotrauma
    Cresol
    Paraquat
    Portland Cement
    Talc
19. Carbon Black
    Coherent Energy (Laser Radiation)
    Ethylene Oxide
    Impact Noise
    Proteolytic Enzymes

1. Criteria now developed.
2. Under intense investigation for causing peripheral neuropathy.

Source: Report to the Committee on Labor and Public Welfare, United States Senate, "Slow Progress Likely in Development of Standards for Toxic Substances and Harmful Physical Agents Found in Workplaces." The 1973 Priority List is very much expanded and contains approximately 475 chemicals.

measures for determining criteria documentation priorities, or that the weight given each of the variables is appropriate. More important, the data to compute the indices is simply not available. NIOSH has stated that the data from its Hazard Surveillance and Illness and Injury Surveillance activities represent "the primary input into [the] priority system."[66] These data samples, however, are very limited, and the problems of managing the collected data have prevented thorough analysis, so that real answers to questions on exposed populations and on illness and injury incidences are not available.* In an effort to work around the lack of data on relative toxicity and incidence rates, NIOSH has asked toxicologists, occupational physicians, and professionals in related areas to establish a severity rating through a Delphi consensus procedure based on days lost from work.[67]

Another major weakness of the priority ranking system is that health and safety hazards have been treated separately. Due primarily to NIOSH's health orientation and lack of expertise in safety areas, the priorities for safety research have been determined under an outside contract.[68] While NIOSH proposes eventually to integrate the priority lists for health and safety, it currently lacks the knowledge and expertise to do so.

The objectives and conclusions of the study of needed *safety* research were summarized in a policy analysis by the Government Research Corporation as follows:

> The objectives of this study . . . were to assist NIOSH in the development of an occupational safety research program by:
> 1. summarizing the present status of occupational safety research;
> 2. assessing the need for developing additional research;
> 3. developing a priority rating system for occupational safety research projects; and
> 4. preparing a priority list of specific research projects for NIOSH safety research.
>
> The study focused on the development of criteria and standards, and psychological, behavioral, motivational, engineering, equipment and educational research.
>
> In line with its objectives, the results of the study were fourfold:
> 1. the study concluded that safety research in the nation as a whole was limited and inadequate;
> 2. the study demonstrated the need for a balanced program of safety research in three principal areas: 1) accident causative factors, accident investigation and statistical studies; 2) general occupational safety research in psychology, motivation, education, training and management safety practices; and 3) physical safety, particularly the development of criteria and standards.
> 3. the study developed a priority rating system to rank occupational safety research projects; and
> 4. the study prepared a priority list of specific occupational research projects [Table 6.15].[69]

*One of the problems encountered has been identifying the chemical components of the trade name products that are used in industry. Until this has been accomplished, neither the quantities used nor the population exposed can be determined. In an interview on February 21, 1973, Richard Witmer of the Office of Health Surveillance and Biometrics estimated that 75% of the exposures to chemicals come from nonlabeled chemicals. Even where exposure to known materials is found, no quantitative measurements are taken in the surveys.

Table 6.15
Relative Rankings of Occupational Safety Research Projects by the Government Research Corporation (1973)

Project	Relative Ranking
Working Surfaces	100
Safety Equipment Footwear	46
Lifting and Moving Aids	40
Packages and Containers	37
Lifting/Moving Job Requirements	33
Loading and Unloading Operations	32
Vehicle Design, Movement and Alarm System	30
Containers and Bulk Storage	28
Machine Guards	28
Atmosphere Monitoring Systems	26
Industrial Stairs	23
Nonelectric Hand Tools	23
Personal Emergency Breathing Apparatus	21
Personal Gas Warning System	21
In-Process Maintenance and Cleaning Tools	21
Lifting/Moving Bulk Materials	21
Machinery Interlocks	18
Lifting/Moving Miscellaneous Common Objects	18
Stationary and Portable Ladders	16
Hand Tools and Overexertion Injuries	14
Scaffolding and Staging	14
Vehicle Working Surfaces	14
Objects Discharged from Equipment	12
Personal Protective Equipment for Major Body Parts	12
Personal Protective Warning Markings	12
Machine Repair and Maintenance to Reduce Overexertion Injuries	11
Portable Power Equipment Guards	11
Protective Clothing for Temperature and Chemical Exposure	11
Injuries in Which Other Persons Are the Accident Agent	9
Containers for In-Plant Use of Hazardous Chemicals	9
Vehicle Laddering, Railing, Foot and Hand Holds	7
Protective Head Gear	7
Hoisting Apparatus and Elevators	7
Gloves and Hand Protection Equipment	5
Safety Devices for Electrical Hand Tools	4
Lifting and Moving Attachment Points for Miscellaneous Equipment	4
Design, Test, and Maintenance of Hooks, Rings, and Other Hoisting and Lifting Apparatus	4
Effects of Blind Corners	4
Excavation and Tunneling	4
Temporary Protective Measure for "Unprotected Openings"	2
Vehicular Equipment–Rollover, Enclosure, and Operator Protection	2
Electrical Protective Equipment	2

Source: Government Research Corporation, "Occupational Safety and Health, A Policy Analysis" (1973), Appendix B.

The order of development of occupational *health* criteria documents is determined more by the availability of scientific information than by the priority rating obtained by applying the indices. This is true because once the priority list has been developed, a literature review is carried out to identify all available information on the substance. Then the necessary additional research is scheduled and completed before the criteria document is prepared. The order of completion of the criteria documents is thus dependent on the timing of the research.* The contribution of the priority ranking system is identification of substances for research, and this process should not be misconstrued as establishing priorities for the development of standards.

Research

A review of the research activities supported by NIOSH, both in-house and under contract, clearly indicates that NIOSH is concentrating on health as opposed to safety hazards. The emphasis on health is needed to close the gap in knowledge between safety and health. NIOSH's research program at the Division of Laboratories and Criteria Development, Cincinnati (with a fiscal 1972 budget of $6.4 million and a staff of 162) covers the following areas:

- toxicology
- physical and chemical analysis
- physiology and ergonomics
- engineering
- behavioral and motivational factors
- physical agents.

In addition, industrywide epidemiological studies are carried out at the Division of Field Studies and Clinical Investigations, Cincinnati (fiscal 1972 budget, $2.2 million; staff, 48).

Safety research is mentioned, in the 1973 *President's Report on Occupational Safety and Health*, only under behavioral and motivational factors. This research is directed at determining accident proneness, a now much discredited concept that is opposed by labor as an attempt by government and industry to blame the worker rather than correct the safety problems. The OSHAct provides in Section 6(b)(5) that the Secretary of Labor, "in promulgating standards dealing with toxic materials or harmful physical agents . . . shall set the standard which most adequately assures, to the extent feasible, on the basis of the best available evidence, that *no* employee will suffer material impairment of health or functional capacity even if such employee has regular exposure to the hazard dealt with by such standard for

*This process is reflected, to some degree, by dividing the list into three parts: a list of those substances and physical agents for which criteria for recommendation of a standard have been developed; a list of substances and physical agents for which criteria documents are in process; and a list of nineteen groups of toxic substances whose relative priority has been established by the rating system.

the period of his working life." Research to shift responsibility onto the employee seems counter to the intent of this section. NIOSH's toxicology research has also taken this approach by emphasizing "hyperreactors," and by proposing standards that "prevent allergic respiratory sensitization for all *but* the most highly sensitive."[70] *

An evaluation of the research done at NIOSH itself and done externally on contract will be found in Section 6.4.2.

The Development of Criteria Documents

Aside from the question of what research is done, it is important to determine how quickly NIOSH is developing criteria documents, since this is where the emphasis has been placed in budget allocations. NIOSH has published a list of 57 substances as the "Hazards for Development of Criteria Documents, FY 72, 73 and 74." Consistent with their health orientation, only health hazards were included on this list. By May 1975 NIOSH had prepared twenty criteria documents on substances for which OSHA has not yet promulgated standards; ten of these had been prepared prior to 1974. OSHA has acted on only one criteria package (asbestos) and on the recommended criteria for standards for 14 carcinogens and vinyl chloride. NIOSH has admitted that it is off to a slow start but expects to increase its pace to a rate of twenty criteria packages per year.[71] This rate increase has not occurred; and even if it did, it would hardly make a dent in the more than 12,000 toxic substances included in NIOSH's 1973 Toxic Substances List.

The Toxic Substances List is a moving target in itself, for according to NIOSH's former Director Marcus Key, 3000 new chemicals are synthesized each year and about 500 of these find some use in industry; thus, "the backlog is increasing faster than our ability to develop criteria for standards."[72] In another speech, Key stated:

As we in NIOSH have attacked this problem of essentially unknown chemical substances, we have come to two important conclusions:
1. Development of comprehensive criteria documents is expensive and time consuming. The cost is approximately $200,000 to $300,000, even when the dose-effect relationship is known. The length of time is 12 to 18 months. . . .
2. We cannot be so specific in the proposed standards as to include every eventuality and contingency.[73]

The process NIOSH has been using for preparing the criteria documents needs to be changed to facilitate more rapid development of meaningful standards that do not attempt to be overly specific. To speed up the process, NIOSH has contracted

*The head of toxicology at NIOSH is Dr. Herbert Stokinger, who also heads the American Conference of Governmental Industrial Hygienists' TLV Committee for Airborne Contaminants. Dr. Stokinger not only believes in hypersusceptibility, but also holds that workers can be biologically exposed to safe levels of carcinogens, and that no material should be banned from the environment until it is *proven* harmful.

with the Stanford Research Institute to prepare eighteen criteria documents by June 1977. In addition, NIOSH has awarded $1.5 million for two contracts to upgrade the TLVs by recommending appropriate control procedures and sampling and analysis methods to be added to the already established toxicity tolerances.[74]

NIOSH's review procedures claim to use an "advocate-adversary" system, involving three rounds of review for every document. The first review is conducted by NIOSH staff members, the second is performed by a group of recognized experts, and the third is undertaken by selected professionals, scientific societies, and government agencies whose activities are directly involved in the specific problem. The appointment of each group of outside reviewers delays the completion of the document, and then progress is subject to the rate at which they act to complete the review. Since OSHA submits all criteria documents to the appointed advisory committees, the reviews by NIOSH seem excessive and subject to too many delays. The goals of its review process, NIOSH states in the 1973 *President's Report*, are to "insure objectivity and comprehensiveness of the final project," yet labor is clearly underrepresented in the three review groups since scientific expertise rather than affiliation is the criterion for participation. Labor gets its chance after the "objective" facts are settled by the "experts"—who are usually drawn from industry and government.

The value of NIOSH's review procedures must thus be considered in the context of the potential for bias in the composition of the review panels. The criteria documents developed by NIOSH are supposed to provide the scientific support required for standard setting, and it is much more difficult to challenge a statement or finding after this scientific document has been finalized than while it is in process. The OSHAct does not require representation of all affected parties on NIOSH review panels, but guarantees should be established to ensure that all *scientific* challenges are heard equally before the document is finalized.

The Society for Occupational and Environmental Health announced on January 2, 1974, that NIOSH would soon send it criteria documents for review on ammonia, chloroform, nitric acid, and hydrogen fluoride. The Society has a more or less equal balance of management, labor, university professionals, and "interested persons," and could provide a less biased mechanism for criteria document review.

Surveillance Activities

NIOSH's surveillance activities have already been mentioned in relation to determining the priorities for criteria documentation. More specifically, NIOSH has undertaken two surveillance projects: the Hazard Surveillance and the Illness and Injury Surveillance. The Hazard Surveillance was designed to identify and define potential occupational health and safety problems. The first phase involved walk-through inspections and interviews with plant managers at some 5200 workplaces, selected in cooperation with the Bureau of Labor Statistics to provide a representative sample of all nonfarm industries covered by the Act. The workplace surveys were started in 1971 and not completed until 1974. The second phase of the project involves computerizing the data to provide information on an industry, job, and exposure specific basis. Several problems have arisen in managing the data, in identifying chemical components of trade-name items, and in coding the occupational categories, and these have further delayed the project's completion.

The Illness and Injury Surveillance project does not involve the collection of any new data. The purpose of this project is "to compile data describing the magnitude and extent of occupational illness and disease." The primary sources of data are the injury statistics developed by the Department of Labor, based on employer-reported injury and illness data. Although NIOSH states that this project "is necessary to establish priorities for research and criteria development," it is not really clear how the data is used—especially since the data has only been cross-tabulated on an occupation vs. disability/mortality basis that does not identify the causal agents.[75]

Another source of data available to NIOSH is the National Surveillance Network (NSN). These data are collected by state inspectors who are asked to fill out the special NSN forms while conducting their own compliance, surveillance, or consultative services. The data collected provide information on in-plant environmental conditions which augments the data from the Hazard Survey, but the NSN data "cannot stand alone for national estimates."[76]

Industrywide Studies

While the surveillance activities report on the statistical relationship between specific occupations and illnesses, NIOSH's industrywide studies are specifically required by Section 20(a)(7) of the Act to investigate "the effect of chronic or low-level exposure to industrial materials, processes, and stresses on the potential for illness, disease, or loss of functional capacity in aging adults." The choice of which industries to study is based on three factors:

1. the level of suspicion concerning possible adverse health effects resulting from occupational exposures;
2. the feasibility of recovering information of the required validity on occupational exposures and health effects either by study of existing records or by detailed environmental, medical, and biostatistical studies; and
3. the size of the population that is potentially affected.[77]

In the 1973 *President's Report*, NIOSH reported on 31 studies that were either underway or planned. The studies were to provide essential information for the "development of sound criteria for control of industrial exposures to noxious agents which encompass a recommended exposure standard, procedures for environmental evaluation, methods of medical surveillance, and engineering control measures."[78]

The findings of these studies have also been used by OSHA to review petitions for variances. For example, in denying a petition by the American Iron and Steel Institute for a variance from the standard for coal tar pitch volatiles, OSHA cited NIOSH's research on the high rate of lung cancer among coke-oven workers.[79] Following each study, the report, sampling data, and summaries of the medical data are forwarded to the appropriate companies and state occupational health agencies.[80] The availability of this information to unions as well as companies might be expected to provide an incentive for corrective actions even in the absence of the development of federal standards. *This route for action may be particularly important for those industries where an abnormal death rate is observed but the particular agent has not been identified.* In this case, a standard cannot be developed, but the industry could adopt *engineering controls* to reduce worker exposure to all possible agents.

The industrywide studies are conducted under the direction of Dr. Joseph Wagoner, who heads the Division of Field Studies and Clinical Investigations. Dr. Wagoner, who was previously affiliated with the National Cancer Institute, is a firm believer that no exposure to a carcinogen is safe, and he is one of the few outspoken NIOSH scientists who have shown a high degree of concern about workers' health. His rapid response to the revelation of a close connection between the use of vinyl chloride in the rubber industry and a high incidence of liver cancer among rubberworkers led to the adoption of the third OSHA health standard.

Technical Assistance

Information on specific health hazards is available to industries through NIOSH's Division of Technical Services. This program provides five categories of services:

1. *Health Hazard Evaluation Services:* provides on request of employer or employee an on-site analysis, following prior notice, of a substance to determine its potentially toxic effects in the concentration as used or found.
2. *Industrial Hygiene Services:* provides on-site consultation, technical assistance, and other services related to engineering and industrial hygiene issues.
3. *Accident Prevention Services:* provides technical and consultative services on various approaches to prevent or reduce in severity injuries associated with machines, equipment, devices, and appliances used in the occupational environment.
4. *Medical Services:* provides medical and nursing occupational health consultation, and provides assistance on health hazard evaluations.
5. *Technical Information Services:* maintains a technical library and information system and responds to requests for information.

Most significantly, these services are provided on-site at the request of either an employer or a representative of the employees. Given that industries, and small businesses in particular, have voiced a need for both information on toxic substances and help in determining whether they are in violation of the Act, one would expect NIOSH's Division of Technical Services would be inundated with requests. The experience of the first years, however, has been the opposite. In the 1973 *President's Report*, NIOSH reported that, during the first year of operation of the Hazard Evaluation Services, 149 requests for hazard evaluations were received. NIOSH only responded to 75 of these requests; the others were not found to be "valid requests as defined under Section 20(a)(5) of the Act."[81]

The 75 requests acted on involved over 200 substances. Determinations had been made for only 22 (under one-third) of these requests, and the remaining 53 were in various stages of completion at the time of the publication of the report. Dr. Bobby F. Craft, Director of the Division of Technical Services, stated in an interview in June 1972 that he expected about 100 requests for hazard evaluation per year. Approximately one year later, another member of that Division stated that there hadn't been a great increase in the rate of requests since the program was initiated, and that he did not think the program was meeting its potential because little had been done to promote use of its services.

A dramatic example of how important a hazard evaluation can be was provided in the determination by NIOSH that nine deaths at Goodrich Rubber Company were caused by exposure to vinyl chloride.[82] Having found vinyl chloride to constitute an imminent danger, NIOSH devoted intensified efforts to the industry and its workers through the Division of Technical Services.

Bobby Craft indicated in 1972 that his limited manpower resources did not allow him to initiate hazard evaluations, but only to respond. Thus, only on OSHA inspections are health hazards likely to be found, and, as discussed previously, inspections are *unlikely*.

A NIOSH staff member summed up the situation in a confidential letter to the author dated March 2, 1973:

> A major problem in NIOSH is that we as health professionals (nursing, medicine) not only have to create the projects (surveys) but are then required to monitor a contract, to spend hours of professional time attempting to complete a statistical survey, to ask for assistance from other divisions for statistical help, to write this up, to edit it for publication, to get it ready for distribution, etc.
>
> In other words, we are told to create projects relative to PL 91-596, and then when we do, we must 'hand carry' this every inch of the way, *in areas we know nothing about, nor have had training to do, nor direction for seeing that it gets done.*

The reports of findings from each investigation, and recommendations for controlling observed hazards, are sent to the employer, a representative of the employees, and to the Department of Labor and the appropriate state agency. The NIOSH investigator does not have the authority to issue citations for violations of a specific standard or the general duty clause; even in a case of imminent danger, he can only report to the Department of Labor so that the Secretary of Labor can take action.

NIOSH, however, hopes that the information gained through the technical service programs will provide valuable inputs for criteria documentation.

Training

The budget allocations for technical services, surveillance activities, industrywide studies, and research and criteria documentation have all increased, while the budget for manpower development was cut by approximately 80% in fiscal 1973 and 1974. The manpower training activities have gone from 11% to only 2% of NIOSH's annual budget (see Table 6.4). The significance of the changes is that they represent a *shift* of resources since the total budget has grown by only 4% during this time period. This shift was required by an Office of Management and Budget policy decision to cut back *all* federal agency training activities, even though the manpower shortages recognized by the OSHAct were still critical. In fact, the budget cuts in manpower training were announced by NIOSH in an article which also reported the finding of a NIOSH working group that the "long-range deficit in occupational safety and health personnel may be as high as 50,000 (professionals)."[83] This same article quoted NIOSH director Marcus Key as warning, "if the problem of the manpower deficit is not resolved, the Occupational Safety and Health Act and its purposes will be jeopardized." Due to the interagency aspects of the training programs, NIOSH having responsibility for training technical manpower and OSHA having responsibility for explaining to employers and employees their rights and responsibilities under the Act, both OSHA and NIOSH have been affected by the training budget cuts. It is sufficient to state here that after a high-priority start, NIOSH's short-term courses have been cut back and are now offered only on a reimbursable basis. In addition, the long-term training courses will be phased out by 1976. (See Chapter 9 for a description of the NIOSH manpower study.)

Problems within NIOSH

Aside from cuts affecting NIOSH's ability to train occupational health manpower, its own manpower has been cut (see Table 6.3). The Public Health Service Corps had been a prime source of professional staff, but the abolition of the draft has affected the number of qualified people made available through that mechanism.[84]

The final comments on NIOSH to be raised in this section are related to its public image. NIOSH's image was initially very poor, according to Ralph Nader's Task Force on Disease and Injury on the Job. The Nader Report makes several suggestions on how this situation might be corrected:

> NIOSH itself has to bridge the gap between scientific professionalism and the workplace. Additional funds and more scientists will not help unless people at NIOSH are able to communicate with workers. To this end, NIOSH must cultivate in the mind of employees the image of a willing listener and a source of help. . . . Finally, NIOSH should view itself as more than a research organization, and assume more

responsibility for urging enforcement of health standards.[85]

NIOSH's image problem was not limited to its interactions with the public; it also suffered from not being able to attract the attention and support of the upper levels of the HEW bureaucracy. Plagued by frequent reorganizations, and by such a low average GS level (7.6) that it could not offer competitive salaries, NIOSH lost important staff members. Dr. Marcus Key, then Director of NIOSH, summed up these administrative difficulties in 1973 by asserting that "NIOSH is not a viable organization."[86]

Dr. John Finklea, formerly Director of the National Environmental Research Center (EPA), has recently assumed the directorship of NIOSH. In the short period of time since his appointment, he has taken encouraging steps toward making NIOSH a more effective force in occupational health. He is working to develop a closer working relationship with other agencies involved in environmental health sciences, to encourage increased use of the hazard evaluation service, to develop criteria in as timely a manner as possible, and to expand research efforts in mortality and morbidity studies as well as in work performance under varying working conditions.[87]

6.3.3 The Review Commission

The Occupational Safety and Health Review Commission holds hearings on all contests of citations and proposed penalties requested by employers, and on all contests of abatement periods requested by either employers or employees. The hearing is held close to the place of employment, and is presided over by a hearing officer. The hearing officer makes a report of his determinations which, after thirty days, becomes the final order of the Commission unless one of the three Commissioners has requested that it be reviewed by the entire Commission.

Jurisdiction

The OSHAct specifically gave the Review Commission jurisdiction to hear only contests of citations, proposed penalties, and abatement periods. Section 6(f) of the Act states that the judicial review of a standard shall be within the jurisdiction of a circuit court of appeals. All other agency actions are reviewable in the federal court system, as provided for under the Administrative Procedure Act.

With such limited jurisdiction, one of the problems of the Review Commission has been to determine which contests presented before it raise issues that are within its authority to decide. The decisions of the Review Commission have been discussed extensively in Section 4.3.2, but some representative decisions will be discussed here briefly to illustrate the Commission's function.

The question has arisen of the extent to which the Commission should conduct itself as though it were a court rather than a more traditional administrative agency.

The Commission cannot look to other independent agencies in the Executive Branch for a definitive resolution of this problem because its duties and its legislative history have little in common with the others. For instance, the rulings of the Review Commission apply only to the individual case before it, and therefore the Review Commission's powers are very different from those of agencies such as the Federal Trade Commission which can issue regulations that are binding on businesses generally. The Review Commission cannot conduct investigations, initiate suits, or prosecute, and therefore the authority of the Commission must be distinguished from agencies such as the National Labor Relations Board which can engage in such activities. The Review Commission is best understood as an administrative agency with the limited duty of adjudicating those cases brought before it by employers and employees who seek review of the enforcement actions taken by OSHA and the Secretary of Labor.[88]

Given that the Review Commission is an administrative agency, the question that must be answered is whether the Commission has any administrative responsibilities under the Act that would distinguish its operations from those of any traditional court. Another way to state the question would be to ask whether or not the Secretary of Labor relinquishes his authority and responsibility for seeing that the Act is complied with once a case comes to the Commission. This issue was dealt with by the Commission in the case of the *Wetmore and Parman, Inc.*, in which the Commission had to decide whether it had the authority to raise the classification of a citation from serious to willful during the review process. The Commission decided that the congressional intent had been to separate the investigatory (inspection), prosecution, and enforcement functions from the adjudicatory function. The first three were given to the Secretary of Labor, and only the adjudicatory function was given to the Commission, so that the Commission lacked the authority to raise the character of a violation in the absence of any motion to do so by the Secretary of Labor. In so deciding, the Commission stated that it was meant to act "as a court within the traditional sense."[89]

Another issue involving the Commission's authority arises in cases in which a limited notice of contest is filed. In *Florida East Coast Properties, Inc.*, the Commission held that when an employer files a notice of contest that is limited to the appropriateness of the Secretary's proposed penalty, the citation itself becomes a final order pursuant to Section 10(a) of the Act.[90] On the other hand, any notice of contest of the citation automatically places into issue the appropriateness of the penalty.[91]

The Commission's Decisions about Coverage under the OSHAct

Section 4(b)(1) of the Act provides that the Act does not apply to the working conditions of employees over which other federal agencies exercise statutory authority to prescribe or enforce standards or regulations affecting occupational health and safety. In *Mushroom Transportation Co.*, the Commission held that a common carrier subject to the Motor Carrier Safety Regulations of the Department of Transportation was exempted from compliance with OSHA standards pertaining to wheel chocks. The Commission stated further that in order to be exempt from OSHA coverage, it is not necessary that the other agency's regulations be similar to or as stringent as the OSHA standards.[92]

In *Fineberg Packing Co.*, the Commission held that a meat processor subject to the provisions of the Wholesome Meat Act was *not* exempted from compliance with OSHA standards. The Commission stated that in order to be exempt under Section 4(b)(1), the policy or purpose of the other Act must be to protect employees. The purpose of the Wholesome Meat Act is to protect consumers.[93]

In *Southern Pacific Transportation Co.*, a majority of the Commission held that the railroad industry is not exempted from OSHAct coverage by virtue of the Federal Railway Safety Act. Working conditions that are covered by the Department of Transportation's regulations, however, including record-keeping, are exempt.[94]

Petitions for Modifications of Abatements (PMAs)

Section 10(c) of the OSHAct provides that if an employer is unable to comply with the abatement requirements of either an uncontested citation or a Review Commission order resulting from a contested citation because of factors beyond his control, the employer, after an opportunity for a hearing, shall be issued an order affirming or modifying the abatement requirements. In *H.K. Porter Co., Inc., Thermoid Div.*, a majority of the Commission held that the Commission and not the Secretary had the power to rule on PMAs. After reviewing the legislative history of Section 10(c), the Commission concluded that Congress had inadvertently used the word "Secretary" in Section 10(c) in place of "Commission" in assigning authority for holding hearings on PMAs.[95] On February 18, 1975, however, the Commission amended its own Rule 34 to authorize the Secretary to grant PMAs unless he objects or there is an objection from an employee or authorized employee representative.[96]

Settlements

In *Thorleif Larsen & Son of Indiana, Inc.*, the Commission held that even though it is the final arbiter of penalties assessed under the Act, when an employer stipulates that the Secretary's proposed penalty is acceptable, the Commission will abide by that penalty unless it is "clearly repugnant" to the purposes and policies of the Act.[97] In *Shaffer Construction and Engineering Co.*, Commissioner Cleary indicated that in his view a penalty is clearly repugnant when there is "overwhelming evidence that indicates a significant difference between the stipulated penalty and a reasonable assessment."[98]

In *Dale M. Madden Construction, Inc.*, the Ninth Circuit Court of Appeals held that the Secretary of Labor has discretion to compromise penalties assessed by the

Review Commission and that the Commission may not object to a settlement after the case has become a final order of the Commission.[99]

Administration

The case load of the Commission has proved to be much greater than originally anticipated, resulting in a backlog of cases awaiting decision. When the Commissioners were first appointed on March 19, 1971, their staff estimated that 100 to 200 cases would be appealed during the first year of operation. Two months later, in May 1971, the number of appeals had reached 890, of which only 88 had been reviewed. Experience has shown that some 5 or 6% of OSHA's inspections lead to Review Commission cases. Recently, the average number of contests filed each month has been between 300 and 400. As of November 30, 1973, the Commission had received a total of 5484 cases, of which only 3399, or 62%, had been brought to final disposition. While a backlog of cases was a constant problem at first, the Commission is catching up. As of March 1975, the Commission had received 12,417 cases, of which 10,070 had reached final disposition. Of the 465 decisions issued by the three-commissioner panel, 232 judge's decisions were affirmed, 132 were modified, 39 were affirmed in part and vacated in part, and 62 were vacated. Of the 4831 petitions for modifications of abatements, 4350 have been granted and 8 have been denied.

The first three members of the Occupational Safety and Health Review Commission were appointed by President Nixon on March 19, 1971:

Robert Moran, J.D., was appointed Director for a six-year term. He had previously served as Director of the Wage and Hour Division of the Labor Department from 1969 to 1971 and as Associate Commissioner of Labor and Industry and Chairman of the Board of Conciliation and Arbitration for the Commonwealth of Massachusetts.

James F. Van Namee was appointed for a four-year term. He had served as Director of Accident Prevention for Westinghouse Electric Corporation from 1961 to 1971 and as a supervising safety engineer for the Liberty Mutual Insurance Company for twenty years. (Frank R. Barnako has been nominated to fill the vacancy created by the expiration of Commissioner Van Namee's term in April 1975.)

Alan F. Burch was appointed to a two-year term. He had been Director of the Department of Safety and Accident Prevention for the International Union of Operating Engineers, AFL-CIO, from 1963 to 1971 and Director of Research and Education for the union in 1961-1962. Prior to that, he had been an operating engineer.

On the second anniversary of the OSHAct, April 28, 1973, Timothy F. Cleary, J.D., was appointed to a six-year term to replace Commissioner Burch, whose term had expired. Cleary had previously worked in the Office of the Solicitor of Labor, where he specialized in legal problems arising under the Fair Labor Standards Act and federal workmen's compensation laws, and he had served as chief legal counsel to Commissioner Burch since July 1971. Cleary's nomination was supported by Secretary of Labor Brennan and by Hunter P. Wharton, President of the International Union of Operating Engineers.

6.3.4 The National Advisory Committee on Occupational Safety and Health

The National Advisory Committee on Occupational Safety and Health (NACOSH) was established by Section 7(a)(2) of the Act to consult with and advise the Secretaries of Labor and HEW on the administration of the Act. With only these broad guidelines for direction, the importance of the Advisory Committee must be assessed on the basis of both the specific issues it chooses to consider and the composition of its membership.

The areas to which NACOSH has given the most attention are state plans, the *Compliance Operation Manual* developed by OSHA, and the fiscal and manpower needs of NIOSH. Among the important issues NACOSH has *not* dealt with are the emphasis that has been placed on safety as compared to health and the delays in the development of health standards.

State Plans

In mid-1972 NACOSH prepared a report on the following six issues:
1. the procedures for postapproval evaluation of state plans;
2. OSHA's experimental and demonstration grant policy;
3. coverage of public employees and employers under state plans;
4. preemption of state activity in those states without approved state plans by December 31, 1972;
5. first-instance citations, penalties, and employee sanctions; and
6. the federal role in granting variances under approved state plans.

This report and its recommendations were prepared by the Subcommittee on State Plans; it was then approved by NACOSH and "referred to the Occupational Safety and Health Administration for whatever use OSHA may be able to make of it."[100]

The recommendations were stated in very general terms, and tended at that time to be supportive rather than critical of OSHA's policies. Neither the controversial aspects of the issue nor the criteria applied in arriving at the recommendations were made clear. For example, the Act specifically requires that the development and enforcement of standards by the states must be "at least as effective as" the federal programs. However, under issue 5, the NACOSH Subcommittee recommended that first-instance *citations* be provided for in state plans, but did not recommend that first-instance *penalties* be required. The Act, on the other hand, had made first-instance penalties mandatory for serious violations and discretionary for nonserious violations—provisions which were reported by an OSHA staff member to comprise one of the two critical authorities that had been lacking from most state occupational safety and health programs in the past. (The other was the right to enter any workplace without advance notice.)[101]

Apparently OSHA did not accept the report's findings under issue 4, for it issued temporary orders continuing state enforcement activities after the December 31, 1972, deadline set in the Act for those states lacking approved state plans. In this case, NACOSH was correct in suggesting that such an action would not be consistent with the intent of the Act. OSHA rescinded the temporary order after the AFL-CIO obtained a preliminary injunction.

NACOSH developed additional recommendations on state plans at its meeting on April 19, 1973, in response to issues that had been raised by speakers before the meeting of the Subcommittee on State Plans.[102] The issues dealt with were more specific than those covered in the previous report, but again the Committee made only general suggestions. For example, with respect to "benchmarks" for the number of compliance officers and industrial hygiene personnel, the Committee stated that the minimum number of inspectors should be determined not on the basis of the number of inspectors the federal government could maintain in the state, but rather on the basis of what was *needed* to do an effective job. OSHA had developed the original benchmarks as the minimum number which would meet the Act's "as effective as" criteria, and those benchmarks were generally only one-third the current number of state personnel.

Another of NACOSH's recommendations was that OSHA's approval policy be reviewed with respect to those elements of a state plan which should be, but are not, in place at the time of approval. (It should be noted that this very general recommendation replaced a defeated motion to recommend that OSHA require enabling state legislation to be enacted before approving a state's plan.)

At a meeting of NACOSH's Subcommittee on State Plans on December 21, 1973, Richard Wilson, OSHA Deputy Director for Regional Programs, reported that of all the Subcommittee's recommendations, OSHA had acted on only one. Wilson said OSHA was developing new benchmarks and instituting a federal variance procedure for employers operating in several states.[103] The purpose of the NACOSH meeting, in addition to receiving Dick Wilson's report, was to refocus attention on issue 1 of the first report: the procedures for postapproval evaluation of state plans. OSHA's policy on this issue was "to shift to monitoring of state plans rather than combining federal enforcement activities upon commencement of enforcement by states with operational approved plans."[104] NACOSH's Subcommittee recommended that federal enforcement in states with approved plans continue until an initial evaluation of actual operations is made. Federal enforcement activities should then be increased or decreased based on the evaluation of the following factors:

1. What is actually operational in the state plan?
2. How effective is the state plan?

3. How large and how competent is the state staff (including industrial hygienists)?
4. Are the standards adopted by the state "as effective as" the federal standards.[105]

The Subcommittee also recommended that the public be provided opportunities to comment on the evaluation, that the Department of Labor's decisions should not be made *in camera.* The Subcommittee's recommendations were included in a report submitted to NACOSH on February 12, 1974. OSHA's response to these NACOSH recommendations was a proposal to investigate *complaints* for the first six months after a state plan is operational.[106] This response is essentially a compromise between NACOSH's recommendations and the states' opposition to dual jurisdiction after a plan has been approved. (See Chapter 5 for a more detailed discussion.)

In sum, NACOSH has been critical of OSHA's policy on the withdrawal of federal enforcement, but it has not influenced OSHA policy significantly because its recommendations have been compromised or ignored. In a suit filed on March 12, 1974, the AFL-CIO asked that the Department of Labor be prohibited from "approving state plans that do not meet the statutory requirements" of the OSHAct and "from giving continuing effect to the orders previously issued approving state plans."[107] The District Court for the District of Columbia dismissed this suit, upholding the challenged state plan's regulations, and the case is now on appeal. It is possible that NACOSH's criticisms will help the plaintiffs in this suit.

Fiscal and Manpower Needs of NIOSH

Other actions at the April 19, 1973, meeting included the development of a recommendation urging action to correct the loss of NIOSH personnel and program curtailment. A subcommittee was appointed to develop information on "current and projected fiscal needs for NIOSH and the significance of its role in HEW so that the adverse impact of grade point averages and the consequent bumping of professional employees can be avoided."[108]

The Compliance Operations Manual

In addition to the subcommittee formed to deal with NIOSH's budget, a second subcommittee was formed at the same meeting to review changes in the OSHA *Compliance Operations Manual.* Later, the changes to the manual proposed by OSHA were presented to this subcommittee—covering wilful, repeated, and serious violations, general duty, follow-up inspections, abatements, and OSHA forms. While they were discussed by the subcommittee, no recommendations were made.[109]

On-Site Citations

The issue of on-site citations was specifically presented for NACOSH's review by the Department of Labor. In October 1973 OSHA authorized senior compliance officers to issue on-site citations during the closing conference of an inspection. This policy produced so much opposition that OSHA's Washington office sent a letter to the Regional Administrators suspending the authorization pending the results of a discussion with NACOSH at their meeting on December 6, 1973.[110] After specifically requesting NACOSH's review, it is interesting to note that OSHA did not ask the Committee to vote on the issue, but the "consensus [of NACOSH] was that most of the controversy stemmed from misunderstanding of the issue." The Committee concluded, however, that the transcript of its discussion of the policy would make good input for consideration by the Department of Labor. The directive authorizing on-site inspections has since been reissued.[111]

Committee Membership

NACOSH has twelve members appointed by the Secretary of Labor, four of whom are to be designated by the Secretary of HEW. These members are to be chosen as "representatives of management, labor, occupational safety and occupational health professions, and of the public" (Section 7(a)(1)). One of the public members is to be appointed as Chairman. The Nader Task Force Report *Bitter Wages* summarized the first appointments, made in July 1971, as follows:

> Not surprisingly, the deck was loaded in favor of business. The lineup on paper is six industry or industry-oriented members, with two labor people, and four members nominally not affiliated with industry (although two had industry backgrounds). Howard Pyle, President of the National Safety Council, is the Chairman, even though he comes from a recognized safety and industry-oriented group, not a public organization as required by the law. . . . In sum, NACOSH is a group of predominately likeminded men and women more likely to applaud than give critical advice to the Labor Department.

These first members were

Howard Pyle, Chairman
President
National Safety Council
Chicago, Ill.

Dr. Richard Sutter
Medical Director
Sutter Clinics, Inc.
St. Louis, Mo.

John J. Sheehan
Legislative Director
United Steelworkers of America
Washington, D.C.

Frank R. Barnako
Manager, Safety and Workmen's Compensation Division
Bethlehem Steel Corp.
Bethlehem, Pa.

Barry C. Brown
Director, Department of Labor
Michigan State
Lansing, Mich.

Donald Peyton
Managing Director
American National Standards Institute
New York, N.Y.

Dr. John V. Grimaldi
The Center for Safety
New York University
New York, N.Y.

Sara P. Wagner
Director of Nurses
Medical Department
Standard Oil Co. (N.J.)
New York, N.Y.

George H.R. Taylor
Executive Secretary
AFL-CIO Standing Committee on
Occupational Safety and Health
Washington, D.C.

Lief Thorne-Thompson
Public Health Administrator
Department of Health and Welfare
Juneau, Alaska

Roscoe W. Botts
Vice President
Industrial Relations
International Harvester Company
Chicago, Ill.

Roger H. Wingate
Vice President
Liberty Mutual Insurance Co.
Boston, Mass.

Three new members were appointed to NACOSH, and nine were reappointed, in November 1973. The new members were

Willis A. Noel
Former Personnel Manager
Reynolds Metals Co.
Longview, Wash.

Charles S. Morgan
President
National Fire Protection Assn.
Boston, Mass.

Paul D. Halley
Director
Environmental Health Services
Standard Oil
Chicago, Ill.

Willis Noel replaced Roscoe Botts, Charles Morgan replaced Donald Peyton, and Paul Halley replaced Sara Wagner. The new appointments resulted in removal of the only woman, and the only representative of occupational nurses, on the Advisory Committee. The new appointees did not result in a shift in the industry/nonindustry split, but replaced industry by industry and standards organizations by standards organizations in representation.

6.3.5 The National Commission on State Workmen's Compensation Laws

The National Commission on State Workmen's Compensation Laws was established by the OSHAct to undertake "an effective study and objective evaluation of State workmen's compensation laws in order to determine if such laws provide an adequate, prompt, and equitable system of compensation for injury or death arising out of or in the course of employment." To meet the responsibilities mandated in the Act, this study had to cover 56 jurisdictions and the sixteen topics specifically mentioned in the Act—all within fifteen months from the effective date of the Act. What made the task even more difficult was the fact that the Commissioners were not appointed until July 15, 1971—six months after the Act was passed and two months after the effective date of the Act. This delay left the Commission only thirteen months to carry out the study and finalize its report. During those months, the Commission met eleven days to develop the plan and review the substance of the report. In addition to the meetings, the Commission held nine public hearings, three in Washington, D.C., plus regional hearings in Chicago, Boston, San Francisco, Dallas, Atlanta, and New York. The meetings required 32 days, and the hearings 18 days, for a total of 50 days of active participation for the Commissioners.

The task of compiling information, conducting surveys, preparing reports, and generally assisting the Commissioners was performed by a full-time staff of thirty employees. The fifteen members of the professional staff included economists, lawyers, physicians, statisticians, and others specializing in workmen's compensation and rehabilitation.

In addition to this staff, the Commission let contracts to 31 private firms, institutions, and individuals for research, surveys, and reports to be published by the Commission. The publications included a *Compendium on Workmen's Compensation* and a series of *Supplemental Studies.* The *Compendium* is a survey of the important issues and the available information. The *Supplemental Studies* examine selected issues in detail. The *Report of the National Commission on State Workmen's Compensation Laws* and the transcript of the public hearings complete the documentation of the Commission's study. The *Report of the Commission* alone expresses the official position of the Commission; since the *Compendium* and the *Supplemental Studies* "were prepared and edited by independent scholars, some of the publications contradict one another or are inconsistent with the collective judgment of the Commission."[112]

The findings of the Commission, as expressed in the report, are as follows:

> We have concluded that there is a significant role for a modern workmen's compensation program and that *the State's primary responsibility for the program should be conserved.* We also agree that the *protection furnished by workmen's compensation to American workers presently is, in general, inadequate and inequitable.* Significant improvements in workmen's compensation are necessary if the program is to fulfill its potential. (Emphasis added)

The report identifies five objectives for a "modern" workmen's compensation program. They are:

1. Broad coverage of employees and of work-related injuries and diseases.
2. Substantial protection against interruption of income.
3. Provision of sufficient medical care and rehabilitation services.
4. Encouragement of safety.
5. An effective system for delivery of the benefits and services.[113]

Section 27(d)(1) of the OSHAct listed sixteen topics to be covered in the Commission's study:

A. The amount and duration of permanent and temporary disability benefits and the criteria for determining the maximum limitations thereon.

B. The amount and duration of medical benefits and provisions insuring adequate medical care and free choice of physician.

C. The extent of coverage of workers, including exemptions based on numbers or type of employment.

D. Standards for determining which injuries or diseases should be deemed compensable.
E. Rehabilitation.
F. Coverage under second or subsequent injury funds.
G. Time limits on filing claims.
H. Waiting periods.
I. Compulsory or elective coverage.
J. Administration.
K. Legal expenses.
L. The feasibility and desirability of a uniform system of reporting information concerning job-related injuries and diseases and the operation of workmen's compensation laws.
M. The resolution of conflict of laws, extraterritoriality and similar problems arising from claims with multistate aspects.
N. The extent to which private insurance carriers are excluded from supplying workmen's compensation coverage and the desirability of such exclusionary practices, to the extent they are found to exist.
O. The relationship between workmen's compensation on the one hand, and old-age, disability, and survivors insurance and other types of insurance, public or private, on the other hand.
P. Methods of implementing the recommendations of the Commission.

The joining of the topics and the objectives is accomplished in the recommendations. The priorities set by the Commission for workmen's compensation programs were identified as follows:

1. Compulsory coverage.
2. No occupational or numerical exemptions to coverage.
3. Full coverage of work-related diseases.
4. Full medical and physical rehabilitation in services without arbitrary limits.
5. Employee's choice of jurisdiction for filing interstate claims.
6. Adequate weekly cash benefits for temporary total, permanent total, and death cases.
7. No arbitrary limits on duration or sum of benefits.[114]

The Commission recommended that "compliance of the States with these essential recommendations be evaluated on July 1, 1975, and, if necessary, Congress with no further delay in the effective date should then guarantee compliance with these recommendations."[115] This recommendation was most controversial among the Commissioners themselves, and three Commissioners submitted supplementary statements to the report suggesting that Congress act immediately rather than wait until 1975 to make the recommendations mandatory. The position of the Commission seemed to be that the passage of the OSHAct had created a new atmosphere in which the states would take voluntary action to improve their workmen's compensation programs. This position seems rather naive in face of the Commission's own finding that even with 300 state bills enacted during 1971, the "encouraging burst of activity nevertheless failed to satisfy many of the basic needs. Of sixteen recommendations for workmen's compensation published by the Department of Labor, the average State meets only eight. The wide variation among the States [is also] disturbing. While nine States meet at least thirteen of the recommendations, ten states meet four or fewer."[116]

Despite strong opposition from both Congress and the labor movement, the Commission's recommendation to give the states until July 1, 1975, to improve their laws was accepted. Senator Harrison Williams (D-New Jersey) stated in late June 1975 that not a single state was yet in compliance with the basic elements of an effective compensation system. Together with Senator Jacob Javits (D-New York), Williams has introduced federal workmen's compensation legislation in the 94th Congress (as well as in the 92d and 93d).[117]

The Secretaries of Commerce, Labor, and HEW served as ex officio members of the Commission. The thirteen appointed Commissioners were:

John F. Burton, Jr., Chairman
Graduate School of Business
University of Chicago

M. Holland Krise, Vice-Chairman
Ohio Industrial Commission

Melvin B. Bradshaw
Executive Vice President
Liberty Mutual Insurance Co.

Clarence E. Carothers
Administrator of Workmen's Compensation
Ford Motor Co.

Daniel T. Doherty
Chairman
Maryland Workmen's Compensation Commission

James L. Flournoy
Commissioner
California State Workmen's Compensation Appeals Board

John A. Greenlee
President
California State University
Los Angeles, Calif.

Samuel B. Horowitz
Attorney
Boston, Mass.

Henry F. Howe
Associate Director
Department of Environmental, Public, and Occupational Health
American Medical Association

Andrew Kalmykow
Counsel
American Insurance Association

Henry H. Kessler
Director for Professional Education and Research
Kesseler Institute for Rehabilitation

Marion E. Martin
Maine Commission of Labor and Industry

William J. Moshofsky
Vice President
Georgia-Pacific Corporation

6.4 EVALUATION OF THE FEDERAL ROLE

This section evaluates primarily the agencies and programs of the governmental units responsible for the implementation of the OSHAct. This evaluation will be made with respect to (1) the stated goals and purposes, and (2) the appropriateness of the goals and achievements for meeting the perceived needs in occupational health and safety.

Problems of interaction between the agencies responsible for carrying out the various mandates of the Act are often caused by the narrow views taken by each agency. However, (1) Nixon administration policy underemphasizing occupational health problems and (2) budget allocation decisions by the Office of Management and Budget (OMB) have seriously affected the ability of the agencies to coordinate their activities. To cite the most serious example, the phasing out of NIOSH's manpower training programs seriously affects OSHA's ability to use its substantially increased allocations for standard setting, enforcement, and state assistance most effectively. As will be discussed at length in Chapter 9, the lack of competent occupational health and safety manpower is the critical resource limitation in the improvement of job health and safety.

The critical comments made in this section are presented in full recognition of the fact that this new federal involvement has been in effect for only four years. The most serious deficiencies, however, arise not so much because the government is not moving "fast enough" as because it is moving in some wrong directions and with the wrong emphases.

While nowhere in the Act is it stated that "economic feasibility" should be taken into account in carrying out the mandates of the Act, it is clear that economic pressures have constrained and influenced the setting and enforcement of standards and general policy. If occupational health problems are to be dealt with to the extent required by the letter of the law, there may be immense economic consequences for some industry, especially in the short run (see Chapter 7). In fact, occupational health problems are being ignored. (The Workmen's Compensation Commission hardly took notice of the occupational health issue; see Chapter 8.)

The general policy to disregard occupational health is due to ignorance, reluctance to act on available data, and general attitudes toward health care and welfare, as well as to the realization that "internalizing" the costs of occupational disease may have serious economic consequences. Specific decisions affecting occupational health hazards have, of course, been subject to political pressure arising from economic consequences—as, for example, in the setting of a "safe" noise level.

A comparison of the situation in the United States with that in eight Western European countries will be found in Chapter 11. The conclusion reached there is that the United States has not shown as strong a national commitment to workers' health and safety as have the European countries. Examination of the foreign experience indicates clearly that productivity is not inconsistent with a policy of concern for job health and safety.

6.4.1 The Occupational Safety and Health Administration

An evaluation of OSHA is most conveniently approached from three points of view: (1) OSHA's own perception of its primary responsibilities and its operations in the federal framework; (2) the extent of OSHA's interactions with other agencies, especially NIOSH, NACOSH, and the Bureau of Labor Statistics; and (3) the role of OSHA in the return of enforcement authority to the states.

Inasmuch as the day-to-day operation of OSHA and OSHA's interactions with other agencies will be greatly influenced by the ultimate fate of the federal program, the return of enforcement authority to the states will be taken up first.

State Preemption of Occupational Health and Safety Authority

Many of the problems arising from the return of enforcement authority to the states under the Nixon administration's "New Federalism" were articulated in Chapter 5. Until recently, OSHA's future as an important federal agency was in doubt because of the extent of the anticipated state takeover. The withdrawal of plans by several large industrial states has drastically changed the situation, but it is still questionable whether OSHA has the ability to monitor and evaluate the 22 approved state programs in order to guarantee adequate protection as mandated by the OSHAct.

As has already been indicated, the *federal* performance on health problems is seriously deficient. Since so few health standards have been adopted by OSHA, and the enforcement of the general duty obligation to provide a safe and *healthful* workplace is not likely to be actively pursued, workers remain essentially unprotected from health hazards. The use of the general duty clause in the original federal framework was difficult enough. Use of the various state general duty clauses and precedents (where they exist) cannot be effectively monitored by OSHA for the purpose of ensuring that they remain as effective as the federal program. Alternatively, it is unrealistic to assume that the states will enforce standards while OSHA will enforce the general duty clause. To the extent that it occurred, the phasing out of the federal role in occupational health and safety has not helped to foster a spirit of innovation in OSHA. Most of the early increase in OSHA's budget went to support the state takeover.

Interactions with Other Agencies

Support for the belief that OSHA will not monitor the state programs effectively, especially in the occupational health area, comes from the observation that OSHA's interaction with NIOSH has been quite poor. If OSHA cannot maintain a close working relationship with NIOSH, how will it ever be able to monitor the occupational health efforts of nearly half the states?

OSHA's relationship with NACOSH has also been poor. Angered with the positions taken by the Subcommittee on State Plans, OSHA has sought to eliminate the subcommittee and to have NACOSH give advice only when asked and only on prespecified topics.

Although OSHA has expressed its intention to give top priority to the issuance of health standards, NIOSH has submitted many more criteria documents than OSHA has acted upon. OSHA has not taken any initiative to act on NIOSH's priority list of 113 toxic substances or on the 1000 to 2000 toxic substances and agents NIOSH deems significant enough to require federal standards. OSHA has regarded NIOSH as a "poor cousin," and since OSHA has virtually no health expertise in-house, the problems of occupational health remain largely ignored.

The lack of concern for occupational health issues in OSHA is further felt in the underreporting by the BLS of statistics for job-related diseases and deaths. OSHA has not emphasized the importance of uncovering occupational health statistics. Until more accurate data is available, the standard-setting process will continue to be seriously biased and highly adversary.

OSHA's Operations within the Federal Framework

While former Assistant Secretary of Labor for Occupational Safety and Health George Guenther has stated that the heart of the OSHAct is the development and enforcement of standards, OSHA has fared poorly in its attempts at standard development. Although OSHA acted rapidly to promulgate consensus standards under 6(a) procedures, some of these standards had little relation to more important occupational safety and health issues. If OSHA had taken a little more time, it could have avoided some damaging criticisms. The initial standards were predominantly safety-oriented, and some were obsolete before they were issued. In promulgating a number of obsolete and/or relatively unimportant consensus standards, OSHA exposed itself to vigorous criticism by industry. Further, OSHA has had to devote significant time to revise, revoke or modify standards which should never have been issued at all. On the whole, however, OSHA's safety standards represent a useful initial attempt to gear up a new federal effort.

Perhaps the most important criticism to be levied against the consensus standards is that many of them do not, in fact, represent a true consensus of all interested parties. In the first place, there was very little labor participation in the original consensus process, and thus few of labor's views are represented in the standards. Second, the consensus organizations are entirely voluntary on the part of employers, and therefore only a limited number of businessmen ever take part in the consensus process. In a 1972 interview, Richard Berman, Labor Relations Counsel for the U.S. Chamber of Commerce, stated that the ANSI standards are not really consensus standards at all, because there were simply no incentives for *industry* to participate fully in the consensus process prior to the OSHAct.

The federal health standards included in the original 6(a) package by reference were never meant to be anything more than guidelines. These threshold limit values do not attempt to describe scientifically proven relationships between exposure and disease. They merely define the limits beyond which exposure is *known* to be hazardous. Because TLVs are often based on poor data and are not designed to protect all those exposed, they afford minimal protection for workers. Only about 400 TLVs were included in the initial standards package, even though it has been estimated that literally thousands of agents and substances found in the workplace pose health hazards. OSHA has not revised nor attempted to revise most of the original TLVs. On March 18, 1974, Assistant Secretary of Labor for Occupational Health and Safety John Stender announced a $3.5 million joint OSHA-NIOSH project to "complete" the TLV standards.[118] Monitoring requirements and other details were to be added, but no real reexamination of the toxicity data was to be done. The standards will be issued as proposed regulations and will be made available for comment upon completion of recommendations by the two agencies. A package of proposed standards for ketones was recently published in the *Federal Register* as the first product of this effort.

While it may have been reasonable for OSHA to promulgate the TLVs as standards under Section 6(a), OSHA explicitly excluded TLVs for carcinogens, thus depriving workers of even recommended guideline protection against known cancer-causing substances. As a result of this policy, workers were left unprotected from 14 carcinogens for an additional *three years.*

OSHA officials have repeated at regular intervals that it is OSHA policy to place increased emphasis on the development of health standards. However, the record shows that OSHA has been grossly deficient in protecting employees' health through the development of new health standards. Not only has OSHA failed to adopt needed occupational health standards, it has done so only under the duress or fear of law suits.

That OSHA has failed to protect workers against known dangerous health hazards is particularly inexcusable given that Section 6(c) of the OSHAct allows the Secretary of Labor to issue temporary emergency standards if a determination is made "that employees are exposed to grave danger from exposure to substances or agents determined to be toxic or physically harmful." It has been known since the 1930s that asbestos is an extremely dangerous health hazard. NIOSH has developed a list of the most serious health hazards to which employees are exposed, yet OSHA issued only the vinyl chloride emergency standard "voluntarily." Even though the deaths of workers at a B.F. Goodrich plant were linked by the company itself to exposure to vinyl chloride, and even though NIOSH determined that the danger involved mandated that worker exposure to vinyl chloride be reduced to levels too small to be measured, OSHA issued an initial temporary emergency standard for vinyl chloride that permitted exposure at 50 parts per million. (It is encouraging to note that in October 1974 OSHA issued a much more protective *permanent* standard for vinyl chloride of 1 ppm.)

Prior to the vinyl chloride standard, the two permanent health standards that had been set were not adequate to protect workers fully. The asbestos standard is currently set at a TLV of 5 fibers per cubic centimeter and will only be reduced to the safer level of 2 fibers per cc in 1976. While in issuing the carcinogen standard OSHA stated that a safe level of human exposure to any of the 14 carcinogens could not be established by application of present knowledge, OSHA was "not prepared to draw from this state of knowledge the conclusion that such levels do not exist." Thus, the carcinogen standard issued by OSHA *does* allow *biologically significant exposure* of workers to the 14 cancer-causing substances.

Economic and other political pressures clearly influence the setting of standards. As stated in the 1973 *President's Report on Occupational Safety and Health*, "Impact studies are, nonetheless, valuable safeguards against adopting standards that might otherwise result in *unnecessarily high compliance costs to employers.*"[119] The standards have certainly taken into consideration the possible financial burden to employers. Even the *temporary* emergency standard for vinyl chloride was what the industry considered economically attainable and not what NIOSH scientists thought imperative for the protection of workers.* The National Academies of Science and Engineering advise that

> if we are running the risk of going beyond points of no return, common sense certainly urges that *it is better to err on the side of caution.* It would not seem to be wise to place too much of a burden of proof on those who contend that use of a certain substance has harmful effects. The object of the game is *not* to completely avoid needless restrictions in use or consumption but to avoid those failures to restrict that will involve us in major or even catastrophic penalties.[120]

OSHA has not been quick to adopt standards once it has decided to act. In the case of the noise and heat-stress criteria documents prepared by NIOSH, it took OSHA over six months just to establish standards advisory committees once the documents were received. OSHA has had numerous disagreements with its advisory committees, and this has resulted in lengthy delays in the promulgation of new standards. To avoid these delays in setting standards, OSHA is turning increasingly to the management-dominated national consensus organizations in which labor has had little participation. OSHA's attempts to issue temporary or permanent pesticide standards, discussed in Section 12.3, represent a mixture of unnecessary delay and lack of the necessary expertise.

Not only has OSHA had a poor track record in setting standards, it has failed to use its standard-setting capability fully. While in 1972 OSHA was authorized a staff of 130 in its standard-setting office, its actual staff numbered between 80 and 90, and half of these people were nonprofessionals. Even though OSHA has increased its budget authorization for standard-setting activity, it has done so at the expense of the training section of the budget (see Table 6.1). As was mentioned in Section

*The adoption of a more stringent permanent standard for vinyl chloride at 1 ppm appears to be more protective of the worker, but other weaknesses exist with the standard.

6.3.1, most of the budget increase for standard setting went not for an increase in professional staff, but rather to support an increased number of standards advisory committees.

In its *enforcement* of standards, OSHA has exhibited a significant *safety bias.* The special Target Health Hazards Program accounted for only 2.8% of all inspections made during fiscal 1974. In that it is reasonable to say that few health hazards were found by the predominantly safety-trained OSHA compliance officers, this figure is a fairly accurate estimate of OSHA health standards enforcement activity. In addition, OSHA has not adequately enforced the 400 TLVs that are on the books. This failure to enforce is in part a result of the small number of industrial hygienists on the OSHA compliance force, and in part a result of OSHA's unwillingness to enforce the permissive TLVs as even minimally protective standards.

In choosing industries to be included in its Target Industries Program, OSHA excluded agriculture, which has one of the highest injury frequency rates of all industries (see Chapter 12). In fact, for the period July to December 1972, OSHA inspected only 80 workplaces classified as agricultural out of a total 16,756 inspections.

OSHA has concentrated less than 2% of its enforcement activity on the 90% of all workplaces which have fewer than 25 employees, even though smaller operations may be relatively more dangerous.

OSHA continues to ignore employers' general duty obligations and thus leaves workers virtually unprotected from serious health hazards not covered by the three promulgated health standard packages.

The "voluntary compliance" which OSHA had hoped to encourage on the part of industry has not been effective because (1) regulations and standards do not exist for most serious health hazards; (2) the inspection force is not large enough to present a significant chance that the employer be randomly inspected; (3) significant penalties are not being levied; (4) the inspectors are poorly qualified; (5) employees are not well informed of their rights; and (6) employers generally did not regard the initial package of safety standards as being reasonable. (Compare the conditions listed in Section 6.3.1 for a successful program of voluntary compliance.)

Recent statistics published by the Bureau of Labor Statistics indicate some improvement in the conditions of the American workplace. However, this improvement has been confined almost entirely to the industries in the Target Industries Program. OSHA citations have been overwhelmingly for conditions that are nonserious in nature (98%), indicating that OSHA inspections are neither strict nor comprehensive.

Industry has been critical of OSHA for the quality of its compliance officers. OSHA's response to this criticism has been to *reduce* the amount of money available for the training of compliance personnel. In fiscal 1973 OSHA training activities received \$6.5 million, while in fiscal 1974 OSHA budgeted less than \$4.5 million for the training of compliance officers.

OSHA has not undertaken a significant program of information dissemination to teach employees about the Act. Whatever union activity exists is directed toward union membership and does not reach unorganized workers who constitute 75% of the workforce. The absence, for example, of nationwide television public service announcements about the OSHAct is a serious omission if the American worker is to be informed. The Government Research Corporation policy analysis on occupational safety and health put it quite well:

> In spite of the controversy which has been raging the last 20 months, occupational safety and health in general and the Act in particular, still do not have wide public exposure, although many more people are now interested and concerned. The public media has largely ignored the issue and no national political figure has jumped to the role of spokesman (George McGovern made a half-hearted attempt to make it an issue in [the 1972] Presidential campaign).[121]

It is clear, then, that OSHA is doing only a fraction of what needs to be done, or what it could do *even with its present resources.* Measured against the serious efforts to reduce occupational injury *and* disease in Western Europe (see Chapter 11), the present U.S. effort is disgracefully inadequate and probably ineffective. Radical policy changes are required if the OSHAct and its partial preemption by state programs are not to result in a reversion to pre-OSHAct inaction and a serious increase in job-related injury and disease.

The lack of administrative commitment to improved job health and safety is seen not only in the actions of the Office of Management and Budget, but also in the appointment of the former Assistant Secretaries of Labor for Occupational Safety and Health. Both men have been relatively unqualified for the job when compared to the technical and medical professionals heading the counterpart agencies in Western Europe.

With the appointment of John Dunlop as Secretary of Labor in April 1975, there is some hope for an increased emphasis on occupational health. Since OSHA is to be reorganized and its previous policies evaluated, a new commitment coupled with adequate funding and staffing and new priorities may yet emerge.

6.4.2 The National Institute for Occupational Safety and Health

An assessment of NIOSH's activities can be conveniently divided into two parts: (1) the emphasis and programs in the three main areas of its specific responsibilities: (a) priority setting, research, and the development of criteria documents, (b) technical assistance, and (c) manpower training; and (2) the relationship between NIOSH and OSHA and the ability of NIOSH to focus attention on occupational health problems. The most often heard explanations of NIOSH's small influence in the solving of occupational health problems are NIOSH's low position in the

HEW hierarchy, the low GS levels of its employees, manpower cuts, and insufficient funding for research and especially for manpower training. While these factors contribute to NIOSH's problems, they hide the more serious systemic deficiencies which increased funding alone will not correct.

Hazard and illness/injury data upon which research priorities are established are inadequate; problems exist both in the management of the data and in the use of the Delphi technique. NIOSH uses the Delphi technique to determine severity ratings for hazardous materials by asking plant physicians to rate severity *based on lost workdays.* Because lost workdays do not adequately reflect the incidence of long-term chronic disease, such as cancer, this kind of information-gathering method does not provide a true picture of severity. Furthermore, the NIOSH priority list does not reflect the seriousness of the problem because surveys conducted to collect data on worker exposures do not include exposure *levels* (except for noise).

NIOSH's excuse of having insufficient manpower in-house to monitor external contracts is difficult to understand. The point is that, for the most part, *new* substances and physical agents suspected of being harmful to workers are not researched. The toxicology branch, headed by Dr. Herbert Stokinger, is not encouraging the really critically needed research, such as investigation of synergistic effects or the effects of stress on toxicity. The toxicology branch also sponsors experiments to determine if safe levels of *already known human carcinogens* can be found in animals. The research done by the toxicology branch thus has little operational relevance to pressing occupational health problems. New engineering solutions, investigation of combined effects, and the role of stress as a cocausative factor of disease initiated by toxic materials are largely ignored.

An examination of the research done at NIOSH itself and done externally on contract reveals the following facts:

- The allocation of research funds among the various areas does not appear to have been decided as a result of any discernible scientific policy, or in a coordinated way.
- Many of the research projects, though labeled "applied research," have only a tenuous relationship to improving worker health and safety.
- Some research, but by no means a large percentage, has important implications for occupational health—especially the epidemiological studies.
- Too many projects concentrate on now-discredited concepts in occupational health and safety, such as those emphasizing accident proneness or hypersusceptibility as a major source of injury and disease.
- There is virtually no support of labor groups or public-interest groups for research or demonstration projects.
- Little research on combined hazards is performed.
- Stress as both a causative and cocausative factor of disease does not receive enough emphasis.

• Too many irrelevant animal studies are done.

The development of criteria documents is a slow and cumbersome process. Some of the fat needs to be trimmed from the various review stages so as to speed up the process. A study by the General Accounting Office concluded that "NIOSH has been making slow progress in developing and recommending comprehensive standards for toxic substances and harmful physical agents."[122] More specific findings of the study were described in Section 6.3.2. I am in essential agreement with these findings, especially with the need for more and higher-level manpower and funding. These deficiencies, however, are only part of the problem.

In addition, the standard recommendations are often politicized, a sphere reserved for OSHA; NIOSH is to arrive at the "best scientific judgment," objectively determined, on what constitutes safe exposures. However, in two cases NIOSH recommended criteria at variance with what was suggested by the scientific evidence alone. NIOSH recommended 90 dBA as an eight-hour noise exposure limit, eventually to go down to 85 dBA. Either level clearly causes a fair amount of hearing loss and takes no account of nonauditory effects. The NIOSH recommendation that the asbestos standard be set at 5 fibers per cc until 1976 (when it will be lowered to 2 fibers per cc) is another example of how politicized the "objective" recommendations of NIOSH are (as a result of economic pressure). NIOSH's recommendation for a standard allowing *no* measurable exposure to vinyl chloride may be viewed as an indication that the NIOSH leadership is finally awakening to the serious protective measures that need to be taken.

The industrywide studies have produced valuable epidemiological results, despite their understaffing and underfunding. The Division of Technical Services is not able to keep up with even the limited number of requests for hazard evaluations that it receives. The program is insufficiently utilized and it has a very low visibility among employers and employees. NIOSH's new Director, Dr. John Finklea, plans to promote an increase in the use of the technical assistance program as an important step in promoting job health and safety.

An unfortunate limitation on the NIOSH manpower training program has resulted from OMB training budget cuts. As a result, NIOSH has been prevented from contributing to the creation of a pool of occupational health and safety professionals necessary to staff in-plant health and environmental services, maintain a high level of quality enforcement, and conduct necessary research.

Finally, an important reason why occupational health problems remained, until recently, underemphasized in the implementation of the OSHAct and the reform of the workmen's compensation system was a lack of real commitment to occupational health problems on the part of NIOSH leadership. Although it is true that OSHA has not acted on all the criteria documents submitted by NIOSH, NIOSH's leadership has been overly cautious and has maintained a low profile in interacting with OSHA and other agencies involved with environmental health. NIOSH's position on vinyl chloride is the first real demonstration on the part of the leadership of a concern for critical occupational health problems.

As of this writing, NIOSH appears to be charting a new course under Dr. Finklea's leadership, with a posture more protective of worker health.

6.4.3 The Review Commission

The Review Commission has conducted itself as an independent quasi-judicial agency. The decisions of the Commission, discussed at length in Section 4.3.2, appear to be relatively free from external political pressure. Decisions favoring workers include a liberal interpretation of the term "recognized hazard," while the decisions imposing a heavy burden of proof on OSHA in cases alleging violation of the general duty clause favor employers. On balance, the Commission maintains the degree of competence one would expect on the basis of its high GS level.

6.4.4 The National Advisory Committee on Occupational Safety and Health

The membership of NACOSH has a heavy safety and industry bias. The Committee has been used as a forum for labor's representatives John Sheehan and George Taylor to raise issues related to OSHA's implementation of the Act, but for the most part neither OSHA nor NIOSH takes NACOSH very seriously. Thus, the effectiveness of NACOSH in ensuring the accountability of OSHA and in criticizing OSHA, NIOSH, or the administration remains generally minimal.

The recommendations of NACOSH have tended to be very general, calling for agency review or suggesting alternatives for action, but rarely presenting a thorough analysis or even a statement of findings of the Committee's review. The Subcommittee on State Plans has taken a strong stand, but its recommendations have not been adequately supported by the entire Advisory Committee.

In contrast, NACOSH did develop a strongly worded statement of its reaction to the small-business exemption when it was passed by the House of Representatives in 1972. Unfortunately, this statement was never transmitted to the members of Congress as requested by NACOSH.

The future role of NACOSH is currently under review, primarily because the strong stand taken by the Subcommittee on State Plans angered Assistant Secretary of Labor John Stender.[123] There is a clear need for such an independent watchdog agency. Moreover, NACOSH meetings are open to the public and could thus provide an informal opportunity for public participation in the development of OSHA policy.

6.4.5 The National Commission on State Workmen's Compensation Laws

The area in which the report of the National Commission on State Workmen's Compensation seems most deficient is the coverage of occupational disease. Workmen's compensation laws initially had no specific provisions for occupational disease. Some states now provide coverage for "all" diseases, while others provide coverage only for specific listed diseases. In both cases, the worker must meet stated time limitations for initiating a claim. The Commission recommended that all diseases be covered, and that the time limit be two years for all impairments, but they did not address the problems involved in identifying the cause of a disease before it can be proved to be work-related. What is the proper role of workmen's compensation, for example, when no agent can be identified, yet a statistical study shows that a set of symptoms is overrepresented in a certain group of employees?

The Commission did mention the problem of multiple causation of diseases. However, the wording of the recommendation, requiring that a work-related factor (once identified) be proved a "significant cause," almost ignores the medical problems of multiple causation and synergism—and the recommendation therefore does not address these problems.

The second-class status of health problems is apparent from the Commission's discussion of how to determine when an impairment due to emotional strain or heart attack is health-related:

> As the purpose of workmen's compensation is to protect the employee, we believe in the traditional practice of resolving doubt in favor of the employee. At the same time, we do not believe that workmen's compensation should be converted into a general insurance scheme: its function is not to protect against all sources of impairment or death for workers. One of its objectives is to provide incentives for employers to improve their safety record.[124]

This emphasis on the role of workmen's compensation as an incentive for the employer to prevent injuries is very strong throughout the report. This is surprising because the relationship between accident rates and insurance costs could not be established in the studies reported on by the Commission.[125] The Commission did not address the possible impact of insurance costs as an incentive to reducing occupational *disease* by providing better preventive health care.

In summary, the report provided too little real analysis of the problems involved in the coverage of occupational diseases. In addition, the Commission favored voluntary action on the part of the states over federal action to mandate the essential elements of a compensation program—even though the states have shown no real indication that they are more willing to change now than they have been for the last fifty years.

REFERENCES

1. House Report No. 92-1341 (August 15, 1972), 92d Congress, 2d Session, p. 12.

2. *Occupational Safety and Health Reporter* (the Bureau of National Affairs, Inc.), Vol. 3, No. 3 (December 27, 1973), p. 941. [Hereinafter cited as *OSHR*, BNA]

3. *The President's Report on Occupational Safety and Health*, G.P.O. Document No. 2915-0011 (May 1975), p. 15.

4. *OSHR*, BNA, Vol. 2, No. 47 (April 26, 1973), p. 1371.

5. *Ibid.*

6. *The President's Report on Occupational Safety and Health*, G.P.O. Document No. 4000-00288 (December 1973), p. 90.

7. *OSHR*, BNA, Vol. 2, No. 47 (April 26, 1973), p. 1346.

8. George C. Guenther, "The Significance of the Occupational Safety and Health Act to the Worker in the United States," *International Labour Review*, Vol. 105, No. 1 (January 1972), p. 62.

9. Government Research Corporation, *Occupational Safety and Health, A Policy Analysis* (1973), p. 6. [Hereinafter cited as GRC]

10. GRC, p. 33.

11. GRC, p. 29.

12. *President's Report*, 1973, p. 1.

13. *Ibid.*, Appendix C in the section on OSHA, p. 60.

14. GRC, p. 30.

15. GRC, p. i.

16. GRC, p. 10.

17. GRC, pp. 12-13; emphasis added.

18. *President's Report*, 1973, p. 12.

19. *OSHR*, BNA, Vol. 2, No. 7 (April 26, 1973), p. 1372.

20. GRC, p. 34.

21. *Slow Progress Likely in Development of Standards for Toxic Substances and Harmful Physical Agents Found in the Workplace*, Report to the Committee on Labor and Public Welfare, U.S. Senate, by the Comptroller General of the United States, General Accounting Office Document No. B-163375 (September 28, 1973), p. 3. [Hereinafter cited as GAO]

22. GRC, p. 33.

23. *Ibid.*, p. 52.

24. GAO, p. 59.

25. GAO, cover letter.

26. *Ibid.*, pp. 16, 19.

27. *OSHR*, BNA, Vol. 3, No. 36, p. 1131; GAO, p. 67.

28. *Wall Street Journal*, March 25, 1974, p. 19.

29. *Industrial Union Department v. Hodgson*, No. 72-1713 (April 15, 1973).

30. GAO, p. 53.

31. GRC, p. 35.

32. *President's Report*, 1973, p. 14.

33. GRC, p. 30.

34. *Ibid.*

35. National Safety Council, *Safety Newsletter* (September 1973).

36. *President's Report*, 1973, p. 33.

37. Joe Collier, "Inspection and Enforcement in the Workplace," *Monthly Labor Review* (August 1973), p. 39.

38. J. M. Stellman, and S. M. Daum, *Work is Dangerous to Your Health* (New York, Vintage Books, 1973), p. xiv.

39. *Ibid.*, p. 42. Frank Errera, Chief of OSHA's Office of Management Data Systems, press briefing, May 3, 1973.

40. *OSHR* BNA, Vol. 2, p. 1214.

41. Collier, "Inspection and Enforcement," p. 35.

42. *OSHR*, BNA, Vol. 4 (1975), p. 1375.

43. *President's Report*, 1973, p. 39; *OSHR*, BNA, Vol. 4 (1975), p. 1556.

44. *American Smelting and Refining Company*, OSHRC Docket No. 16 (August 17, 1973).

45. Earl Heath, "William Steiger's Occupational Safety and Health Act," *University of Florida Law Review*, Vol. 25 (1973), p. 251.

46. *OSHR*, BNA, Vol. 2, p. 1246.

47. *Ibid.*, p. 1214.

48. *President's Report*, 1973, p. 19.

49. *Ibid.*, p. 26.

50. Robert McKeand, Director of OSHA Training Center, interview with author, June 13, 1972.

51. *President's Report*, 1973, p. 33.

52. *Ibid.*

53. *International Labour Review* (International Labour Organization, Geneva), Vol. 105, No. 1 (July 1972), p. 63.

54. Maurice F. Bresnahan, "Availability of Occupational Injury and Illness Data and Future Plans," speech presented before the National Symposium on Occupational Safety and Health, Washington, D.C., June 4, 1973.

55. *OSHR*, BNA, Vol. 4, No. 30 (December 26, 1974), p. 867.

56. *President's Report*, 1972, p. 111.

57. GAO, p. 1.

58. *Ibid.*, p. 2; emphasis added.

59. Mike Causey, "GAO Cites Safety Hazards in Agencies," *Washington Post*, March 24, 1973.

60. GAO, pp. 3, 4.

61. *Ibid.*

62. *OSHR*, BNA, Vol. 4, No. 18 (October 3, 1974), p. 492; Vol. 4, No. 41 (March 13, 1975), p. 1346.

63. *President's Report*, 1973, p. 103.

64. GAO, p. 3.

65. *The Plans and Programs of NIOSH, 1973*, unpublished document.

66. *President's Report*, 1973, p. 105.

67. *Ibid.*, p. 106.

68. Arthur D. Little, Inc., *The Present Status and Requirements for Occupational Safety Research* (1972).

69. GRC, p. 38.

70. *President's Report*, 1973, p. 114.

71. *OSHR*, BNA, Vol. 4, No. 39 (February 27, 1975), p. 1277.

72. Marcus Key, speech presented before the 45th Annual Maine State Safety Conference, September 21, 1972.

73. Marcus Key, speech presented before the American Public Health Association, Atlantic City, New Jersey, October 14, 1972.

74. *OSHR*, BNA, Vol. 4, No. 36 (February 6, 1975), p. 1192.

75. *President's Report*, 1973, p. 105.

76. *Ibid.*

77. *Ibid.*, p. 133.

78. *Ibid.*, p. 131.

79. *President's Report*, 1972, p. 128.

80. *Ibid.*, p. 135.

81. *President's Report*, 1973, p. 125.

82. *OSHR*, BNA, Vol. 2, No. 7 (April 26, 1973), p. 1372.

83. *News from NIOSH*, Vol. 1, No. 9 (August 1973), p. 31.

84. GAO, p. 43.

85. Joseph Page and Mary-Win O'Brien, *Bitter Wages* (Grossman, New York, 1973), pp. 244-245.

86. *OSHR*, BNA, Vol. 2, No. 47 (April 26, 1973), p. 1371.

87. *OSHR*, BNA, Vol. 4, No. 43 (March 27, 1975), pp. 14-15.

88. John Tavela, "If a Citation is Appealed," *Monthly Labor Review* (August 1973), p. 46.

89. OSHRC Docket No. 221 (January 17, 1973).

90. OSHRC Docket No. 2534 (February 5, 1974), 1 OSHC 1532.

91. *Ibid.*, p. 1533. Cf. *Bill Echols Trucking Co.*, 487 F.2d 230 (5th Cir. 1973).

92. OSHRC Docket No. 1588 (November 7, 1974), 1 OSHC 1390.

93. OSHRC Docket No. 61 (February 22, 1974), 1 OSHC 1598.

94. OSHRC Docket No. 1348 (November 15, 1974), 2 OSHC 1313, petition for review docketed, Nos. 74-3981, 75-1091 (5th Cir., November 29, 1974).

95. OSHRC Docket No. 1210-P (March 22, 1974), 1 OSHC 1600.

96. 40 *Fed. Reg.* 3594 (January 23, 1975).

97. OSHRC Docket No. 370 (October 11, 1974), 2 OSHC 1256.

98. OSHRC Docket No. 675 (December 24, 1974), 2 OSHC 1449.

99. 502 F.2d 278 (9th Cir. 1974).

100. Howard Pyle, address before the banquet of the International Association of Governmental Labor Officials Convention, July 18, 1972.

101. Collier, "Inspection and Enforcement."

102. *Report of the State Programs Subcommittee to NACOSH*, p. 6.

103. *Current Report*, BNA, Vol. 3 (December 27, 1973), p. 933.

104. *Ibid.*, p. 932.

105. *OSHR*, BNA, Vol. 3, No. 37 (February 14, 1974), p. 1166.

106. *OSHR*, BNA, Vol. 3, No. 35 (January 31, 1974), p. 1102.

107. *OSHR*, BNA, Vol. 3, No. 41 (March 14, 1974), p. 1275.

108. *OSHR*, BNA, Vol. 2, No. 47 (April 26, 1973), p. 1371.

109. *Ibid.*, p. 1372.

110. *OSHR*, BNA, Vol. 3, No. 28 (December 13, 1973), p. 891.

111. OSHA Program Directive No. 200-28 (April 22, 1974), superseding Program Directives No. 200-20 (September 7, 1973) and No. 200-20A (September 28, 1973).

112. National Commission on State Workmen's Compensation Laws, *Compendium on Workmen's Compensation*, G.P.O. Document No. 496-632 (1973), Foreword by John F. Burton, Jr., Chairman.

113. *The Report of the National Commission on State Workmen's Compensation Laws*, G.P.O. Document No. 477-053 (July 1972), p. 125.

114. *Ibid.*, p. 26.

115. *Ibid.*

116. *Ibid.*

117. *OSHR*, BNA, Vol. 5, No. 5 (July 3, 1975), p. 162.

118. *OSHR*, BNA, Vol. 3, No. 42 (March 21, 1974), p. 1325.

119. *President's Report*, 1973, p. 14; emphasis added.

120. *Man, Materials and Environment*, A Report for the National Commission on Materials Policy by the National Academy of Engineering and the National Academy of Sciences (March 1973); emphasis added.

121. GRC, p. ii.

122. GAO, p. 45.

123. *OSHR*, BNA, Vol. 3, No. 42 (March 21, 1974), p. 1324.

124. *Report of the National Commission on Workmen's Compensation*, p. 51.

125. *Ibid.*, p. 97.

7 Economic Issues. I. The OSHAct and Its Consequences

After reading about the difficulties and inadequacies of using the law (in this case the OSHAct) to solve occupational health and safety problems, the reader might be tempted to conclude that other policy instruments—particularly those designed to affect the market—would be more successful mechanisms for achieving an increase in worker health and safety. Indeed, for the problems of the general environment and consumer safety, two blue-ribbon studies have concluded that less rigid standard enforcement and more "attempts to raise the value of the social product in the market" is the preferable policy.[1,2] The discussion of broad economic issues in this chapter, of workmen's compensation in Chapter 8, and of manpower issues in Chapter 9 will hopefully convince the reader that, whatever the case for the general environment, the market cannot be relied upon to solve the health and safety problems in the work environment.

7.1 INTRODUCTION

7.1.1 Why Economic Analysis Is Important for Occupational Health and Safety

An understanding of the economic issues is important because

- economic forces explain both past and present behavior, in part;
- economic mechanisms and policies can be used to change future behavior, in part;
- there are limitations to the use or effects of economic mechanisms or policies, i.e., there are both imperfections in the "market" and "nonmarket" forces at work;
- to the extent that the law is used as an important mechanism for allocating (and internalizing) costs and liability, the law affects the workings of the economic system.

Occupational disease and injury are part of the human and social costs of production. (For a full discussion of the present costs of occupational injury and disease, see Section 3.4.) As such, the level of workplace-related injury and disease is influenced by the economic forces that come to bear on firms and their employees. Companies, for example, are in business to make profits for their owners, and must produce an adequate rate of return if they are to remain economically viable. Competition among firms all striving to make profits results in constant pressure to reduce costs—often a socially desirable objective in itself. Unfortunately, it is also the case that many ways of reducing costs also increase workplace hazards, especially in the short run. Workers, for their part, would of course prefer a safe and healthy workplace, yet the securing of this objective may often result in reduced wages or even the loss of jobs through plant closures or automation. Government is particularly sensitive to economic forces in the occupational health and safety area. OSHA's former standards chief Gerald Scannel stated that OSHA should and would consider economic feasibility in the development of new standards (see Section 6.3.1).

Since the turn of the century, it has been recognized that placing sole reliance upon the unregulated free market leads to a socially unacceptable level of workplace injury, illness, and death. Political pressure, especially at the federal level, led first to the establishment of state workmen's compensation laws and health and safety codes, and more recently to the OSHAct. While these and other laws have helped improve working conditions, their design and application have always been greatly influenced by their anticipated economic consequences. Even after laws have been enacted to improve work safety and health, constant political pressure has been applied to limit the application of the laws and even to amend them or keep appropriations so low that they remain ineffective.

Although private and governmental behavior affecting occupational hazards is in some sense a reaction to strictly economic forces, custom, tradition, and a host of other social, cultural, and attitudinal characteristics mediate the influence of these forces upon behavioral outcomes. The role of such considerations is perhaps most apparent in, though scarcely confined to, the political arena. For example, unions, employers, and other groups often put great pressure upon legislators to enact what they view to be favorable legislation, with neither the pressuring groups nor the legislators first properly calculating the likely costs and benefits of the desired legislation or fully weighing its economic and social consequences. Such behavior is a reflection of the difficulty of predicting even beneficial outcomes with certainty, or of comprehending their full complexity. Thus, mine owners' fears of the economic consequences of proposed coal mine health and safety legislation led them to exert great pressure against its passage, despite evidence that improved mine health and safety conditions could result in benefits for employers even in the narrow economic sense. Supporters of better safety conditions in the coal mines were unable to overcome political opposition until after the tragic deaths of 78 men in a mine explosion near Farmington, West Virginia. The public's reaction to the Farmington disaster was a highly emotional one, yet decisive in securing the passage of an act with significant economic consequences.

Workers themselves may refuse to wear safety equipment or oppose technological improvements in working conditions which they (sometimes incorrectly) believe may threaten their jobs or reduce their control over their tasks.

Much of the behavior of firms, unions, government, and individual working people is indeed guided by economic forces. However, government is often forced to react to the pressures exerted by interest groups for programs that will maximize their own benefit at the expense of another's or of society's. This can occur with regard to a broad range of activities relevant to occupational safety and health. Among these are (1) research, education, and information transfer; (2) technology development and transfer; (3) manpower development and its utilization for research, technology, training, and enforcement; (4) maintenance of full employment; (5) rehabilitation; (6) tax incentives; (7) financial, technical, and medical assistance to small businesses and marginal firms, displaced workers, universities and research institutes, and to industry and labor unions in general; (8) enforcement of health and safety standards; and (9) workmen's compensation.

To the extent that our institutions (government, labor, and management) do not react optimally (from a social point of view), the design of economic and legal mechanisms and policies will have to be modified. Therefore, an appreciation of both the economic issues and forces and the behavioral responses and motivations of man, individually or collectively, is necessary.

7.1.2 Market Imperfections and Economic Analysis

Economic theory helps us to understand why the unregulated market does not function efficiently as an allocative mechanism in many cases. In order for the free market to function efficiently, a number of stringent conditions must be met, including the absence of monopoly power, free and perfect information, full employment of all scarce productive inputs, and no errors or omissions in the allocation of costs and benefits. The latter condition—what the economist calls the absence of externalities—is especially significant, and means that the price of each product must reflect *all* the costs incurred in its production, both private *and* social. In the presence of externalities or other market failures, government regulation is necessary to ensure the smooth functioning of the market mechanism.

One familiar example of an externality is the case of water pollution. A factory may pollute a river and make it unusable to people downstream. If the firm is not held accountable for the damage so inflicted, it will pollute too much from society's point of view, especially if the people downstream are uninformed or unorganized. Government can intervene to correct this situation by establishing a "pollution market" through the assignment of (tort) liability, by directly ordering the firm to pollute less, or by compelling the firm to pay effluent charges on its untreated waste.

With regard to occupational health and safety, the problem of market imperfections is indeed a serious one. Among the most important market imperfections are problems of (1) the unavailability of information; (2) "indivisibilities" in the transfer of new technology, i.e., scale problems; (3) the training and utilization of technical expertise; and (4) disparity between the social and private costs of injuries and disease.

7.1.3 Other Economic Problems

Attempts to diminish occupational hazards are often complicated by other kinds of economic problems, not necessarily identifiable as "market imperfections." Safe conditions are often more expensive to achieve for smaller firms than for larger ones. The cheapest control technologies often require a minimum size which is larger than the operations of many production units. These firms also often find it difficult to locate and pay for the expert advice or information which they may require.

Many firms—often the ones with the worst health and safety problems—may also be economically marginal. Such firms are often found in highly competitive industries and are under constant threat of failure. Especially when these marginal firms have more profitable domestic competitors with substantial financial resources or foreign competitors not forced to internalize the costs of health and safety, they may be faced with a choice between hazardous working conditions and bankruptcy. However, in specific instances there are often many ways of improving health and safety which are not costly, and in such cases the marginal nature of the firm need not be a reason for ignoring workplace hazards. In general, it has been found that the most profitable firms in an industry are also the safest. This may be explained by many factors, such as better management (for both production and safety), the utilization of longer time horizons over which to evaluate the investment in good safety practices, and better access to information and expertise.

Time horizons can differ within the firm as well as among firms. Top management may believe in principle that it is consistent with the long-run economic interest of the firm to adopt good health and safety practices. Nonetheless, it may at the same time judge the performance of its middle management in the short run with the result that disease and injury may be higher than they need be even on the basis of narrow economic considerations.

Chronic disease which becomes manifest only after a long period of time presents a "discounting" problem which is difficult for even the most farsighted management to perform. The discounting problem for future harm (costs) is paralleled by a discounting problem for future benefits as well.

Even if a firm thought it profitable in terms of conserving its scarce human resources to reduce chronic disease, the uncertainty of future benefits makes it unlikely that long-term action would be taken.

Finally, the introduction of new chemical substances into the workplace is proceeding at a rapid rate, while an understanding of the potential hazards associated with their use grows much more slowly. Assessment of the damage which these substances may cause several decades in the future (or mutations in the next generation) becomes enormously difficult. There is thus an increasing divergence between what is known and what needs to be known. The dynamic nature of increasing problems means that economic solutions based on static assumptions will be of minimal use.

7.1.4 Cost-Benefit Analysis and Economic Efficiency

Economic analysis not only helps to describe many issues in occupational health and safety, it provides tools such as cost-benefit analysis for helping to evaluate the consequences of decisions. There are several levels of possible application for cost-benefit analysis: (1) estimation of the immediate economic consequences of a project which have established market values; (2) inclusion of monetarily quantifiable costs that the market does not normally capture (because of imperfections); (3) at a yet more sophisticated level, inclusion of those effects that have no established market value, but are clearly part of the overall "social cost" of the project.

7.1.5 The Issue of Equity

Often, decision-making has economic efficiency as its only objective. However, the question of *who* pays the cost and *who* reaps the benefit is also important. Minimizing nonrandom victimization through a concern for individual justice is a legitimate social goal which may at times conflict with attainment of economic efficiency. Society may prefer to move away from an economically efficient point in order to have a *fairer* distribution of costs and benefits. Of course, different people view what is fair differently—but this fact makes the consideration of equity no less important. Whatever the alternative value judgments as to what is fair, we should know what the costs are for those alternatives we wish to consider. This fourth level of application of cost-benefit analysis is rarely considered; it differs from the level (3) because it recognizes that one person's cost cannot simply be evenly traded for another person's benefit if the risk of harm falls on a selected group of people in an industry.

7.1.6 Questions This Chapter Attempts to Answer

The following questions are offered, not as being all-inclusive, but as representative of the multifaceted economic problems that need to be addressed in occupational health and safety.

- To what extent are costs presently being borne by the worker, by industry and the employer, and by the society generally?
- How has the new federal legislation shifted the allocation of these costs?—or provided economic incentives to reduce injury and disease?
- To what extent does the workmen's compensation system act as an economic incentive to reduce accidents and disease?
- Are good health and safety policies profitable in the short run? in the long run?
- How would full enforcement of occupational health and safety standards affect the mix of small and large firms? the displacement of workers?
- How do alternative health and safety policies affect the competitive position of

the United States in the world market?
- To what extent is selective enforcement and voluntary compliance effective?
- Can a worker be "economically rational" in taking risks on the job?
- What are the present and projected manpower needs? What manpower policies have been adopted by management, labor, government, and educational institutions?
- What are the present and projected research and technology needs? What policies have been adopted by management, government, research institutions, and industry?
- How serious is the information transfer problem? To what extent is ignorance a factor in an unsafe and unhealthy work environment?
- How serious is the technology transfer problem, both vertically within the society and horizontally at the national and international level?

7.2 ELEMENTS OF THE COST AND BENEFIT OF COMPLIANCE WITH OCCUPATIONAL HEALTH AND SAFETY LEGISLATION

7.2.1 Possible Costs of Improved Workplace Health and Safety

One possible consequence of the OSHAct, the Coal Mine Health and Safety Act, and pending reform of the workmen's compensation system is increases in certain types of costs for many firms, their employees, and society at large. Such changes in costs can be expected for a broad range of items. Among the more important of these are the following:
- Increased in-plant services, in the areas of industrial safety, industrial hygiene, and occupational medicine.
- Process changes such as the implementation of closed manufacturing systems to minimize or eliminate exposure to toxic or carcinogenic substances, improved ventilation to reduce exposure to potentially harmful industrial dusts, and acoustical redesign to reduce or eliminate noise pollution.
- Design of new equipment to meet the objective of improved occupational safety and health, such as the development of a new, quieter loom that generates less cotton dust.
- Development and implementation of substitutes for hazardous and carcinogenic materials, such as the use of fiberglass in place of asbestos.
- Changes in the organization of work to reduce the level of occupational hazard, such as institution of longer rest periods, shortening of the working day, reduction in the pace of work, and implementation of job rotation to reduce the possibility of chronic or long-term exposure to hazards such as noise.
- Education of workers and employers as to their rights and duties regarding the provision of improved working conditions; improved orientation of newly hired

workers; technical training to enhance the capability of working people to monitor the hazards of their job environment and take the necessary steps to bring about alleviation of hazardous conditions.

- Health research on the broader implications of occupational hazards in terms of workers' physical and mental well-being; increased research effort regarding the *preventive* (as opposed to the curative or treatment) aspects of occupational medicine.
- Manpower development for the detection and prevention of occupational hazards.
- Increased workmen's compensation premiums, especially for firms and industries which generate high levels of presently unrecognized occupational *disease.*
- Costs of displacement, relocation, and retraining of workers who lose their jobs as a consequence of their employer's compliance (or inability to comply) with the OSHAct.
- Relative worsening of the competitive position of small firms, so that one consequence of compliance with the OSHAct may be an alteration of the mix of small and large firms in the economy in favor of the latter.
- Worsening of the relative position of less profitable firms within an industry, with the result that the rate of disappearance of marginal firms may be increased.
- Worsening of the relative position of less profitable industries (e.g., coal mining versus oil extraction).
- Possible deterioration of the *international competitive position of U.S. firms* whose foreign rivals are not subject to the costs of complying with the OSHAct.

In regard to all of the above items of possible cost increases, it is expected that the magnitude of change is potentially greatest with respect to health-related (as opposed to safety-related) hazards.

7.2.2 Estimates of the Costs of Compliance with the Legislation

Impact on the Relative Competitive Position of the United States

In an interview, a representative of the National Association of Manufacturers expressed his feeling that a deterioration in the relative international competitive position of American firms would perhaps be the most adverse consequence of compliance with the OSHAct.

A report for the National Commission on Materials Policy concludes that

> there are only a few econometric estimates of the impact of environmental control on the short-term balance of payments and level of aggregate income. One, using an econometric model of the U.S. economy and projected future American EC measures (assuming other nations do *not* follow suit), emerges with the following results: Over the 1972-1980 period, pollution control and compensatory macroeconomic policy measures are expected to raise the U.S. unemployment rate by 0.3%,

raise the average annual rate of inflation by 0.26%, reduce fixed investment not related to pollution control by an average of $2.3 billion annually, and exert a negative impact on the U.S. balance of trade in the amount of $1.9 billion. (Council on Environmental Quality, 1972). If, on the other hand, pollution control costs are 50% higher than current estimates, or standards are raised correspondingly, the average annual 1980 negative trade balance effect is estimated to be in the neighborhood of $3.2 billion or 6-7% of current level of exports.

The meager empirical evidence available at this time suggests the balance of payments, employment, and national income effects resulting from domestic environmental controls are likely to be relatively small for the United States. This is true even if the trend toward indirect subsidization is reversed toward a full "polluter pays" principle. Thus, a tentative conclusion is that the *aggregate impact* of domestic environmental controls should not be a matter of deep concern for policy-makers. Of more crucial importance are the short-term impacts on particular firms and industries. Substantial additional research is needed on these questions.[3]

However, the President's Science Advisory Committee takes into account the expectation that foreign rivals would be adopting similar regulatory controls:

The United States has consistently been a net exporter of chemical products. It is also an exporter of technology and research and development but with repatriation of profits through multinational firms. The regulatory process does not seem to have been a major force in this trend. Rather it simply reflects business opportunities and the regulations prevailing in other nations.

The Panel is persuaded that much or all of the industrialized world is moving towards a pattern of regulation which resembles that of the U.S.[4]

Thus, the notion that the United States would suffer a competitive disadvantage internationally by enforcing strict occupational health and safety regulations is open to much question. Perhaps what is of concern to American businessmen active in international markets is the opportunity cost of losing the competitive *advantage* which they would enjoy by operating under a system which does not have to adopt or enforce the regulations other trading nations expect to adopt.

National Association of Manufacturers

The National Association of Manufacturers recently conducted a survey to determine the impact which the OSHAct was having on U.S. companies. NAM found that 84.4% of companies responding to its questionnaire had "initiated new safety and health programs or activities to implement the provisions of the Act."[5] Companies in the sample were also queried about the cost of complying with OSHA standards. Estimated cost of compliance according to firm size is given in Table 7.1. Note the decreasing cost per worker as firm size increases.

The Conference Board

The Conference Board has published a study based on a survey of over 800 companies and extensive interviews with representatives of top management and corpo-

Table 7.1
Estimated Cost of Compliance with OSHA Standards (in Dollars)

Company Size (No. of Employees)	Estimated Total Expense (Weighted Average)	Estimated Expense per Worker*
1-100	33,000	660
101-500	104,000	400
501-1000	212,000	270
1001-2000	372,000	250
2001-5000	863,000	245
over 5000	7,146,000	--

*Author's calculation, using midpoints of firm size ranges.

Source: National Association of Manufacturers, "Where We Stand," p. 21.

rate medical and personnel directors. One purpose of this study was to investigate the response of firms to the OSHAct at a time when the law had been in effect for one full year. A large proportion of the firms surveyed (47%) reported that at least some of their employees were presently subjected to hazardous working conditions. Of these firms, a substantial proportion (83%) had initiated changes in one or more corporate activities in an effort to comply with the provisions of the OSHAct. A large proportion of these companies reported changes in methods of monitoring the work environment (71%), occupational safety and health staffing or organization (48%), and research activities directed toward alleviation of hazardous working conditions (27%).

Among companies changing their method of monitoring the work environment, a large proportion reported having introduced new or improved measures to monitor specific job environment hazards such as noise, fumes, dust, radiation, and exposure to other toxic substances. Other companies reported improvements in their self-inspection practices, such as increasing the frequency and thoroughness of inspections.

A large proportion of companies reported making additions to their occupational health and safety staff. Sometimes these staffing changes involved hiring additional fully trained occupational safety and health professionals; more typically, the duties of personnel already employed were augmented to include occupational safety and health responsibilities. A number of firms reported the creation or expansion of occupational safety and health committees or departments.

Medical programs have likewise been created or expanded, according to many firms. Often these involve intensification of health screening efforts, or utilization of medical staff in giving first-aid instruction to employees. Significantly, increased utilization of medical expertise for *preventive* purposes—such as consultation regarding possible health implications of new equipment or processes—was not mentioned.

McGraw-Hill

The first economywide survey of corporate capital investment plans regarding employee health and safety was undertaken in 1973 by McGraw-Hill.[6] Highlights of the survey's results were as follows:

1. American industry planned to spend $3.16 billion for employee safety and health in 1973, a 26% jump above 1972. Preliminary plans for 1976 totaled $3.56 billion.
2. Manufacturing industries expected to spend $1.26 billion in 1973 for upgrading safety and health conditions on the job. This represented a 34% increase over 1972. Preliminary plans for 1976 totaled $1.83 billion.
3. Projected investment for employee safety and health accounted for 3% of total planned capital investment in 1973, compared with 2.8% in 1972. Manufacturing companies' projected share was 3.1%, up from 3% in 1972.

More detailed information on corporate investment plans for employee health and safety care are contained in Tables 7.2 and 7.3.

It can be concluded that the magnitude of planned investment in this area, though significant in absolute terms, is relatively small as a proportion of total capital investment (only 3%). Moreover, investment in employee health and safety is far less than investment in general environmental protection in many industries. Though the employee health and safety component of capital investment is growing at a fairly rapid rate, it appears to be growing no faster than total investment, remaining roughly stable as a proportion of total planned investment through 1976.

Dun's Review

According to *Dun's Review*, an influential business publication, costs of compliance with the OSHAct are likely to prove substantial: "While no one has attempted an exact accounting, estimates of OSHA's total cost to U.S. industry run into the tens of billions of dollars. A few industries will get off lightly, but costs of operations in many could rise by as much as 5% to 10%."[7]

Dun's Review predicts that costs of compliance will be especially burdensome in the reduction of noise levels, and in the improvement of air quality by means of ventilation. Marginal firms operating with old, fully capitalized equipment may be forced out of business. *Dun's* also mentions several industries which may find compliance particularly costly:[8]

- *Metal-stamping machines.* The companies that produce the units that form sheet metal into everything from car bodies to bottle tops will be spending more than $6 billion over the next five years to comply with OSHA, according to Jefferson Keith.
- *Lumber and wood products.* With its excessive noise, the use of large, unguarded saws, flying wood chips, dust, and manual handling of huge logs, the lumber industry will be taking on a massive dose of safety under the Act. Observers think the cost will go into the billions of dollars.

Table 7.2
Plans for Investment in Employee Safety and Health (Millions of Dollars)

Industry	Actual 1972	Planned 1973	% Change 1972-1973	Planned 1976
Iron and steel	$ 193	$ 215	11%	$ 478
Nonferrous metals	37	46	24	97
Electrical machinery	57	64	12	105
Machinery	86	131	52	213
Autos, trucks, and parts	88	74	–16	65
Aerospace	14	26	86	16
Other transportation equipment	6	15	150	18
Fabricated metals	20	29	45	32
Instruments	12	21	75	23
Stone, clay, and glass	30	87	190	93
Other durables	37	66	78	54
Total durables	580	774	33	1194
Chemicals	72	96	33	146
Paper	50	66	32	71
Rubber	15	35	133	72
Petroleum	68	99	46	142
Food and beverages	71	95	34	107
Textiles	58	67	16	67
Other nondurables	24	25	4	29
Total nondurables	358	483	35	634
All manufacturing	938	1257	34	1828
Mining	84	116	38	86
Railroads	31	34	10	44
Airlines	54	55	2	28
Other transportation	70	66	–6	73
Communications	404	569	41	509
Electric utilities	203	370	82	445
Gas utilities	23	26	13	31
Commercial*	702	663	–6	519
All nonmanufacturing	1571	1899	21	1735
All business	2509	3156	26	3563

*Figures based on large chain, mail order and department stores, insurance companies, banks and other commercial businesses.

Source: McGraw-Hill, "First Annual Survey of Investment in Employee Safety and Health," Table 1, p. 5.

Table 7.3
Employee Safety and Health Investment as Percent of Capital Spending

Industry	Actual 1972	Planned 1973	Planned 1976
Iron and steel	12.3%	10.6%	17.4%
Nonferrous metals	3.1	2.4	5.2
Electrical machinery	2.4	2.3	3.4
Machinery	3.0	3.2	5.4
Autos, trucks, and parts	4.8	3.1	2.4
Aerospace	3.3	4.1	2.4
Other transportation equipment	2.4	3.4	5.0
Fabricated metals	1.3	1.5	1.6
Instruments	1.7	2.3	2.3
Stone, clay, and glass	2.5	4.3	4.7
Other durables	2.3	2.7	2.1
Total durables	3.7	3.6	5.2
Chemicals	2.1	2.2	3.1
Paper	3.6	3.3	4.1
Rubber	1.4	2.3	4.6
Petroleum	1.3	1.7	1.9
Food and beverages	2.8	3.2	3.5
Textiles	7.9	8.1	8.8
Other nondurables	1.9	1.8	2.6
Total nondurables	2.3	2.6	3.1
All manufacturing	3.0	3.1	4.2
Mining	3.5	4.4	2.8
Railroads	1.7	1.9	2.1
Airlines	2.2	2.1	2.1
Other transportation	4.8	3.6	3.6
Communications	3.4	4.2	2.8
Electric utilities	1.4	2.2	2.1
Gas utilities	0.9	0.9	0.8
Commercial*	3.5	2.9	2.1
All nonmanufacturing	2.8	2.9	2.3
All business	2.8	3.0	3.0

*Figures based on large chain, mail order, and department stores, insurance companies, banks, and other commercial businesses.

Source: McGraw-Hill, "First Annual Survey of Investment in Employee Safety and Health," Table 2, p. 6.

• *Housing.* A prominent member of the National Association of Home Builders predicts that OSHA will add $2000 to $3000 to the cost of each new home.
• *Paper.* An industry executive predicts that OSHA's cost to the pulp and paper industry will run at about $200 per worker over the next year or two—for a total cost of $140 million.

Noise

As has been pointed out elsewhere in this report, noise is perhaps the most pervasive occupational health hazard in American industry today. Controversy surrounding the setting of appropriate standards for regulating the exposure of workers to this hazard has been intense, with affected industries applying considerable political pressure to prevent imposition of standards which they feel are too strict.

Against this background, estimates have recently been prepared for OSHA by the consulting firm of Bolt, Beranek and Newman regarding the cost to industry of compliance with proposed noise standards. Their report concludes that the cost of compliance with an 85-dBA noise standard would be in excess of $31 billion for 19 major industries.[9] (The report was unavailable at the time of this writing; thus, no evaluation of the study could be made.)

According to the same report, the cost of complying with the 90-dBA noise standard favored by business would be substantially less, though nevertheless considerable—$13.4 billion. Over one-fifth of this cost of noise abatement would be borne by the utilities industries, which would have to spend $6.3 billion to comply with an 85-dBA standard, and $3.2 billion to comply with a 90-dBA standard.

Estimated costs of compliance with the 90-dBA and 85-dBA standards are shown in Table 7.4. Though these costs would be spread out over a period of three to five years, it is nevertheless clear that the required outlays (if the estimates are correct) will be large relative to sales and profits in a number of industries. (The authors of the study acknowledge that the estimates may indeed be high because of their assumption that the cost of noise-control devices will remain constant.)[10] However, these figures by themselves do not tell the whole story. Benefits of reduced noise exposure, such as reduced absenteeism and workmen's compensation costs saved, would result in lower *net* costs to the industry.*

Asbestos

Controversy has surrounded the establishment of an appropriate standard for controlling employee exposure to this hazardous substance. Much of the controversy has revolved around the cost to corporations of complying with standards of varying degrees of strictness. Indeed, the prestigious consulting firm of Arthur D. Little concluded in a report to the Department of Labor that costs of compliance with a proposed two fiber per cc asbestos standard were so high as to warrant adoption of a significantly less stringent standard.[11] According to ADL's report, "a two-fiber

*For an attempt to estimate net costs, see D. Hattis et al., *Some Considerations in Choosing an Occupational Noise Standard* (EPA, 1975).

Table 7.4
Projected Industry Costs to Comply with 85- and 90-dBA Noise Levels

SIC Code	Industry	Costs to Comply with Present (90-dBA) Standard[1]	Costs to Comply with Proposed (85-dBA) Standard[1]	1971 Sales[2]	1971 Profits[3]
20	Food and kindred products	590	2600	103,631	5141
21	Tobacco products	45	90	5528	1217
22	Textile mill products	1100	2700	24,030	1068
23	Apparel and related products	0	10	25,025	1025
24	Lumber and wood products	160	650	14,931	927
25	Furniture and fixtures	190	580	9754	424
26	Paper and allied products	140	500	25,458	951
27	Printing and publishing	870	1000	26,887	1996
28	Chem. and allied products	110	1400	51,873	6700
29	Petroleum and coal products	210	260	26,935	6825
30	Rubber and plastic products	300	500	17,044	1208
31	Leather and leather goods	0	8	5218	317
32	Stone, clay, and glass	290	520	18,535	1483
33	Primary metals	900	1900	53,067	2055
34	Fabricated metal products	1100	3200	42,026	2103
35	Nonelectrical machinery	1400	4200	55,560	4869
36	Electrical machinery	780	2300	49,168	4705
37	Transportation equipment	1100	2900	86,920	6893
49	Utilities	3200	6300	22,922	N.A.
	Total	13,480	31,618	664,512	49,763

1. Expressed in millions of 1973 dollars.
2. Expressed in millions of 1971 dollars. From Department of Commerce, *Annual Survey of Manufacturers*, 1971.
3. Corporations only; expressed in millions of 1971 dollars. From Federal Trade Commission, Securities and Exchange Commission, *Quarterly Reports.*

Source: *OSHR*, BNA, Vol. 3, No. 37 (February 14, 1974), pp. 1183-1184. Estimates prepared by Bolt, Beranek and Newman, Inc.

standard, if uniformly enforced, would require capital investment on the part of industry that is at least twice that required to meet the five-fiber standard."[12] In dollar figures, according to the report, capital investment required over a two-year period to meet the two-fiber standard would be approximately $192 million, compared to $80 million to meet the five-fiber standard. (Of course, these estimates ignore possible economic *benefits* to be derived from the substitution of other materials for asbestos; cf. the experience of the Swedish shipbuilding industry, discussed below.)

Economic Impact of the Black Lung Benefits Program

It has been estimated that 200,000 or more active and former coal miners suffer to some degree from pneumoconiosis,[13] and the incidence of this debilitating and often fatal disease has probably been on the upswing (at least until recently) because of technological advances in the extraction of coal which have tended to increase dust levels in the mines. In a recent study of over 9000 miners conducted by NIOSH and the Bureau of Mines, it was found that nearly one-third had black lung, and that 2.5% had the "severe, advanced form that can cause disability and death."[14] The death rate among coal miners from chronic respiratory disease is five times greater than among a matched cohort drawn from the general population.[15] Yet, prior to 1969, victims of black lung disease were eligible for workmen's compensation benefits in only one state.[16]

One consequence of the Coal Mine Health and Safety Act of 1969 and subsequent amendments has been to extend black lung benefits to disabled miners and their surviving dependents. Commencing January 1, 1974, the law provides that employers must assume liability for these disability payments, either directly or through the purchase of insurance.* In view of the large proportion of miners eligible for compensation, the economic impact of this program could be substantial. This program comes at a time when the energy shortage makes it likely that there will be significant increases in the demand for coal in the years ahead.[18] It should be clearly recognized that the assumption of liability by the coal producers is a payment for "past sins" and, while significant, it should not influence the decision to invest *now* in abatement (mandated by the legislation) in order to reduce *future* compensation payments. (See Section 7.2.3 for a discussion of possible benefits of the mining legislation.) However, because the compensation premiums for past harm are levied on wages paid to those presently employed, the method of production may be influenced.

*Presently the federal government is still absorbing the cost of black lung disability payments, due to the vigorous legal efforts of the coal mine operators to avoid liability. The companies are basing their legal argument on the contention "that Congress had no constitutional right to presume that miners necessarily contract black lung in coal mines."[17] However, the mine operators recently lost a court battle and may soon have to begin payment of benefits.

Taking as given the projected premium rates of 11% of the wage bill in underground mining and 7% in strip mining, it is predicted that the black lung benefits program will cause modest increases in price and decreases in the quantity of coal produced; but, more importantly, there is likely to be a "significant shift from underground to strip mining."[19] As a result, employment in coal mining may decline between 7.4% and 9.6%–including a 9% to 13% decline in underground mining employment, which implies a loss of 10,000 to 14,000 jobs. Moreover, Morris Goldstein and Robert Smith, in their study of the economic impact of the black lung benefits program, doubt whether the 11% (underground) and 7% (strip) insurance premium rates mentioned earlier will be sufficient to generate the necessary revenue. They predict that the premium rate required in underground mining will be 22.5% of the wage bill (though only 5.5% of the wage bill in strip mining).[20] If these estimates prove correct, the impact on employment and the underground mining-strip mining "mix" will be even greater than the significant impact predicted on the basis of far lower premium rates.

The authors point out that an output tax generating the same revenue will have far smaller adverse consequences for employment than the proposed *payroll* tax. If insurance premiums are levied on the output of coal rather than payrolls, the predicted job loss will be held to only 2.6% to 4.9%.[21] Nevertheless, it seems clear that even the partial internalization of the social costs of occupational disease in the coal mining industry entailed by the black lung benefits program could have substantial economic consequences.

Unfortunately, Goldstein and Smith fail to consider that one of these economic consequences in the long run is likely to be a reduction of dust levels in underground coal mines and, hence, a reduction in the incidence of black lung disease among underground miners. Coal mine operators faced with the prospect of stiff black lung disability payments to miners harmed in the future may find it cheaper to initiate changes in their production process–such as improved ventilation and methods of wetting the mine face–which will significantly reduce coal mine dust levels. In the long run, modification in equipment design may bring about further reductions in the level of dust generated, without undue sacrifice of productive capacity,[22] and perhaps with an increase.[23]

7.2.3 Possible Benefits of Compliance

One operative principle underlying OSHA's enforcement philosophy of "voluntary compliance" is the notion that corporations can be convinced that investment in improved workplace health and safety is in their long-term economic self-interest. In order for voluntary compliance to work, companies must become convinced that expenditures on employee health and safety "pay." In fact, OSHA has thus far met with relatively limited success in convincing firms that such investment is justified on the basis of the same cost-effectiveness criteria that are employed in the evaluation of other potential investment programs.* Nevertheless, there are a number of reasons to suppose that investment in significantly improved workplace health and safety, though expensive, will ultimately yield substantial benefits, both to society and the firms which undertake it. Among these private and social benefits are the following:

- Improved workplace health and safety would lead to increased well-being of employees and their families.
- Productivity may increase as a result of improved employee health and morale and the reorganization of work necessary to achieve this end (job rotation, increased relief time, shorter workday, etc.).
- Development and implementation of nontoxic substitute materials which may turn out to be more profitable in the long run than the hazardous substances being replaced. (E.g., the Swedish shipbuilding industry found that the substitution of fiberglass for asbestos, originally undertaken for reasons of industrial hygiene, has in fact been accompanied by higher profits.)[24]
- Changes in the "mix" of firms resulting from compliance with the OSHAct could prove beneficial to the economy, in the long run, in terms of increased productivity. Hastening the disappearance of badly managed, unsafe, marginal firms may confer a dividend on the entire economy. Smaller firms which find the cost of compliance prohibitive in their present lines of activity may shift to product lines for which modes of production are less hazardous to employee health and safety.
- New industries and product lines may be developed as a result of increased demand for health and safety technology. The United States could become an international leader in the production of occupational health and safety equipment and the provision of expertise, opening up the possibility of a significant growth of exports in these areas.
- Compliance with the OSHAct may stimulate innovation in technology and job redesign and new capital investment to improve productivity and economic efficiency.
- A gain in overall economic efficiency could be achieved to the extent that a socially desirable redeployment of resources results from the internalization of the costs of occupational injury and disease brought about by OSHAct compliance. To the extent that the prices of products produced under hazardous conditions will more nearly reflect the true costs of occupational injury and illness, the efficiency

*George Guenther, then Assistant Secretary of Labor for OSHA, stated in a 1973 lecture that OSHA had not really been successful in selling the idea that "safety pays." Though he personally believed that improved workplace health and safety would pay in the long run, due to improved employee morale, reduced workmen's compensation costs, lower turnover and downtime, it had not been possible to convince employers that this was in fact the case. (Lecture at the Kennedy School of Government, Harvard University, April 26, 1973.)

of our economic system will be improved by OSHAct compliance.

In studying the economic effects of regulation in the chemical, pharmaceutical, and pesticide industries, the President's Science Advisory Committee stated that not only is there "little evidence of capital flight having taken place,"[25] but also

> a Government regulatory agency, if it performs appropriately, represents in the form of its approval or certification for marketing a *capital asset value* to the industries petitioning to it. It is in the industries' best interest to assure that the best scientific information is obtained for decisions.[26]

In the coal mining industry, Louise Langlois estimates that adherence to the standards imposed by the Coal Mine Health and Safety Act of 1969 will not only reduce black lung significantly, but will also be profitable for the industry.[27]

A report by the National Academies of Science and Engineering to the National Commission on Materials Policy states,

> Moreover, we question whether production cost in order to comply with federal health and safety provisions has increased as much as recently claimed. The ease with which most coal mining firms have complied with former "unattainable dust standards" suggests that technology can usually reduce the "higher costs" of health and safety standards. . . . Reaching of these standards will probably prove good business for the mining industry.[28]

In this section, I have attempted to indicate the elements of cost and benefit that are possible consequences of compliance with occupational health and safety legislation. The various estimates of the costs are only estimates. The benefits are much harder to quantify. It is my best guess that in the short run, the costs will be high and unequal in different industries, a different mix of large and small firms may result, marginal firms may disappear, some workers may be displaced, and severe pressures to resist the expansion of enforcement of the legislation will arise. If the enforcement persists, longer-range adjustments and the encouragement of new technology, substitute materials, and the redesign of jobs should result in higher total productivity, decreased worker absenteeism, and improved job health and safety.

Finally, it should be noted that many of the costs of improving workplace health and safety are one-time expenditures, which should properly be compared with the permanent or long-term benefits of better occupational health. The initial fixed costs of improved working conditions will no doubt be substantial in many industries; once made, however, these expenditures should continue to accrue benefits over a long period of time, requiring only a relatively modest yearly outlay for maintenance and upkeep. Viewed on an appropriate annualized basis (i.e., averaged over an appropriate time period), therefore, the costs of improved occupational health and safety are probably far smaller than the figures presented in the preceding pages. The latter reflect the large initial costs of overcoming decades of neglect.

7.3 COST-BENEFIT ANALYSIS

7.3.1 What Is Cost-Benefit Analysis?

Cost-benefit analysis is a methodology for evaluating and comparing the consequences of different policy alternatives. Even though cost-benefit analysis is sometimes thought of as a single "technique," there are, in fact, a variety of approaches to using it. For the purposes of clarifying the decision-making process, cost-benefit analysis can be of value to individuals, private corporations, labor unions, public institutions, and government.

A user of cost-benefit analysis is obliged to specify his goals (or alternative sets of goals) as precisely as possible. He must understand the alternatives open to him in meeting his goals, as well as their consequences. He can then use the results of this understanding to find the "best" way of achieving his objectives. The "best" method is often taken to be the one that is most economically efficient.

Suppose that a decision-maker had as his only goal the reduction of occupational injury and disease. He could employ one of the three modes of calculation to efficiently meet this goal. First, given a fixed budget, he could choose actions that would minimize injury and disease subject to the constraint of this fixed expenditure. Second, he could set a predetermined target level of injury and disease incidence and seek a way to achieve this goal at minimum cost.

These two methods of evaluation are limited by somewhat arbitrary (although often very real) constraints. A third method would be to attempt to find the "best" combination of expenditure and incidence of harm. The costs of damage and prevention at different incidence rates are shown in Figure 7.1 for a simple case in which all the costs have been expressed monetarily. Suppose a certain program will reduce the accident rate from one level to another (for example, from point B to

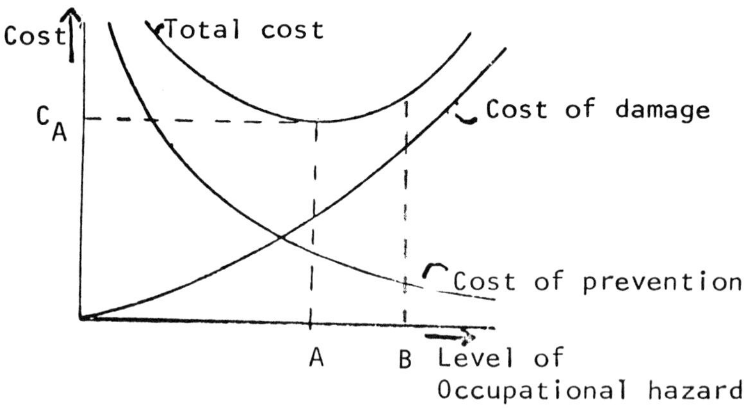

Figure 7.1 Determination of an economically efficient level of expenditure for the elimination of an occupational hazard. From Craig Sinclair, "The Incorporation of Health and Welfare Risks into Technological Forecasting," *Research Policy*, Vol. 1 (1971-1972), p. 52.

point A in Figure 7.1). If, as in the case shown, the value of the reduced accident rate (damage) is greater than the costs of the program (prevention), the program would be beneficial if undertaken. While Figure 7.1 shows only one "best" (optimal) level of workplace hazard corresponding to a minimization of the social costs associated with occupational injury and disease, in reality there may be many such levels.

The first two methods are the ones most likely to be used by decision-makers concerned with occupational health and safety, since no direct attempt need be made to quantify in economic terms the value of reduced injury or disease. To be sure, the value of human life and suffering is established *implicitly* when one decides upon the constraints facing the decision-maker, i.e., the amount of money to be spent (Case 1) or the maximum tolerable incidence rate (Case 2). For many decision-makers the third method is too theoretical, in that it requires a difficult-to-arrive-at *explicit* monetary equivalent of human life or suffering. One caveat, however, is necessary. If the decision-maker limits the value of reduced accidents and disease to the more readily quantifiable benefits such as reduced insurance premiums, absenteeism, and measurable production losses, then a determination by the third method can be readily made. However, such an approach overlooks a broad range of difficult-to-quantify benefits to the worker, his family, and society. Hence, the "optimal" point level is a fiction (i.e., will in fact be suboptimal) as far as the employee and society are concerned. Other difficulties in applying cost-benefit analysis to job safety and health will be dealt with throughout the remainder of this chapter.

7.3.2 The Limitations of Cost-Benefit Analysis

It is useful to consider the framework in which cost-benefit analysis can properly be used as a rational basis for action. Setting aside for the moment the question of whether cost-benefit analysis can be used in a given situation by a particular actor, the theoretical conditions which must be met are as follows: (1) market imperfections must be identified and properly accounted for (e.g., the costs of acquiring information must be accurately estimated); and (2) the effect of "nonmarket" forces such as custom, attitudes, and legal restrictions on the economic system must be accounted for.

A more detailed examination of the problems inherent in meeting these conditions is contained in later sections. There are, however, further difficulties which limit the practical usefulness of cost-benefit analysis as a tool for rational decision-making.

In an organization with only one goal, the problem of decision-making is relatively straightforward. One such example is the economist's model of the competitive, profit-maximizing firm. Although insufficient knowledge may complicate decision-making, the best alternative is always the one with the largest difference between estimated revenue and estimated costs—the alternative that gives the greatest expected profit. Most organizations, however, have a number of goals and may have a great deal of difficulty choosing among them. Decision-makers faced with multiple goals often try to reduce their diverse objectives to a single composite goal. For example, a safety department with a fixed budget may decide to try to minimize fatalities, or lost-time accidents. The department can thus simplify its objectives—but in doing so it ignores other goals, such as the reduction of disease.

For this reason, economists often approach the problem of decision-making by putting a market value on as many program consequences as possible. Even where there is no monetary value on a particular outcome, it is often possible to impute a market value by looking at similar situations in other parts of the economy. If it is possible to use both actual and imputed market values, the total costs of the program can be summarized by a single dollar value—as can the total benefits.

Often, however, cost-benefit studies stop at this point, and are thereby misleading in their conclusions. If objectives are not readily quantifiable, they are rarely given proper weight in the analysis. The most important examples of this drawback with reference to occupational health and safety fall into three categories, all of which will be discussed in more detail in subsequent sections. (1) There are consequences of planned activities for which there is no obvious market analogue. The value of human life is an important example. Sometimes, the issue is inadequately dealt with by equating the value of a person's life with his earning capacity. (2) Another set of problems involve comparing events in the present with those in the future. How should one compare $100 in benefits (or costs) today with $200 in benefits (or costs) ten years hence? This decision involves what economists call the selection of an appropriate discount rate. (3) Perhaps an even more difficult problem is that of equity (or fairness). The issue of who pays the costs and who reaps the benefits is all too often overlooked in the process of evaluating the two sides of the cost-benefit equation.

Cost-benefit analysis is a method of clarifying choices among alternative programs by evaluating their consequences. Of necessity, the analysis uses market prices as a measure of value, and these are often a poor index of social value. However, even if we accept market prices (where they exist) as an adequate measure of value, we cannot avoid the inherent conflict present in many situations involving decision-making. In Chapter 2, five kinds of conflicts characteristic of the work environment were discussed. Cost-benefit analysis should not be thought of as a means of resolving such workplace conflict. At best, it can serve to improve our understanding of some of the factors inherent in these conflicts. Even in those situations where cost-benefit analysis is appropriate, value judgments and compromises are usually essential steps between the analysis and its application.

The 1973 *President's Report on Occupational Safety and Health* notes that

> in 1972, the National Safety Council, under contract with OSHA, completed a study to determine whether cost-benefit analyses could be made at the establishment level to show the benefit of effective safety and health programs in economic

terms. The Safety Council concluded, after surveying responses from over 4200 firms, that information on items of direct costs of accidents and the costs of safety programs is generally available at the establishment level, but that the indirect costs of accidents generally are not. Previous accident cost studies indicate that indirect costs are at least as great as direct costs, probably greater, and the lack of this information means that *it is not feasible to perform valid cost-benefit analyses at the establishment level at this time.*[29]

The report also discussed OSHA-sponsored economic impact studies:

In 1972, OSHA began a program to evaluate the cost of compliance of major new standards. When a proposed standard appears likely to have significant economic consequences, an economic impact study is undertaken. The study weighs the cost of compliance against the safety and health benefits provided, assesses the extent of the problem, looks at alternative methods of easing the problem, and provides an objective body of information to be considered in deciding whether to adopt the standard.

Economic impact studies are not traditional cost-effectiveness evaluations. They contain inherent limitations–the difficulty of economic prediction; the paucity of reliable injury and illness data; the understandable inability to objectively quantify pain, suffering, and loss of bodily function and loss of life. Impact studies are, nonetheless, valuable safeguards against adopting standards that might otherwise result in unnecessarily high compliance costs to employers.[30]

7.3.3 Who Uses Cost-Benefit Analysis and How They Use It

The description of cost-benefit analysis so far might lead the reader to believe that, despite its limitations and the methodological problems associated with its use, this technique is at least employed in good faith, solely as a technical aid, by decision-makers. In practice, unfortunately, this is often not the case. Cost-benefit analysis is often used in an attempt to convince other parties that a course of action, predetermined on other grounds, is justified. Value judgments are often hidden in the assumptions on which the calculation is based.

For example, one prominent investigator, Paul Brodeur, suspects bias in the Arthur D. Little report on the economic impact of the proposed two-fiber asbestos standard (see above, Section 7.2.2). In his article Brodeur gives the following description of how estimates of costs and benefits were in fact established:

It turned out that the firm had solicited "guess-estimates" on the biological effects of a five-fibre standard, as compared with a two-fibre standard, from an eleven-man "expert health panel" that included only one medical doctor who had not been a paid consultant of, or whose investigations into asbestos-related disease had not been supported by, some segment of the asbestos industry, and that it had solicited "guess-estimates" on the economic impact of the respective standards from a panel of twelve men representing various asbestos-producing companies and from a committee of thirteen men representing private shipbuilding companies.[31]

Even when the analysis is undertaken with sincerity, its conclusions may remain confidential if their disclosure is felt to be in conflict with other organizational objectives. In a recent conference, one participant pointed out that

> although his company performed a number of fairly sophisticated benefit-risk analyses, they were used only for internal decision-making. The company lawyers felt that disclosure of finite risks associated with company decisions would possibly increase the organization's cost in liability actions.[32]

Finally, even if cost-benefit analysis indicates that investment in improved workplace safety and health would be a profitable venture, management may fail to take action on this finding. Such a study of the cost and prevention of black lung in the coal industry[33] has never been utilized, although clear profits are predicted if the report's recommendations are adopted.

Attempts at formal cost-benefit analysis have been far less influential in management decisions to improve job health and safety practices than the threat of fines imposed under the OSHAct for violations found during an inspection, or the pressure of collective bargaining. Cost-benefit analysis is even less likely to be effective in altering the behavior of workers; nor does the OSHAct impose enforceable duties on workers. Modification of workers' behavior must await a major educational effort to raise the level of their consciousness of the health and safety issues.

When I asked an economist with the U.S. Chamber of Commerce why the Chamber had not undertaken cost-benefit analysis related to occupational health and safety, he replied that American business is worried more about red tape and government forms than about cost. Another official of the Chamber stated that he has never had a letter from a businessman saying that he could not afford to comply with the OSHAct.

7.3.4 Inherent Differences in Cost-Benefit Analysis for Injury versus Disease

Estimating costs and benefits in the case of job-related injury, though difficult, is far simpler than comparable estimations in the case of occupational disease. Many of the problems discussed earlier apply with particular force to occupational disease. The occupational etiology of many illnesses is often unknown or difficult to ascertain. Synergistic effects further complicate the problem of identifying separately the relative importance of various individual factors in the workplace and the general environment. Illnesses are sometimes ascribed to causes outside of employment when workplace conditions in fact play an important contributory role. The difficulty of determining the causation of disease is reflected in the slowness of the state workmen's compensation system to recognize the occupational origin of many diseases for compensation purposes (see Chapter 8).

The delayed effects of many workplace exposures and the chronic nature of much occupational disease make identification still more difficult. A further distinction between injury and disease can be drawn in the case of mutagenic hazards, where ill effects are not manifest until one or more generations after exposure.

Detecting many health hazards is a more difficult technical problem than detecting safety hazards. Many causes of accidents are relatively easy to identify. An unguarded blade, holes in the floor, or oil spills are often plainly visible, even to an untrained observer. Highly toxic carbon monoxide, on the other hand, is odorless, colorless, and tasteless. It is just as dangerous as the more obvious hazards, but sophisticated instrumentation is required for its detection.

Safety risks are more easily evaluated by workers than risks to *health.* Not only are safety hazards more easily recognizable, but when accidents do occur, they are perceived to be a direct consequence of these hazards, often spectacularly so. Not surprisingly, people generally find it more difficult to relate to long-term, low-probability "invisible" health risks than to the more obvious and immediate risks of serious injury on the job.

These differences between safety and health hazards make it apparent that while estimating costs and benefits is an arduous task for job safety, it is much more difficult in the case of occupational health.

7.4 MARKET IMPERFECTIONS, NONMARKET FORCES, AND THEIR IMPLICATIONS FOR ECONOMIC DECISION-MAKING

7.4.1 Introduction

Many economists are accustomed to thinking about the problem of providing a safe and healthful workplace in terms of market imperfections and methods of overcoming them. According to this view, the reason why the free market results in too much workplace injury and illness is that market "signals" or incentives are "wrong" from society's point of view. The problem of public policy then becomes one of finding ways to make the market more effective in obtaining the socially desired level of job-related hazard. This approach suggests that public policy should be geared toward intervention in the price system to make it function more smoothly in such a way that all prices reflect true social cost and all "externalities" are "internalized."

This way of conceptualizing the problem of occupational health and safety is helpful in contributing to our understanding of the implications of market imperfections for economic decision-making. However, this analysis suffers from the drawback of making the problem of appropriate policy design appear too simple. The pervasiveness of market imperfections in regard to occupational health and safety serves to make the problem of constructing sound social policy in this area extremely complex and, from a practical point of view, precludes our reliance upon the market mechanism as the only or predominant method of social intervention. Particularly with regard to the question of occupational *health*, public policy which relies solely upon market incentives for its implementation is likely to be seriously deficient. Preoccupation among many analysts with problems of occupational *safety*, as opposed to *health*, has led to the creation of a climate in which economic solutions to the problem of job-related hazards appear to be seductively straightforward. Unfortunately, economists' emphasis upon the correction of market imperfections through economic incentives may be deflecting or delaying us from the important task of finding other solutions.*

This section will have the following goals. Market imperfections, other economic problems, and nonmarket forces which are important for understanding the failure of the free-market mechanism to generate a socially tolerable level of workplace health and safety will be enumerated and discussed. An attempt will be made to illustrate both their implications for private economic decision-making and the limitations they pose for a public policy which is restricted to indirect attempts to alter these private decisions.

Market imperfections will be discussed with specific reference to problems of information, technology, and manpower as they relate to occupational health and safety. *Other economic problems* which seriously complicate the process of private and public decision-making will also be discussed, including problems associated with market structure, economies of scale, divergent time horizons, externalities, and the operation of the labor market.

These market imperfections and other economic problems enormously complicate the task of the policy-maker, even if all relevant actors in society can be assumed to be "economically rational." Departures from economic rationality are likely to be substantial, however, in view of a wide variety of *nonmarket forces* and institutional considerations which impinge upon economic actors in our society.

*For example, economic theory suggests at first glance that it would be better to "tax" employers for the results of industrial hazards than to enforce specific workplace standards such as TLVs, because the former approach leaves employers free to choose the "cost-minimizing" method of hazard reduction. In the words of one economist,

> The government should consider placing a "tax" on work injuries. . . . The evidence presented here suggests such a "tax" would be effective. The "tax" would also be superior to present safety legislation in that it would leave the firm free to choose its own best strategy to reduce injuries (thus encouraging innovation), eliminate costly inspections and be more easily adjusted over time.[34]

This view that we should rely solely on tax incentives to solve our occupational safety problems overlooks the important fact that society may not be indifferent as to the various strategies open to the employer in reducing the accident rate of his firm. Instead of reducing the level of physical hazard, the employer may choose to implement postemployment physicals and other screening techniques which threaten the job security of his employees. If the employer does not have to pay the full social cost of dislocation associated with the dismissal or transfer of workers considered to be "accident-prone" or "hypersusceptible," then society would be most unwise to leave employers free to choose their own cost-minimizing strategy of accident reduction. Yet economists seem to have a trained incapacity to accept the necessity of government intervention going beyond economic incentives, even when the proper application of their own analysis suggests the limitations of the latter approach.

These departures from "economic rationality" associated with nonmarket forces cast additional doubt on the effectiveness of a public policy based solely upon economic incentives, and further limit the usefulness of techniques such as cost-benefit analysis in aiding decision-makers.

One qualifying remark does need to be made, lest the reader feel that too negative a view of the usefulness of economic analysis has been taken. Environmental impact statements, often containing extensive economic analysis, should generally be required prior to the construction or relocation of technology-based operations. For example, the building of a new oil refinery in the northeastern part of the United States should not be undertaken without full analysis of its probable effects on the general environment. Though economic analysis may not be very useful for dictating policy in the work environment, this does not mean that such analysis is not a valuable tool in the preparation of analogous work-environment impact statements. In fact, economic analyses can greatly aid in the recognition of problem areas and could facilitate the operation of the administrative law process.[35]

7.4.2 Information

Problems of inadequate and imperfect information are pervasive with regard to occupational health and safety. These problems exist at every stage of the information process, including its generation, dissemination, and utilization. A massive research effort would be required merely to *generate* an adequate data base regarding the level of workplace hazard which we presently face, and the nature of the alternatives available to us in reducing it.*

The phenomena of multiple etiology, synergistic effects, long time lags between exposure to occupational health hazards and the onset of (chronic) disease, and the divergence between the rate of introduction of new substances into the occupational environment and the growth of our understanding of their long-run physiological implications, all serve to make the technical and scientific problems of information generation enormously complex. Yet it is obvious that in the absence of sufficient or unequivocal information about workplace hazards and their associated costs, private individuals and government cannot begin to make informed judgments that will lead to a minimization of the social cost of these hazards. This problem has been pervasive in the setting of health standards. Even if adequate information could somehow be generated, we would still be faced with a host of problems regarding the *dissemination* of information. Information does little good if it is not

*For example, it was recently reported in a study prepared for *Work in America* that known environmental and physiological factors, such as cigarette smoking, excess weight, serum cholesterol, and hypertension account for only 25% of the risk of heart attack. Occupational factors such as job-related stress–note that noise is a stressor, for example,–may well account for much of the remaining 75%.[36]

made widely available to decision-makers and to parties who bear the consequences of the decision-makers' actions—or inactions. Beyond the logistical problems, there is a broad range of social, political, and institutional considerations which enormously complicate difficulties of information dissemination, and these would still face us even if the problems of information generation could be overcome.

In large measure, however, the problems of hazards on the job are attributable to the nonutilization of information that already exists. A great deal of useful and pertinent information has already been generated and fairly widely disseminated, but decision-makers have been painfully slow in utilizing this information. A tragic example of this situation is provided by the case of America's uranium miners, a large proportion of whom have or soon will die prematurely from lung cancer of occupational origin.[37] It has long been known that exposure to airborne radiation carries greatly increased risk of lung cancer, and that the airborne radiation to which our underground uranium miners were being exposed was well in excess of levels proven to be unsafe. It had, furthermore, long been known, based upon European experience, that this hazard can be substantially alleviated by the simple expedient of installing adequate ventilation in uranium mines. Failure of public and private decision-makers to utilize this widely disseminated information has exacted a terrible toll among America's uranium miners.

Problems of the generation, dissemination, and utilization of information are further complicated by *inequality of access to information.* Inequality of access, as between management and labor or between large firms and small ones, creates incentives for special interests to withhold or distort potentially damaging (or beneficial) information. Differential access converts information into a bargaining advantage for the more knowledgeable, and compounds the difficulties of public and private decision-makers faced with the problem of evaluating the merits of a bewildering variety of conflicting claims. Thus, the phenomenon of differential access to information removes the problem of improving our understanding from the purely scientific and technical realm of information generation, dissemination, and utilization, to the political arena. Moreover, the pervasiveness of patent rights, licensing arrangements, and the proprietary nature of much of the information regarding newly introduced substances and processes mean that the problem of differential access to information regarding workplace hazards is endemic to our economic system, not a mere superficial imperfection (see Section 10.6).

The problems of inadequate information are illustrated by the controversy surrounding the industrial use of asbestos. Our information about the existing level of workplace hazard associated with current rates of asbestos use is woefully inadequate as a basis for informed decision-making. In part, the problem is one of *generating information.* Carcinogenicity and other morbidity and mortality risks of asbestos use are only imperfectly known at present. Scientific findings are in dispute, and it is indeed difficult to evaluate the bewildering variety of conflicting claims.* The fact that asbestosis and other asbestos-related disorders are seriously aggravated by environmental factors such as cigarette smoking adds yet another dimension to the problem of generating information, especially in view of the widespread tendency to ascribe such cases of multiple etiology to the environmental rather than the occupational causes.

It is apparent that in the absence of perfect information, the level of workplace hazard which results from private and public decision-making is likely to be too high from society's point of view. Without full information, economic actors—firms, employees, unions, and government—will be unaware of or misperceive the true costs of occupational hazard; they may underestimate the true costs of damage of occupational illness and injury, overestimate the costs of reducing the level of occupational illness and injury, or both. Such miscalculations are readily illustrated in terms of a simple diagram. In Figure 7.1 the cost of damage resulting from a workplace hazard is assumed to rise with increasing levels of hazard, while the cost of reducing the level of workplace hazard falls with increasing levels of hazard. The socially optimal level of workplace hazard is the point at which the total social cost of the phenomenon is minimized—the minimum point of the total cost curve which represents the vertical summation of thc cost-of-damage and cost-of-prevention curves (denoted by point A in Figure 7.1).

If, instead, the cost of damage associated with each level of occupational hazard is underestimated as a result of imperfect information, the perceived socially optimal level of workplace hazard will be higher than the true social optimum. This situation is illustrated in Figure 7.2, in which the true cost-of-damage curve lies above the cost-of-damage curve perceived by economic actors on the basis of imperfect information. Clearly, the socially optimal level of occupational hazard will be exceeded by economic actors in this case, as a result of imperfect information. (In terms of Figure 7.2, the socially optimal level of occupational hazard is perceived to be A′ instead of A.)

Similarly, the level of workplace hazard reached will exceed the socially optimal level if economic actors overestimate the cost of prevention of this phenomenon (as may well be expected for a variety of reasons).

One plausible example of such a situation is provided by the synergistic interaction between noise levels and visual acuity. It is a physiological fact that noise reduces visual acuity, and hence could be partially responsible for a considerable proportion of industrial accidents, poor workmanship and quality control, and other labor standards problems in noisy workplaces. Furthermore, it has been re-

*The opinions of medical experts vary enormously concerning the incidence of asbestosis, lung cancer, and mesothelioma among workers exposed to different levels of asbestos dust. Experts predict that anywhere between zero and sixty deaths would occur from asbestos-related causes among 100 randomly selected workers exposed for a forty-year period to an asbestos level of five fibers per cc.[38]

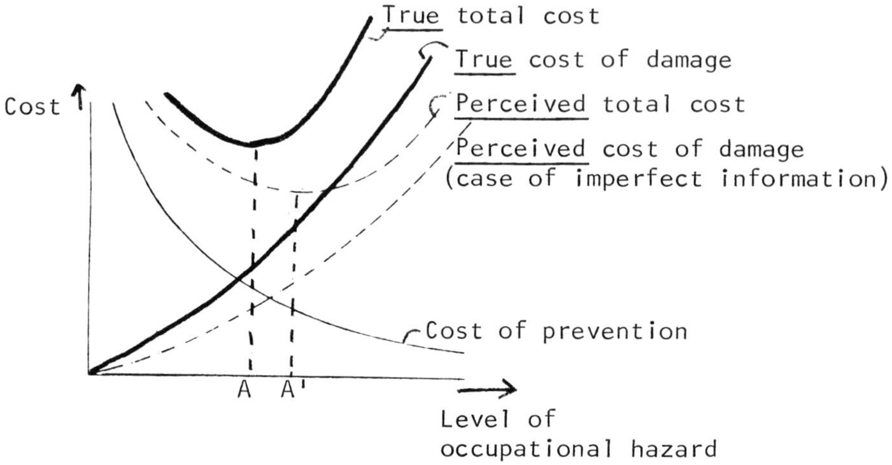

Figure 7.2 Effect of imperfect information on the determination of an optimum level of expenditure on occupational health and safety.

ported in a study commissioned by NIOSH that the rate of absenteeism is higher in noisy workplaces, as is the accident rate, the rate of somatic complaints, and the incidence of diagnosed medical problems.[39] If management is unaware of this fact and bases its calculations regarding investment in noise abatement solely on possible savings in workmen's compensation claims, training costs, and other costs associated with direct loss of hearing, then too little will be spent on noise abatement, even from the firm's point of view. In such a situation, managerial decisions will be based upon an erroneous conception of the "cost of noise" curve which it faces, as illustrated in Figure 7.2 (For an extended discussion of the magnitude and pervasiveness of synergistic effects in the workplace of the type just discussed, see Chapter 3.)

7.4.3 Technology

The dynamic process of technological change makes reliance on a static model inappropriate as a basis for decision-making. Decision-makers who do not take proper account of the process of innovation are likely to overestimate the long-run costs of preventing occupational injury and disease. Firms will invest too little in occupational health and safety if they fail to consider the possible increases in benefits which innovation may bring. Yet they will not themselves undertake adequate expenditures on the research and development required to bring about such innovations, because they will not be able to capture all of the social benefits that result, and because such investment may appear too risky from a purely private point of view. Having failed to make an adequate initial outlay of resources on innovation, firms will miss further opportunities to reduce the cost of preventing accidents and illness afforded by the "learning curve" and by economies of scale in the production of new, safer equipment and processes. If such initial outlays are undertaken, however, it is likely that as firms gain experience in the prevention of occupational hazards, they will learn how to engage in such activity more cheaply, with the result that the socially optimal level of workplace hazard—and the associated total cost—will shift downward over time.* This learning-curve effect is illustrated in Figure 7.3. The cost-of-prevention curve shifts downward over time as firms become more proficient in reducing the level of workplace hazard. As a result, the total cost curve also shifts downward and to the left, and with it the level of workplace hazard at which the total cost of this phenomenon is minimized.

Market imperfections also impede the *transfer* and *diffusion* of the new technology that results from innovation. Among firms in the same industry, and among industries, the transfer of technology is hampered by secrecy, patent rights, licensing agreements, and the proprietary nature of newly developed equipment and processes. Rather than being made available to everyone in the interest of social welfare, new technology for improved occupational health and safety confers a competitive advantage on the firm responsible for its development. Indeed, collaboration among firms in the dissemination of safer and more healthful workplace technologies may, at present, subject them to the risk of antitrust law violation (see Section 4.4.4). Other institutions, including government, organized labor, and the universities, have heretofore played little part in the process of innovation and diffusion of new technology for a safer and healthier workplace, with the result that the void left by the inadequate activities of private corporations in this area has largely gone unfilled. Even under the OSHAct, the level of resources allocated by government to promote the development and transfer of new health and safety technology has been far from sufficient to meet the need. Not surprisingly, in view of the level of resources presently devoted to the subject, universities and research institutes have placed a low priority upon occupational health and safety. Organized labor, the only other institution in society potentially capable of substantial commitment of resources to this question, has, for a variety of reasons, largely been preoccupied with other concerns, though this is beginning to change (see Chapter 10). Indeed, a prime need of organized labor, and of working people generally, is the transfer of sophisticated technology developed in the private sector and in government to the development of simple, durable, and inexpensive equipment which would aid working people to acquire the self-help capability needed to monitor their own occupational environment.

Nor does the process of technology transfer among nations proceed any more smoothly at present, if the transfer of occupational health and safety technology from other countries to the United States is any indication. The failure of the United States to adopt a simple method of reducing the hazards of uranium mining,

*The learning-curve phenomenon, with regard to cost reduction over time, has been documented in manufacturing processes such as the airplane industry.[40]

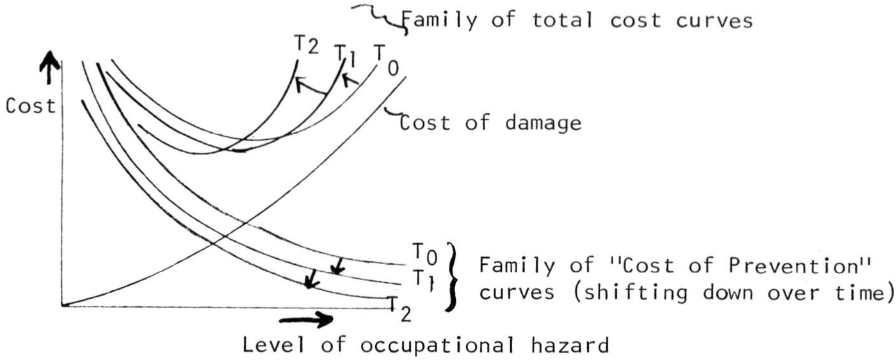

Figure 7.3 Learning-curve effects.

developed in France and Czechoslovakia prior to World War II, has already been noted. Similarly, the debate surrounding the industrial use of asbestos has proceeded largely on the assumption that asbestos is an essential raw material without close substitutes for many uses such as shipbuilding.[41] Given this assumption, the terms of the debate were confined to the choice of an appropriate asbestos exposure standard and the costs of meeting it in terms of filtration and ventilation equipment and the like. Yet, while this controversy raged, the Swedish shipbuilding industry had for some time been using a substitute material in place of asbestos which completely eliminated the necessity of exposing shipyard workers to the hazardous substance, at least in the construction of new ships.[42]* It seems clear that we have many lessons to learn from other countries in regard to superior technologies for improving the level of occupational health and safety. An important objective of public policy must, therefore, be to facilitate the international technology transfer process.

Beyond the problems of technology development and transfer, there are a variety of barriers which act to retard technology *utilization.* Little concern has been devoted to the problems faced by organized labor in its attempts to utilize health and safety technology. Indeed, it has typically been beyond the means of organized labor to utilize the sophisticated, relatively expensive means of monitoring the industrial environment that have been developed thus far. Small firms and older firms with expensive capital equipment already in place have similarly been at a disadvantage in utilizing new technology. Often, newer, less hazardous modes of production require substantial initial outlays or can only be introduced economically in firms which have a large volume of production. Firms with sizable investment in existing capital equipment may undertake relatively inexpensive, largely ineffective, stopgap measures to improve workplace health and safety, when from a social point of view

*Of course, even Swedish shipyard workers are still exposed to asbestos hazards in the repair and reconstruction of older ships.

it would be more desirable that existing equipment or processes be scrapped and replaced by new machinery of less hazardous vintage. Once noisy machinery is in place in a plant, for example, it is far more expensive to reduce the noise level than would have been the case if the facility had been properly designed acoustically in the first place. Once financially committed to noisy equipment, management is likely to pass out ear plugs and absorb the cost of increased workmen's compensation, turnover, absenteeism, and accidents, rather than pay for a proper job of acoustical engineering. From society's point of view, however, the latter course of action may in fact be less costly.

7.4.4 Manpower*

Market imperfections seriously impede the development and utilization of adequate manpower to meet our occupational health and safety needs. Properly trained manpower in sufficient numbers is essential if we are to have the medical and technical expertise to bring about a socially desirable reduction in the level of workplace injury and disease. Moreover, the problems of insufficient numbers and inadequate training are compounded by that of poor distribution of the existing manpower. The safety bias that pervades the field is especially serious in the case of manpower, as manifest in a serious imbalance in the mix of safety versus health professionals at the present time, and in the composition of facilities for their training. This imbalance is in large measure attributable to the rational response of firms and insurance companies to economic forces and government regulations. Workmen's compensation, insofar as it acts as an incentive at all, primarily influences firms' demand for *safety* as opposed to *health* manpower, as few diseases are recognized as being of occupational origin for compensation purposes. The OSHAct with its emphasis on safety has, if anything, worsened the manpower imbalance, as firms demand additional safety personnel to comply with the Act and government demands additional safety personnel to enforce it.

Labor faces a serious shortage of the skilled health and safety manpower needed to perform the essential function of serving as an independent check on management. In view of the limited resources available to it for paying a full-time professional health and safety staff, organized labor is in urgent need of manpower on a part-time or consulting basis, especially for the purpose of conducting training programs which could begin to provide working people with the self-help skills to monitor their own job environment.

Industry faces a variety of market imperfections which are impeding the development and utilization of occupational health and safety manpower. Firms are reluctant to pay training costs of health and safety professionals, both because the pro-

*See Chapter 9 for an extensive discussion of the manpower problem.

fessional, once trained, can sell his services to another employer and because, even while employed by the firm which has paid for his training, the firm does not capture the full social value of his services. The result is that firms pay too little to train health and safety manpower. Furthermore, the interests of the firm which hires the manpower may diverge from the public interest, as the case of industry's safety bias illustrates. In the absence of sound public policy, there is little basis for hope that the market will induce private industry to utilize properly the health and safety manpower that it employs.

Small firms face special problems in regard to the development and utilization of occupational health and safety manpower. Often, the large fixed cost of setting up a full-time in-house manpower capability is prohibitive for the smaller firm. Yet the market has proven extremely sluggish in providing the kind of ancillary consulting services which could be a useful substitute for full-time staff to the smaller firm. (As will be discussed in Chapter 9, uncertainty about the future needs for occupational health and safety manpower is partly responsible for the present unresponsiveness of the market.)

Nor has government been playing its proper role in the occupational health and safety manpower field; indeed, it has recently made several decisions resulting in staffing cutbacks (see Chapter 6 and Section 9.5). Government has not been sufficiently active in promoting the training and development of health and safety manpower, nor has public policy been geared toward achieving the objective of proper manpower utilization by the private sector. Indeed, the activities of government have served to worsen certain imbalances in the supply and demand for health and safety manpower, directly through insufficient government demand for occupational *health* manpower* and indirectly through enforcing incentives on business which make them respond in a similar manner.

Likewise, universities and nonprofit research institutions have not been playing their potential effective role in the training of occupational health and safety manpower. In view of the low level of support for health and safety manpower developed by government and business, universities and nonprofit research institutions have lacked the funds to place sufficiently high priority on this important objective. As a result, our society is currently underutilizing a potentially important resource—our universities and research institutions—in the development of occupational health and safety manpower.

In conclusion, market imperfections have been partly responsible for a serious shortage of qualified occupational health and safety manpower. Insufficient resources are devoted to the training of such manpower, and the cost-effectiveness of presently committed training resources (in terms of *social* rather than private priorities) is exceedingly low.

*State government, in particular. For the manpower implications of safety bias in the implementation of the OSHAct, see Section 9.4.2.

7.4.5 Problems of Market Structure

As has already been mentioned, new and potentially harmful substances and manufacturing processes are constantly being introduced into our industrial environment at a rapid rate which far outstrips increases in our knowledge of their long-run consequences. The commercial life of products in the chemical industry has been estimated to be on the order of five years.[43] Hence, by the time a chemical substance has been proven harmful, it may have long since fallen into disuse and been replaced by newer and just as hazardous substitutes about which there will also be little or no health information until it is too late. Moreover, for every new chemical product reaching the market, *workers* (often in laboratories) are exposed to countless substances which never reach this stage:

> Most of the products that the laboratories of pharmaceutical houses initially take up for investigations, or firms making pesticides, prove to be commercial failures. Five to ten thousand chemicals are initially studied as serious possibilities for every single product that makes its successful way clear through the R&D process to the market.[44]

Problems of market structure lie at the root of this phenomenon, which has such adverse consequences for the level of workplace hazard. In many cases, being first to introduce a new, potentially harmful substance or process contributes to a firm's market advantage. Being first may also mean that the firm has at least a temporary monopoly of information about the potential deleterious effects of the substance or process in question upon workers and the environment, information which the firm will generally find in its interest to withhold. Sometimes, as in the case of a beryllium compound used in fluorescent lighting which led to the production of a brighter, longer-lasting light bulb, the new substance proves useful to consumers but hazardous to workers, and the producer's monopoly of information keeps workers ignorant of the hazard they are encountering.[45] * In other cases, the hazardous new substance is a marketing ploy of unquestionable utility to the consumer, which serves to secure for the firm a temporary advantage through product differentiation[47] —as illustrated by the recent use of enzymes in the manufacture of detergents.

Regardless of the potential usefulness of new substances or processes from the consumer's point of view, however, the following conclusion seems to be justified.

*According to Public Health Service estimates, as many as 30,000 workers are presently at risk of disabling disease and early death as a result of occupational exposure to beryllium. At the Conference on the Current Status of Beryllium as an Occupational Hazard, sponsored by the Society for Occupational and Environmental Health, scientists speculated that "as many as one out of nine workers in the beryllium industry may get some form of beryllium disease"[46] – despite the fact that the adverse health effects of beryllium exposure have been known to at least some occupational physicians for years.

Considerations of market structure pervasive in the American economy today have resulted in a chronic tendency among many firms to introduce new substances and processes for short-term marketing advantage before the longer-term implications of their use are properly understood.

7.4.6 Economies of Scale

The existence of significant economies of scale underlie many of the difficulties encountered by public and private decision-makers in improving the level of occupational health and safety. It has already been noted that there are a number of market imperfections and other economic problems which are preventing us from generating an adequate stock of properly trained health and safety manpower. These problems, with regard to both the development and the utilization of the necessary expertise, are partly attributable to economies of scale. Training of safety and health professionals is most cheaply undertaken in large numbers, yet few firms could utilize a sufficient number of professionals to realize economies of scale in training. Thus, few firms develop the in-house capacity to train health and safety professionals. This task is left by default to government, universities, and research institutes. Yet these institutions often lack sufficient financial resources, with the result that they are also unable to capitalize on economies of scale in training. A major national commitment to improved occupational health and safety would be likely to generate sufficient demand for skilled health and safety professionals to reduce significantly the long-run average cost of their training, since educational institutions could then take advantage of economies of scale in this regard.

Serious problems also exist with regard to economies of scale in manpower utilization. Retention of an in-house health and safety staff often involves a substantial fixed cost which must be spread over a large volume of production in order to meet the criterion of economic feasibility. Often, the complementary nature of services provided means that the average cost of manpower services falls as firm size increases. For example, the acquisition of an adequate health and safety library may be no more expensive for a large firm employing several health and safety professionals than for a smaller firm with a full-time staff of one.

Problems presented by economies of scale may be even more serious with regard to the process of technological change. Equipment and processes designed to improve workplace health and safety are presently expensive, in part, because so few firms have been in the market for such "esoteric" items. If a large number of firms began to demand such equipment and processes, it is likely that significant economies of scale could be realized in their production, thereby lowering their long-run average cost. A wider market for less hazardous equipment and processes would induce suppliers to adopt the techniques of mass production. These firms, and the firms that wish to utilize the less hazardous equipment and processes, would also begin to increase the share of their research and development budget allocated to inventive and innovative activity with respect to improved occupational health and safety, with the result that the long-run average cost of reducing workplace hazard would quite likely fall even further.

Presently, the existence of what economists call "indivisibilities" limits the institution of safety and health practices and controls. For example, it makes little sense to insulate acoustically a single machine or process that is surrounded by many other noisy machines. Thus, firms are often faced with a choice between total inaction and the considerable expense involved in implementing acoustical redesign throughout the entire plant. Such indivisibilities are pervasive in occupational health and safety, and have heretofore provided a serious impediment to investment in the reduction of workplace hazard.

7.4.7 Problems of Divergent Time Horizons

Many of the problems of decision-making regarding occupational health and safety are related to divergent time horizons among different economic actors and institutions, which lead to several types of inconsistency between private decisions and social objectives. Among the inconsistencies associated with divergent time horizons for decision-making are the following:

- Divergence between top management's cognizance of the benefits of a safer workplace and the incentives and evaluation structure facing lower and middle management.
- Divergence of time horizons for decision-making between small and large firms.
- Divergence of time horizons for decision-making between marginal firms and the rest of society.
- Divergence between private and social time horizons regarding prevention, causation, and manifestation of occupational disease.
- Divergence between corporate management and society regarding tolerance for the risk and uncertainty of investment in a more healthful workplace.

Top management may be cognizant, in principle, of the long-term benefits of a safer workplace, and may have the good-faith intention of directing corporate policy toward the achievement of this objective. Yet, perhaps unintentionally, top management often evaluates subordinates on the basis of criteria which force them to have an extremely short time horizon with respect to this issue. The incentives operative on middle- and first-line supervisory personnel typically lead them to emphasize smooth, uninterrupted production at the maximum feasible rate, often to the detriment of their subordinates' health and safety. Thus, there is a widespread tendency for managerial *practice* in regard to occupational health to be geared toward the short time horizon imposed upon middle- and first-line supervision, even when top management understands in principle the beneficial consequences of improved workplace safety.

Differential availability of the capital needed for investment in a safer workplace also affects the time horizon for decision-making. Large, profitable firms can rely on internal financing out of retained earnings, or can utilize their extensive capability to raise capital externally, and are therefore able to undertake the kind of large-scale investment project with long gestation period which is often required for significant improvement in workplace health and safety. Small firms, in contrast, are less able to raise the necessary capital, and are often insufficiently diversified to shoulder the additional risk associated with such a major undertaking.

The problem of marginal firms is even more serious in this regard. Firms on the edge of bankruptcy clearly are not likely to choose the alternative of current safety outlays if this would spell certain economic disaster. Insofar as continued survival is an objective of the firm, marginal firms are likely to opt for deferring the risk of bankruptcy as long as possible, electing to forego expensive investment in improved workplace health and safety in the short run and take the chance of higher disability payments in the future.

There is also the more general problem of divergence between private and social evaluations of the time profile of the relative costs of occupational hazards and their prevention. There is often a long time lag between exposure to an occupational hazard and the manifestation of disability. Firms may well find that it is cheaper to risk payment of disability benefits 20 or 30 years hence than to invest in a safer workplace today—even when, from a social perspective, the reverse may be true. The discount problem is especially acute in the cases of mutagenic or teratogenic hazards, in which the potential harm does not become manifest for one or more generations. Private companies may use a discount rate which is too high from society's point of view in evaluating the trade-off between long-run costs of occupational hazard and the short-run costs of hazard prevention.*

Furthermore, there is the possibility that individual firms may be unwilling to assume privately the burden of uncertainty associated with investment in the reduction of workplace hazard, even though the expected value of such investment may be high. Firms may prefer relatively certain and calculable workmen's compensation outlays to the potentially smaller but more uncertain costs associated with reduction of occupational hazard. From society's point of view, such investment would not appear so risky and should perhaps be undertaken.

7.4.8 Internalization of the Social Cost

Perhaps the most serious impediment to improvement in the level of workplace health and safety is the fact that employers presently are not held financially accountable for the human and social consequences of their failure to provide safe

*The possibility of divergence between private and social rates of discount in the evaluation of investment projects has been widely debated by economists. It is beyond the intended scope of this work to contribute to the controversy.

and healthful working conditions. A disproportionate share of the damage associated with occupational illness and injury befalls working people, their families, and society at large, without ever directly entering a corporate profit-and-loss statement. As a result, business has insufficient incentive to improve job health and safety.

It is true, of course, that exposing employees to hazardous and unhealthful working conditions does impose substantial costs on the firm even at the present time. These costs to the employer are of two types: (1) direct, calculable, and quantifiable costs of occupational injury and illness which are presently accounted for in the corporate income statement; and (2) indirect costs which presently elude the income statement and therefore do not properly serve as incentives to improve working conditions. At the present time, corporate expenditures on hazard *prevention* (investment in occupational health and safety manpower and equipment, etc.) typically must be justified on a cost-effectiveness basis relative to the *direct* costs of damage to the firm associated with occupational injury and illness. Yet it may well be the case that the *indirect* costs to the firm, which presently do not influence corporate decisions regarding investment in workplace health and safety, are substantially larger than the direct costs which do influence decisions. (See Figure 7.2 and the discussion of the implications of imperfect information in decision-making in Section 7.4.2. See also the 1973 *President's Report on Occupational Safety and Health*, p. 22, for a statement that the indirect costs of accidents are probably larger than the direct costs.)

Direct costs to the firm of occupational injury and illness are, of course, very substantial in some industries. Among these direct costs are (1) workmen's compensation premiums; (2) payroll and fringe benefit costs such as sick pay, medical and life insurance, and other payments to employees who are unable to work on account of occupational illness or injury; (3) direct medical expenses such as the salaries of the company doctor and nurse and the cost of equipping the plant dispensary; (4) costs of recruiting and training replacements for employees who are unable to work because of job-related injury or illness; (5) costs of impairment to capital equipment and goods in process which are damaged as a result of industrial accidents; and (6) the "value-added" of production which is lost on account of interruptions associated with industrial accidents.

These and other direct costs to the firm of occupational illness and injury can be substantial, and have been estimated by the National Safety Council to amount to roughly $10 billion per year—around 1% of GNP. In 1970 workmen's compensation cost employers $4.9 billion or 1.13% of covered payrolls, and workmen's compensation costs are considerably above this level in some industries.[48] Fred Foulkes cites an example from the longshoring industry, where workmen's compensation costs were a significant enough factor to influence the location of economic activity.[49]

Nevertheless, the fact remains that these direct costs of occupational injury and illness are not a sufficient lever to bring about needed improvements in working conditions. The indirect costs to industry presently elude economic calculation, and the social costs, which are perhaps even greater, do not befall business at all.

The *Compendium on Workmen's Compensation* contains the following discussion of indirect costs to the employer:

Less apparent to employers are the other losses, often called indirect losses, associated with industrial accidents. One source lists ten examples of such losses:
1. Cost of wages paid for working time lost by workers who were not injured.
2. Net cost to repair, replace, or straighten up material or equipment that was damaged in an accident.
3. Cost of wages paid for working time lost by injured workers, other than workmen's compensation payments.
4. Extra cost due to overtime work necessitated by an accident.
5. Cost of wages paid supervisors while their time is required for activities necessitated by the accident.
6. Wage cost due to decreased output of injured worker after return to work.
7. Cost of learning period of new worker.
8. Uninsured medical cost borne by the company.
9. Cost of time spent by higher supervision and clerical workers on investigations or in the processing of compensation application forms.
10. Miscellaneous unusual costs (e.g., loss of profits on contracts canceled, public liability claims, and cost of excess spoilage by new employees).

According to H.W. Heinrich, a pioneer in developing scientific approaches to industrial accident prevention, on the average these indirect costs are about four times the employer's share of their employees' earnings losses and medical expenses.

In a more recent study, Simonds and Grimaldi have argued that, although Heinrich should be commended for dramatizing the importance of indirect costs, the ratio of these costs to direct costs varies markedly among employers and over time and on the average may be less than four to one.[50]

The indirect costs to industry of reduced worker morale and efficiency due to the effects of occupational illness and injury on worker productivity may well be substantial. Employees suffering from occupational diseases or in constant fear of harm due to the hazardous nature of their employment are likely to function less efficiently as a result, and may also adversely affect the morale and efficiency of their coworkers. It has been estimated, for example, that the debilitating effects of black lung disease are responsible for reducing the productivity of labor in underground coal mining by 20-30%.[51] Similarly, it appears likely that absenteeism, turnover, and quality of workmanship are adversely affected by hazardous working conditions, leading to further substantial hidden costs on the firm.[52]

On average, each employee is absent from work roughly six days per year, according to the BLS, and it has been estimated that reduction of one day per year in the annual rate of absenteeism among the U.S. labor force would add $10 billion to GNP.[53] In view of the recent NIOSH findings, it is certainly plausible that unpleasant working conditions such as excessive noise may be responsible for at least *one* of these six days of absence per year. Moreover, respiratory diseases, which account for more days away from work than all other disease combined and which cause more than one billion days lost from work yearly, have been shown to be closely related to industrial air pollution.[54] Since the level of air pollutants to which workers are exposed on the job is often several orders of magnitude greater than that to which they are exposed in the general environment, it is likely that a substantial proportion of respiratory disease-induced absence from work is of occupational origin. Therefore, it would not be surprising if inclusion of the hidden costs of occupational hazard due to excess absenteeism alone would result in a doubling of present estimates of the cost of occupational illness and injury.

A wide range of synergistic effects which are only beginning to be understood multiply the hidden costs of workplace hazards to the firm. For example, the previously mentioned NIOSH noise study found that the rate of seemingly nonoccupationally related medical complaints is higher in plants with hazardous (e.g., noisy) working conditions, with the result that the costs of equipping and staffing the plant dispensary are higher than they would otherwise be. Lack of awareness of such complex interactions among workplace hazards leads companies to underestimate the benefits associated with investment in improved workplace health and safety.

Noise, perhaps the most pervasive occupational hazard in industry today, provides a number of striking illustrations of hidden costs of occupational hazard due to synergistic effects. Noise is a stressor, for example, and may therefore be responsible for increased incidence of heart disease among employees in noisy workplaces. Noise has been found to increase the toxicity of amphetamines,[55] a significant finding in view of the unknown but possibly large proportion of young blue-collar workers who take pep pills in an effort to maintain the rigorous pace their job often demands. Noise may very well be a cocausative factor in diseases associated with toxic material in general. Noise also reduces the level of visual acuity,[56] with the result that the cost of accidents and shoddy workmanship is likely to be higher in noisy shops. Since all of the foregoing effects of noise are poorly understood and not well known at present, it is unlikely that they are properly accounted for in business decisions regarding investment in noise abatement. It seems clear that many companies could save more than the workmen's compensation premiums associated with claims relating to occupational deafness (which in any case do not fully internalize even the direct economic costs of deafness), but such firms presently may not realize this fact and may forego potentially profitable investment in a quieter workplace.

If the full extent of such synergistic effects and the broader implications of hazardous and unpleasant working conditions for employee physical and mental wellbeing ever became fully known, it is quite likely that the presently hidden indirect costs that they impose upon firms will prove to be substantial.

Nevertheless, it remains the case that the bulk of the costs associated with occupational injury and illness do not presently befall the firm, directly or indirectly; and herein lies the primary explanation for the failure of the market mechanism to provide a socially tolerable level of workplace health and safety.* Most of the costs of occupational illness and injury are directly borne by working people and their families. Workers who become injured or ill as a result of their job, and their dependents, suffer a considerable loss of income and fringe benefits such as group medical and life insurance, and a general diminution of economic security and well-being. The removal of the right to sue in tort since establishment of the system of workmen's compensation has prevented disabled employees or their dependents from recovering money damages for pain and suffering, with the result that employers presently do not have to pay for the full consequences of their negligence (see Section 8.2.5).

Workmen's compensation payments do not reflect even the direct *wage* losses of disabled employees, much less such indirect economic losses as fringe benefits and the incalculable intangible costs of pain, suffering, and bereavement. Moreover, since most illness of occupational origin is not presently recognized as compensable, many disabled workers do not even enjoy the inadequate measure of income security which workmen's compensation provides.

Those victims of occupational injury and illness who are fortunate enough to receive workmen's compensation will typically find that benefits represent less than 50% of their prior earnings. Pay raises received for any reason by coworkers remaining on the job– such as promotion, cost-of-living adjustment, or union negotiation–are not reflected in the level of workmen's compensation benefits received by the previously disabled. The relevant factor in determining the impact of occupational illness and injury upon a victim's income is not his earnings prior to disability, but rather the level of income he would currently be earning had he not been disabled on the job.

Even if workmen's compensation payments fully restored the income losses sustained by the victims of occupational illness and injury and their families, the costs in terms of the value of human life and the pain and suffering inflicted would still not be fully internalized. As previously mentioned, industrial injury claims over 14,000 lives annually in this country, while industrial disease may claim as many as 100,000. Each year there are an estimated 390,000 new cases of occupational illness, and roughly one employee in eight suffers a disabling injury at work; 100,000 per year are disabled permanently as a result. The individual and social costs of this enormous amount of occupational illness and injury is in no sense limited to the

*Many economists will disagree with this statement, on the grounds that "hazard pay" compensates workers for assuming differential job-related risk, and acts as an incentive for employers to reduce the risk to which their workforce is exposed. For reasons that will be discussed in Section 7.5.3, I do not agree with this contention.

attendant losses of income or production, though these are far from negligible. Though some economists equate the value of a human life with the person's income for purposes of cost-benefit analysis (as do many courts in assessing damages in civil suits), there is no justification in economic theory for employing such a procedure.[57]

Furthermore, much of the cost of occupational illness and injury befalls, not the employer or worker, but society at large. The social cost of maintaining and rehabilitating victims and providing for their dependents is by no means entirely paid by the victims themselves, or by the firms which employ them. Victims of occupational illness and injury and their surviving dependents place an enormous burden on social-welfare services in terms of the costs of medical care, rehabilitation, retraining, and direct income maintenance. Family breakup sometimes ensues upon the death or disability of the primary breadwinner, with the result that surviving dependents may become wards of the state. Discouragement and disillusionment attendant upon the trauma experienced by the victim and members of his family may trigger or exacerbate emotional problems that require the publicly subsidized assistance of social workers, psychiatrists, or other costly professionals.

As victims of occupational illness and injury, large numbers of working people and their families are converted from economically self-sufficient taxpayers to economically dependent welfare recipients. The enormous social costs of this sad transition are not internalized by their employers, but are diffused more generally throughout society—ultimately, in large measure, in the form of higher taxes on those working people who remain healthy enough to work. Management, secure in the knowledge that it will probably not be held financially accountable for the costs of pain and suffering that accompany its decisions, may therefore not attach sufficient weight to maintaining the health and safety of the workforce. Thus, it could be said that the employer "free-rides" on the health and safety of his workers.

Furthermore, even if these measurable social costs could somehow be properly assigned to the responsible parties, there would still remain the problem of the intangible costs that befall individuals other than those directly involved in the decision-making process. People value the lives and well-being of their friends, relatives, and associates, and suffer intangible but nonetheless very real psychic loss if any of these people become victims of occupational illness or injury. Though workers may in principle receive compensation in the form of hazard pay or damage award commensurate with their personal assumption of job-related risk, there is no comparable mechanism through which potentially aggrieved friends and loved ones can receive adequate compensation.

The question remains whether recent and proposed federal legislation (such as the OSHAct, the Coal Mine Health and Safety Act as amended by the Black Lung Benefits Act, and proposed reforms of the workmen's compensation system) is likely to prove effective in internalizing the social costs of occupational injury and disease.

Preliminary experience under the OSHAct with its attendant philosophy of voluntary compliance does not give cause for optimism in this regard. Inadequate compliance manpower has meant that the probability of a firm's being inspected is relatively low. Laxity and misplaced emphasis in the setting of standards, coupled with inadequate training among compliance officers, have meant that many serious occupational hazards—especially health hazards—are overlooked when firms are inspected. Moreover, even if cited, employers are relatively unlikely to be fined; and fines, even if imposed, often are far too low to serve as much of an incentive regarding improvement of working conditions.* For these and other reasons, the OSHAct has failed thus far to live up to its potential as a significant policy lever for the internalization of the social costs of occupational illness and injury.

Proposed reforms of the workmen's compensation system, though necessary and desirable, are likewise unlikely to provide a significant policy lever for the internalization of these social costs. Proposed reforms do not presently include restoration of the right to sue in tort for disabled workers and their surviving dependents, nor do they contemplate any requirement that people be compensated for the pain and suffering that result from the disability.

In several respects, of course, the proposed changes in the system of workmen's compensation do represent steps in the right direction. Disease of occupational origin may become more easily compensable; the level of benefits as a proportion of previous wages may be increased; eligibility and scope of coverage may be extended to a larger proportion of the labor force; and statutes of limitation which presently prevent the granting of compensation in many cases of occupational disease with a long gestation period may be modified (see Chapter 8).

It has been predicted that these and other proposed reforms will lead to a doubling or even a tripling of workmen's compensation premium costs to employers. If so, this may provide some incentive for improving working conditions. Nevertheless, a variety of difficulties still remain which preclude sole or primary reliance upon workmen's compensation as a policy instrument to achieve internalization of the social costs of occupational illness and injury. For example, there is the inherent difficulty of assigning to the "right" employer the costs of occupational disease which do not become manifest until many years—and several employers—after a worker's exposure to hazard. Likewise, there are a range of conceptual and administrative problems involved in assigning costs in cases where illness is of multiple or uncertain etiology and where nonoccupational factors may play a contributory

*From April 28, 1971, through December 1973, OSHA made 117,745 inspections resulting in 76,533 citations alleging 386,000 violations, with proposed penalties totaling $10,118,451. Of that amount, $7,225,597 was remitted to the Secretary of the Treasury. The balance represents funds not yet collected, plus penalties proposed by OSHA that were reduced or vacated by order of the Occupational Safety and Health Review Commission after employer appeals.[58]

causal role. The costs of litigation and administration associated with placing full reliance upon the workmen's compensation system for internalizing the social cost of occupational illness and injury would no doubt prove considerable.* These and other limitations on the usefulness of workmen's compensation in bringing about an internalization of costs remain; workmen's compensation is a necessary palliative, but it should not be used as a substitute for adequate standards and strict enforcement.

7.4.9 Labor Market Problems

In economists' abstract models, price and wage flexibility and perfect factor mobility guarantee full employment of all scarce productive inputs, including labor. One characteristic assumption of such models is that workers are indifferent between continuing to work at their present job and being fired or laid off, since they can always find a similar job at the same wage with another employer without costly transition. Job security, a dominant real-world concern of working people, has no place in this abstract model. Recent findings, however, indicate that job loss ranks as the third most serious form of stress which people experience during their lives, surpassed only by death of a loved one and the stress associated with family break-up and divorce.[59] Compared to the world of economists' abstract models, then, our real world—characterized by substantial unemployment and immobility of labor, both geographical and occupational—is one in which the private decisions of both firms and their employees are likely to result in too much work-related injury, illness, and death. Working people faced with the prospect of serious economic harm in the event of job loss are willing to tolerate a too heavy burden of workplace hazard rather than "rock the boat." Firms, for their part, do not place a sufficiently high value on the lives and health of their employees if there are unemployed workers eagerly waiting to take the places of those who are injured or become ill.**

Therefore, government policy to reduce unemployment and the occupational and geographic immobility of labor would be likely to have the beneficial side effect of reducing the magnitude of our occupational health and safety problem. Conversely, any direct government policy measures to reduce the hazards of the workplace will have to allay workers' fears of job insecurity if they are to be effective.

Indeed, the traditional fear of job insecurity among working people, coupled with the emergence of certain important economic trends, may well be partly responsi-

*For example, litigation involving automobile accidents—which poses far fewer conceptual problems in the assignment of liability than does occupational disease—has proven to be extremely costly, providing the stimulus for "no-fault" automobile insurance.

**Of course, the existence of some types of labor market friction has the opposite tendency, making employers more safety-conscious than they would be if workers really were like homogeneous, interchangeable parts. Yet the dominant effect, pending further investigation, is probably the one reported above.

ble for recent increases in the magnitude of America's occupational health and safety problem. In a large segment of the labor market, termed the "primary sector" by dual labor market theorists, the interaction of institutional and economic forces has led to a secular reduction in interfirm mobility among a substantial group of workers.[60] Employee compensation, retirement benefits, and career advancement opportunities have become increasingly linked for a large segment of the workforce to stable, continuous, and uninterrupted service with the same employer. Thus, at a time when rapid technological change is multiplying the hazards of the workplace, chronic exposure to these hazards may be on the increase due to decreased interfirm and interindustry mobility among precisely those groups of workers who are most subject to risk of harm.

Extensive reliance upon job tenure (seniority) as a determinant of job security, pay, pension rights, and intrafirm promotion and training opportunities has led to a situation in which many working people find the "opportunity cost" of voluntary midcareer changes of employer prohibitively high. Such employees are increasingly locked by "golden handcuffs" into what one scholar has characterized as our system of "industrial feudalism."* Since the risk from many occupational hazards is primarily associated with—or is greatly increased by—chronic long-term exposure (many industrial dust diseases are of this type, for example), the spread of labor market "imperfections" of the type just described may well be partly responsible for the increasingly serious nature of our occupational health problems.[62]**

7.4.10 Nonmarket Forces†

Clearly, the pervasiveness of market imperfections makes it unlikely that the outcome of private decisions based solely upon market forces will be a socially optimal level of workplace health and safety, even if all relevant actors in society can be assumed to respond rationally to economic stimuli. But the problems in attaining a safe workplace are not those of market imperfections alone. Departures from "economic rationality" are likely to be substantial in view of the wide variety of nonmarket forces and institutional considerations which impinge upon economic actors in our society. In such circumstances, the effectiveness of economic incentives as a major (or sole) instrument of public policy comes greatly into doubt.

It is customary in economic theory to assume that individuals know what is best for themselves in all matters and always act accordingly. Economists who subscribe

*The term "industrial feudalism" was coined by Arthur Ross, although he disputed the existence of the phenomenon.[61]

**The term "market imperfection" is used advisedly in this context, and is not meant to imply that things would be "better" if job security were not protected by seniority or other means.

†I am not unmindful of the anathema placed on this term by economists. The reader, if he wishes, may substitute "forces (cultural or social) which prevent an individual from examining a risk situation in detail."

to this view believe that it is not ordinarily within their professional competence, or the competence of government, to question the wisdom of individual decisions which do not interfere with the rights of others when these are freely made on the basis of full and complete information. In particular, it is not within the professional competence of economists to question the wisdom of an individual's expenditure decisions, even when consumption of the commodity in question may pose serious health risks to the individual himself. The conventional economic wisdom on this subject is well illustrated in the following statement regarding the proper application of the techniques of cost-benefit analysis to an evaluation of the net social benefits of tobacco consumption:

If smoking tobacco causes 20,000 deaths a year, no subtracting from the benefits, on account of this risk, need be entered in a cost-benefit analysis of the tobacco industry inasmuch as smokers are already aware that the tobacco habit is unhealthy. And if, notwithstanding their awareness, they continue to smoke, the economist has no choice but to assume that they consider themselves better off despite the risks. Indeed, the benefits to smokers, net of risk, are reflected in the demand schedule for tobacco. Once the area under the demand curve has been estimated and used as an approximation of the benefit smokers derive from the use of tobacco, any further subtraction for such risks would entail double counting.[63]

A further application of this mode of thinking is reflected in a recent analysis of the use of seat belts in automobiles. The authors of the study estimated the time and expense involved in using seat belts, compared these magnitudes with the incremental reduction in risk of death associated with seat-belt use, and concluded that individual decisions to forego the use of seat belts need not be viewed as inconsistent with "economic rationality":

The National Safety Council estimates that highway deaths would be reduced by about 10,000 per year if all automobile users wore lap safety belts. Assuming the estimate to be correct, seat belts would reduce the probability of dying in an automobile accident from about 25 per hundred thousand (25×10^{-5}) per year to about 20 per hundred thousand per year (20×10^{-5}). . . . [We] estimate that the average person in our sample would be willing to pay at least $8.80 per year (in 1967 dollars) for a seat belt for themselves. The cost of seat belts includes not only the purchase price and installation costs, but also costs associated with using the belts, including the bother and time spent buckling and unbuckling. *Clearly, it is easily within the realm of possibility that decisions not to purchase seat belts prior to the law were rational.* We can make some more back-of-the-envelope calculations. How much would the time and bother costs have to be in order to justify not using seat belts even after they are mandatory? In our sample the mean hourly wage was about $3.50. Using that as an estimate of the value of time, time spent buckling and unbuckling would have to be about 2.5 hours per year to cost as much as $8.80. Assuming 500 trips per year, this amounts to about 18 seconds per trip in time equivalent costs of using seat belts. . . . We leave it to the reader to experiment with other possibilities.[64]

It must be conceded that economists' concern for safeguarding the individual's right to make decisions is often justified. It cannot be denied that powerful interests often wish to substitute their own decision-making authority for that of people most directly affected by the outcome of those decisions. Nevertheless, there are serious reasons for questioning the notion that the existing level of workplace hazard should be tolerated because it represents working peoples' "choice" regarding the assumption of job-related risk. Inasmuch as decisions are often made by trial and error, there are serious reasons for doubting whether people really *do* know what is best for them in regard to their initial decision to accept hazardous employment. One important reason why this is so is the irreversibility of such decisions.

Irreversibility of Decisions

In the standard static theory of consumer choice, people are assumed to know what they want. In allocating their income between apples and nuts, for example, they always choose that combination of apples and nuts which yields them the greatest satisfaction. Though the theory requires that people enter into "economic adulthood" with their preferences fully formed, in fact it does little harm if in the case of choosing between apples and nuts people must expend some resources in forming their preferences through experience. They try various combinations of apples and nuts, perhaps several times each, in the process of deciding which combination suits them best. Though economic theory overlooks this process, no major distortion results, because the process of preference formation does not involve any serious *irreversibilities* (if we rule out such unlikely contingencies as lethal allergies to either commodity). People need have no cause to regret mistakes in allocating their budget between apples and nuts because they are not permanently "locked in" to a particular pattern of consumption, and can easily shift to any pattern that comes to appear more desirable.

When we leave the static model of consumer choice and begin to consider problems of decision-making over time (intertemporal allocation), the fact that people's preferences are formed through experience becomes much more consequential. In deciding among jobs which differ only in regard to safety and pay, for example, it is not possible to ascertain one's preferences by moving reversibly among the jobs: an injury or disease sustained in this process will not "go away" once preferences have been fully formed. A worker can therefore come to have serious cause to regret an earlier decision regarding the assumption of workplace risk, but at that point it may be too late to move to a safe job. The hands of the clock cannot be turned back, nor can the slate be wiped clean. A worker cannot "try" black lung disease for a week or so in order to discover whether he really thinks that it is worth the risk of becoming a coal miner. Having once been contracted, the disease will be with him for the rest of his (shortened) life, and cannot be cast aside when he moves to a less hazardous occupation.

Attitudes of Workers

A wide variety of nonmarket forces–including social, cultural, psychological, and environmental factors–influence workers' decisions regarding the assumption of job-related risks. Inability to assess or relate to low-probability, large-harm contingencies is a behavioral trait common to many–if not most–individuals. Everyone is familiar with the "it can't happen to me" syndrome, whether in himself or among friends or associates. The pervasiveness of this attitude often renders individual judgment suspect with regard to occupational health and safety.* As one panel of experts put it,

> When the probability of injury or death is seen as low and the individual has had no personal experience with others who have suffered the consequences, his awareness of the risk may not be incorporated into his behavior. This phenomenon might explain the remarkably high incidence of cigarette smoking by the general population compared with the low incidence among radiologists. For the latter group, exposure to the possible consequences is a matter of everyday experience.[65]

"Neurotic risk-taking"–compulsive assumption of risk–is a related and widespread behavioral pattern which reduces the desirability of exclusive reliance upon individual decisions regarding occupational hazard. It is not unreasonable to argue that society should protect "neurotic risk-takers" from themselves and from the adverse consequences of their choices.

Of course, there is a possibility that a term such as "neurotic risk-taking" may appear subjective, imprecise, and value-laden to some. However, investigators have already made the attempt to distinguish between "healthy" and "neurotic" risk-taking, on the basis of the following criteria:

> constructive or healthy risk-taking has the following characteristics: (1) conscious deliberation; (2) adequate preparation; (3) positive rather than negative aims; and (4) personal responsibility for the outcome. Neurotic risk-taking, on the other hand, is characterized by maladaptive qualities, repetitiveness, and inflexibility in the face of reality considerations.[66]

On the basis of these criteria, a panel of experts concluded that cigarette smoking comes under the category of "neurotic risk-taking." If so, it may well be asked what proportion of individual decisions regarding assumption of job-related risk falls under the same heading.

The pervasive phenomenon of machismo is another important factor retarding improvement in the level of occupational health and safety. Male workers are often socialized to believe that engaging in risky or hazardous employment is a mark of

*Inability to relate to risk is well illustrated by the following anecdote told to the author by a member of the United Steelworkers of America: A meeting was convened by union safety stewards to present members with information on the special health hazards of working near coke ovens. Upon hearing this information, the assembled coke oven workers were momentarily appalled, and then proceeded to express their discontent regarding the lack of overtime opportunities.

"manliness" or "virility," and that using safety equipment is not compatible with this ideal. Machismo also leads to the creation of a climate in which supervisory and peer-group pressure precludes the emergence of safety and health consciousness, and workers who do nevertheless become safety-conscious may face the disdain and ostracism of their coworkers. The effect of machismo in inhibiting aggressive union bargaining over improvement in the level of job health and safety has recently been mentioned with reference to the trucking industry:

> It's something of a machismo approach to the idea of being a truck driver. It comes down to this:
>
> The money's good so you have to accept these dangers the way other real tough men accept the dangers of their occupations. In effect, the man is told: What kind of a chicken-heart are you anyway? It's like the situation with the United Mine Workers, which also stresses the point that the money's good but to get it, you have to go down into those bloody mines for 20 years. The idea that comes out is this: Any red-blooded male confronted with the choice of being a truck driver despite all the dangers isn't going to let himself be typed as the gutless type who would back away from such an assignment.[67]

A wide variety of related personality and attitudinal considerations influence the way in which an individual views a particular risk, including (1) familiarity with the risk (which may make it seem less threatening); (2) the time lag between exposure to a hazard and its possible effects; and (3) the level of sophistication required for understanding the basis of the benefit-risk calculation.[68]

In view of all these intervening factors, it may at least be questioned whether sole reliance upon individual judgment is a sound basis for policy-making with regard to setting an acceptable level for a workplace hazard.

There are also a variety of sociological and cultural factors which influence the formation of individual preferences regarding the assumption of job-related risk. Many workers are socialized to accept the hazardous nature of jobs and are convinced of the necessity of performing them in order to earn their livelihood. The Puritan work ethic and the notion that work has to be hard and unpleasant helps induce a willingness to accept hazardous conditions. Family and community norms and traditions further contribute to the resignation with which many workers accept the fate of a hazardous job. Many workers live in "company towns" in which there is, in fact, no viable alternative to hazardous employment. Faced with such conditions, it is hardly surprising the many workers don't *want* to know (or admit) the risks they are facing on the job.

"Rugged individualism" and an antipathy to being told what to do no doubt contribute to workers' failure to follow safety practices, which is responsible for a small portion of job injuries. Workers resist being told how to do their job. (This situation is not to be confused with workers taking hazardous shortcuts in order to meet production demands or to improve piecework incentive earnings. In these situations where economic disincentives prevent safe and healthful work practices, neither "carelessness" nor "ignorance" are suitable descriptions of risk-taking.)

Finally, the typically American inability to relate to statistics and to death itself is in no small way responsible for behavior toward risk. Moreover, hazard pay differentials clearly are not arrived at by rational decision-making processes.

Attitudes of Businessmen

Nonmarket forces also significantly influence the attitudes of businessmen toward occupational health and safety. Businessmen have an antipathy verging on the irrational to government regulation in this area (perhaps even stronger than their antipathy to general environmental controls). OSHA's record-keeping requirements, inspections, and standards have aroused the strong opposition of a large segment of the business community, despite considerable evidence that at least some firms will ultimately benefit financially from the resulting improvements in workplace health and safety.

It would appear that many businesses are extremely reluctant to alter customary hazardous production practices, even if an appeal is made in terms of the increased profits associated with improved levels of job health and safety. Once habituated to the practice of making business decisions without due consideration for their consequences in terms of employee health and safety, it is very difficult indeed to change such managerial behavior. Attempts to do so by external means such as government intervention tend to arouse considerable hostility.

Even in those firms which do make active decisions regarding job health and safety, those decisions are usually reserved for management alone. Reluctance to permit labor participation through joint safety committees or by other means is rooted in a firmly held conviction that such decisions are the sole prerogative of management. This posture is often maintained at great economic cost to the firm (as in the case of the strike and boycott precipitated by Shell Oil Company's refusal to negotiate safety and health issues with the Oil, Chemical, and Atomic Workers Union).

7.5 EQUITY

7.5.1 The Issue of Equity As It Affects Workers

Considerations of equity and concern for individual justice are of fundamental importance for decisions regarding the socially acceptable level and distribution of workplace hazard. Individuals presently subjected to job-related health risks are hardly a randomly chosen group representative of the total population. Though occupational hazards are pervasive throughout American workplaces, a disproportionate bulk of the burden of this risk is in fact shouldered by individuals employed in a variety of particularly hazardous occupations and industries. In addition, workers exposed to the most hazardous substances frequently belong to the lowest socioeconomic group in the industrial process and live closest to the plant, where the external air and water pollution is especially high.[69] These workers are subjected to a form of double jeopardy—a disproportionate risk of harm both off and on the job. One could also argue that middle-level managers are subject to stress both on the job and off the job driving home or commuting. In the case of the blue-collar worker, physical hazards need to be removed, while stress for the white-collar worker is more related to the characteristics of the job task. Noise, of course, affects both groups adversely.

As a result of the special job-related risks which these individuals face, life is made easier and more comfortable for all of us. Farmworkers face a large number of occupational hazards, contributing to an exceptionally high accident and fatality rate, so that the rest of society may enjoy relatively abundant food supplies (see Chapter 12). Coal miners pursue their hazardous and unpleasant trade in order that the rest of society may have available to it abundant supplies of electricity, steel, and other commodities fundamental to our standard of living. Asbestos workers currently experience tremendous excess mortality in order that our ships and buildings may be properly insulated and the brakes of our automobiles may function efficiently. (Of the 500,000 individuals alive today who have ever worked with asbestos, as many as 34% may die from asbestos-caused disease.)[70]

Though it may be *expedient* for the rest of society to subject these and other selected groups of workers to special occupational risks, it can by no means be considered *just*. As John Rawls has put it, principles of equity and fairness "rule out justifying institutions on the grounds that the hardships of some are offset by a greater good in the aggregate. It may be expedient but it is not just that some should have less in order that others may prosper."[71]

There is little justification for society to be less protective of the man in his work environment than it is of the man in the general environment. It is now recognized that caveat emptor (let the buyer beware) is *not* an equitable principle in product safety and imposes an unconscionable contract upon the buyer. Yet the principle of "let the worker beware" in the manufacturing of that good is retained. The argument that wage differentials in hazardous industries incorporate the appropriate level of workplace risk is about as convincing as the argument, no longer made, that the price the consumer is willing to pay incorporates the risk he is willing to take on being injured by the product he buys.

It may, of course, be the case that the pursuit of equity is in conflict with the attainment of other social goals such as economic efficiency. In economic terms, there may be a "trade-off" between equity and efficiency implicit in decisions regarding the level of workplace health and safety. If so, and if society places positive value on equity per se, it may be sound social policy to forego maximum economic efficiency in order to obtain increased equity. Of course, this procedure may involve a larger allocation of social resources to improved workplace health and safety than would be dictated by efficiency considerations alone. Rather than attain the chosen target level of improvement in workplace health and safety in the most efficient way possible—by means of the smallest total expenditure—it may well be preferable to distribute this reduction in job hazard more equitably among workers, even at an increased cost. Society must decide, as a long-term goal, how far to travel down the road toward equalizing the level of hazard in all occupations and industries. The *distribution* among individuals of reduction in the level of workplace hazard is an important policy consideration which must be addressed.

Some have argued against allocating additional resources to reduce the level of workplace hazard on the grounds that risk is inherent in all activities. Driving an automobile, flying in an airplane, and breathing polluted air all entail health risks. Why should work be any different? Such reasoning is fallacious, however, since it overlooks three important elements of risk:

1. The degree to which the decision to assume risk is *collective* (vs. individual).
2. The degree to which there can be *voluntary choice* in opting out of taking the risk.
3. The extent to which there is randomization (vs. selective victimization) of harm resulting from the risk.

These principles may be illustrated by comparing the risks associated with automobile driving, breathing polluted air, and hazardous work. The decision to drive an automobile is, in principle, voluntarily made on an individual basis, in view of the existence of less hazardous alternative modes of transportation. Under particularly adverse driving conditions, such as bad weather, the only other cars on the road are those driven by individuals who have similarly elected to assume the higher-than-normal burden of risk involved. Victims of automobile accidents represent in many respects a cross section of individuals randomly selected from the driving population, and there is near equality in the level of risk assumed. Though drivers of small cars face a higher probability of serious harm if they are involved in an accident than do drivers of large cars, the probability of being involved in an accident is greater for a large car than for a small one. Thus, the probability of harm is roughly the same for drivers of both sizes of car.

The risk situation in the workplace might be thought to be much closer to the analysis of hazards in the general environment than to the risk associated with automobile driving. However, a careful examination of hazards in the general environment (such as air or water pollution) will show that the two situations are significantly different. If society *collectively* decides to tolerate a certain sulfur content in its heating fuels to save costs and is willing to suffer an increase in respiratory disease as a result, there is a certain equity which gives us comfort. All citizens save money, and the risk of disease within an urban area is randomly distributed, more or less. Unfortunately, "opting out" of this risk situation is more difficult than choosing not to drive an automobile, since moving to a less-polluted location is more difficult.

However, if society *collectively* decides to use asbestos wherever it will save money, and a selected group of asbestos workers pay for this decision with their lives and health, there is a strong argument that the situation is not equitable, *even if more lives are saved than lost.* For example, by the use of asbestos brake linings, more lives may be saved among the general public on the highways than are lost by workers in the plant. Yet the nonrandom selection of the workers who bear the risk—a distinction which is often lost—deserves special attention. It is often the case that changing jobs is difficult for people who initially enter a hazardous occupation because no other jobs are available to them. Moreover, once the decision is made to enter hazardous employment, the fact that compensation, retirement benefits, and opportunities for training and advancement are often linked to continued, uninterrupted employment makes subsequent job changes too costly for many workers.

There is little awareness of these issues of *who benefits* and *who bears the risks*, and of the differences in risk analysis for the work versus the general environment. Though it might be expected that these two issues would be underemphasized by management and research scientists, I have also found insurance companies, labor, public-interest groups, and legal practitioners to be almost totally unaware of these issues (see Chapter 10).

At the present time, nonrandom victimization is the norm for exposure to occupational hazard. With the rise of economic forces and institutions which place a substantial premium on job tenure as a determinant of pay, pensions, and advancement in a large segment of the labor market, selective nonrandom exposure to occupational hazards may well be on the rise—that is, the "population at risk" may come to represent a less and less accurate cross section of the general population.

Many working people—especially the poor and unskilled, members of discriminated-against minority groups, and the geographically and occupationally immobile—are presently subjected to the cruel dilemma of a very sharp trade-off between their health and the economic well-being of their families. Considerations of equity and social justice dictate that a daily assumption of undue risk not be a prerequisite for their claim to reasonable job security and a decent wage. Of course, different people view what is fair differently, but this fact makes the consideration of equity no less important. Whatever the alternative value judgments as to what is fair, we should know what the costs are for those alternatives we wish to consider.

7.5.2 The Issue of Equity As It Affects Firms

Equity considerations also enter into the differential impact of occupational safety and health regulations upon small and large firms. It has long been a recognized objective of government policy in this country to promote and preserve small business, and to influence the "mix" of small and large firms in such a way as to result in a larger share of small firms than would be dictated by the operation of economic forces alone. Without special provisions for small firms, it is likely that public policy regarding occupational safety and health will further accentuate economic forces tending to undermine the viability of small business. *If* it is determined that continued promotion of small business should remain as an objective of social policy—a constraint imposed by political reality—then perhaps a larger fraction of social resources should be expended to obtain a more standardized level of workplace safety and health than would result from "least-cost" methods which gives large businesses a competitive advantage (see Section 6.6).

Although it may be argued that a marginal increase in the resources devoted to improving job health and safety would be allocated more efficiently to the larger and safer firms, following this policy would continue to improve the health and safety of those workers already better off without benefiting those workers who face the worst working conditions. Thus, equity considerations might justify deviating from maximum efficiency in the allocation of resources for improving working conditions.

7.5.3 Hazard Pay and the Assumption of Risk

A 1973 study on *Chemicals and Health* concludes that

> another example of voluntary acceptance of high risk is the field of occupational exposures. Some occupations can represent very high probability of death and injury when compared to other involuntary activities. Of course, it can be argued that voluntarism does not completely describe occupational choices.[72]

This last comment may indeed be an understatement. While jet test pilots may in fact be risk preferrers, the ability of many working people to make "free choices" in their selection of jobs is highly questionable. The employer's defense of "the employee's assumption of risk" to defeat a suit for damages by an injured employee has long disappeared from our system of workmen's compensation.

One mechanism for compensating workers who are subject to job-related risk is *hazard pay.* Hazard pay is relatively common in many blue-collar occupations, and has customarily been sought by labor unions for members who are subjected to unusually dangerous working conditions. In some industries, hazard-pay differentials have even been embodied in collective-bargaining contracts or formal joint labor-management job-evaluation plans.*

*Perhaps the best-known formal joint labor-management job evaluation plan is in the steel industry. Hazardous and unpleasant working conditions have long been an explicit factor in determining employee compensation in this industry.[73]

When interviewed, corporate medical directors often voiced criticism of using hazard pay as a mechanism to allocate job-related risk, though the reasoning by which they reached this conclu-

Economists have further generalized the notion of compensating wage differentials which reflect differences in hazard among occupations.[74] In the abstract world of the economist's model, individuals have perfect information, are perfectly mobile (between jobs or employers), know what is best for themselves in all matters and always act accordingly. Employers are analogously situated in this make-believe world, and choose among alternative technologies and prospective employees so as to maximize profit. Externalities and other market imperfections are assumed not to exist. In such a model, hazard-pay differentials fully compensate individuals for the assumption of job-related risk of injury or disease; thus, other things being equal, the more hazardous the job, the more highly paid are those who perform it. Individuals evaluate their job safety-expected income-opportunity locus (set of best attainable points) and choose the point they most strongly prefer (the one that yields them the greatest expected utility or satisfaction). Firms search among existing technologies which vary as to level of workplace hazard (given their level of output and the prevalent range of risk premia in the labor market) and then choose a mode of production and associated level of workplace hazard which minimizes the expected cost of production. In this best of all possible workplace worlds, the operation of the free market establishes the socially optimal level of workplace hazard. Note that compulsory insurance schemes such as workmen's compensation and the right to sue in tort—based on legal liability for the costs of damage associated with occupational hazard—are largely irrelevant in this model to the level of workplace hazard that is established. In the absence of the right to sue in tort, workers as a group receive hazard pay in lieu of recovering specific damages from their employer. If individual workers are "risk-averse," they can buy an insurance policy with their risk premium to protect against the contingency that they may become injured or ill on account of their job. (If employers are assigned legal liability, then *they* can buy the insurance instead of their employees.) Thus, in this model, hazard-pay differentials are assumed to compensate individuals fully for the expected costs involved in the assumption of workplace risk.

In the real world, it is questionable whether hazard pay should continue to be relied upon as the primary instrument for allocating the burden of job-related risk. Hazard pay is an inequitable substitute for the correction of dangerous working conditions, and should not be as widely utilized by business as it presently is. For a variety of reasons, it is doubtful whether hazard pay really compensates workers for the job-related risk they assume. In view of the pervasiveness of imperfect information, and of employers' information advantage, it is unlikely that employees have sufficient knowledge at the present time to make proper evaluation of the risks

sion varied greatly from company to company. Some medical directors felt that hazard pay was a morally indefensible substitute for improvement in working conditions, and represented an abdication of managerial responsibility for employee health and well-being. Other medical directors, by way of contrast, opposed hazard pay on the grounds that it was utilized by labor unions as a mechanism to extort higher wages.

facing them. Furthermore, both workers and employers saddle society at large with much of the cost burden associated with occupational injury and disease. Costs of medical care, therapy, and rehabilitation in such cases often befall society at large, as do social welfare services consumed by the surviving dependents of workers killed by occupational diseases or accidents (see Section 7.2.2).

Fears of job loss and unavailability of alternative employment opportunities by reason of occupational or geographic immobility or lack of training often make hazard pay an unconscionable bargain to those individuals who are forced by circumstances beyond their control to place an inordinately low implicit valuation upon their own lives. Nor is it appropriate to rely upon hazard pay if the distribution of income and opportunities is inequitable to begin with. Individual decisions about the amount of job-related risk to assume are not independent of society's distribution of nonhuman wealth and such associated factors as home environment, educational opportunities, "connections," etc. An individual in need of $100 per week to support his family, when choosing between two jobs, one paying $125 but with high risk of illness or death, another paying $75 per week but safe, may choose differently if he has wealth bearing $50 interest per week than if he has no wealth at all. Hazard pay is *too* attractive to such individuals from society's point of view, and in any case doesn't come close, typically, to compensating them for the true costs of the assumption of risk involved.

A variety of other considerations make the presumption that workers always "know what is best for themselves" regarding their balancing of hazard pay and workplace risk even more questionable. The prevalence of "inability to relate to risk" and "neurotic risk-taking," as well as the influence of social and cultural conditioning and group pressure, have been discussed previously in terms of their implications for individual behavior toward risk. Many workers opt for hazard pay under the mistaken notion that the potential adverse consequences of so doing "can't happen to them." Feelings of invincibility or invulnerability, and the notion that all that is required to avoid harm is the exercise of proper caution, are prevalent among many people. Other individuals may subsequently regret decisions to undertake hazardous work, but the assumption of risk is an irreversible process once the unwanted contingency has occurred. If an individual "voluntarily" decides to accept a high-risk, high-paid job and becomes injured or ill, it may be too late for him to reconsider this decision, as irreparable harm may have been done.

Clearly, therefore, it must be concluded that hazard pay is not an equitable or socially acceptable substitute for general improvement in the level of workplace health and safety. (In saying this, I am not unmindful of the existence of inherently dangerous occupations such as fire-fighting, police work, and aircraft testing for which hazard pay is an accepted tradition. Perhaps the distinction lies in the extent to which the employer can make the job safer by his own affirmative efforts.)

7.6 THE SPECIAL PROBLEMS OF THE SMALL FIRM

7.6.1 Introduction

The social benefits of improved health and safety in the American workplace can hardly be questioned. However, unless policies for achieving a safer workplace are carefully chosen, many workers and employers could be hurt in the process of attaining this desirable objective. Smaller firms may find it particularly difficult to pay for the information, new technology, health and safety manpower, and other production changes that will be required. Compliance with strict health and safety regulations will generally be relatively more expensive for the smaller firm. Even though the small firm may eventually be able to maintain a good health and safety program, the initial costs of implementing such programs may be prohibitively high.

Moreover, these costs seem to be coming at a time when many smaller firms find themselves in a precarious competitive position. If left unassisted, a number of these firms may be forced to close their doors, while still more will curtail production and lay off workers. When we recall that nearly two-thirds of the U.S. labor force is employed in firms of under 500 employees (and 30% in firms of under 25 employees), the importance of designing policy to deal with the special problems of small business becomes clear.

The loss of *some* jobs and firms is probably unavoidable if we are to reach an adequate standard of workplace health and safety. In the long run, however, adjustments by employers and workers will probably involve substantial benefit to both. But in the short run, some people who have invested their savings and effort in a particular endeavor may find their lives seriously disrupted. No health and safety policy is fully adequate unless it addresses the problem of mitigating the harm to such individuals and firms. Thus, a major concern must be that of equity—i.e., the extent to which we can avoid placing the costs of social policies unfairly on the shoulders of a select few employers and their employees.

7.6.2 The Marginal Position of Many Small Firms

As has been mentioned, small firms are often in a precarious competitive position.*

*On the basis of a study of regulation in the chemical, pharmaceutical, and pesticide industries, the President's Advisory Science Committee reported that

> after examining a variety of economic issues concerning the industries which manufacture chemical products, the economic health of these industries remains high. Sales of chemical products continue to increase. The value of shipments of chemical items continues to rise with a doubling time of something over ten years. . . . [It] is the Panel's understanding that profits generally remain high and commensurate with an industry where risks are considered to be correspondingly high. There is some indication of the departure from the field of some small manufacturers and of secondary manufacturers (formulators of pesticides, distributors, etc.).[75]

They exist in many different kinds of markets, producing many different types of products, but all share the likelihood that small economic setbacks will lead to failure. This is true whether (1) they are in an industry composed entirely of small firms, each with little market power (as in apparel and construction), (2) they compete with larger firms within the same industry, but without the market power, diversity, or cash reserves available to their more sizable competitors, or (3) they compete in international markets, with firms that are not subject to American regulations. All these types of market conditions present their own difficulties for small firms faced with the additional cost burden of improved occupational health and safety.

Larger firms can often absorb such costs in a number of ways. For example, a firm with diversified operations can (temporarily) offset increases in cost in one area by expanded production and sales in another. Moreover, a firm with large resources and available credit will be able to withstand short-term losses in the anticipation of future profits. A smaller firm faced with similar short-term increases in costs may have no choice but to retrench or close down.

For small firms facing international competition, the problems can be even more severe. Such firms, with their relatively modest profit margins, are in competition with foreign producers who may not have to pay the costs of compliance with strict health and safety standards. In such cases, American companies may no longer be able to compete in international markets for goods which are most cheaply produced when dangerously produced. (There are really two ways of looking at this. One is that we are losing jobs to foreign competition. The other is that we are exporting injury and disease, as in the case of South African asbestos workers or Zairian uranium miners.)

7.6.3 Costs of Prevention and Firm Size

Not only are many small firms in a difficult competitive situation, there is reason to believe that reduction of workplace hazard is more expensive in small firms than in large ones. (It should be noted, however, that the small firm may find it easier to adopt new work practices, since it is less likely to be committed to expensive hazardous methods of production than its larger competitors.) It may well be the case that the least expensive way of protecting workers' health in many industries is available only to production units above a minimum size. If so, all firms below this size would be forced to pay more per worker to achieve a given level of protection (see Table 7.1). Small firms may not be able to spread the costs of hiring a full-time safety and health specialist over a sufficient volume of production to pay for his services. Yet such a full-time person is often much more effective than an employee with only a partial commitment to this issue—or a consultant who is unfamiliar with the specific conditions facing the particular firm.

The problems of small firms with regard to occupational health and safety are likely to be similar to problems that they face with regard to environmental pollution control. A recent study of the costs of pollution control in the fruit and vegetable canning and freezing industries concluded that there are substantial differences in costs to large versus small firms. This study presents the estimate that the average increase in prices necessary to offset the costs of meeting federal pollution standards is 5.5% in large plants, 6.4% in medium-sized plants, and 9.6% in small plants.[76]

Such cost differentials have undoubtedly been reflected in the relatively poor performance of small firms in providing occupational health and safety medical services to date. It has been pointed out that large firms began to establish elaborate health and safety programs before 1950. On the other hand, "Small industries could not afford these sophisticated programs. Their per-worker costs for comparable services proved to be hundreds of dollars per worker, as contrasted with $25-$30 in plants of over 500 employees."[77]

As a result of the marginal nature of many small firms and their high costs of compliance, full enforcement of adequate standards—whether through the OSHAct or some other statute—may well place many small firms in an economically untenable position. Both workers and owners who made the decision to enter a particular industry long before the establishment of new, tougher health and safety standards may be forced to pay a high price for their enforcement. Such workers face both the loss of their jobs and the possibility of reduced earnings in the future. Owners can lose their substantial investment of savings and effort.

7.6.4 Possible Programs to Reduce the Burden on Small Firms

There are two issues to be faced: (1) Are we willing to allow small firms to disappear from an industry in order to reduce accidents and disease in that industry, or should we pay the additional costs for making small industry safer in order to maintain a certain mix of small and large firms? (2) Given that we desire to preserve some small firms, what are the best mechanisms for achieving this objective?

There are many possible methods of reducing the burden of economic dislocation on the workers and owners of small firms. Small firms may eventually bear the full cost of health and safety expenditures; however, without long-term subsidy to facilitate the adjustment of small firms to new health and safety requirements, it is likely that the rate at which small firms are disappearing from the U.S. economy may greatly increase.

Aid to Owners

Attempts have been made to exempt small firms from the effects of the OSHAct. Indeed, an amendment exempting small firms from being inspected under the

OSHAct was actually passed by Congress and only incidentally vetoed by the President because it was attached to an appropriations bill. An OSHA regulation was subsequently issued exempting most firms with seven or fewer employees from record-keeping requirements under the Act (see Section 4.3.4). Certainly, a small-business exemption does solve the problem of the dislocation of some workers, but at the unacceptable cost of sacrificing adequate protection for millions of other workers.

Exemption of small business from OSHAct coverage is a particularly unwise policy in view of the extremely high concentration of smaller firms in hazardous industries. For example, in the construction industry, 90% of companies employ 20 or fewer workers.[78] Though construction activity accounts for only 5% of total employment, 20% of work-related accidents occur in this industry. In 1970 the rate of lost time due to industrial accidents was 2100 per million employee-hours worked in construction, compared with a rate of 759 in manufacturing. Similarly, in logging, the accident rate is three times the general rate which prevails in manufacturing, and 96% of the logging companies in the United States employ fewer than 20 workers.[79] Other industries designated as "targets" by the Department of Labor due to their extremely high injury rates include longshoring, roofing, meatcutting, and mobile home construction—all of which are industries with relatively high concentrations of smaller firms.

Moreover, a large proportion of companies engaged in the manufacture, processing, or use of particularly hazardous substances are smaller firms. Benzene and other hazardous chemical substances are widely used in industries such as printing and furniture refinishing, activities which have high concentration of smaller firms. In the chemical industry as well, there is a significant proportion of smaller firms: 35% of all OCAW locals, the members of which are exposed to a wide range of harmful and often toxic chemicals, consist of 25 or fewer employees.[80] In agriculture, where a large proportion of the labor force is regularly exposed to potentially harmful pesticides, employment is overwhelmingly concentrated among small employers. Asbestos is a particularly hazardous substance which is conservatively estimated to be responsible for 3000 premature deaths annually among asbestos workers—1000 each from asbestosis, mesothelioma, and lung cancer.[81] In several activities which rely heavily upon asbestos use, including construction insulation and automobile brake relining, smaller firms predominate.

It is furthermore the case that *within* industries, smaller firms tend to be more hazardous than large ones. Prior to the new record-keeping requirements imposed under the OSHAct, "the likeliest employers to be underrepresented in BLS [injury] statistics [were] small employers with bad safety records."[82] Despite this fact, the data indicate that *within industries*, small firms above a certain minimum size are significantly more hazardous than large ones (see Figure 7.4). It is interesting to note, however, that the very smallest firms (those with 49 or fewer employees) are

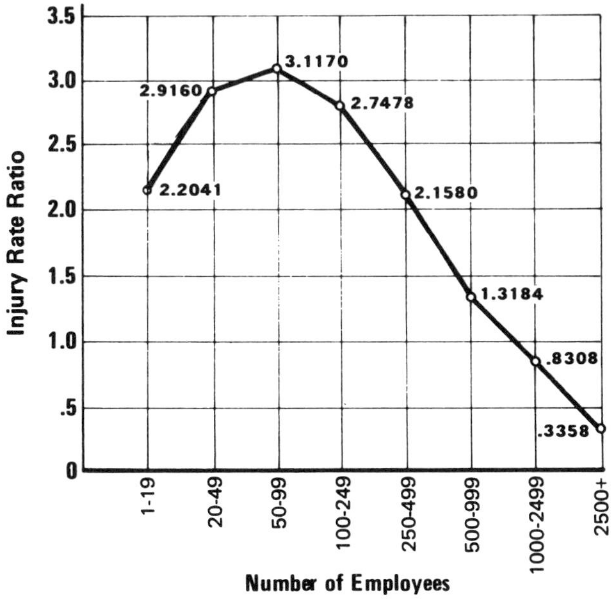

Figure 7.4 Ratio of the injury rates for firms of various sizes to the average injury rates in their industry. Ratios are weighted averages for 39 manufacturing industries. From *The Report of the National Commission on State Workmen's Compensation Laws*, p. 92.

slightly less hazardous than firms with 50-99 employees. Of course, this may well be a statistical artifact, attributable to the chronic underreporting and unrepresentative nature of the BLS small-firm sample mentioned above. Moreover, the data presented refer only to *injury* rates—there is no comparable information on the relationship between occupational *disease* rates and firm size.

It may also be the case that in the very smallest firms, the employer is himself often a worker and subject to the same health risks as his workforce, which provides considerable incentive to alleviate hazardous conditions. However, this explanation is somewhat implausible in view of the fact that such employers are likely to be every bit as ill-informed regarding the nature and level of occupational hazard as are their employees. In any case, firms with 99 and fewer employees are clearly far more hazardous than their larger competitors.

It is thus clear that complete exemption of smaller firms from the requirements of the OSHAct is not the correct solution to their special problems of compliance, for this would leave unmet the needs of a substantial group of employees most subject to excessive workplace hazard at present. As so large a proportion of the U.S. labor force is employed in small firms, the "population at risk" is far too large to be exempted from the protection provided by the OSHAct. The fact that the incidence of occupational illness and injury is presently so great among this substantial fraction of the labor force makes the exemption of small business particularly unwarranted.

The Occupational Safety and Health Act amends sections of the Small Business

Act to provide for financial assistance to small businesses for the purpose of altering equipment, methods of operation, or facilities to comply with standards and duties required by the Act. In practice, however, this provision of the OSHAct has hardly been utilized to date. During congressional hearings on "Small Businesses and the OSHAct of 1970," held in June 1972, it was pointed out that in the two years since the OSHAct had passed, the Small Business Administration had granted only eight loans totaling $596,400 to small companies for the purpose of investing in improved workplace health and safety.[83] Partially as a result of the stimulus provided by these congressional hearings, the rate at which SBA loans were granted for OSHAct compliance purposes increased significantly during the next year. However, by the end of 1973 only a total of 70 such SBA loans had been granted.[84]

Another method of aiding small firms could be to subsidize their expenditures on occupational health and safety, either through direct payments or through special tax credits for health and safety investment. This seems to be a reasonable short-run solution, but it would mean that the market price of goods in more dangerous industries would not adequately reflect all costs of production. As a result, too large a quantity of these goods would be produced from a social-efficiency point of view, but this may properly reflect our commitment to maintain the economic viability of some small firms.

In order to help small firms deal with the problems of international competition, special tariffs could be levied on goods produced under dangerous conditions. Or the importation of these goods could be restricted, on grounds similar to those involved in forbidding the importation of products made from animals on the "endangered species" list. This might mean, for example, that South African asbestos and asbestos products would be highly taxed.

Finally, programs to increase the availability of the services of health and safety experts would greatly assist small firms. As was pointed out earlier, the costs of expertise are often much higher for small firms than for large ones. Moreover, small businesses in some cases cannot presently obtain expert assistance even if they are able to pay for it. If contacts could be facilitated between health and safety professionals and small firms—through governmental or private means—this would prove a major benefit (see Section 7.6.5).

Aid to Displaced Workers

Despite government programs to assist small business, some workers are likely to lose their jobs as a consequence of the transition to a safer and healthier workplace. The loss of one's job is often a source of major trauma, both because of lost income and because of a diminished sense of self-worth; hence, programs to minimize loss of jobs and to ease the transition to new ones are of great importance.

The most obvious policy to reduce the adverse employment effects of workforce reductions associated with improved occupational health and safety is a national commitment to the maintenance of full employment. This, of course, is a goal worth pursuing for reasons quite apart from occupational health and safety. But maintenance of full employment is nevertheless perhaps the single most important policy area affecting displaced workers.

In addition, special training and relocation programs could ease the problems of workers who are laid off. Such programs *must* be linked to the development of employment opportunities if they are to prove effective—a lesson which has been learned from the manpower programs of the 1960s. Of course, income supplements to trainees are often necessary. Note here that efficiency would suggest that relatively more training be made available to displaced young workers who will have longer to use their skills and to build on them. However, other considerations—e.g., the fact that the unassisted labor market mechanism is likely to do a better job in relocating a younger than an older worker—would suggest an emphasis on training and relocating older workers, since they have invested more of their lives to now useless skills, and since their alternative employment opportunities may be severely limited.

7.6.5 Group Occupational Health Services and the Small Firm

Group occupational health services are one important way of attempting to overcome scale problems in the provision of occupational health and safety services. Several European countries have substantial experience with these services, and their experience has generally proven encouraging (see Chapter 11). The rationale for such programs is straightforward: since efficient provision of preventive and medical services requires a large minimum size, several small firms are able to support a joint program which none of them would be able to afford individually.

The actual form of these group services varies from country to country, depending upon the type of labor-management relationship and the legal system.

In England there are six cooperative group occupational health services, with combined coverage of about 100,000. There is, as yet, no organized pressure for such programs, so all have found financial support to be a continuing problem. Moreover, "Employers are more concerned to maintain their labor force at work through prompt, efficient, on-the-spot care for injuries and ailments than with environmental protective measures."[85] The *treatment* aspect of these services have thus been emphasized, to the exclusion of potentially beneficial *preventive* service.

In France such group cooperative occupational health and safety services are numerically much more important. There are about 630 of them, covering approximately six million French workers. The difference between the experiences of these two countries is the result of two factors. First, the French occupational health services do not compete with a National Health Service, as in Britain. Second, enrollment is legally compulsory for all firms regardless of the number of employees on the payroll.[86]

The Swedish case is strongly conditioned by the important roles played in the country's political life by the National Swedish Employers' Confederation and its labor counterpart, the Federation of Trade Unions. These two groups have a joint agreement through which a program of health services has been established.[87]

The renewal of interest in occupational health and safety in this country has led to at least one interesting and innovative suggestion for a group health service in Philadelphia. The Franklin-Hahnemann Institute of Occupational and Environmental Safety and Health is presently well into the planning stage.[88] Using the facilities of a group prepaid health insurance plan, a hospital, and a research and consulting organization, this service would attempt to integrate occupational health services with a total health-care program, including prevention. It would also have an integrated research component. This proposed institute is an exciting concept that should be watched closely by everyone involved in occupational health and safety issues.

7.6.6 Consultation for Small Firms

Another policy of potential importance to small business would be provision of technical consultation services. Thus far the market has been relatively sluggish in responding to the needs of small business in this regard. Congressman William Steiger, coauthor of the OSHAct, has proposed an amendment to the Act which would extend OSHA's responsibilities to the provision of such consultative services.[89] This proposed amendment does have the drawback of possibly diverting OSHA compliance officers from their enforcement duties under the Act.[90] However, in principle, the idea is a sound one, subject to the proviso that sufficient funding is allocated to enable OSHA to discharge properly both its consultative and enforcement functions.

7.7 THE SPECIAL PROBLEMS OF UNORGANIZED WORKERS

7.7.1 Introduction

Unorganized workers constitute 75% of the American labor force, a proportion which is not likely to change in the years ahead in the absence of considerable organizing effort. Of course, there is considerable variation among industries and occupations, both with regard to the proportion of the workforce organized and the level of workplace hazard faced. Some particularly hazardous industries, such as underground coal mining and construction, are heavily unionized. On the other hand, some relatively safe occupations, largely white-collar, are relatively unorganized. Yet it is also the case that a number of very hazardous industries remain predominantly unorganized, agriculture being a prime example. Moreover, the most dangerous firms within those industries that are heavily unionized are often the unorganized ones, again as in the case of underground coal mining. In sum, a great many workers in hazardous industries, and especially in the most hazardous firms, do not enjoy the protection associated with membership in a labor organization.

While many of these unorganized workers are employed by large, progressive firms with relatively good health and safety practices which have managed to avoid unions by keeping one step ahead of them, not all unorganized workers are so fortunate. A large portion of the unorganized workforce is employed by smaller, marginal firms or consists of traditionally difficult-to-organize low-wage groups such as female and minority workers. Even in the larger firms, workers without unions to back them up often find that they lack access to information regarding possible workplace hazards and that, historically, attempts to obtain such information have often resulted in summary discharge without recourse as well as a host of lesser discriminatory actions. Thus, unorganized workers quickly learn, even in the larger and relatively more progressive nonunion firm, that workplace health and safety is a firmly held management prerogative over which workers have very little influence. Yet working people are the ones who are most affected by managerial decisions in this regard. The special problem of unorganized workers is thus quite simply that they have very little control over these decisions which can have momentous consequences for them.

Unorganized workers do, of course, benefit directly and indirectly from the activities of labor unions in improving workplace health and safety. Indirectly, unorganized workers benefit from improved health and safety practices "voluntarily" initiated by management in an attempt to forestall unionization. Thus, gains extracted by unions at the bargaining table or through health and safety strikes will ultimately "filter down," at least in part, to some unorganized workers. Directly, unorganized workers theoretically benefit from state and federal legislation to improve workplace health and safety, such as the OSHAct, which was enacted largely due to the political influence and strong backing of organized labor. The legal and political activities of unions to ensure proper implementation of the OSHAct also benefit unorganized workers. For example, OCAW activities leading to the adoption of carcinogen standards theoretically resulted in equal protection for all workers exposed to those materials. It is particularly evident that unorganized workers will benefit directly from planned improvements in the level of payments and scope of coverage of the workmen's compensation system which, if enacted, will be largely the result of political lobbying efforts by the AFL-CIO and various international labor unions. Similarly, the new federal black lung screening and compensation program, attributable in substantial measure to the political efforts and activities of the United Mine Workers of America, is beneficial to all coal miners, organized or not. Pension reform legislation, passed in 1974, may also serve in a significant way to increase the mobility of many unorganized workers who are presently tied to hazardous jobs in part through fear of losing their pension.

Nevertheless, it is clearly the case that the activities of unions, however beneficial, hardly solve the special problems of unorganized workers in obtaining a safer and healthier environment. This is especially so in view of the fact that the threat of competition from the massive nonunion sector of the workforce seriously limits the advances that can be made by unions in this area. The present willingness of non-union workers to assume a heavy burden of job-related risk, whether due to a lack of information about the hazards facing them, or to a belief that there is no alternative if they are to earn a living, undermines the ability of organized labor to obtain a safer and healthier workplace for all. An understanding of the special problems faced by unorganized workers in attaining a safe and healthy workplace is therefore an essential ingredient in the design of policy alternatives affecting both organized and unorganized labor.

7.7.2 The Nature of the Problems

Information

Unorganized workers suffer with particular severity from a lack of information regarding (1) the nature and seriousness of workplace hazards which they face, and (2) their rights under the OSHAct and the Act's provisions for improving working conditions. In the absence of a union, workers must depend on the goodwill of their employer for information whose disclosure is likely to be of serious financial detriment to the firm and to cause labor-relations problems. Hence, it is not surprising that employees receive little information about occupational hazards when they are not backed by a union. When employers do choose to impart some information to their workforce, moreover, its reliability may be suspect in view of the substantial costs which disclosure of reliable information could impose. And since even the good-faith employer (especially in the smaller firm) may in fact be ignorant of health and safety hazard, the information problem is particularly acute.

Though these problems of ignorance and information control exist to a certain extent with regard to occupational safety, they are particularly serious in the case of occupational disease. Employers will in some cases actually attempt to persuade workers that their disease is in no way connected with the work environment. The case of Southern textile workers faced with employer assertions that byssinosis doesn't exist is instructive in this regard.* As recently as 1969, unorganized textile workers could expect to receive the following sort of information from their employers:

*The employers often defend their views by using expert scientific or medical testimony to draw fine lines, often verging on obfuscation. For example, Dr. Henry Howe of the AMA has said "there is no such thing as byssinosis," that the "disease" is really emphysema, rather than a diagnostically different disease.[91] The determination that textile workers have a medical problem should not turn on resolution of the question of whether they have "byssinosis" or occupationally caused emphysema.

We are particularly intrigued by the term "byssinosis," a thing thought up by venal doctors who attended last year's ILO [International Labour Organization] meetings in Africa, where inferior races are bound to be afflicted by new diseases more superior people defeated years ago.

As a matter of fact, we referred to the "cotton fever" earlier, when we pointed out that a good chaw of B.L. dark would take care of it, or some snuff.[92]

It may of course be the case that such positions are publicly taken on the advice of counsel, for the purpose of avoiding financial liability rather than for the purpose of actually persuading anyone. Nevertheless, this case serves to illustrate the quality of information unorganized workers can expect to receive from employers bent on denying responsibility for the occupational illnesses of their employees.

Manpower

The special problems of unorganized workers are compounded by their abysmally low level of access to expertise and manpower in regard to occupational safety and health. Though union efforts in this area have been far from adequate, at least the industrial hygienists and other health and safety professionals in those unions which have them can provide some semblance of an independent check on the activities of business. Organized workers, moreover, are just beginning to have available to them training programs taught by outside professionals. Such programs have proven to be of considerable value in raising consciousness and in equipping working people with some of the skills needed to monitor their own job environment.* Unorganized workers, by contrast, are forced to rely on the services of health and safety professionals on the company payroll (if such exist), and have little access to independent training programs of the type which have proven so successful in unionized settings. Company-supplied manpower and training, though often well-intentioned, has its primary obligation to the employer, with the result that unorganized workers can be shortchanged when improved safety and health conditions would result in a significant increase in operating costs to the firm.

7.7.3 The Implications of the OSHAct

One may well ask at this point if there isn't anything in the new law that is specifically addressed to the special problems of unorganized workers. It is of course true

*The activities of activist public-interest groups such as the Boston-based Urban Planning Aid have met with considerable success in educating unionized workers regarding the occupational health hazards that they face, in appraising them of their rights under the OSHAct, and in suggesting creative collective bargaining strategies for use by their local union.[93] Similar efforts directed toward organized workers are underway at the University of Wisconsin's School for Workers and have recently begun at the University of California's Labor Extension Program. For an assessment of the effectiveness of these programs at labor studies centers in universities, heretofore directed primarily toward organized workers, see Chapter 10.

that unorganized workers now have the right to file complaints under the new law, and can theoretically accompany OSHA inspectors on plant tours. Yet these rights are rarely exercised by unorganized workers. Workers are often not aware of their right to file a complaint, and are not likely to be aware that their names must be withheld from the employer if they so request. Anonymous complaints do not receive attention from OSHA, and at any rate they are often of little value, since the hazard to which they refer in many cases cannot be located by the OSHA inspector if the worker who filed the complaint is unwilling to risk being seen talking to him. In nonunion plants, there is no mechanism for designating an "employee representative" to accompany the OSHA inspector, with the result that management has virtually a free hand in conducting the plant tour.

There are a host of other rights which are available to the worker under the OSHAct. Senator Harrison Williams, coauthor of the OSHAct, has referred to certain provisions of the Act as a "worker's environmental bill of rights."[94] (For a list of these provisions, see Section 4.2.14.)

Why have unorganized workers been so slow to exercise their rights under the OSHAct? On one level, fear of employer reprisal is responsible. Unorganized workers do not have the job security afforded by union membership, and they have little recourse in practice from managerial decisions regarding discipline and discharge, although the OSHAct does theoretically protect them from discriminatory action taken because they file an OSHA complaint. Although unorganized workers enjoy a certain amount of formal legal protection (apart from the OSHAct) against improper or arbitrary actions of their employer, these remedies available to them are not widely known, are infrequently utilized, are weak to begin with, and have largely been eroded over time through a series of adverse court decisions (see Section 4.5).[95] Even when unorganized workers are successful in legal action against their employer, legal remedies can be slow in coming, often do not make the aggrieved worker "whole," and are applied in a highly capricious and uneven fashion. In general, the unorganized worker can only assume that any management decision regarding discipline or discharge will not be reversed. When management is free, for example, to dock employees' pay for time spent accompanying OSHA inspectors on their rounds, even under the OSHAct, it becomes clear why the "walkaround right" of unorganized workers exists largely on paper. (See Section 4.3.2 for a fuller discussion of the walkaround right and its limitations.)

Yet there are also deeper and more fundamental reasons why unorganized workers have been so unassertive of their newly acquired rights. Many unorganized workers genuinely fear, in some cases with considerable justification, that forcing their employer to comply with the provisions of the OSHAct could seriously undermine the economic viability of the enterprise, thereby placing their jobs in jeopardy.* In this respect, the *special problems of the unorganized worker are often*

*This fear has been referred to as "environmental blackmail."[96]

much the same as the special problems of the small firm. The costs associated with plant shutdowns or layoffs are often substantial for unorganized workers, since they typically do not enjoy such benefits as vested pensions, severance pay, transfer rights, and supplemental unemployment compensation which ease the economic burden of job loss for many unionized workers. Retirement benefits, in particular, are often tied to the continued economic success of the enterprise in the case of unorganized workers. Years of accumulated benefits may be completely wiped out if the plant is forced to close its doors. In some cases, unorganized workers live in "company towns" where the lack of alternative employment opportunity makes the prospect of unemployment especially serious. Such workers, moreover, are unlikely to develop an awareness that things could be "different"—that there actually *could* be jobs at which a safe and healthful work environment is not inconsistent with earning a decent living. Thus, unorganized workers are particularly susceptible to the problems of occupational and geographic immobility which can make unemployment a harrowing experience. Often, unorganized workers may feel that their present job is the only one available to them which offers the opportunity to utilize their painstakingly acquired skills, or gratify their need for self-fulfillment.

Unorganized workers are also particularly susceptible to the pressure of ridicule and rejection by their peers if they show excessive concern with an "unmanly" subject like occupational health and safety. In a unionized plant, local union leadership is often able to use its influence to overcome such attitudes, but in a nonunion plant management is able to play upon them to its own advantage, spurring employees on to assume even greater burdens of risk. Thus, unorganized workers are socialized to accept the hazardous nature of jobs and are convinced of the necessity of performing them in order to earn their livelihood. Family and community norms and traditions also contribute to the resignation with which unorganized workers accept the fate of hazardous work. Thus, it should hardly prove surprising that unorganized workers—dimly aware of the hazards they face, faced with the necessity of earning a living, and not cognizant of the existence of less risky employment alternatives—are not likely to assert their rights and raise the issue of occupational health and safety.

7.7.4 Some Possible Approaches to the Problems of Unorganized Workers

What means are available for addressing the special problems faced by unorganized workers in attaining a safe and healthful workplace? Mechanisms need to be designed and implemented to increase the awareness and utilization among unorganized workers of already existing rights and self-help avenues. Where existing avenues are too narrow, additional policies must be designed to broaden them.

Even the most well intentioned, best organized, and most concerned labor unions are making relatively slow progress in stimulating employers to reduce the level of workplace hazard. Thus, the process of "filtering down," though helpful, will take far too long to be of significant benefit to the 75% of the nation's labor force which remains unorganized. Nor can the Occupational Safety and Health Administration be relied upon to bridge this gap: of a total budget of $134 billion allocated by Congress during fiscal 1975 to all programs of the Departments of Labor and HEW, a mere $102 million was budgeted for administration of the OSHAct—less than two dollars per worker covered by the Act.

One indirect method of assisting unorganized workers would be to increase the speed of the filtering-down process by which gains won by organized workers are eventually enjoyed by unorganized workers as well. Facilitation of the efforts of *organized* workers to achieve improved workplace health and safety would therefore redound to the ultimate benefit of some unorganized workers. Successful efforts in the unionized sector of the economy could provide a beneficial "demonstration effect" which would lead to an increased awareness of and demand for health and safety improvements elsewhere. Similarly, promotion of union efforts to use occupational health and safety issues as an organizing device would benefit unorganized workers.* Insofar as occupational health and safety issues prove successful in stimulating the spread of unionization, some of the health and safety problems facing previously unorganized workers will automatically be addressed. Indirectly, other unorganized workers will benefit as employers accelerate their efforts to alleviate hazardous working conditions in an attempt to forestall unionization. Furthermore, political lobbying efforts of organized labor on behalf of legislation of potential benefit to *all* working people, such as the pending reform of the workmen's compensation system, could be supported.

More generally, policies and programs tending to promote increased *public* awareness of the magnitude of the nation's occupational health and safety problems, and the methods available for overcoming them, could prove of significant indirect benefit to unorganized workers.

More importantly, it is imperative that programs and policies be designed which are of *specific* assistance to unorganized workers. Among the types of programs which may prove useful in addressing the specific needs of unorganized workers are the following:

1. Free legal counsel to assist unorganized workers in exercising their rights under the OSHAct and in filing compensation claims and damage suits, independently or by class action, against negligent employers.
2. Education programs to assist unorganized workers in learning about their rights under the OSHAct, such as the details of filing a complaint with OSHA. It may be especially useful in this regard if the types of labor education programs at universities and elsewhere which are proving so successful among organized workers could

*For example, the United Farm Workers have used occupational health and safety extensively as an organizing tool, with considerable success. For the activities of this union in regard to pesticide regulation and other health and safety related matters, see Chapter 12.

also be made available on an extensive basis to unorganized workers.

Such programs, though of great potential value, would hardly be sufficient in and of themselves. Unorganized workers are urgently in need of full and accurate information concerning the level of workplace hazard they face, and they are just as urgently in need of mechanisms to act upon such information in bringing about improvements in the level of workplace health and safety. Briefly, unorganized workers are in need of access to

1. full and accurate information about materials they are working with, in regard to toxicity and other health and safety characteristics;
2. company accident and health records, employee medical records, etc.; and
3. the resources necessary to monitor working conditions in order to provide an independent check on whether health and safety standards are being violated. This implies access to outside expert assistance and employee health and safety training, among other things.

One important mechanism for attaining these objectives which has met with considerable success in Western Europe is a public policy mandate for the creation in all places of employment of enterprise-level health and safety committees and/or works councils with worker-elected representatives (see Chapter 11). In the United States, such committees are relatively common at present in the unionized sector, where they have typically emerged as a result of union collective-bargaining proposals, but they are virtually nonexistent in the unorganized sector.*

An important condition for the successful operation of such health and safety committees is access to information and technical expertise. Thus, an essential component of the successful operation of these enterprise-level committees is the availability of medical and environmental health services that are directly accountable to the committee. It might be mandated that such professional expertise be made available to the health and safety committee, with the further stipulation that these professionals be required to report to the committee at specified intervals.

Another important condition for the successful operation of health and safety committees is that they be given sufficient authority to implement decisions regarding the improvement of working conditions. Past attempts of companies to use the mechanism of rank-and-file safety committees have floundered because of the unwillingness of top management to give them enough authority to do their job.[98]

*A recent analysis of major collective-bargaining agreements by the Bureau of Labor Statistics revealed that of 6.3 million workers covered by the collective-bargaining agreements under study, 29.7% were covered by agreements providing for the establishment of joint labor-management safety committees.[97] In several particularly hazardous industries, the proportion of organized workers covered by agreements containing safety committee provisos was in excess of 50%. It would, of course, be interesting and worthwhile to assess the activities of such committees, which have by now been in existence for quite a number of years in some industries. Such an analysis could provide an important input into the process of formulating policy alternatives with respect to mandatory enterprise-level health and safety committees with worker-elected representatives.

Stiff managerial resistance to the granting of any decision-making rights to safety committees, even where there is a powerful union to insist upon it, makes it likely that strong legislation would be required to bring such committees into being in a meaningful way.

REFERENCES

1. *Man, Materials and Environment*, A Report for the National Commission on Materials Policy, by the National Academy of Engineering and the National Academy of Sciences (March 1973).

2. *Chemicals and Health*, Report of the Panel on Chemicals and Health of the President's Science Advisory Committee (September 1973) Science and Technology Policy Office, National Science Foundation.

3. *Man, Materials and Environment*, p. 172.

4. *Chemicals and Health*, p. 115.

5. National Association of Manufacturers, "Where We Stand," p. 13.

6. "First Annual McGraw-Hill Survey of Investment in Employee Safety and Health," mimeograph (May 1973), p. 1.

7. "The Crushing Cost of Safety," *Dun's Review* (January 1972), p. 53.

8. *Ibid.*, p. 54.

9. *Occupational Safety and Health Reporter* (the Bureau of National Affairs, Inc.), Vol. 3, No. 38 (February 21, 1974), p. 1187. [Hereinafter cited as *OSHR*, BNA]

10. *OSHR*, BNA, Vol. 3, No. 37 (February 14, 1974), p. 1163.

11. Arthur D. Little, Inc., *Impact of Proposed OSHA Standard for Asbestos*, First Report to the U.S. Department of Labor (1972), unpublished.

12. Government Research Corporation, *Occupational Safety and Health, A Policy Analysis* (1973), p. 46.

13. *Man, Materials and Environment*, p. 75.

14. "News from NIOSH," *Job Safety and Health*, Vol. 2, No. 1 (January 1974), p. 33.

15. Edward Baier, deputy director of NIOSH, quoted in Agnes Fahy, "Enemies in the Dust: Occupational Respiratory Diseases," *National Tuberculosis and Respiratory Disease Association Bulletin* (June 1972), p. 15.

16. Morris Goldstein and Robert S. Smith, "The Predicted Impact of the Black Lung Benefits Program on Prices, Output, and Employment in the Bituminous Coal Industry," Technical Analysis Paper No. 14, Office of Evaluation, Office of the Assistant Secretary for Policy, Evaluation, and Research, Department of Labor (November 1973), p. 1.

17. United Mine Workers *Journal*, Vol. 85, No. 6 (March 1-15, 1974), p. 3.

18. Dennis Meadows et al., "Miner Health Benefits and the Potential for Long-Term Shortages in the Coal Mining Industry," unpublished paper (June 1973).

19. Goldstein and Smith, "Predicted Impact of Black Lung Benefits," p. 59.

20. *Ibid.*

21. *Ibid.*, p. 60.

22. See Lucille Langlois, *The Cost and Prevention of Coal Workers' Pneumoconiosis* (Appalachian Regional Commission, Washington, D.C., June 1971), for an excellent discussion of the costs of reducing dust levels in underground coal mines.

23. *Man, Materials and Environment*, p. 75.

24. H. Sundstedt, "Introduction of Asbestos Substitutes in a Swedish Shipyard," *Swedish Journal of Occupational Medicine* (1973), p. 27.

25. *Chemicals and Health*, p. 109.

26. *Ibid.*, p. 115; emphasis added.

27. Langlois, *Cost and Prevention.*

28. *Man, Materials and Environment*, p. 75.

29. *The President's Report on Occupational Safety and Health*, G.P.O. Document No.

4000-00288 (December 1973), pp. 22-23; emphasis added.

30. *Ibid.*, p. 14.

31. Paul Brodeur, "Annals of Industry," *New Yorker* (November 26, 1973), p. 131.

32. National Academy of Engineering, *Perspectives on Benefit-Risk Decision Making* (Washington, D.C., 1972), p. 9.

33. Langlois, *Cost and Prevention.*

34. Robert S. Smith, "Intertemporal Changes in Work Injury Rates," Industrial Relations Research Association *Proceedings* (December 1972), p. 174.

35. See, for example, Michael O. Finkelstein, "Regression Models in Administrative Proceedings," *Harvard Law Review*, Vol. 86, No. 8 (June 1973), pp. 1442-1475; F. M. Fisher and G. Kraft, "The Effect of the Removal of Firemen on Railroad Accidents, 1962-1967," *Bell Journal of Economic and Management Science* (fall 1971).

36. *Work in America*, Report of a Special Task Force to the Secretary of Health, Education, and Welfare (MIT Press, Cambridge, Mass., 1973), p. 79.

37. H. Peter Metzger, *The Atomic Establishment* (Simon & Schuster, New York, 1972); U.S. Department of Health, Education, and Welfare, Public Health Service, *Radon Daughter Exposure and Respiratory Cancer: Quantitative and Temporal Aspects*, Report from the Epidemiological Study of United States Uranium Miners, National Institute for Occupational Safety and Health, Joint Monograph No. 1 (1971).

38. Arthur D. Little, Inc., *Impact of Proposed OSHA Standard for Asbestos.*

39. Raytheon Service Company (Burlington, Mass.), *Industrial Noise and Worker Medical, Absence and Accident Records*, G.P.O. Document No. 1972-759-570/1311 (June 1972).

40. See, for example, Winfred B. Hirschmann, "Profit From the Learning Curve," *Harvard Business Review* (January-February 1964). The seminal theoretical article in the economics literature on this subject is Kenneth J. Arrow, "The Economic Implications of Learning by Doing," *Review of Economic Studies*, Vol. 29, No. 80 (June 1962).

41. Arthur D. Little, Inc., *Impact of Proposed OSHA Standard for Asbestos.*

42. Sundstedt, "Introduction of Asbestos Substitutes," p. 27.

43. *Chemicals and Health*, p. 108.

44. *Ibid.*, p. 107.

45. See Harriet L. Hardy, "Beryllium Poisoning—Lessons in Control of Man-Made Disease," *New England Journal of Medicine*, Vol. 273 (November 25, 1965), pp. 1188-1199.

46. Phyllis Lehmann, "Job Health Hazard: Beryllium," *Job Safety and Health*, Vol. 2, No. 1 (January 1974), p. 27.

47. See David H. Wegman, Leslie Boden, and Charles Levenstein, "Health Hazard Surveillance by Industrial Workers," *American Journal of Public Health*, Vol. 65, No. 1 (January 1975).

48. National Commission on State Workmen's Compensation Laws, *Compendium on Workmen's Compensation*, G.P.O. Document No. 496-632 (1973), p. 6.

49. Fred Foulkes, "Learning to Live with OSHA," *Harvard Business Review* (November-December 1973), p. 67.

50. National Commission, *Compendium*, p. 6; sources omitted.

51. Langlois, *Cost and Prevention.*

52. See, for example, Raytheon, *Industrial Noise.*

53. Quoted in *Man's Health and the Environment—Some Research Needs*, Report of the Task Force on Research Planning in Environmental Health Science, National Institute of Environmental Health Science, HEW (March 1970), p. 73; the rate of six days per year refers to 1969. The estimate of the GNP value of decreased absenteeism is from *Protecting the Health of Eighty Million Americans: A National Goal for Occupational Health*, Special Report to the Surgeon General of the United States Public Health Service, HEW (November 19, 1965), p. i.

54. *Man, Materials and Environment*, p. 75.

55. Lennart Levi, *Society, Stress and Disease*

(Oxford University Press, New York, 1971), Vol. 1, pp. 212-213.

56. Dr. R. van der Vlist, Netherlands Institute for Preventive Medicine, interview with author, Leiden, Netherlands, September 13, 1972.

57. See E. J. Mishan, "Evaluation of Life and Limb: A Theoretical Approach," *Journal of Political Economy*, Vol. 79, No. 4 (1971), pp. 687-705.

58. *OSHR*, BNA, Vol. 3, No. 39 (February 28, 1974), p. 1219.

59. See Sidney Cobb and Stanislau Kasl, "Some Medical Aspects of Unemployment," *Proceedings of the Industrial Gerontology Seminar*, Annapolis, Maryland, April 28, 1970, for one description of the deleterious physical and mental consequences of involuntary job loss among middle-aged blue-collar workers.

60. See Peter Doeringer and Michael Piore, *Internal Labor Markets and Manpower Analysis* (D.C. Heath, Lexington, Mass., 1971).

61. Arthur M. Ross, "Do We Have a New Industrial Feudalism?" *American Economic Review*, Vol. 48, No. 5 (December 1958), pp. 903-920.

62. See Michael Piore, in "Upward Mobility, Job Monotony & Labor Market Structure," MIT Department of Economics Working Paper No. 90 (September 1972).

63. Mishan, "Evaluation of Life and Limb," p. 696.

64. Richard Thaler and Sherwin Rosen, "The Value of Saving a Life: Evidence from the Labor Market," unpublished discussion paper, Department of Economics, University of Rochester (November 1973), p. 58; emphasis added.

65. *Perspectives on Benefit-Risk Decision-making*, Report of a Colloquium Conducted by the Committee on Public Engineering Policy, National Academy of Engineering (1972), Summary and Recommendations, p. 4.

66. *Ibid.*, pp. 5-6.

67. *Daily Labor Report*, No. 2 (January 3, 1974), p. C-3.

68. *Perspectives on Benefit-Risk Decision-making*, p. 4.

69. *Man, Materials and Environment*, p. 73. Richard England and Barry Bluestone, "Ecology & Class Conflict," *Review of Radical Political Economics*, Vol. 3, No. 4 (fall-winter 1971).

70. Data from Irving Selikoff, quoted in Jeanne Stellman and Susan Daum, *Work is Dangerous to Your Health* (Vintage Books, New York, 1973), p. xiv.

71. John Rawls, *A Theory of Justice* (Harvard University Press, Cambridge, Mass., 1971), p. 15.

72. *Chemicals and Health*, p. 136.

73. United Steelworkers of America, *Job Description and Classification Manual* (Pittsburgh, 1963).

74. Thaler and Rosen, "The Value of Saving a Life," The notion of compensating wage differentials among occupations which vary in their degree of pleasantness is as old as Adam Smith.

75. *Chemicals and Health*, p. 113.

76. *The Economic Impact of Pollution Control*, Study Prepared for the Council on Environmental Quality, Department of Commerce, and the Environmental Protection Agency (March 1972), p. 105.

77. Henry F. Howe, "Preventing Occupational Health and Safety Hazards in Small Employee Groups," *American Journal of Public Health*, Vol. 61 (1971), p. 1581.

78. Statement of Peter Fosco, General President, Laborers International Union of North America, AFL-CIO, Hearings Before the Subcommittee on Labor of the Senate Labor and Public Welfare Committee Concerning Oversight Hearings on Administration of the Occupational Safety and Health Act, 92d Congress, 2d Session (July and September 1972), p. 530. [Hereinafter cited as *Oversight Hearings*]

79. Testimony of Jacob Clayman, Administrative Director, Industrial Union Department, AFL-CIO, *Oversight Hearings*, p. 262.

80. Testimony of Dr. Sidney Wolfe, Health Research Group, *Oversight Hearings*, p. 297.

81. *Ibid.*, p. 296.

82. Statement of James F. Bailey, Legislative Advocate, United Brotherhood of Carpenters and Joiners of America, *Oversight Hearings*, p. 292.

83. "Small Business and the OSHAct of 1970: A Report of the Subcommittee on Environmental Problems Affecting Small Business," H. Rep. No. 92-1341 (August 15, 1972), 92d Congress, 2d Session, p. 12.

84. *OSHR*, BNA, Vol. 3, No. 30 (December 27, 1973), p. 941.

85. W. Jones, "Occupational Health in Small Industries in the United Kingdom," report presented at the Consultation on Occupational Health in Small Industries, Geneva, Switzerland, October 1970.

86. A. Rothan, "Occupational Health in Small Industries in France," report presented at the Consultation on Occupational Health in Small Industries.

87. Sven Forssman, "Occupational Health in Small Industries in Sweden," report presented at the Consultation on Occupational Health in Small Industries.

88. Franklin-Hahnemann Institute of Occupational and Environmental Safety and Health, "The Development of Comprehensive Safety and Health Service Concepts," report presented at the Consultation on Occupational Health in Small Industries.

89. *OSHR*, BNA, Vol. 3, No. 37 (February 14, 1974), p. 1184, and No. 41 (March 14, 1974), p. 1277.

90. *OSHR*, BNA, Vol. 3, No. 37, p. 1185.

91. Henry Howe, interview with author, Chicago, June 21, 1972.

92. *America's Textile Reporter* (July 10, 1969).

93. Wegman, Boden, and Levenstein, "Health Hazard Surveillance by Industrial Workers."

94. *Trial* (July-August 1973), p. 15.

95. Sheldon Friedman, "The Right to Refuse Abnormally Dangerous Work Under Section 502 of the Taft-Hartley Act," unpublished paper, Massachusetts Institute of Technology (1970); and Leslie Boden and Sheldon Friedman, "The Right to Refuse Hazardous Work: Managerial Prerogatives vs. Workers' Health," paper in progress, Massachusetts Institute of Technology (1975).

96. Leonard Woodcock, "Labor and the Economic Impact of Environmental Control Requirements," paper presented at the Conference on Jobs and the Environment–Whose Jobs? Whose Environment?, sponsored by Institute of Industrial Relations, University of California, Berkeley (San Francisco, November 28, 1972).

97. U.S. Department of Labor, Bureau of Labor Statistics, Bulletin No. 1784, "Characteristics of Agreements Covering 1000 Workers or More," Table 15, p. 18.

98. See Russell De Reamer, "Are Safety Committees Useful?" *Job Safety and Health*, Vol. 2, No. 3 (March 1974), p. 23.

8 Economic Issues. II. Workmen's Compensation

8.1 INTRODUCTION*

The National Commission on State Workmen's Compensation Laws (hereafter called the Commission) estimates that in 1970, "14,200 workers died from a work-connected injury or disease; another 90,000 were permanently disabled."[1] As discussed in Section 3.4.6, these figures probably grossly underestimate the numbers suffering from occupational disease. These workers—and millions of others with less serious disabilities—often find themselves damaged in three ways: (1) the injury or disease itself is often painful and emotionally trying; (2) the worker's family is often faced with the burden of paying for medical and rehabilitative services; and (3) the family's income is often drastically reduced. The injured worker may be forced to accept a lower-paying job because of the disability; or, even worse, he may be completely unable to work for an extended period of time.

Until the beginning of the twentieth century, the worker's only recourse was to sue his employer, claiming liability based on negligence. However, the employer's legal defenses were so powerful that compensation through the courts was unlikely, uncertain, and often inadequate. The three defenses available to the employer and their implications were:

1. *Contributory negligence.* If the worker, even in a minor way, contributed to the injury through his own negligence, he could not collect.
2. *The fellow-servant doctrine.* The injured employee could not collect if a fellow worker's negligence contributed to the injury.
3. *Assumption of risk.* If a cause of the injury was an inherent hazard of the worker's occupation, and the worker could be expected to know of the hazard, he could not recover. Under the law, he had knowingly assumed the risks of his occupation.

Through the beginning of the twentieth century, the courts interpreted these defenses so liberally that it was almost impossible for injured workers to collect. In case after case, the courts left disabled workers and their families with neither income nor the means of obtaining medical care or rehabilitation. Finally, around 1910, the National Civic Federation and the National Association of Manufacturers (both with members representing large corporations) began to lobby for workmen's compensation legislation to replace the common-law systems. The new statutes were supposed to provide adequate benefits to injured workers, covering lost wages and medical and rehabilitation expenses. These payments were to be predetermined and delivered promptly in order to relieve employees and employers from uncertainty and to eliminate wasteful litigation. The legal principle was liability without

*This chapter bases much of its factual material on the publications of the National Commission on State Workmen's Compensation Laws—especially the *Report* and its *Compendium.* (The *Supplemental Studies* were received too late for adequate consideration.) The chapter is, in part, a critical analysis of the *Report*, although it only covers those areas considered most in need of comment. The reader is referred to Chapter 11 for a general description of the workmen's compensation systems in Europe.

fault. Thus, the costs of work-related injuries were to be allocated to the employer not because of any presumption that he was personally to blame, but rather because industrial employment was inherently hazardous.

The actual workings of the compensation system fell far short of its stated goals. To begin with, the early statutes provided neither adequate income supplements or adequate medical payments. For example, most states initially set a limit of 50% recovery of lost income for permanent total disability—up to a maximum of $10 to $12 per week.[2] (In 1920 the average weekly earnings of production or nonsupervisory workers in manufacturing industries was $26.02.)[3] No state allowed more than ten years for recovery, and most had a six year maximum.[4] In addition, no state allowed payment of more than 90 days of medical benefits—and some allowed none at all.[5]

Even though they were low, these benefits were often an improvement over the settlements likely under court interpretation of the common law. However, there is good evidence that the probability of high payments under the common law was growing over time. A recently completed study indicates that defendants were more and more regularly losing negligence suits during the period from 1875 to 1905.[6] The study focuses on negligence suits in general, but includes work-related negligence suits. Litigation was becoming much more common in the years after 1875, and more plaintiffs were recovering damages. These two facts indicate that employers were increasingly vulnerable, and it was therefore in their economic interest to lobby for the passage of compensation legislation which established payments.

Workmen's compensation legislation was an effective response to the increased risk of large losses by business to injured workers. It gave injured workers the right to collect payments from their employers without proving negligence; but it also took away from workers their right to sue under the common law. Although it guaranteed payment for injury, the cost to employers was small (because the benefits were low) and predictable (because it eliminated the possibility of large court judgments). In addition, the legislation seemed progressive and humanitarian, thus improving the public image of those employers who favored it.

8.2 A BRIEF DESCRIPTION OF THE GOALS OF WORKMEN'S COMPENSATION

8.2.1 Income Maintenance

In the course of their workday, many workers are faced with the risk of substantial disability or death. A large number of workers do not even know the extent of the hazards they are exposed to, and others, while understanding the risks, do not have access to relatively safe employment. One of the original purposes of workmen's compensation was to provide such workers and their families with sufficient resources to maintain an adequate standard of living despite disability or death of the primary breadwinner. (Promptness of payment is, of course, an important aspect of income maintenance. The *Report* deals adequately with this issue.)

8.2.2 Insurance*

Another original and continuing goal of workmen's compensation is to provide insurance for employers. By taking away the employee's right to common-law recovery, workmen's compensation statutes protect the employer from the risk of costly negligence suits. Since compensation payments are generally both small and relatively fixed, financial risks are negligible for the employer. In addition, compensation statutes led to the organization of a market in workmen's compensation insurance, so that employers have the opportunity to pool their risks.

Workmen's compensation "insures" employees as well. By supplementing income during periods of disability, it reduces the worker's risk of losing income because of hazardous employment. This part of the insurance function is accomplished in several ways: (1) it establishes the employer's liability without fault for work-related disability; (2) it attempts to assure the employer's ability to pay; and (3) it establishes legal minimum rates of recovery. Even if payment levels are low, the guarantee of payment may justify a particular statute.

8.2.3 Minimizing the Costs of Workplace Hazards (Efficiency)

Prevention of Accidents and Disease

Since workmen's compensation makes employers liable for the costs of accidents and disease, it can theoretically act as an incentive for them to decrease workplace hazards. Section 7.5 dealt with the many market imperfections that lead to employers' paying less than their share of the full social costs of workplace hazards. The result is an unnecessarily high level of injury and illness. (As was noted in that section, preventing *all* preventable accidents may not be economically efficient, but may be justified on equity grounds.) It has been said that the most effective incentive to employers to improve workplace health and safety conditions is to make "safety pay." If employers were liable for a larger proportion of the costs of accidents and disease, it would be more profitable for them to reduce workplace hazards. Thus, to the extent that workmen's compensation increases employers' share of the burden of the costs of disability, it increases their interest in preventive activities.**

*Here, insurance means more than insurance policies. It refers to any method of reducing uncertainty (risk).

**It is recognized that the same argument applies to workers as well. As long as they are able to avoid accidents and disease, paying the costs of disability will encourage workers to decrease

Efficient Administration and Low Legal Costs
One of the weaknesses of recovery through common-law negligence suits is the high cost of legal services inherent in a system based on litigation. Workmen's compensation is meant to reduce these costs drastically without substituting high administrative costs for the high costs of litigation.

8.2.4 Broad Coverage

To meet its basic goals adequately, workmen's compensation must be broad and uniform in its coverage. It cannot omit large groups of workers from coverage, or offer different amounts of income protection to different groups. Moreover, it must treat all injury and disease in as uniform a manner as possible.

One widely held legal objective is "equality under the law"—in this case, the same level of protection for all workers. This is a normative basis for uniform and broad coverage. But there is a pragmatic basis as well. Since broad labor markets make workers in different occupations, industries, and areas all competitive with one another, lack of coverage in one area can undercut attempts at coverage in another. While noting that such attitudes may be irrational, the Commission reports that

> it seems likely that many states have been dissuaded from reform of their workmen's compensation statute because of the specter of the vanishing employer, even if that apparition is a product of fancy not fact. A few states have achieved genuine reform, but most suffer with inadequate laws because of the drag of laws of competing states.[7]

8.2.5 Just Compensation of Accident Victims (Equity Considerations)

In the law of products liability, caveat emptor (let the buyer beware) is no longer the guiding doctrine. Manufacturers are responsible for the harm that their products might cause in their intended use. As Justice Traynor stated:

> The purpose of such liability is to insure that the costs of injuries resulting from defective products are borne by the manufacturers that put such products on the market rather than by the injured persons who are powerless to protect themselves.[8]

This opinion mirrors the moral value that people should not be forced to bear

dangerous activities. However, as noted throughout this report, management has traditionally had more control over hazards. Workers often do not know of the hazards they face, or are too weakly organized to take action, or are afraid of losing their jobs in a world of high unemployment. Also, worker responsibility comes most effectively with worker control of health and safety conditions; and management traditionally wishes workplace control to rest in its own hands. As the *Report* states (p. 39): "Because the employer's control over working conditions far exceeds that of the employee, we believe that assigning to the employer the largest portion of the costs of work-related injuries and diseases will best serve the objective of safety."

harm done them by events outside their understanding and control. The consumer is often not in a position to be aware of the dangers in a product he buys. Since control lies in the hands of the manufacturer, the consumer may therefore sue a negligent manufacturer to recover damages incurred in using an unsafe product.

Workers are often in a similar position in their places of employment. They are faced with many risks over which they have little effective control. Especially when working with disease-producing agents, the employee may be unaware of the danger. Workmen's compensation therefore places the burden of worker's economic loss on his employer.

However, compensation laws often completely disallow payment for nonincome losses. As noted above, when workers gained the right to workmen's compensation, they lost the right for recovery through negligence suits. In theory, they gained a guaranteed (although perhaps small) payment in place of an uncertain recovery through the courts. But settlements under tort law commonly include payment for both loss of income *and* "pain and suffering." Workmen's compensation statutes include payment for the former, but provide only limited payment for the latter. Looked at in this way, the worker gave up his right to sue for lost income in return for workmen's compensation income payments; but he gave up his right to sue for pain and suffering at the same time and got little in return.

At the turn of the century, when recovery under the common law was unlikely, this trade-off might not have been a substantial loss. But now that the establishment of negligence has become relatively easy, why should the worker "be obliged to surrender the blessings of a full and generous recovery in tort?"[9] The enactment of the OSHAct has further strengthened the position of those workers who have retained the right to institute negligence suits against their employers. Those covered by the Longshoremen's and Harbor Worker's Compensation Act are such a group. In a recent ruling, *Arthur v. Flota Mercante, Gran Centro Americana, S.A.*, the U.S. Court of Appeals for the Fifth Circuit affirmed that a violation of safety and health regulations is one of the elements sufficient to establish negligence per se on the part of the employer, thus shifting the burden of rebutting onto the employer. The claimant was thus able to collect damages more easily than would have been possible without the OSHAct.[10]

8.3 THE GOALS AS VIEWED BY THE COMMISSION

The National Commission on State Workmen's Compensation Laws lists five objectives for a workmen's compensation program:

- Broad coverage of employees and work-related injuries and diseases.
- Substantial protection against interruption of income.
- Provision of sufficient medical care and rehabilitation services.
- Encouragement of safety.

- An effective system for delivery of the benefits and services.[11]

These objectives are consistent with the ones we described earlier, but omit insurance and equity as explicit goals. An important weakness of the *Report* is that the Commission focuses on each goal individually and omits discussion of the interrelationships and conflicts between the goals. The problem of conflict is hidden in its statement of goals by the use of words like "substantial," "sufficient," and "effective." Much important debate surrounds the interpretation of, for example, "*substantial* protection against interruption of income." Such words only have meaning when interpreted administratively or by the courts, or when given substance by a well-defined statute. It is precisely on the meaning of such words that the effectiveness of workmen's compensation is based.

8.4 CONFLICT AMONG GOALS*

Why not "full protection against loss of income," instead of "substantial protection"? If income maintenance were the only goal of workmen's compensation, full protection might be the appropriate objective. However, compensation for *all* lost wages can come into conflict with other goals. Every day that the injured worker is out of work, society loses his potential contribution to production. Many fear that, "As the proportion of wages replaced is increased, the worker will . . . have less incentive to return to work."[12] Production will be lost, and thus the cost of lost-time accidents will rise. If this is true, then the efficiency objective of workmen's compensation is in conflict with its income-maintenance objective.

In addition to this conflict in societal goals, the interests of workers and management differ with respect to the level of income maintenance to be provided. The higher the level of compensation, the higher the costs of premiums to the firm. Therefore, irrespective of the goal of income maintenance, employers want to keep payments relatively low, while employees and their unions will generally fight for maximum benefits.

There are several other important points of conflict among the social goals of workmen's compensation (income maintenance, insurance, efficiency, broad and uniform coverage, and equity), and these will be discussed later. In addition, the private interests of insurance companies, physicians, and lawyers, as well as those of management and labor, are often in conflict with each other. These conflicts lead to political struggles over the form of workmen's compensation programs. Especially when the social goals are not internally consistent, the relative power of private interests may determine the outcome.

*For a more general discussion of conflict in occupational health and safety, see Chapter 2.

8.5 HOW WORKMEN'S COMPENSATION MEETS ITS GOALS

8.5.1 Income Maintenance

Lost Wages

Present state workmen's compensation programs have provided generally inadequate levels of wage replacement. Most estimates of wage loss tend to understate workers' foregone earnings,* but even these report abysmally low levels of replacement. A 1967 article reports, "In the majority of workmen's compensation jurisdictions the actual rate of replacement lies between 10 and 40 percent, depending on the extent of disability."[13] The Commission estimates that, over all United States jurisdictions in 1972, the median percentage of wage loss replaced in temporary total disability cases was 40-44%.[14]

The low rate of recovery of lost income is caused by several factors. First, most states limit recovery to two-thirds of previous wages. Second, about one-half of all covered workers face an absolute maximum on their weekly benefit (for temporary total disability) of $80 or less.[15] This makes the two-thirds maximum effective only for those earning less than $120 per week—not a high income by today's standards.** It should also be noted that benefits for permanently disabled workers are often limited in duration or dollar amount.

Another measure of low recovery is that most workmen's compensation laws have ignored the fact that a person's future earning ability is not accurately measured by his income at a particular point in time. A 30-year-old worker can expect his income to rise over much of the rest of his working life, as he moves to better jobs or receives raises on the basis of his seniority or merit. Compensating him on the basis of his present rate of pay surely underestimates his total wage loss. (This factor is equally important to the families of workers who die.)

In all but six states, state workmen's compensation statutes do not consider the effects of inflation. Especially in today's economy, with prices rising at approximately 10% per year, benefits of a fixed dollar amount soon lose much of their purchasing power. This is especially serious in long-term disability cases, such as are characteristic of chronic disease. Laurence Locke illustrates the effect of fixed dollar rates on women who contracted beryllium poisoning in the early 1940s:

> The average wage of these girls was between $15 and $20. The compensation rate then was $12 to $14 per week. Even though a victim might first [have] become

*These estimates use gross earnings at the time of injury as a measure of loss. This is not a good measure for several reasons. As the Commission points out, it tends to overestimate workers' available income because taxes are taken from normal earnings, but not from workmen's compensation payments. On the other hand, there are several important ways in which this measure is an underestimate. First, it does not include the value of fringe benefits, which are becoming increasingly important in the total remuneration of workers. Second, it does not recognize the fact that people's incomes rise over much of their lives. On balance, gross earnings at the time of injury seem to be smaller than income lost, especially for permanently disabled workers.

**These statistics are timely as of January 1, 1972.

disabled in 1952 or 1962 [and no matter how many years she survived], a maximum compensation would still be limited to that low rate.[16]

The Commission recognizes the gravity of this problem, and recommends "that beneficiaries in *permanent total disability* cases have their benefits increased through time in the same proportion as increases in the State's average weekly wage."[17] Such a policy should surely be adopted for *all* types of disability.

In summary, wage losses seem to be widely undercompensated for three reasons: (1) relatively low maximum payments, (2) disregard for the individual's rising income potential, and (3) the problem of inflation.

In addition to its recommendations that would increase payments for wage loss, the Commission also suggests that social security payments for death and total disablement be offset by workmen's compensation payments.[18] It reasons that people with the same disability should have equal compensation. Since those who were injured off the job would not receive workmen's compensation, the Commission argues that equity considerations should lead us to reduce social security disability payments to those covered by workmen's compensation. The Commission does not seem to consider the fact that the social security system is an insurance system, in which half of the premiums are paid for by workers. The Commission would not deny recovery under private insurance plans, yet it would deny recovery under this public plan.

The social security offset has been federal policy since 1965. In that year, social security benefits to permanently, totally disabled workers were reduced in cases where the sum of disability insurance benefits and workmen's compensation benefits exceeded 85% of earnings prior to disability.[19] The offset now reduces the already insufficient rate of recovery for over 10,000 permanently disabled workers.

This policy overlooks entirely the fact that inflation destroys much of the value of the workmen's compensation payments—and the usefulness of the preinjury wage as a basis for compensation. But, more important, it overlooks the fact that the worker's pain and suffering may be a great part of the full cost of the injury to him, and that this part is denied him, or his family, under compensation statutes. Social security benefits might help to offset nonpecuniary losses, yet the Commission would deny these benefits to workers so badly injured that they are permanently and totally disabled.

Given the widespread undercompensation of accident and disease victims, it seems regrettable that any energy should go into attempting to keep workers with permanent and total disability from collecting as much as 90% of their predisability earnings. Our legislators' time would be better spent adjusting all payments to a higher level, rather than denying high payments to those who qualify under several programs. By denying disabled workers income from both workmen's compensation and social security, we are using the concept of equity to make our laws *equally unfair.* On the whole, then, neither the Commission's arguments concerning equality of treatment nor that involving removal of an incentive for staying away from work is very convincing.[20]

Medical and Rehabilitation Expenses

Workmen's compensation jurisdictions have a better record in providing the injured worker with medical and rehabilitation payments than in paying him lost wages.* Most jurisdictions provide for payments for "all reasonable and necessary medical, surgical, and hospital care," although fifteen did not as of January 1, 1972.

Complete coverage is the Commission's goal. As the *Report* puts it, arbitrary limits on medical payments,

> while they may be intended to achieve the worthy objective of high quality care at a reasonable cost, are almost invariably inappropriate to that end. In essence, these limits can be self-defeating because they may force a disabled worker to be unproductive indefinitely because his medical rehabilitation is incomplete.[21]

A part of the worker's cost of rehabilitation is the income he loses by not working during that period. While the decision to work rather than receive adequate professional aid is less elective for medical rehabilitation, it is a primary consideration for workers considering retraining. Immediate financial problems may push the worker to return to a job for which he is no longer suited or to take an unsatisfactory job, whereas his long-run interests might better be served by vocational rehabilitation. In order to encourage workers not to return to work prematurely, the Commission suggests that workers be paid a special maintenance benefit in addition to normal recovery during rehabilitation. This seems to be an excellent suggestion.

Legal Expenses

Workmen's compensation was designed to avoid litigation, but, especially in the case of serious disease of occupational origin, many cases are contested by the employer or insurer. In all but a handful of jurisdictions, the employee's lawyer is paid out of the settlement. Thus, especially in the case of a large total loss, lawyers' fees often reduce an already inadequate settlement.

The Commission's response to this is to suggest the regulation of attorneys' fees.[22] Although the Commission recommends regulating fees charged by both claimants' and defendants' lawyers, most states will probably follow only half the recommendation—regulating only claimants' attorneys' fees. This is several states' current practice. Moreover, since many defendants' lawyers are paid a salary by the insurance carriers, it is difficult to earmark and regulate their fees for a specific

*Peter Barth, of the Department of Economics, University of Connecticut, points out that "the record in terms of vocational and nonmedical rehabilitation [as opposed to rehabilitation payments] could hardly be worse in most states" (personal communication). Testimony before the Commission in Texas indicated that the Director of the State Vocational Rehabilitation Agency felt that his program had nothing to do with workmen's compensation.

case. The effect of too stringent regulation will be a reduction in the number and quality of lawyers who plead compensation cases for workers. Insurance carriers already have superior legal and medical resources on which to draw. The Commission's recommendation would exacerbate this inequality.

Litigation is often excessively advantageous to carriers, especially in the case of serious, long-term disability (see Section 8.7.3). The time delay of litigation is favorable to the carriers, who collect investment earnings on the unpaid claim. On the other hand, if the victim's family is without adequate income, they will often settle for a smaller sum rather than go through a prolonged legal battle.

In summary, allocating the burden of legal expenses, rather than the dollar amount, seems to be a major weakness of many workmen's compensation statutes. Limiting legal fees for the worker without limiting them for the carrier is clearly discriminatory. Making the carrier pay for the victim's legal fees when the worker wins the suit seems a more adequate and equitable procedure. (Of course, such a procedure carries with it the possibility that the courts or awarding agencies would adjust downwards to compensate for the added cost to the insurer.)

8.5.2 The Insurance Function (Reduction of Uncertainty)

For the Employer

Perhaps the single most effective achievement of workmen's compensation has come in spreading and reducing the risks to employers. This is first accomplished through the employee's loss of his common-law right to sue his employer. The statutes thus replace an uncertain and perhaps very high court settlement with low and certain payments. In addition, compulsory workmen's compensation insurance coverage led to the establishment of a market in workmen's compensation insurance, through which insurance companies further reduce and spread risks. In some ways, many employers may now be overinsured, a factor affecting their incentive to provide safer workplaces (see below).

For the Employee

Covered employees are guaranteed minimum wage recovery and usually full reimbursement of medical expenses. Moreover, by requiring most employers to buy insurance, workmen's compensation statutes eliminate the risk of an employer's not being able to pay the legal benefits. This reduces the uncertainty of recovery as compared with a tort liability system. However, the income guaranteed is often a low one. Moreover, it only holds for covered employees–and many are not afforded the protection of workmen's compensation. In addition, some uncertainty remains where states do not have funds to protect against uninsured firms or bankrupt insurance companies.

8.5.3 Incentives for the Prevention of Accidents and Disease

The fact that employers (and their insurance companies) are legally liable for the costs of accidents and disease can act as an incentive for them to introduce more effective hazard controls. But employers' assumption of the costs of workmen's compensation will only lead to an efficient result if they bear the *full social costs* of their actions. To the extent that employers' compensation premium costs are relatively small, or relatively unresponsive to changes in injury and disease rates, this incentive is diminished.*

The Size of Payments

Noted above was the fact that workmen's compensation only pays disabled workers a small portion of their actual lost incomes. A greater share of the burden of medical and rehabilitation costs is covered—but sometimes by public funds. Most vocational-rehabilitation activities take place through state departments of vocational rehabilitation. "These departments largely are funded by federal money and often are associated with education programs or other activities of the state government with little formal connection with the workmen's compensation agency."[23]

Even more important is the fact that difficult-to-evaluate costs, such as those of pain and suffering, are not counted in indemnity to the worker. These are certainly true social costs, but an inability (or unwillingness) to evaluate them seems to bar their consideration. These are often a substantial portion of the settlement in tort cases involving bodily injury in those areas where it is recognized that personal damage is not restricted to loss of income and payment of medical expenses. Yet workmen's compensation settlements often only consider monetary losses, and covered workers are denied recourse to the courts. Under this system, it is impossible for workers to recover full damages, or for employers to pay them.

The Commission makes a distinction between two types of possible cash benefits:[24]

1. *Impairment benefits* are paid to a worker with an impairment caused by a work-related injury or disease (whether or not he suffers a wage loss).
2. *Disability benefits* are paid when an employee has impairment and wage loss, both due to a work-related injury or disease.

The Commission in one place states:

Impairment benefits are justified because of losses an impaired worker experiences

*A demonstration of the small incentive effect of workmen's compensation payments can be found in the *Report*, pp. 96-97. The Commission reports that "there does not appear to be a systematic relationship . . . between the level of benefits and the safety record in the State." Furthermore, there is a surprisingly poor correlation between premium rates and average benefits across states (Peter Barth, private communication).

that are unrelated to lost remuneration. The impairment may, for example, have lifetime effects on the personality and normal activities of the worker.[25]

However, the Commission also states that

the total cost and the imbalance of permanent partial benefits tend to undermine the entire workmen's compensation program. Employers, disturbed by what they consider excessive payments for minor injuries, have refused to support general increases in benefits. Labor spokesmen oppose surrender of the substantial awards for minor injuries because they fear that any general benefit increases given in exchange would soon be eroded by the passage of time and the assault of inflation.[26]

The major function of workmen's compensation has traditionally been payment for disability, not for impairment. Generally, the states have given a secondary role to benefits based on impairment. Most jurisdictions have not dropped such benefits entirely, but tend to be arbitrary and inconsistent in their awards for impairment. Impairment is often important in permanent partial injury (e.g., loss of a hand, an arm, or an eye). Workers receive payments for these injuries equal to a specified number of weeks of workmen's compensation benefits. These payments, however, do not adequately reflect both impairment and disability. The actual length or importance of disability does not at all enter into the determination of a settlement. (Loss of a thumb would be treated equally for a pianist and for an economics professor.) In addition, the differences among states in the major "scheduled" payments is much more important than the difference in weekly payments for more minor injuries. Some examples of the variation of these benefits among jurisdictions are shown in Table 8.1.

Some permanent partial benefits are established by the state workmen's compensation boards rather than by schedule. These benefits are for cases such as back injury and disfigurement, where a schedule is difficult to apply. Many employers believe that these payments are excessive, and base their opposition to increases in all benefits on the fear that these permanent partial benefits will become even more burdensome.

Because of the importance of nonwage losses in such cases, the Commission has stated:

Permanent partial benefits are . . . the most controversial and complex aspect of workmen's compensation. We were impressed during our hearings and meetings that for no other class of benefits are there more variations among the states or more divergence between statutes and practices.[27]

Impairment benefits would partly "internalize" the social cost to workers. They would consequently also provide additional incentive for employers to reduce workplace hazards since they would increase the size of payments.

The impression that impairment benefits are used to inflate artificially and exorbitantly the size of payments for injuries should be judged with reference to a

Table 8.1
Specified Maximum Amounts of Benefits Provided for Loss of Arm, Foot, or Eye in Various Jurisdictions, 1972

	States (50)	Other "States" (6)*	Federal (2)
Arm at Shoulder			
Over $25,000	6	1	1
$20,000-24,999	1	1	0
$15,000-19,999	14	1	0
$10,000-14,999	22	2	0
Under $10,000	7	0	0
Foot			
Over $20,000	2	1	1
$15,000-19,999	3	0	0
$10,000-14,999	14	2	1
$6000-9999	26	2	0
Under $6000	5	0	0
Eye			
Over $20,000	1	1	1
$15,000-19,999	4	0	0
$10,000-14,999	8	2	1
$6000-9999	27	1	0
Under $6000	10	0	0

*No set amount in Puerto Rico

Source: *The Report of the National Commission on State Workmen's Compensation* (July 1972).

much more serious problem—that of occupational disease. Impairment *without* disability is a major consequence of much chronic occupational disease. The onset of disease may be slow, and through most or all of the course of the disease, the worker may suffer some impairment without any consequent loss of income. Occupational respiratory diseases and occupationally induced deafness fall into this category. It is thus difficult to agree with the Commission which, while stressing the importance of the problem of impairment, concludes that "disability should be the primary basis for permanent partial benefits."[28] Impairment may in fact be more important than disability if the "disability" does not occur until after retirement, as is often the case with chronic disease.

If impairment benefits are to be removed from the compensation system, one possible method of "internalizing" the nonincome losses is to permit negligence suits. As the *Report* discusses, there is much weakness to litigation, especially when it is the *only* means of recovery.[29] However, the courts could be made available to *supplement* the statutory disability payments when obvious employer negligence or indifference is a contributing factor. This may both be fair to the worker and provide the necessary incentive for the employer to remove workplace hazards. This revision of the system would preserve the insurance and income-maintenance functions without substantially reducing the costs of negligence to the employer. Such an additional remedy now operates in England.[30]

The British have yet another method of capturing the value of nonincome losses. One part of their system compensates the worker for income losses and expenses incurred. The second part establishes disability not on the basis of earning power, but as a loss "of the power to enjoy the life of a normal healthy person of the accident victim's age and sex. . . . A disability can be found to exist even if preaccident employment and earnings remain unaffected or, after an interruption, are resumed."[31]

There are four other ways in which workmen's compensation leads to employers' not absorbing the full social costs of accidents and disease. First, diseases in which occupational etiology is difficult to establish are often undercompensated. In fact, workmen's compensation was not originally set up to include payments for occupational illness, and its history reveals a singular difficulty in establishing adequate criteria and modes of payment for the compensation of victims of occupational disease (see Section 8.7).

Second, payments are often limited in total amount or duration. In 19 states, payments to permanently and totally disabled workers are not made for life. In 15 of these states, payments are limited to 10 years, and in 11 states the gross sum is $25,000, less than four years' wages for the average full-time worker.[32] Death benefits are similarly limited, with only 15 states paying benefits to the widow until her death or remarriage and to dependent children until age 18.[33]

Third, plaintiffs' legal fees often come out of court settlements. These fees should be part of the total costs of the injury prevention and compensation system. If an employer attempts to avoid payment of a just claim through the courts, claimants' legal fees should rest with the employer.[34]

Finally, some administrative and rehabilitation costs are paid for by state and federal government.

The Responsiveness of Payments to Costs of Disability

In order to act as an incentive, payments must not only be high, but must also be sensitive to the level (and costs) of accidents and disease. An employer with a rising accident rate must see that fact reflected in his insurance costs if he is to be adequately motivated to undertake a program of hazard reduction. However, workmen's compensation being essentially a risk-pooling insurance scheme, it is designed to avoid large fluctuations in employers' payments. To the extent that insurance means the pooling of risks—both among employers and over time—employers are rarely forced to bear the full costs of increased accident or disease rates.[35] Most employers pay for workmen's compensation insurance at rates determined by the experience of a large class of similar establishments. Their own safety record does not influence their insurance rate, unless it is so bad that they become assigned risks. However, many of the larger firms (and one-fourth of all firms covered) are merit-rated. Merit rating changes firms' premiums to reflect their loss records. Merit rating as currently practiced has some faults,[36] but even a perfect merit system would probably not be an adequate incentive to safety unless benefits were raised to reflect the full social costs of accidents (see Sections 8.5.1 and 8.3). As the *Compendium* notes,

A saving in premium costs could be a significant reward for success in accident prevention if benefits were higher, because premiums would then more truly measure the costs of accidents at work. The higher the benefit levels, the larger the premium costs to be avoided and the larger the incentive for prevention.[37]

Increasing the extent of merit rating and increasing benefits can obviously encourage preventive programs. Another way of increasing the ability of workmen's compensation to act as a safety incentive is through the use of deductible amounts of coinsurance. Presently, most covered firms must fully insure against workmen's compensation losses. Only some of the larger firms who set up their own internal insurance programs have a real option of partial insurance. An alternative to this system is one presently practiced in automobile insurance. Like many motorists, a firm would choose a deductible amount, so that it paid the full cost of any accident up to that level. The costs of insurance would be greatly reduced for those employers choosing a substantial deductible amount. (This amount could be limited to a small percentage of the net worth of the firm, to avoid the threat of insolvency.) The deductible would establish a closer relationship between employers' payments and the costs of accidents or disease.

In an effort to make compensation payments more responsive to health and safety hazards, the state of Wisconsin provides that additional compensation (15%) is payable *from the employer* if he is found to have violated a safety standard, and, conversely, the compensation is reduced if the worker has violated a safety practice. In fact, what results is an underreporting of violations, stimulated by the inspector's compassion for employers.[38] And, naturally, any underreporting decreases the incentive to improve workplace conditions.

Incentives to Insurers

If the insurance carrier ultimately bears the cost of compensation, there would seem to be a substantial incentive for the carrier to engage in loss-prevention activities. However, prevention activities amount to only 1.3% of premiums collected.[39] Even this amount does not go entirely to prevention activities. George Guenther, now Senior Vice President of the Insurance Company of North America and former

Assistant Secretary of Labor for Occupational Safety and Health has stated that this money goes to the underwriters "for inspection to determine the nature of risk and assist them in setting prices."[40]

Carriers' costs are more responsive than the employer's to injury and disease rates. However, as noted earlier, payments to injured workers are relatively low, so that the carrier's cost incentive for increased safety is still small. In addition, the insurance industry is not very competitive. The larger firms, who are generally insured by the larger carriers, are not likely to change insurance companies for small differences in available services, especially given today's changing compensation picture. The smaller firms are often insured by small local carriers who maintain a local monopoly and supply little loss-prevention service anyway. (Self-insured firms may, of course, be more responsive to their own accident history.)

Rates are carefully regulated by the states. One of the generally accepted guidelines of regulation has been that rates "should not be so low as to threaten the solvency of the insurer."[41] While it is certainly important to assure that carriers are solvent, such regulations clearly reduce competition among insurance companies. Since rates are often regulated to assure an adequate average rate of return on premiums written, the higher the total losses (and cash flow), the higher the profits will be in the industry in the *aggregate.* Some have assumed that this fact provides a sufficient disincentive for individual carriers to expand their loss-prevention activities. However, this is not true if the carrier's loss-prevention activities are effective. As long as the regulating agencies permit only small deviations from average rates, any insurance company with a good history of losses will be making extra profits because of its low loss rate. However, an individual company would nonetheless encourage the setting of high premium rates for the industry as a whole.

In spite of this chance to increase profits through loss prevention, competition in the insurance industry has largely been based on selling effort or administrative efficiency. Employers do not generally view workmen's compensation as a safety tool. They are likely to see the insurance inspector as a nuisance, and a potentially expensive one at that. Often, the inspector (whether for fire insurance or workmen's compensation) is regarded as outside interference, in spite of the fact that he can help the firm to reduce costs. Many insurance companies recognize that low losses are advantageous. However, rather than using tactics that would pressure their customers, they seem more inclined to contest the more costly cases. This situation has led to insurance companies' spending four times as much on selling commissions and brokerage fees as on inspections.

Even if the carriers had sufficient incentive to minimize their losses, reducing hazards at all the workplaces insured would be a difficult job. Each workplace has its own peculiar hazards, which might be more difficult to locate for the carrier's inspectors than for those familiar with the day-to-day operations of the firm. Also, employers who are hesitant to correct hazardous conditions can "clean up" for the insurance inspector just as they "clean up" for the government inspector. Often, the easiest way for insurance companies to minimize losses is to delay payments and to contest claims in court.

Adequate Flow of Information

In order for economic incentives to work properly, there must be adequate generation and communication of information about occupational injury and disease. It is necessary to understand both the causes of impairment and how to control them. However, the same incentive that can lead employers and insurance companies to attempt to reduce workplace hazards—i.e., liability for the costs of accidents and disease—can also lead them to hide information that could increase their liability. For example, information on the toxicity of a chemical could help prevent severe illness; yet an employer or carrier who discovers this illness in workers may prefer to disguise the cause rather than face the costs of workmen's compensation. As Harriet Hardy notes in an article on beryllium poisoning,

> Fear of compensation costs . . . had great influence on recognition of beryllium disease and its proper study. An insurance company handling workmen's compensation claims is a business existing to earn money. Cost of litigation was and still is influential in withholding accurate knowledge of how many workers were exposed to beryllium-using operations.[42]

Minimizing Administrative and Legal Costs

Between 1950 and 1970 loss payments for private insurers as a percentage of workmen's compensation premiums has fluctuated around 60%.[43] During this same period, about 9% of total premiums went to commissions and brokerage fees. Even these figures underestimate the true costs of insurance, since investment earnings on premiums are both a major source of income to the carriers and an additional cost to the insured firms. In comparison, payments from state workmen's compensation funds during this period averaged better than 80% of premiums; and those of the Federal Employees' Compensation Act averaged over 95%.

An important reason for the rather low ratio of loss payments to costs is the low level of competition among carriers, especially for small- and moderate-sized accounts. Oligopoly in this industry is generally strengthened by the influence of public regulatory agencies, which establish uniform "manual rates" which tend to make price competition less likely. The National Council on Compensation Insurance, which has traditionally been the filing agency for rates, has stated clearly that it wishes "to establish compensation rate-making on a scientific and noncompetitive basis."[44]

In addition, many insurance companies rely on independent agents to sell their policies. The agents' commissions and brokerage fees account for about one-fourth of the total expenses of private insurance.[45] And, as Arthur H. Reede has pointed out,

Clearly, the agents are a powerful interest group capable of fighting vigorously any efforts they think will reduce their income. . . . Individual stock companies are particularly vulnerable to opposition by their agents, for the latter can easily retaliate against companies which reduce rates of commission by offering their clientele to companies which continue to pay the old rates.[46]

Litigation costs can also be substantial, especially in the case of occupational disease. Even though negligence no longer needs to be established for recovery, eligibility for compensation benefits often turns on whether an injury or disease has occurred "out of or in the course of the employment." As discussed in greater detail below, this distinction is difficult to make for many occupational illnesses. The *extent* of disability can present another problem of quantification, as in the case of back injury. Many cases are contested for these reasons.

High legal costs have led the Commission to recommend that plaintiffs' legal fees be strictly regulated. This may not be advisable. As we have pointed out above (Sections 8.5.1 and 8.5.3), limiting plaintiffs' legal payments may lead to limitations on the adequacy of the legal aid they can secure. A more effective lawyer may charge his clients more, but increased settlements may still leave them better off. Moreover, carriers may be more willing to contest cases when faced with less effective attorneys. Thus, total litigation could well rise if the fees of claimants' lawyers were strictly regulated. However, if carriers were required to pay the costs of the employee's counsel in cases decided in favor of the employee, they might decide to contest in fewer instances where the claimant's case is sound. The greater potential cost to the carrier would partially offset the advantages to be gained by postponing payment. Thus, total legal expenses might well fall.

The Relationship between Payments and Benefits

The discussion of this section has been written as if benefits to workers were directly related to costs to employers and insurance companies. This assumption is warranted neither in reality nor in principle. The relationship between benefit levels and workmen's compensation payments vary among jurisdictions depending, for example, on local administrative, medical, and legal costs.

Moreover, workers receiving workmen's compensation payments do not pay their share of the costs of public services, since benefits are generally not taxable. Thus, other taxpayers subsidize disabled workers' use of these services, and employers' insurance premiums do not reflect these costs. This subsidy is potentially regressive. Workers with higher incomes normally pay a larger proportion of their incomes in taxes. If injured workers all receive the same proportion of their lost wages from the workmen's compensation system, the *net benefit* to higher-income workers is greater than that to lower-income workers.

The Commission suggests that this problem be handled by payments based on aftertax earnings.[47] This would reduce the problem of the regressiveness of payments based on pretax wages. However, it would not deal with the fact that the employer, the insurer, and the disabled worker presently do not contribute to the provision of public services while workmen's compensation is being paid. A better solution might be to tax compensation payments and to increase them accordingly.

8.5.4 Uniform and Broad Coverage

Broad Coverage of Employees

One of the basic objectives of a workmen's compensation system is the broad coverage of employees. State and federal programs, according to the Commission, reach about 85% of all employees. The coverage is inadequate and inequitable,[48] inadequate because many employees are not covered and inequitable because the employees that are now covered usually are not those most in need of protection—low-wage workers (such as farm help, domestic and casual workers) and employees of small firms. Further inequity exists because of wide variations among the states in the proportion of their workers protected by workmen's compensation. Thirteen states cover more than 85% of their workers, while fifteen states cover less than 70%.

The lack of coverage is to a major extent caused by statutory limitations in the coverage of classes of employees and specific occupations. Some of these limitations are as follows:

1. *Elective rather than compulsory:* Over a third of the states retain the elective feature of workmen's compensation, by which an employer may choose not to be covered.
2. *Numerical exemption:* More than half of the states do not extend coverage to firms with a limited number of employees. Some states also exempt certain classes of employers (e.g., charitable organizations).
3. *Exclusions:* Certain categories of employees are also excluded in many states. Farmworkers are covered on the same basis as other workers in only one-third of the states. Although some casual and domestic workers are covered partially, no states use the same basis for coverage as for all other workers. Many government employees do not have coverage due to elective laws in state and local governments.

Broad Coverage for All Forms of Disability

Uniformity in workmen's compensation laws should extend to types of disabilities as well as groups of workers. (Sometimes, a type of disability is limited to a particular group of workers—e.g., black lung.) The coverage is especially limited in the case of occupational diseases, many of which remain uncovered. A more detailed discussion of this problem is found in Section 8.7.

8.6 CONFLICT ANALYSIS

The goals of workmen's compensation have thus far been discussed individually, as if they were not significantly interrelated. In fact, not only do the public goals of workmen's compensation overlap in their projected effects, they are often inconsistent with one another. Furthermore, private interests are in conflict at almost every turn. These conflicts are inherent in the workmen's compensation system, and their resolution determines much of the content and substance of compensation statutes.

The first of these conflicts is the trade-off between the insurance function and the incentive to reduce hazards. As discussed above (Sections 8.5.2 and 8.5.3), the function of insurance is to reduce the variability of employers' payments. On the other hand, the incentive to reduce injury and disease depends on payments' fluctuating with the costs of accidents and disease. A common attitude toward this problem is exemplified by the following statement:

> Sound insurance principle requires [that the objective of prevention] be kept clearly secondary to those of equity and adequacy. . . . When safety is made the main end of the rate structure, this end, as in workmen's compensation schedule rating, may indeed be achieved but at the price of so great a failure in rate equity that (with rare exceptions) it has had to be abandoned. On the other hand, if the safety objective is kept subordinate [sanctions against the employer] will not be large enough to influence his attitude toward his risk.[49]

This conflict is inevitable, but the balance could be shifted in the present system by providing for partial insurance—for example, in the form of a substantial deductible amount.

A second, much-discussed conflict is that between adequate income replacement and an adequate supply of labor. The question here has been raised in many contexts—for example, in determining an adequate rate for unemployment compensation or welfare benefits. Some people fear that if payments are "too high," people will choose to live on the payments rather than go back to work. In the case of workmen's compensation, it has been argued that high payments will discourage workers from recuperating.

Historically, this fear has been overrated. When workmen's compensation was first being implemented, many feared "that compensation would undermine the morals of the workers and encourage them to injure themselves for the purpose of collecting compensation."[50] Recent large-scale social experiments with a "negative income tax" have demonstrated the exaggeration involved in the fears of those who believe that adequate income guarantees invariably lead to "loafing." Even if it is true that some people take advantage of offers of income without work, society must decide whether to punish the many for the vices of the few.

This issue is also a source of conflict between employers and labor unions. The unions want to see their members compensated as fully as possible for any work-related injury. When payments are high, the law requires the employer to provide a benefit that unions would otherwise be forced to bargain for. This leaves them free to gain additional concessions for their members. Also, by requiring high disability payments for *all* workers, generous workmen's compensation laws reduce the likelihood that lower benefits in the unorganized sector will undercut unions' abilities to win benefits for their own members.

A third source of conflict arises from employer liability for the full social costs of accidents and disease as an important incentive for reducing these costs. One way of reducing costs is to control workplace hazards. However, costs to the employer can also be reduced by controlling the flow of information that might lead workers to realize that they have contracted a work-related disease.

It has already been mentioned that employers have traditionally had almost exclusive access to medical and scientific data. In order to recognize that a certain disease is occupationally related, one must know both the medical histories of workers and the hazards they face. Without both sets of facts, it is generally impossible to draw conclusions about the etiology of disease. Employers have been in the best position to detect and publicize many disease problems. However, such publicity might well leave the employer open to substantial workmen's compensation claims and to new demands by labor. Thus, the employer's liability may lead to the suppression of information crucial to the detection and prevention of disease.

Yet another way of avoiding payment while not correcting hazards is to make legal recovery of lost income difficult. This can be done either by lobbying for weak laws (e.g., ones which do not recognize particular diseases as compensable) or by contesting claims. Especially when an employee's family is in critical financial condition, simply contesting the claim can result in a lower settlement.

The political and economic environment of workmen's compensation is itself an area of conflict. Employers and workers are constantly bargaining over the conditions of work and workers' benefits. Workmen's compensation influences both these aspects of labor-management relations. For this reason, management will generally speak out on the side of restraining benefits, while labor unions will generally be in favor of the most liberal construction of the statutes. High benefits, broad coverage of workers and of disease, etc., are all benefits to workers and costs of production to employers.

The Commission seems to overlook the difference between private and social benefits when it discusses the estimated "cost of adopting benefit recommendations."[51] Table 8.2, reproduced from the *Report*, lists increases in workmen's compensation rates and calls these increases the costs of suggested improvements in the system. But these are costs *only in the eyes of employers.* To the extent that they increase the income of the disabled worker, they represent a benefit to these workers; and to the extent that they help to reduce the number of work-related disabil-

Table 8.2
Estimated Increase or Decrease in Cost, Expressed as a Percentage of Current Costs, of Incorporating the Recommendations of the National Commission on State Workmen's Compensation Laws into State Laws in Effect January 1, 1972

	All Recommendations		Essential Recommendations	
Jurisdictions	With 1973 Maximum Weekly Benefits	With 1975 Maximum Weekly Benefits	With 1973 Maximum Weekly Benefits	With 1975 Maximum Weekly Benefits
Alabama	48.2	64.2	37.8	48.6
Alaska	64.7	80.0	35.8	44.5
Arizona	1.8	7.1	−1.5*	2.4
Arkansas	19.1	30.1	13.6	21.3
California	22.1	30.2	13.6	19.2
Colorado	36.9	50.5	25.6	34.6
Connecticut	24.6	33.5	−0.4*	4.5
Delaware	38.4	51.9	28.4	37.1
Florida	32.7	43.5	24.0	32.0
Georgia	51.4	67.0	39.5	49.9
Hawaii	8.8	16.9	3.2	9.1
Idaho	11.5	20.9	8.3	15.2
Illinois	21.1	26.3	10.1	17.3
Indiana	37.3	50.3	26.4	35.3
Iowa	36.4	48.7	27.7	36.0
Kansas	47.2	61.1	38.4	48.6
Kentucky	46.5	62.0	31.2	42.3
Louisiana	64.9	80.4	48.9	60.1
Maine	−9.8*	−2.4*	0.5	7.7
Maryland	18.6	27.6	13.5	20.7
Massachusetts	8.8	17.9	4.2	11.3
Michigan	24.3	34.8	17.1	25.1
Minnesota	21.4	32.3	14.7	22.7
Mississippi	48.3	63.8	40.7	51.2
Missouri	40.0	52.0	33.6	42.6
Montana	58.9	81.7	49.1	62.4
Nebraska	20.6	33.2	11.7	19.9
Nevada	17.9	25.4	2.4	10.0
New Hampshire	7.2	16.8	2.8	10.3
New Jersey	11.1	32.0	3.6	20.6
New Mexico	47.2	62.2	32.5	42.0
New York	19.7	28.2	12.3	18.6
North Carolina	30.0	42.0	27.9	36.8
North Dakota	34.5	51.6	28.4	39.1
Ohio	54.1	72.0	14.9	25.7
Oklahoma	53.3	67.8	42.8	53.1

Oregon	25.7	36.7	11.7	19.2
Pennsylvania	44.0	60.4	27.4	37.8
Rhode Island	10.1	16.7	6.3	11.8
South Carolina	33.2	45.1	26.6	35.2
South Dakota	33.3	48.5	27.4	38.1
Tennessee	30.7	42.5	23.4	31.8
Texas	50.0	64.8	38.3	48.9
Utah	42.2	56.8	31.4	40.3
Vermont	19.8	28.6	13.7	20.0
Virginia	35.5	48.1	24.0	32.4
Washington	22.6	33.1	–0.4*	6.5
West Virginia	21.0	35.6	16.4	27.9
Wisconsin	12.9	21.3	8.5	14.8
Wyoming	60.0	80.8	49.4	61.9
D.C.	24.0	34.2	17.8	25.0

*Negative values were entered in calculations where present state benefits exceeded those recommended.

Source: National Council on Compensation Insurance.

ities, they represent a social benefit as well.

Such increased payments might ultimately affect workers themselves, but only if markets pass the costs on to them. One possible result of a revised workmen's compensation system is the moving or closing of marginal employers. In an economy characterized by substantial unemployment and generally inefficient labor markets, the fate of workers is often closely tied to the fate of their employer. If the costs of workmen's compensation were to lead to plant closures, or to the emigration of industries, some people would be left jobless. Even when this possibility is remote, uncertainty can often lead workers to take a surprisingly conservative attitude toward programs that might make their employers less competitive.*

The interests of the professionals and administrators who service the workmen's compensation program should not be overlooked. Many Massachusetts lawyers lost a large part of their practices when no-fault automobile insurance was established. Several areas of administrative and legal inefficiency exist in present workmen's compensation systems. To the extent that the incomes of administrators, insurance carriers, and lawyers depend on these inefficiencies, they will be opposed to change.

*Fred Foulkes writes that "in Philadelphia, the cost of workmen's compensation for longshoremen is about $25 per $100 of wages; but in Newport News, the bill is just $9 per $100 of wages. Increasingly, therefore, maritime companies are having their ships unloaded in Virginia as opposed to Philadelphia."[52]

In addition to the problem of layoffs, the more general problem of the incidence of workmen's compensation payments should be considered. The incidence (or final burden) of these payments may rest with the employer, the employee, or the final consumer of the produced goods. Only under respective conditions will the actual incidence of the tax coincide with the optimal incidence.[53]

8.7 WORKMEN'S COMPENSATION AND OCCUPATIONAL DISEASE

The congressional charge to the Commission states that

> in recent years serious questions have been raised concerning the fairness and adequacy of present workmen's compensation laws in the light of the growth of the economy, the changing nature of the labor force, increases in medical knowledge, change in the hazards associated with various types of employment, new technology creating new risks to health and safety, and increases in the general level of wages and the cost of living.[54]

One of the most important changes in the workplace environment since the passage of the first workmen's compensation laws has been the increasing importance of occupational disease. Every year new and potentially harmful substances are introduced into the occupational environment. Both medicine and law have found it difficult to keep pace with the new industrial hazards caused by changing technology. In the words of the Commission, "we assumed as our responsibility a complete reexamination of workmen's compensation in light of the historical changes noted by Congress."[55] In spite of this, the recommendations of the Commission are weakest in the area of occupational disease. For these reasons, the problems involved in adapting workmen's compensation to cover occupational illness are given special emphasis in this separate section.

Workmen's compensation was originally designed to reduce the burdens of the most obvious and catastrophic consequences of a growing industrial society. Injuries, often serious and permanently disabling, were the object of legislation at the beginning of this century. The designers of the original statutes did not foresee the slow, silent disability caused by many of today's occupational hazards, and the problems of recognizing and understanding such occupational diseases have therefore raised critical issues for the entire workmen's compensation system. While benefits may be inadequate and coverage of employees may be incomplete, workmen's compensation has generally succeeded in providing relatively direct and speedy payments to those injured by safety hazards. Its performance for even the most serious of occupational diseases has been much less adequate. In this section, some of the major problems of occupational disease will be reemphasized and their consequences for workmen's compensation discussed.

8.7.1 Detecting Occupational Disease

Many occupational diseases have difficult-to-detect symptoms. Sometimes, the consequences of occupational exposure coincide with normal signs of aging. For example, people in this society tend to lose hearing acuity as they age. At the same time, they are exposed to noise on the job. This noise will produce partial deafness, but only a worker or observer sensitive to the problem of occupational noise exposure will consider the loss of hearing to be other than "natural." Similarly, people expect to find climbing stairs more difficult with age, but shortness of breath is also a symptom of certain occupational lung diseases.

Some occupational diseases have the same symptoms as illness contracted off the job. Emphysema, for example, can be the result of either occupational exposure or cigarette smoking. A skin rash can be induced by fiberglass or by poison ivy. In this case, again, someone who does not understand the dangers of occupational exposure can mistake the cause of a disease.

Even when the cause of a disease is clearly occupational, the effect may not be felt for ten, twenty, or even thirty years. In this case, a particular worker may not recognize a hazard until it is too late. Also, a dangerous (perhaps carcinogenic) substance may be introduced and widely used for decades before its harmful effect is discovered. Even if its use is curtailed as soon as its effects appear, it may be too late to reverse those effects in workers who have been exposed to it.

8.7.2 The Problem of Multiple Etiology

The causes of accidents are often relatively obvious and proximate. In spite of this, there is still much debate around the question, "Who or what causes accidents?" But the etiology of disease is often much more obscure and complex than that of accidents. Therefore, the problem of whether a disease arose out of or in the course of employment is often difficult to assess. A syndrome can be caused–

- by several conditions all originating in the workplace;
- by several conditions, one or more of which may have nonoccupational origin;
- by both physical and emotional stress;
- cumulatively, by conditions at several different workplaces.

Given both poor data about exposure and the wide latitude for the exercise of nonscientific judgment in medicine, such multiple etiologies often make causation a forensic issue rather than a scientific one.

8.7.3 The Effects of Uncertainty, Multiple Etiology, and Long Latency Periods

The difficulties of assessing causation in occupational disease have weakened workmen's compensation in two ways. First, to the extent that the statutes have not been designed to incorporate these problems, the laws are inadequate. Second, even were there "perfect" compensation laws, employers and workers would still tend to undervalue the true costs of disease.

Present Statutes and Disease

Several aspects of current workmen's compensation statutes make compensation for many diseases extremely difficult. Some problems result from the fact that workmen's compensation was originally designed to cover only injury, and from a certain reluctance to open the door to compensation for "too many" reasons.

For example, some statutes still require that the disability be the result of an "accident," a sudden and unexpected event. Since most diseases are either slow in onset or caused by conditions which are a normal part of employment, or both, they are not compensable by this criterion. One of the Commission's recommendations wisely suggests dropping this "accident" requirement wherever it remains.[56]

As more and more occupational diseases have been recognized as such, states have begun to include them in worker's compensation. Some do so by enumerating those diseases which are considered both occupational and compensable. With this system, many workers are left without coverage, however, both because of time lags from the recognition of the occupational origin of a disease to its inclusion in the compensation system and because of employer pressure to prevent legal recognition. The most tragic examples in recent times are provided by the belated recognition and inclusion of coal miners' pneumoconiosis (black lung) and asbestos-related respiratory disease and cancer. Most states have now decided to cover occupational disease fully.* However, as of 1972, there were still nine which had not done so. We concur with the Commission's recommendation "that all states provide full coverage for work-related diseases."[57] This coverage becomes increasingly important as more chemicals are introduced into the occupational environment each year.

Workmen's compensation statutes have traditionally had a time limit for the filing of claims. This time limit is usually not a serious problem for an employee who suffers a work-related accident. However, if he contracts a disease whose etiology is obscure or whose latency period is very long, a time limit could very well foreclose recovery. As the *Compendium* states, "if the running of the claim period dates from the time of last exposure rather than from the date of disability, the first signs of the illness may not be seen until well after the claim period has run."[58] The U.S. Department of Labor recommends that the time limit on filing should be at least one year after the victim knows of his illness *and* knows that it is work-related, and after he has experienced some wage loss. Of the 58 workmen's compensation jurisdictions, only 26 met this standard in 1972.[59] The Commission further strengthens the Department of Labor's recommendation by suggesting that the time limit be extended to three years.[60] Statutes of limitations in the state workmen's compensation system continue to be a problem.

The problem of diseases with long latency periods (including most occupational respiratory diseases, heart disease, and cancer) is further complicated when the "dis-

*The catch, of course, still lies in the determination that the disease is in fact "occupational" in origin. This determination is not arrived at easily or without much politicizing of the scientific data and use of "experts." In European compensation systems, *sickness* arising from exposure to harmful chemicals is also compensable, thus avoiding the necessity for proving that a "disease" exists (see Chapter 11).

ablement" occurs after retirement. If, as the Commission suggests, workers are not compensated for impairment either through the compensation system or through employer negligence suits in the courts, then the retired employee who can no longer breathe well or exert himself, and who will die early, will receive no payment for his impairment. Since he is no longer "disabled" because he is out of the labor force, he will receive neither workmen's compensation nor social security disability payments. Thus, if an employer can get the "system" to avoid recognizing an impairment until after retirement, both he and the social security system will be "home free." Since long-latency diseases are, in fact, quite common, the social cost of the "system" as it stands today—or as the Commission would have it—puts a heavy burden on the worker and his family. The anticipated costs for revamping the compensation system as exhibited in Table 8.2, does *not* include the real costs which should have been included by "a complete reexamination of workmen's compensation in light of the historical changes noted by Congress."

The problems of multiple and uncertain etiology have been a major cause of contested claims. This has often left seriously disabled workers and their families without financial support for long periods of time. It has made workmen's compensation systems less prompt, less efficient, and less just than they otherwise might be. The Commission proposes two solutions to this dilemma. The first suggests that "full workmen's compensation benefits be paid for an impairment or death resulting from both work-related and non-work-related causes if the work-related factor was a significant cause of the impairment or death."[61] The word "significant" masks some very important problems here. How important must a cause be before it is "significant"? Such vague wording could lead to applications of the law which did not meet the spirit in which the recommendation was given. Who will define "significantly," operationally? Such problems aside, this recommendation is a good one.

The second change in the statutes would establish a "disability evaluation unit." This body would legally determine causation in the case of disability or death. It would be considered a professional unit, and used to establish, factually, the etiology of a worker's disease, since, in the words of the Commission, "the etiology of a disease [is] a medical question."[62]

This recommendation is simplistic, avoiding both the scientific and legal subtleties involved in determining causation. There have been widely different opinions—among professionals—on the health effects of a number of substances. Especially when a large amount of profit depends on the use of the substance, or when large workmen's compensation payments are likely, the experts hired by industry often seem to take a conservative view of occupational hazards.* For example, Paul

*In contested cases, plaintiffs' physicians may be as biased in their views of causation as are those hired by the insurance company or the contesting firm. Certainly, there is lack of objectivity by all parties in this advocacy relationship.

Brodeur has discussed the recent outbreak of lung disease among detergent workers exposed to proteolytic enzymes. In a 1969 conference, several studies were presented, showing severe asthmalike reactions by workers to the enzymes. Detergent industry representatives responded, refusing to acknowledge the importance of the enzyme hazard:

> Many of the doctors who had attended the conference, hoping for the cooperation of industry, were dismayed by the intransigence of the soap companies in refusing even to admit that such a problem might exist. Their spirits were scarcely raised when the medical director of one of the largest detergent manufacturers in the world got up before them and sought to discount the very reason they had come together by stating that "in any case nobody ever died of asthma." It was a remark that his fellow M.D.'s greeted with a gasp of disbelief.[63]

An interesting alternative to the disability evaluation unit has been suggested by Leslie Boden, of the Harvard School of Public Health: this would be an arbitration unit, set up to hear cases in which the etiology of a disease is disputed. The arbitration unit would be composed of experts, as in the case of the disability evaluation unit. However, the fiction of purely "scientific" judgment would no longer be espoused. A particular arbitrator would hear a case only if both parties agreed to choose him. Thus, any person whose labor or management bias is too extreme would be unable to be an arbitrator, no matter how "expert" he might be. Hopefully, such a system would lead to a more equitable resolution of conflicts involving the etiology of a disease.

The liberalization of compensation laws with respect to disease may lead to a sudden rise in the premium payments of employers. For example, as "black lung" became compensable, employers could have faced the burden of paying for former employers' carelessness as well as their own. As an interim measure for any similar situation, it is suggested that (as in the case of black lung) a federal fund be set up to help pay the initial costs of compensation for formerly uncovered disease.

Behavioral Implications of Uncertainty, Time Lag, and Multiple Etiology

Decisions about uncertain events in the distant future are among the most difficult to make. And these are exactly the decisions which must be made by management and workers about occupational exposure.

Unless a condition is particularly evident, workers must make their decisions about occupational exposure on the basis of statistics which indicate that something may happen to them decades from now—and, then again, may not. If workers have any information at all about occupational disease, it is often in an unconvincing form. Because of this, they will commonly underestimate the extent of their risk.

Many management decision-makers will also put a low value on the possible costs of disease. Even if compensation payments are high, they may be both uncertain and far in the future. This uncertainty is a function not only of the uncertain consequences of exposure, but of a general lack of knowledge about occupational disease by both workers and their physicians. Even if workers do become ill, they might not associate their illness with occupational factors.

The fact that much illness occurs years after exposure has two effects. First, if the firm has a high rate of return, profits now (from not spending money to reduce hazards) may be considerably more valuable than losses in the future from compensation costs. Second, many firms are uncertain that they will exist in ten or twenty years. And even if the firm does exist, the manager who makes the decision on abatement will probably not be held responsible then. If decisions to promote the manager are made partly on the basis of production, and if healthful work conditions will reduce productivity (in the short run at least), then foremen and managers will be pushed toward accepting relatively unhealthful conditions.

8.7.4 Summary of Occupational Disease Problems

Accidental injury is a relatively simple event. It is concrete and easy to understand, and its effects are immediately felt. Disease is also a consequence of occupational conditions, but its etiology is often complex and not well understood. Onset sometimes comes decades after exposure. Often, the disease is not even known to be of occupational origin. These problems are exacerbated by workmen's compensation statutes which treat disease as a special kind of injury. The complexity of disease also fosters decisions on the part of management and workers which lead to inadequate abatement of health hazards at the workplace.

These difficulties are further complicated by insurance company lobbyists' attempts to keep the definition of occupational diseases stringent, and by powerful economic interests (e.g., the coal industry) manipulating the legal definition of occupational disease to reduce the costs of disease prevention. All of these factors have combined to keep occupational disease claims well under 1% of all compensation payments. Yet the burden of disability and death from occupational disease may be as great as or greater than the burden from accidents.

8.8 WHAT CAN'T WORKMEN'S COMPENSATION DO?

Workmen's compensation is a single instrument attempting to meet many objectives. It would be unreasonable to expect it to meet all these objectives completely. In fact, as suggested in Section 8.6, some of the objectives are in conflict with each other. I believe that, especially in two areas, the most perfect workmen's compensation system must fall short. However, this does not mean that some of the goals must be sacrificed. Rather, it indicates that we must look for other instruments to achieve them.

8.8.1 The Insurance Function and the Prevention Objective

The concept of insurance carries with it the problem of "moral hazard." Moral hazard describes the phenomenon that leads people to engage more in covered activities when they are insured than when they are not. The problem of moral hazard is usually inherent in the insurance situation, deriving from the fact that insured people pay less for extra use of the covered activities.[64] For example, a person with health insurance will use medical services more often than if he were uninsured, since the extra medical services used cost little or nothing.

This conflict between the prevention objective and the insurance objective of workmen's compensation was discussed earlier, and it was noted that because the employer does not pay the full social cost of each additional disability, his incentive to reduce hazards is too low. Here it must be emphasized that *this problem of moral hazard is inevitable as long as the insurance aspects of workmen's compensation are retained.* Compensation laws can encourage safety more effectively than they do today, but they can never be fully effective by themselves.

The insurance coverage of workmen's compensation should be retained, since it serves a very important function. The implication to be drawn from the conflict between prevention and insurance is that additional mechanisms to reduce workplace accidents and illness are necessary. The OSHAct provides one such mechanism.

8.8.2 Perfect Coverage and the Avoidance Problem

Another inherent dilemma of a well-conceived workmen's compensation program is caused by its success in making benefits high. High benefits mean high costs to employers, and thus incentives to find ways to reduce these costs. One way to reduce costs is, of course, to provide a less hazardous workplace. But, unfortunately, there are others. When claims are likely to be high, litigation becomes more profitable. Similarly, when information about possible disease can mean high payments to those affected, this information may be hidden.

Another way of attempting to limit the costs of accidents and disease is to screen out potential problems in the worker population. Medical screening may be advisable, for example, to prevent people with potential back problems from doing heavy lifting. In an economy with very low unemployment, such screening would probably lead to a person's taking jobs better suited to his physical makeup. However, in an economy with very inefficient labor markets, a person could find himself unable to get a good job—or any job at all—if employers strongly emphasized medical screening examinations.

This problem has long been recognized by workmen's compensation in one of its aspects. A previously injured worker may be unable to find employment because of employers' fears that he will be reinjured, leading to a high workmen's compensation claim. In order to give the injured worker a better chance to get a good job, most states therefore have a "second-injury fund." As the *Report* states,

> A second-injury or subsequent injury fund within the workmen's compensation program insures that a handicapped worker who subsequently suffers a work-related injury or disease will receive full compensation to cover the resulting impairment. At the same time, the employer will be charged only for the benefits that are associated with the second injury.[65]

Such funds have some potentially adverse consequences. Workers who are likely to be reinjured may be given dangerous jobs, since the employer will not be forced to bear the full cost of reinjury. Obviously, a better system would reduce occupational hazards in general, and provide adequate alternative employment opportunities to those with medical problems. But until such conditions are met, employees will suffer economically because of unilateral decisions on their employability made by corporate personnel offices or medical departments. Some will not be hired because of medical screening. Others will be refused promotion. Still others may be laid off and not rehired. These problems are likely to become more severe as workmen's compensation payments become higher.

As in the case of the conflict between prevention and insurance, a solution to this dilemma must come from outside the present workmen's compensation—for example, through a National Health Service.

8.8.3 Implications

The implications of the foregoing analysis are clear. Workmen's compensation cannot meet all its goals at once, even if it is perfectly designed. This means that it must be considered as part of an integrated system of occupational safety and health programs. In order for the system to work properly, all its programs must be adequately designed.

8.9 RECENT LEGISLATIVE PROPOSALS

In 1973 several important bills related to the work of the Commission were introduced in Congress. Two of these focused on the special problems of occupational disease, while the third was an attempt to establish minimum standards for workmen's compensation statutes. As of this writing, none of these bills have been enacted.

In February 1973 Senator Taft (R-Ohio) introduced S. 1029 to provide federal workmen's compensation for total disability or death caused by occupationally related lung disease. This bill would provide coverage like that of the black lung legislation to cover a number of other diseases, including asbestosis, berylliosis, byssinosis, bagassosis, diatomite, pneumoconiosis, Shaver's disease, siderosis, silicosis, and occupationally induced lung cancer. As in black lung legislation, initial compensation would be made by the federal government. After 1976 compensation would revert either to state plans or to compulsory employer insurance.

Recognizing that occupational causality is difficult to establish in many chronic lung diseases, the bill would allow work in an identified hazardous industry to "be a rebuttable presumption that such disease arose out of . . . employment in such industry." This reverses the normal burden of proof, leaving the employer with the task of showing that the disease is not occupationally related. Further strengthening employees' normally weak legal position when they have occupational respiratory disease, the bill would make employers liable for disability caused "at least in part" by workplace hazards.

The bill also confronts the problem of employers' firing workers who are ill. It prohibits such discharge or any other discriminatory activity on the part of the employer.

One weakness of the Taft bill is that ten years of service is required for the presumption of work-related disability. Most of the occupational lung diseases can be contracted in a much shorter period of time. However, it is clear that such a presumption is better than none at all. The bill also lists specific occupationally related lung diseases, rather than allowing for all such diseases to be compensated.

In October 1973 Senator Humphrey proposed an Employees Comprehensive Disease Compensation and Registration Act. This bill would require compensation for all workers totally disabled by job-related lung disease, or for their families if they die. A particularly innovative part of this proposed legislation is a national register of employees exposed to respiratory disease hazards. The register would contain data on all workers at risk, giving duration and severity of exposure to disease-producing agents. The information would be available in detail only to the employee or his physician. The registry would provide information to employees so that they can better assess occupational risk. It would also be an excellent data source to increase our knowledge about occupational respiratory disease—including data on particularly elusive synergistic effects.

The broadest of the three bills was introduced by Senators Javits and Williams in June 1973. This bill (S. 2008) attempts to implement many of the recommendations of the Commission by establishing minimum standards to which states would conform. The bill includes the following standards:

1. No maximum limitations on total benefits for death or temporary or permanent total disability.
2. No maximum limitation on the type or extent of medical care or rehabilitation services.
3. Totally disabled workers must be paid not less than two-thirds of their average weekly wage, subject to a maximum of no less than 100% of statewide average

weekly wages during 1975, rising to no less than 200% of statewide average weekly wages commencing January 1, 1978.
4. Appropriate periodic adjustments of benefits for those already receiving benefits to reflect the increases in statewide average weekly wages and benefit levels.
5. Attorney's fees to be added to an award where claimant has been successful in formal adjudicatory proceedings.[66]

In addition, S. 2008 provides for the development of criteria for the work-relatedness of occupational disease by the Secretary of Health, Education, and Welfare. The bill accepts many of the recommendations of the Commission, including the retention of state administration of workmen's compensation, and it attempts to get several of the most important recommendations enacted. The bill is strongly supported by labor representatives, and just as strongly opposed by both insurance associations and some large corporations.

The bill states, "An injury shall be deemed to have arisen out of and in the course of employment if work-related factors were a *contributing* cause of the injury" (Emphasis added). This differs from the Commission's recommendation that work-related factors be a "significant cause of the impairment or death." Insurance carriers seem to feel that the deletion of the word "significant" makes the causal connection too easy to establish.

Leslie Cheek, of the American Insurance Association, suggests that this would lead to compensation when a worker claims "he fell down and broke his leg at home because he was tired from the job."[67] Andre Maisonpierre, of the American Mutual Insurance Alliance, another insurer's organization, mirrors Cheek's concern that "too much" disease will be compensated. Maisonpierre states, "It is hard to think of a heart or ulcer case that is not covered" under S. 2008.[68] Neither of these insurance industry spokesmen addresses the relevant question, "Are these problems actually occupational in origin?" In the HEW report on *Work in America*, the following assessment is made:

> Why is job satisfaction perhaps one of the best ways of extending the length of life? Other factors are undoubtedly important—diet, exercise, medical care, and genetic inheritance. But research findings suggest that *these factors may account for only about 25% of the risk factors in heart disease, the major cause of death.* That is, if cholesterol, blood pressure, smoking, glucose level, serum uric acid, and so forth, were perfectly controlled, only about one-fourth of coronary heart disease could be controlled. Although research on this problem has not led to conclusive answers, it appears that *work role, work conditions, and other social factors may contribute heavily to this "unexplained" 75% of risk factors.*[69]

The Insurance Alliance has asserted that the proposed legislation would be, like the Coal Mine Health and Safety Act, "another billion-dollar fiasco." Its spokesmen point to the Social Security Administration's estimate that black lung payments through 1981 will total $10.6 billion, and that the coal operators' responsibility for all new cases may cost an average of $25 per $100 of payroll.

However, the Alliance fails to note that before the Coal Mine Health and Safety Act, miners themselves were required to bear the full burden of black lung. The $10.6 billion figure is shocking not because the federal government is paying it, but because it is an estimate of the lost income of thousands of disabled coal miners. Black lung is the scandal, not the extent of compensation of already disabled workers. The Alliance thinks of S. 2008 as "a fantastically costly pension program for everyone who can't or won't work."[70] This statement reflects the same distorted perception that once led employers to oppose all workmen's compensation on the grounds that workers would injure themselves to avoid working.

In summary, several important bills have been submitted, but none has passed. Especially in the case of S. 2008, a struggle is brewing between the insurers (and at least some employers) and labor unions. It is in this struggle that the future of workmen's compensation may be determined for the near future.

In an article surveying the changes in state workmen's compensation laws in 1973, it was noted that "[state] legislatures . . . adopted a record number of 400 amendments to their workmen's compensation laws."[71] Although the article states that "full coverage of occupational diseases was newly provided in five states," the provisos of time limitations (statutes of limitations, etc.), narrow interpretations of "occupational," and other caveats make the apparent progressiveness of the new legislation illusory at best. A careful analysis of recent state response clearly indicates that no really meaningful coverage of occupational disease is likely to come from "voluntary" state efforts. Firm national policy and requirements that go beyond the recommendations of the Commission will be necessary if workers are to be compensated fully for occupational disease and injury.

An examination of the workmen's compensation systems in Europe suggests an interesting alternative to the problems of identifying a disease as "occupational" to establish eligibility. We could make *sickness* due to exposure to a large number of chemicals and dusts compensable, thus eliminating the necessity of proving that a "disease" exists. Sickness among workers exposed to acknowledged harmful substances would be presumptive evidence sufficient to establish eligibility for workmen's compensation. Then, anomalies such as having silicosis but not pneumoconiosis compensable for coal miners could not arise.[72]

REFERENCES

1. National Commission on State Workmen's Compensation Laws, *Compendium on Workmen's Compensation*, G.P.O. Document No. 496-632x(1973), p. 3. [Hereinafter cited as *W.C. Compendium*]

2. Arthur H. Reede, *Adequacy of Workmen's Compensation* (Harvard University Press, Cambridge, Mass., 1947), pp. 90-93.

3. *Handbook of Labor Statistics*, U.S. Department of Labor, Bureau of Labor Statistics, p. 185.

4. Reede, *Adequacy of Workmen's Compensation*, pp. 90-93.

5. *Ibid.*, pp. 160-161.

6. Richard Posner, "A Theory of Negligence," *Journal of Legal Studies*, Vol. 1 (January 1972), pp. 29-96.

7. *The Report of the National Commission on State Workmen's Compensation Laws*, G.P.O. Document No. 477-053 (July 1972), p. 125. [Hereinafter cited as *W.C. Report*]

8. Quoted in Milton Katz, "The Function of Tort Liability in Technology Assessment," *Cincinnati Law Review*, Vol. 38 (1969), p. 626.

9. Wex S. Malone, "The Limits of Coverage in Workmen's Compensation: The Dual Requirement Reappraised," *Supplemental Studies for the National Commission on State Workmen's Compensation Laws*, Volume 1, G.P.O. Document No. 521-668 (1973), p. 174. [Hereinafter cited as *W.C. Supplement*]

10. *Arthur v. Flota Mercante, Gran Centro Americana, S.A.*, 487 F.2d 561 (5th Cir. 1973).

11. *W.C. Report*, p. 35.

12. *Ibid.*, p. 56. The *Report* continues, "Of course, if the proportion is too low, a worker may be in such dire circumstances that he may be forced to return to work before he is properly recovered or he may become so demoralized as to be indefinitely disabled." The validity of the conflict between an adequate rate of compensation and the incentive to return to work is an empirical matter, which will be discussed in more detail later.

13. O. E. Williamson, D. G. Olson, and A. Ralston, "Externalities, Insurance, and Disability Analysis," *Economica*, Vol. 34 (August 1967), p. 252.

14. *W.C. Compendium*, p. 115.

15. *Ibid.*, p. 116.

16. Laurence S. Locke, "Environmental Health Hazards in Industry," speech delivered at the meeting of the American Trial Lawyers Association, Denver, Colorado, July 29, 1969.

17. *W.C. Report*, R3.14, p. 64; emphasis added.

18. *Ibid.*, R3.27, p. 73, and R3.18, p. 66.

19. Joseph A. Page and Mary-Win O'Brien, *Bitter Wages* (Grossman, New York, 1973), pp. 67-68.

20. See *W.C. Report*, pp. 65-66, 72-73.

21. *Ibid.*, p. 80.

22. *Ibid.*, R6.15, pp. 109-110.

23. *Ibid.*, p. 81.

24. *Ibid.*, p. 54.

25. *Ibid.*, p. 69.

26. *Ibid.*, p. 67.

27. *Ibid.*, p. 66.

28. *Ibid.*, p. 54.

29. *Ibid.*, pp. 119-120.

30. H. M. Somers and A. R. Somers, *Workmen's Compensation* (John Wiley, New York, 1954), p. 307. See also Chapter 11, this volume.

31. *W.C. Compendium*, p. 66.

32. *W.C. Report*, p. 65.

33. *Ibid.*, p. 72.

34. Locke, "Environmental Health Hazards," p. 11.

35. See, for example, Louise Russell, "Pricing Industrial Accidents," in *W.C. Supplemental Studies*, Vol. 3, p. 27.

36. *W.C. Compendium*, p. 289.

37. *Ibid.*

38. George Hagglund, School for Workers, University of Wisconsin, private communication.

39. C. Arthur Williams, "Insurance Arrangements Under Workmen's Compensation," U.S. Department of Labor, Wage and Labor Standards Administration, Bulletin No. 317 (1969), pp. 74-75.

40. *Occupational Safety and Health Reporter* (the Bureau of National Affairs, Inc.), Vol. 3, No. 1 (June 7, 1973), p. 9. [Hereinafter cited as *OSHR*, BNA]

41. *W.C. Compendium*, p. 254.

42. Harriet L. Hardy, "Beryllium Poisoning–Lessons in Control of Man-Made Disease," *New England Journal of Medicine*, Vol. 273 (November 25, 1965), p. 1197.

43. *W.C. Compendium*, p. 269, Table 16.1. Note also that not all of this 60% is returned to workers, and that any legal costs incurred must be paid for out of the settlement.

44. From the 1952 *Annual Report* of the National Council on Compensation Insurance, p. 3; quoted in Somers and Somers, *Workmen's Compensation*, p. 102.

45. *W.C. Compendium*, p. 269.

46. Reede, *Adequacy of Workmen's Compensation*, pp. 301-302.

47. See *W.C. Report*, p. 56, including R3.1. See also R3.6, p. 60.

48. *Ibid.*, pp. 43-44.

49. C. A. Kulp, "The Rate-Making Process in Property and Casualty Insurance–Goals, Techniques, and Limits," *Law and Contemporary Problems* (1950), pp. 494-495.

50. Abraham Epstein, *Insecurity–A Challenge to America* (Smith and Haas, New York, 1933), p. 593; quoted in Somers and Somers, *Workmen's Compensation*, p. 31.

51. *W.C. Report*, p. 73.

52. Fred Foulkes, "Learning to Live with OSHA," *Harvard Business Review* (November-December 1973), p. 67.

53. Peter Gregory and Micha Gisser, "Theoretical Aspects of Workmen's Compensation," in *W.C. Supplemental Studies*, Vol. 1, pp. 107-127.

54. *W.C. Report*, p. 13.

55. *Ibid.*, p. 34.

56. *Ibid.*, R2.12, pp. 49-50.

57. *Ibid.*, R2.13, p. 50.

58. *W.C. Compendium*, p. 187.

59. *W.C. Report*, p. 108.

60. *Ibid.*, R6.13, pp. 107-108.

61. *Ibid.*, R2.17, p. 51.

62. *Ibid.*, p. 51.

63. Paul Brodeur, *Asbestos and Enzymes* (Ballantine Books, New York, 1972), pp. 81-82.

64. Kenneth J. Arrow, "The Economics of Moral Hazard: Comment," *American Economic Review*, Vol. 53 (1964), p. 53; M. V. Pauly, "The Economics of Moral Hazard," *American Economic Review*, Vol. 54 (1964), p. 531.

65. *W.C. Report*, p. 83.

66. *OSHR*, BNA, Vol. 3, No. 3 (June 21, 1973), p. 68.

67. *Daily Labor Report* (December 28, 1973), p. A-9.

68. *Ibid.*, p. A-12.

69. *Work In America* (MIT Press, Cambridge, Mass., 1973), p. 79; emphasis added.

70. *Daily Labor Report* December 28, 1973), p. A-11.

71. Florence C. Johnson, "Changes in Workmen's Compensation Laws in 1973," *Monthly Labor Review*, Vol. 97, No. 1 (January 1974), p. 32.

72. *Ibid.* See also Chapter 11, this volume.

9 Manpower Problems in Occupational Health and Safety

9.1 INTRODUCTION*

9.1.1 Why the Manpower Issue Should Be Addressed

The field of occupational health and safety requires expertise in many disciplines if the congressional mandate of a safe and healthful workplace is to be achieved. In order to develop a coordinated manpower policy, an understanding of the breadth and depth of the needed expertise in each manpower area is essential. Unfortunately, information on the existing stock of occupational health and safety manpower is seriously deficient. Uncertainty about both the existing stock and requirements for the future makes this preliminary analysis of occupational health and safety manpower needs extremely difficult.

Manpower needs can be classified into four categories, each of which includes a number of different types of professionals:

1. Enforcement personnel are needed to ensure compliance with the OSHAct in both safety and health by inspecting workplaces and citing violations of standards.
2. Qualified researchers, including chemists, physicians, environmental physiologists, epidemiologists, toxicologists, and engineers, are needed to identify potential occupational health and safety hazards, to assess their effects upon those who work and have previously worked under exposure to these hazards, and to determine the requirements for assuring a safe and healthful working environment.
3. Personnel who provide health and environmental services are needed to aid in developing and implementing techniques, methods, and programs for the prevention and treatment of occupational disease and injury. This group includes occupational physicians, industrial nurses, safety professionals, industrial hygienists, and technicians, as well as executives and managers in business, labor unions, and government.
4. Personnel are needed to teach the skills necessary for performing the tasks within each of the above-mentioned groups.

The existing stock of occupational health and safety manpower is seriously deficient in terms of society's needs.** Indirect evidence of insufficient manpower is

*Many of the assertions in this chapter are made without extensive documentation, because of a serious lack of information at the present time. Nevertheless, these observations are based on an extensive series of interviews with health and safety professionals, educators, and industry, labor, and government representatives. Before a comprehensive manpower policy can be formulated, more detailed analysis should be undertaken to confirm the observations made herein.

**An internal NIOSH document, entitled *The Plans and Programs of NIOSH*, circulated in 1973, states,

> Under the Occupational Safety and Health Act of 1970 NIOSH is responsible for developing educational programs which will provide an adequate supply of qualified personnel to carry out the purposes of the Act. Knowledgeable people from industry, labor and government cite the inability to provide full and meaningful protection to workers until adequate numbers of trained professionals become available. There has been little evidence to show that problems of

provided by the recent increase in the incidence of industrial injury and disease. As noted in previous chapters, the industrial accident rate alone increased 29% between 1961 and 1970.[1] It is also clear that much disease is of occupational origin and has not been recognized as such. A further indication of insufficient manpower is the fact that only about 6 out of the approximately 10,000 people who graduate from medical school each year go *directly* into occupational medicine, so that some 75% of the 75 available occupational medicine residency positions go unfilled.[2]* One estimate of the total manpower deficit numbers 50,000 occupational health and safety personnel.[4]

There have also been serious imbalances in the allocation of manpower within the occupational health and safety field. For example, it has been noted that the Bureau of Mines of the Department of the Interior employs 700 inspectors to serve 23,000 mine sites,[5] while OSHA has an inspection force of only 600 with which to serve the approximately 5 million workplaces in all other industries.[6] Another serious imbalance exists in the mix of safety versus health personnel. In June 1973 the OSHA enforcement staff included 456 compliance officers, but only 68 industrial hygienists.[7] Similarly, an estimate of enforcement capability in the states yielded the ratio of 8:1 for safety versus health personnel.[8]

A long-range manpower policy is needed because of the long lead time for the training of specialized Ph.D. scientists and physicians. However, a short-range plan is also needed to deal with the many immediate occupational health and safety problems. Expertise can be developed more quickly by master's degree programs or by retraining baccalaureate scientists or engineers on the job. (Occupational health and safety may thus represent an excellent field of retraining for displaced mid-career professionals such as aerospace engineers, physicists, or chemists.) A less sophisticated four-year undergraduate degree program for safety and health professionals takes longer and develops less expertise, but produces a lower-paid professional who may be more useful in certain types of applications.

Finally, coming to grips with the manpower issue is essential if the Occupational Safety and Health Administration and its research and training counterpart NIOSH are to be effective instruments in improving the work environment. Too often, poor

---

recruitment, curriculum development and funding are receiving the attention they deserve.
Estimates of needs vary and, unfortunately, there is a hidden dimension in the manpower shortage. The average age of available trained occupational safety professionals is 50 years and [that] of available trained industrial hygienists is 41 to 42 years. As the needs increase, the shortage will grow by attrition for age unless training resources are improved through grants to universities and colleges for professionals, through grants to community and junior colleges for para-professionals and through short-term technical training courses to upgrade the competency of personnel now employed in occupational safety and health disciplines to supplement their on-the-job training. Short-term training is also applicable for cross-training safety and industrial hygiene personnel so that each has a better understanding of the overall problem.

*The results of a survey by the Industrial Medical Association indicate that 4100 U.S. physicians practice occupational medicine as a primary or secondary specialty, out of an estimated 335,000 physicians (about 1% of the total).[3]

quality manpower or lack of sufficient manpower can turn a good idea or program into a poor one.* OSHA inspectors who are insufficiently trained and are faced with an overload of inspections to carry out, can produce an enforcement program that is both ineffective and marked by misplaced emphasis. This imbalance can serve as a disincentive for employers to comply with important standards and regulations which are unlikely to be enforced, while forcing compliance with less important ones. The lack of experienced safety and health professionals and scientists (with broad technical, medical, environmental, and statistical backgrounds) can prevent the effective delivery of occupational health and safety services and the collection of sound research data. As former Surgeon General Jesse Steinfeld has noted, "Any large-scale, technologically based activity will succeed only to the extent that qualified people are available."[9]

9.1.2 Major Problems in Manpower Development

A consistent set of policies for developing a sufficient supply of occupational health and safety manpower in each of the relevant professional areas can be derived only after an understanding has been reached of the present manpower stock and its distribution and anticipated future changes in demand. As of this writing, only fragmentary information on the existing stock exists, and only one tentative projection of anticipated manpower requirements has been completed. The following discussion will attempt to describe briefly some of the significant issues that bear on the development of manpower policies to meet the objective needs in the field.

Choosing the Right Kind of Professional

The prospective employer of health and safety professionals must both know what kind of services he requires and be able to place an incremental value on those services in order to make a rational determination of the kind and quantity of professionals to hire. Since the going rate (market wage) for some professionals is determined primarily in areas other than occupational health and safety, the industrial employer often has to take the cost of those professional services as given. For example, the rate of pay of occupational physicians, though often less than that of physicians in general practice, is essentially determined by the overall market wage for physicians. (It should be noted, however, that physicians employed in corpora-

*Indeed, the recent report by the General Accounting Office (B-163375, September 28, 1973) states that

> NIOSH has channeled a portion of its monetary resources into research contracts and agreements with private organizations and other governmental agencies. At the end of fiscal year 1972, however, it had not expended $800,000 of its total appropriation for that year and it had not expended an estimated $600,000 at the end of fiscal year 1973, despite the magnitude of its problems. NIOSH officials said these funds could have been directed into contracting but were not because the lack of personnel prevented any further development and monitoring of contracts.

tions receive other important benefits such as job security, pensions, stock options, etc.) Wages for safety engineers, on the other hand, are determined primarily in the relatively narrow market for industrial safety professionals, which includes specialists in product, public, and job-related safety.

If an employer is not clear about what services various professionals provide, or what services his firm needs, or if there is considerable overlap in the type of services provided, the professional with the lowest market price is often the choice. On the other hand, an understanding of both these factors may not result in the "best" professional being hired, if the perceived incremental value to the organization is below the market-determined wage for that professional. Thus, in cases where the services of an occupational physician are required, a nurse may be hired instead. Finally, the decision to hire a particular kind of professional may be a result of the employer's experience during an OSHA inspection. If safety violations are cited while health hazards are ignored, an employer will be inclined to hire a safety professional rather than a health professional, even if he knows he has more problems with health hazards. New programs designed to train B.A.-level professionals in *both* industrial hygiene and safety engineering may eventually make this "choice" unnecessary.

The problem of hiring safety and health manpower can be even more complicated for small firms, which often do not have adequate resources or an easy enough access to existing information to comply fully with the Act, let alone hire full-time occupational health and safety professionals or technicians as part of the regular staff. Moreover, such firms are often unable to take advantage of health and environmental services from insurance companies, since it is the small insurer—without safety engineers and industrial hygiene professionals on its staffs—who generally insures small firms.[10]

The Time Horizon for Training

As has already been mentioned, there exists a delay between the market's recognition of a need for professional health and safety manpower and the entry of these professionals into the labor market. The training of physicians, nurses, and Ph.D. scientists often requires at least four years, as does that of baccalaureate safety engineers. Planning ahead four years is inherently difficult, and the problem is compounded by philosophical and budgetary changes that result from the four-year political cycle in the executive branch of the government and the two-year cycle in the Congress.

A more quickly available supply of manpower can be generated by introducing and expanding master's degree and associate degree programs, and by retraining chemists or engineers on the job. However, except for the relatively few industrial hygienists, this level of manpower has so far been heavily safety-oriented. Even trained chemists are mostly concerned with fire and explosion hazards and receive little training in toxicology, although they would be ideal candidates for this type of training. The greater availability of safety-oriented manpower in the shorter term has (inadvertently) contributed to the underemphasis on health problems in the workplace.

One further important complication to the development of a consistent manpower policy is the fact that occupational health and safety problems are "moving targets." Rapidly changing technologies and the introduction of new health hazards into the workplace are constantly shifting the needs to be met by the application of expertise. To deal with this situation, professionals must be trained broadly as well as in depth.

Salaries and Status

Occupational health and safety professionals have been the "poor cousins" of the professionals in both the safety field and the health field. Product, highway, and public safety, and surgery and internal medicine, are more prestigious and more remunerative alternatives. With the demand for occupational health and safety manpower on the increase, it appears that qualified professionals will be able to demand higher salaries. While this may be true of industrial hygienists, nurses, and certain safety professionals, the issues of status and salary remain a problem for occupational physicians. Although the salary and status of an occupational physician is high relative to other occupational health and safety professionals, it is not high relative to other medical specialists, for reasons to be discussed later. This fact has, in part, served as a disincentive for medical students to enter the field of occupational medicine.

Another problem of status is found in the job classified as "safety engineer." This category has been characterized by a wide variety of competency levels, ranging from people with minimal technical training who deal with simple injury control (e.g., installing machine guards, using manuals and instruments) to professionals with master's or doctoral degrees who conduct comprehensive control surveys, hazard analyses, and biomechanical studies and operate large-scale programs, usually for larger corporations.[11] Both the public's perception and the small businessman's view of the safety engineer is often that of a limited or older employee operating in a dead-end job. Safety engineers, as a group, have not been highly regarded or highly remunerated. The OSHAct may change this.

Advancement and Evaluation

As a result of the low status generally given to health and safety professionals, especially in smaller firms, neither job mobility "up the management ladder" nor expansion of the job itself into broader or new areas is encouraged. The limited advancement opportunities, in turn, discourage new entries into the field. In the past, therefore, such jobs have often been filled by less capable people who have proven unsuccessful in fields that do provide advancement opportunities. (Plant physicians and resident safety engineers serving smaller companies are often examples, though significant exceptions do, of course, exist.)

The position of nurse within a business firm is also limited from an advancement point of view. However, the chance of advancement outside the firm is often no better; thus, it is usually only the narrowness of the job itself that will discourage the more imaginative nurse from seeking industrial employment.

According to a 1972 British parliamentary report on job safety and health, "The promotion of safety and health is not only a function of good management, but it is, or ought to be, a normal management function–just as production or marketing is a normal function."[12] Awareness and positive interest on the part of top management could promote a sense of importance at all levels of the company, thereby improving both the status of occupational health and safety staff professionals and increasing their desire and potential for advancement.

The report continues:

> It is generally accepted that the primary operational responsibility for ensuring safe working must rest with line management, and here there are two key levels that require particular attention. At the top, at board level, direct responsibility for the general oversight of safety and health matters within the firm should be included in the duties of one of the directors, in the same way that a director may be allocated overall responsibility for production or marketing or exports. In our investigations, we formed the impression that undivided line management responsibility for safety and health matters more often than not stops at some point in the middle-management chain; further up the chain the responsibility tends to become diffused and uncertain. Safety and health should be treated like any other major management function, with a clear line of responsibility and command running up to an accountable individual at the very top.[13]

Problems of advancement within the firm are paralleled by a lack of employer encouragement for the health or safety professional to advance within or beyond his area of specialization. Continuing education, attendance at professional meetings, and other professional activities are neither routinely encouraged nor paid for by management. This is done partly because management's perception of occupational health and safety problems is limited and partly because the value of many of the services performed by such professionals are not included in the balance sheet of the firm and are thus, to some extent, "invisible."

Evaluation of the professional within the firm is limited both by problems of perception and by the ever-present conflict between better working conditions and lower costs (i.e., the better the health or safety professional performs his job, the more costly it may be to the employer). Frequently, then, the professional is instructed not to do "any more than is necessary," i.e., to improve safety and health as long as it doesn't cost the firm more than would the alternatives of increased compensation premiums, absenteeism, or worker turnover.

It has been suggested that the image of the occupational health nurse "as primarily a first-aider" or medical record keeper, has no place in today's industrial health arena.[14] Yet the occupational health nurse is often expected to do no more than that, to refrain from utilizing her abilities in the broader environmental and policy-making area.[15]

Differences in perception and evaluation criteria between an employer and his health and safety staff raise serious questions concerning the objectivity of staff decisions and possible conflicts of professional ethics. The existence and magnitude of this sort of conflict between the industrial physician and his employer is a subject of great controversy (see the discussion of "industrial captives" in Section 9.6).

In sum, then, both the kind and level of evaluation regarding the job being performed has often served to deter potential occupational health and safety professionals from choosing a particular position, from working toward advancement, or from entering and remaining in the profession.

Who Should Pay for Training?

In line with the Nixon administration's "New Federalism," a movement was started to return responsibility for occupational health and safety to the states, including the development of enforcement manpower. This contraction of federal involvement in the enforcement area was paralleled by cuts in occupational health and safety manpower funding in general; the result was an additional burden on the private sector to involve itself in training and to absorb the associated costs (see Section 9.5).

The NIOSH manpower training program (divided into long-term training based on grants to colleges and universities and short-term training, operated by NIOSH itself) originated as a fairly strong program of federal participation in manpower training but has become considerably weakened as a result of substantial and controversial budget cuts. As of July 1973, *all* NIOSH direct funding for the training of occupational health and safety personnel was phased out, including funding grants to universities and colleges and subsidies for NIOSH short-course training, which is now provided on a fee basis.[16] The elimination of free courses—which were oversubscribed by approximately 400%[17]—and stipends offered as a result of direct grants to colleges reflects the present administration's desire not to pay for occupational health and safety manpower training.

Even though the public sector has minimized its level of participation in training, there is no guarantee that labor or industry will take up the cause or the costs. The international unions have strongly opposed both state takeover and federal budgetary cuts for manpower, and are not willing or financially able to assume the burden of training professional manpower.[18]

Since industry has been a primary employer of the existing manpower, it has borne training costs in the past to some degree. While occupational medicine residents must usually pay for their first two years of training (residents in other specialties *earn* between $5000 and $10,000 a year during their residency training), they usually receive salaried third-year clinical training with a company. Only rarely do they receive outright stipends for training in an academic setting or hospital.[19] Many companies have combined on-the-job training with supplemental coursework. Whether or not industry will increase its training expenditures to take up the slack depends on the benefit to be anticipated. This will require a better understanding of the professional's optimal role.

Insurance carriers have traditionally been employers and on-the-job trainers of occupational health and safety manpower for the purpose of monitoring their insurees and providing them with consulting services. Whether they extend their efforts will depend, in part, upon changes in the workmen's compensation system, particularly the recognition of occupational disease (see Chapter 8).

The universities have depended primarily on federal grants, and to a much lesser extent on private support, to initiate occupational health and safety programs. If programs do not attract enough students and funding, they are often abandoned because it is financially impossible to continue support, even if the faculty believes the "need" to train a particular type of specialist exists. Unless government provides incentives to develop or continue such programs, universities, and particularly smaller colleges, will usually be unable to go it alone or to experiment with new programs. John Grimaldi, Director of the Center for Safety at New York University, has expressed concern over the future of occupational health and safety training: "With the reduction this year [1973] in NIOSH training budgets and the planned phase-out of training support, the opportunity for developing competent faculties and constantly improving curricula just about fades completely."[20]

The costs (and thus financial risks) for short-term training (including on-the-job training)—by professional societies, trade associations, universities, and labor unions, as well as by government—are considerably lower than the costs for associate, baccalaureate, and advanced degree training in terms of both absolute costs and immediate benefits from participants' increased knowledge. As with most education, the firm or institution that pays for the training is not easily guaranteed the capture of the product of that education. Thus, long-term professional training has traditionally been supported or supplemented by government aid. The present lack of governmental support leaves the development of occupational health and safety manpower highly uncertain.

The Commitment and Responsibility of the Various Sectors

The level of commitment on the part of federal and state governments, industry, labor, and the universities to the development of occupational health and safety manpower must be assessed if we are to determine the potential role each has to play. As discussed in previous chapters, the states did an inadequate job in protecting the health and safety of workers partly because their enforcement manpower was poorly trained, even though they numbered three times the present federal enforcement staff. The federal government is now endeavoring to limit the OSHA inspection force, in anticipation of the return of enforcement functions to the states. It has similarly restricted expansion of activity in NIOSH, although the states are not expected to provide comparable research or manpower training activities.

Representatives of industry, including the U.S. Chamber of Commerce and the National Association of Manufacturers, fought hard against the passage of the early occupational health and safety bills. Now that the OSHAct is law, industry may hire more staff physicians or industrial hygiene professionals for a variety of reasons, including a desire to pay lower workmen's compensation premiums, to reduce employee absenteeism and turnover, or to increase productivity directly.

Unions have recently been upgrading the priority of health and safety issues in collective bargaining, particularly where the economic climate does not permit demands for substantial wage increases. Such issues might include management paying for an independent health and safety consultant or worker involvement on health and safety committees.

Although medical schools have pursued research applicable to occupational disease, nursing and medical schools have been uninterested or unsuccessful in integrating occupational health into their curricula, and the cries of occupational health practitioners for a change in emphasis have so far gone unheard.

Thus, it appears that in each sector there are conflicting economic or political motivations which serve to mitigate the process of occupational health and safety manpower development. The future level of demand for manpower (as opposed to equipment or materials) is, therefore, quite uncertain.

The Self-Help/Professional Mix

The OSHAct mandates not only the development of professional occupational health and safety manpower, but also the training of employers and employees about the requirements of the Act. Although NIOSH and OSHA sponsor short courses for employers, as do trade associations and professional groups, the level of expertise gained from several weeks of training is not sufficient to monitor or control all the potential hazards of the workplace or to maintain employee health programs. The extent to which an employer relies on professionals once he has received OSHA or NIOSH training and information depends in part on (1) the extent to which he recognizes hazards in his establishment, and (2) what he perceives his needs for professional advice to be. His insight in both these areas is enhanced by education, but short-term training cannot replace the specialized training of a professional. The amount of time an employer can give to occupational health and safety is, by the nature of his diversified responsibilities, far less than what is necessary.

The size of a company also helps to determine the professional/self-help mix. A small employer, who is less able to afford the cost of hiring a professional, will be more inclined to improvise his own methods for compliance, hazard surveillance, and medical care than a middle-sized or large firm, even if he knows that these methods will not be the "best" or most cost-effective ones available.

Although the training of workers may not be thought of as manpower development in the strict sense of the term, the limited access of labor to professional services makes the development of employee self-help training programs and mechanisms very important. The OSHAct mandates the Department of Labor to educate employees as well as employers, but OSHA has done very little in this regard. The needs of both the small employer and his employees will be largely met through the self-help mechanism, if at all.

The Specialist/Generalist Mix

When new demands for manpower reflect steadily expanding problems, a determination of the scope of a training needed by that manpower becomes complicated. This occurs with regard to (1) degree of specialization in a particular profession and (2) breadth of skills and responsibility and overlap with similar and cooperating professions.

For example, at what point do the skills of a technician or a nurse become insufficient? And to what extent is it feasible, or desirable, to have a plant physician filling the role of an industrial hygienist? Determination of the duties a professional should be able to perform outside his clearly defined domain depends on the employer's perception of the job and the need for manpower and the professional's background and interest in expanding the scope of his responsibilities. Manpower development must therefore incorporate a determination of the proper balance between depth and breadth of training within each occupational health and safety manpower category; and this problem is dependent, in turn, on the collection of occupational health and safety professionals to be used in a particular setting—e.g., a physician and a safety engineer or an industrial hygienist and a nurse.

The Discrepancy between Perceived Requirements and Market Demands

Finally, since the market demand for occupational health and safety manpower does not adequately reflect the actual problems and their seriousness, *there is a discrepancy between what manpower is, in fact, needed to solve the health and safety problems and what the market demand implies.* Herein lies the most serious problem in manpower development.

9.1.3 New Demands Created by the OSHAct

In Section 2(b) of the OSHAct, Congress stated its intention to assure safe and healthful working conditions by the following means, among others:

1. [by stimulating] employers and employees to institute new and to perfect existing programs for providing safe and healthful working conditions;
3. by authorizing the Secretary of Labor to set mandatory occupational safety and health standards . . . ;
5. by providing for research in the field of occupational safety and health, . . . and by developing innovative methods, techniques, and approaches for dealing with occupational safety and health problems;
6. by exploring ways to discover latent diseases . . . ;
7. by providing medical criteria which will assure insofar as practicable that no employee will suffer diminished health, functional capacity, or life expectancy as a result of his work experience;
8. by providing for training programs to increase the number and competence of personnel engaged in the field of occupational safety and health;
9. by providing for the development and promulgation of occupational safety and health standards;
10. by providing an effective enforcement program. . . .

These broad directives have had a significant impact on manpower development programs of government, industry, insurance carriers, and labor (see Section 9.5).

New Demands on Government

Under Section 8(a) of the OSHAct, the federal government is authorized to carry out inspections and investigations of places of employment. The states already have enforcement staffs for occupational health and safety inspections, but the federal government has had to develop and train its own enforcement staff since the OSHAct.

Sections 6(a), 6(b), and 6(c) provide that the Department of Labor will promulgate occupational safety and health standards for chemical substances, physical agents, and other health- and safety-related conditions. Moreover, "where appropriate, any such standard shall prescribe the type and frequency of medical examinations or other tests which shall be made available, by the employer or at his cost, to employees exposed to such hazards in order to determine whether the health of such employees is adversely affected by such exposure."

Section 20(a) directs the Secretary of Health, Education, and Welfare to conduct—directly or by grants or contracts—research, demonstrations, and experiments to develop criteria dealing with toxic substances, to describe safe exposure levels, and "to establish such programs of medical examinations and tests as may be necessary for determining the incidence of occupational illnesses and the susceptibility of employees to such illnesses." Medical examinations required for both research purposes and health maintenance under newly promulgated standards will probably create a new demand for industrial nurses, physicians, and medical technicians (e.g., the asbestos standard requires employers to provide periodic lung function tests for employees).

A specific mandate for guaranteeing sufficient manpower to implement the purposes of the Act is given in Section 21(a) and (b):

Sec. 21. (a) The Secretary of Health, Education, and Welfare, after consultation with the Secretary and with other appropriate Federal departments and agencies, shall conduct, directly or by grants or contracts (1) education programs to provide an adequate supply of qualified personnel to carry out the purposes of this Act, and (2) informational programs on the importance of and proper use of adequate safety and health equipment.

(b) The Secretary is also authorized to conduct, directly or by grants or contracts, short-term training of personnel engaged in work related to his responsibilities under this Act.

Thus, to carry out the Act, the federal government must involve itself in standard setting by technical experts, in training and hiring inspectors and other occupational health and safety personnel on a short-term basis, in research of a scientific-medical nature with involvement of the medical-scientific community, and in training other personnel involved in implementing the Act, such as physicians and nurses to perform medical exams and tests under new standards promulgated by the Act.

New Demands on Industry

As a minimum, industry must comply with the federal standards and inspections of the workplace. According to Section 5(a), the employer—

1. shall furnish to each of his employees employment and a place of employment which are free from recognized hazards that are causing or are likely to cause death or serious physical harm to his employees;
2. shall comply with occupational safety and health standards promulgated under this Act.

The employer is thus responsible for compliance with the OSHA record-keeping system (which may include medical exams and tests), specific standards, and the general duty clause, Section 5(a)(1). Furthermore, Section 2(b)(1) recommends that employers and employees be encouraged to set up voluntary health and safety programs. The need for manpower to consult with employers regarding their compliance efforts and health maintenance programs is readily apparent.

New Demands on Insurance Carriers

The more research that is conducted on the connections between various diseases and the work environment, the greater the number of occupational diseases that will be compensable under workmen's compensation. The resulting increase in insurance premiums may cause employers to reassess whether the payment of such premiums is the most efficient mechanism for solving their job health and safety problems. The insurance carriers will have to compete by suggesting programs to help employers reduce illness and injury rates and thus premiums (e.g., offering a reduced premium rate for more comprehensive or frequent medical exams that are required by OSHA, or for a professionally administered health and safety program). Larger carriers may increase their staff of industrial hygienists to consult with clients and inspect their workplaces.

New Demands on Unions

The OSHAct does not make specific requirements upon employees other than to state in Section 5(b) that "Each employee shall comply with occupational safety and health standards and all rules, regulations, and orders issued pursuant to this Act which are applicable to his own actions and conduct." Yet labor wants to be part of the action of implementation. The unions, as primary movers in lobbying for passage of health and safety legislation, have voiced the desire to see the Act used as a viable lever to push industry to action. While not themselves prepared to hire large numbers of professionals, many unions have indicated that they want increased manpower to implement the Act. For example, the Industrial Union Department (IUD) of the AFL-CIO has demanded that OSHA hire more inspectors from union ranks.[21] The Oil, Chemical, and Atomic Workers International Union won 1974 contract clauses with all major oil companies (except Shell and Chevron) providing for periodic plant inspections by independent industrial hygienists and company-paid medical exams.[22] Other unions are negotiating joint labor-management health and safety committees at the plant, local, and international levels. The IUD is calling for more OSHA inspectors, upgrading of NIOSH salary levels to the level of those in OSHA, and more research expertise.[23]

Demands for Manpower to Develop New Technology

The need for engineering controls, such as closed systems for chemical and other types of processes, and equipment for on-site environmental and biological monitoring will require the development of engineering and systems design manpower. The redesign of jobs may be required for the design of new technology, taking into consideration ergonomic factors and organization-of-work issues.

9.1.4 The Role of Government in Manpower Development

Because of all the problems inherent in manpower development and the new demands created by the OSHAct, there is a need for a reexamination of the role of government in occupational health and safety manpower development. As mentioned earlier, the occupational health and safety manpower market is characterized by high risk, long time lags, and sizable capital requirements for even the smallest development program. If a large increase in the manpower stock is perceived as being necessary within a short time period, the market may not be able to meet the need without government assistance. The government is in a position to make both long-term and short-term investment, thereby helping to generate a sufficient stock of manpower. It can facilitate increases in the supply of manpower by establishing a system of incentives to the private sector such as training grants, fellowships, and low-interest loans which encourage students to enter the field and universities to develop educational programs.

A second role which government plays in developing manpower is that of stimulating demand by legislating new programs which require personnel for their implementation (see also Section 9.5). As noted in Chapter 5, the policy of transferring responsibility for occupational health and safety programs from the federal government to the states has further exacerbated the problem by first creating a demand for manpower and then cutting it off, leaving each state to decide its own manpower needs.

9.2 THE EXISTING STOCK OF OCCUPATIONAL HEALTH AND SAFETY MANPOWER

This section will attempt to describe the level, distribution, and characteristics of the existing stock of manpower in each of four major categories: health and environmental services, enforcement, research, and teaching.

9.2.1 Health and Environmental Services

Occupational Physicians

Estimates of the number of U.S. physicians practicing occupational medicine range from 4000 to 10,000, with an average estimate of 8000.[24] The wide variation in the estimates is caused by (1) definitional problems regarding the different settings in which the practice of occupational medicine occurs, and (2) the amount of involvement in occupational medicine necessary in order for it to be defined as a physician's specialty.

A December 1973 survey, sponsored by the Industrial Medical Association, of the distribution of U.S. physicians in all AMA-defined specialties estimates that 4100 physicians practice occupational medicine in a private office or a private clinic, or in teaching, administration, or research.[25] On the basis of an earlier report by the American Medical Association, it is estimated that 2400 physicians are employed in industry and nonprofit organizations and are members of IMA, and of these, 2000 are employed full-time by industry (this estimate does not include 1800 non-IMA members, 1200 IMA members who are not affiliated with a company or institution, and part-time or consultant physicians).[26] Only 450 of the total are certified by the American Board of Preventive Medicine.[27]

A survey of 858 companies conducted in mid-1972 by the Conference Board revealed the following information: "To the extent that it is reflected in changes in expenditures and staffing, industry's commitment to employee health care does not seem to have changed appreciably in the past decade." The report illustrates this static situation with the following data: the ratio of full-time physicians to total employees declined between 1961 and 1971 from 1 in 7600 to 1 in 8800; of part-time physicians, from 1 in 2300 to 1 in 2630; and of nurses (85% of whom are full-time), from 1 in 1200 to 1 in 1300.[28]

The survey estimates that approximately 3200 physicians are employed full-time by companies with 500 or more employees.[29]* Both this estimate and the AMA estimate agree, then, that less than 1% of all U.S. physicians are employed full-time by industry.

The Conference Board study also estimates that 10,600 physicians are employed part-time by companies.[30] However, a definition of "part-time" employment is not clearly made, so that this estimate cannot be easily assessed.

Based on the AMA estimates of the number of physicians involved in occupational medicine, about 50% practice directly in an industrial setting. Estimates of the distribution of the remaining 50% are often unreliable as to whether they practice privately or in clinics, as general practitioners with some emphasis on job-related health problems or as occupational physicians per se. In a survey of 66% of the IMA members, all were employed in 71% of the 200 largest industrial corporations in the United States.[31] The Conference Board survey also found a significant correlation between employment of physicians and company size. They found that 81% of the largest companies (with over 50,000 employees) had at least one full-time physician. Among the smaller companies in their survey (under 1000 employees), 3% employ at least one full-time physician. The ratio of part-time to total physician employment was found to increase as company size decreases; however, in the smallest firms, only 13% employed a physician part-time.[32] Thus, the large industrial companies utilize the majority of occupational physicians, while smaller businesses either hire part-time consultants, utilize nurses or clinics, or use no occupational medical services at all.

The Conference Board study indicated that the employment of physicians in industry varies with several other company characteristics:

1. There is a high incidence of physicians in manufacturing and financial companies.
2. There is a strong relationship between high company earnings and the employment of physicians.
3. Companies located in suburbs of cities with a population over 100,000 have a greater incidence of physicians.
4. There is an inverse relationship between the presence of company physicians and the number of females employed in the company.[33]

*The survey did not include companies with less than 500 employees because a preliminary study indicated that physicians were present in less than 5% of companies with 500-999 employees, and that data on even smaller firms would not affect their estimates.

The mere presence of a company physician is not enough to ensure his accessibility to employees or to provide a medical program that is not bogged down in administrative red tape or burdened with an excess of decision-making channels. The position of the physician or medical director in the organizational structure is, therefore, often a key factor to the success of an in-plant health-care program. The Conference Board survey addressed this issue, and the following data resulted:

> The company physician, or the head of the medical department in multi-physician companies, is typically titled "medical director." Only 11% of medical directors report at a top corporate level—either to the president, chairman, or chief executive. At least 75% report to an executive having personnel, employee relations, or corporate relations responsibilities. Six percent report to a "senior vice president," and one or two percent each say that they report to the executive vice president, the vice president for public affairs, the director of training and education, the vice president for "selection" or the general manager.
>
> To the degree that there has been change in the past decade in the medical department's place in the corporate organization it has been toward consolidation with the personnel function rather than as an independent function reporting to the chief executive.[34]

Thus, the company physician is primarily used for activities closely related to personnel issues—preemployment examinations, promotion and transfer decisions, workmen's compensation, and excessive absenteeism. His potential value in directly improving working conditions through the elimination of hazards and the provision of preventive care to workers is greatly underemphasized.

In 1966 it was estimated that there were about 25 privately operated occupational health clinics providing services for approximately one million employees in over 9000 firms, though the number of physicians in these clinics was not reported.[35] Although the number of such clinics may have increased, they still provide coverage for less than 5% of the working population.

It is estimated that only twelve medical schools include occupational health in their curricula.[36] According to Dr. Henry Howe, of the Department of Occupational Health of the AMA, there are seven educational institutions with AMA-approved two-year residency programs in occupational medicine.[37] An additional 25 residency programs are conducted by industrial firms.[38] Between 25 and 50 physicians complete training in occupational medicine each year, barely sufficient to staff a dozen corporation medical programs.[39]

Occupational medicine has traditionally been held in very low esteem by the medical establishment, partly because it brings no money into medical schools or hospitals, in the way that surgery does, and partly because it has a prevention rather than a treatment orientation. Because of this orientation, the subject, when it is included in the curriculum, is clearly separated from the major clinical courses on which attention is focused.[40] Thus, it is not surprising that medical students receive little encouragement to enter occupational medicine from medical school boards and professors.

Just as labor has historically lacked sources of occupational health and safety information, it has also lacked occupational medicine manpower. Of seven large unions studied, only one, the United Mine Workers, employs a full-time physician. Only a few other large unions have even an industrial hygienist or a safety professional on their staff. While industry has sponsored the majority of medical residency programs in occupational medicine, there are no programs providing opportunities for doctors to work with unions. The Mount Sinai School for Medicine in New York has such a program in the planning stages, to include clinical medicine, courses in epidemiology, biostatistics, and occupational, environmental, and public health, work at a large union's headquarters, and occupational health research.[41]

Occupational Health Nurses

According to Jane Lee, of the Medical Services Branch, NIOSH, and others, there are 20,000 occupational health nurses. This figure excludes over 1000 practical nurses who are hired by management to provide health services at the workplace.[42] According to Jeremiah Lynch of NIOSH, approximately 75 certified occupational nurses and 200 industrial nurses enter the field each year.[43] A comprehensive survey of over 10,000 occupational nurses conducted in 1964 provided the following data:[44]

- One-third had worked in occupational health nursing for 15 or more years. Few had entered the field in the last five years (p. xii).
- The majority of nurses worked in manufacturing industries (as opposed to nonmanufacturing or government) and in workplaces with 500 or more employees (p. xii).
- Only 16.8% of the occupational nurses had taken a college course in occupational nursing (p. 43).
- Less than 10% had college degrees (p. 53).
- Salaries were higher in larger workplaces (p. 55) and government (p. 73).
- Occupational health nurses tend to be older than other nurses, with a median age of 48 in contrast to 36 for other nurses. Fifty percent of the occupational nurses who act as supervisors are over 50 (p. 78).
- Forty percent of the nurses worked alone. Only 25% had full-time medical direction (p. xii).

The 1972 Conference Board survey of 858 companies found that 24% of companies with 500 or more employees have occupational nurses only. Nurses working alone were not found in companies with 50,000 or more employees, while 34% of such nurses were found in firms with 1000-5000 employees. The survey did not show a strong relationship to urban versus suburban locations, as the data on physicians did, but it indicated that nurses working without physicians are found in more rural areas about 50% of the time.[45]

The role of the occupational health nurse has been as varied as the employer for whom she works. Audrey Brenny, Assistant Professor of Nursing at the University of Wisconsin, in a speech before the 1973 American Public Health Association Annual Meeting, expressed the belief that "the image of the occupational health nurse as primarily a first-aider has no place in today's occupational health scene."[46] According to Jane Lee, the traditional role of the occupational health nurse lies in the provision of

> care for illness and injuries occurring at work, in prompt referral for additional and specialized care, in preventive and rehabilitative services, in counseling and providing health and safety education, in maintaining an occupational health record system, in assisting with physical examinations and health screening tests and measurements.[47]

Yet the occupational nurse is struggling with her role and responsibilities as a professional practitioner. As a result of the OSHAct, the industrial nurse is taking on new functions as new requirements for medical exams, monitoring, histories, and records become effective. According to Audrey Brenny, nurses (unlike physicians) are unable to receive any form of insurance payment (except for private duty nursing) and are thus discouraged from forming private or group practices. However, a few occupational nurses have succeeded in forming group practices which offer services to small and medium-sized firms on a part-time contractual basis.[48]

Despite the increasing responsibilities of the occupational health nurse, training in occupational health for nurses has remained virtually unchanged. Nursing schools, like medical schools, have retained a treatment rather than a prevention orientation, with occupational health, when recognized, relegated to the public health curriculum. Professional nursing organizations have also tended to isolate occupational health nurses from the nursing mainstream. Neither industry, labor, nor the insurance carriers have encouraged formal occupational health training for nurses either through stipend or internship programs. As a result, most occupational health nurses enter the field from other nursing areas,[49] and their training usually results from on-the-job experience related to employer requirements.

Mary Louise Brown, Chief of the Health Maintenance Branch, NIOSH, wrote in 1966 that "because little comprehensive job-oriented training at the in-service level has been available for occupational health nurses," short courses and conferences sponsored by nursing organizations and state or federal government have been the primary mechanism for the occupational health education of nurses.[50] Several universities, including the University of North Carolina and the University of Wisconsin, have established master's degree programs in either occupational health nursing or community health with occupational health content.[51] In 1970 the University College of Nursing published a manual for NIOSH, entitled *Occupational Health Content in Baccalaureate Nursing Education*, aimed at developing and expanding occupational health curricula.[52]

Although the 1964 report *Occupational Health Nurses: An Initial Survey* does not contain the most recent data, it is the most comprehensive study done to date.

Since 1964 the number of occupational health nurses has increased by 3000. However, it does not appear likely that the characteristics described in 1964 have changed very much, since the level and type of training and the rate of entry into the field have remained fairly static.

Industrial Hygienists

According to *Health Resource Statistics*, published by HEW in 1971, industrial hygiene activities are defined as "the recognition, evaluation, and control of those environmental factors which have an adverse effect on health and efficiency of workers in places of employment or among citizens of the community."[53] Industrial hygienists are concerned with assessing and controlling environmental hazards and toxic substances such as dusts, gases, vapors, and fumes; physical agents such as excessive noise, heat, and radiation; biological hazards such as enzymes; and other job-related stresses such as monotony.

The HEW study estimates that there were 2500 industrial hygiene personnel in 1971.[54] This figure includes not only industrial hygienists per se, but also technicians and other similar persons performing industrial hygiene activities. The American Industrial Hygiene Association has expressed concern that the title "industrial hygienist" has been defined too loosely to include personnel not meeting AIHA membership requirements, particularly technicians and enforcement personnel with shorter-term training.[55] An estimate of industrial hygienists according to AIHA membership criteria would be 1600-1800.[56] However, this estimate does not include all of the 1200 members of the American Conference of Governmental Industrial Hygienists, whose qualifications are similar to those of members of AIHA but who are employed in the government sector. Dr. Henry Howe of the AMA estimates that there are less than 2000 industrial hygienists in the United States.[57]

The American Academy of Industrial Hygienists awards certification to personnel who meet educational and experience requirements and who pass an oral and written examination. By 1971 the AAIH had certified about 600 persons.[58] No attempts were made to determine the overlap between certification and membership in AIHA or ACGIH. It is estimated that 75 industrial hygienists are produced each year, as a result of M.S. or M.P.H. programs, short-course training, and internal conversion of professionals such as chemists to industrial hygienists.[59]

One estimate of the distribution of industrial hygienists indicates that they have generally been employed by government (about one-third, mostly state) and industry (about two-thirds), with a substantially lower number hired by the insurance companies and labor unions.[60] Data collected by the University of Michigan's School of Public Health in a 1973 survey of their industrial health alumni revealed a slightly different employment picture: one-third of the total of 117 industrial hygiene graduates were employed in industry and insurance companies, one-third in government (equally mixed between state and federal), and the remainder were scattered among universities, armed services, and foreign countries.[61]

As of June 1973 OSHA employed 68 industrial hygienists, thinly spread across the country, or less than 5% of the total estimated number.[62] One major insurance company, Liberty Mutual, employs 18 industrial hygienists (8 of whom are certified).[63] The international unions together probably employ less than that number.

A determination of the distribution of industrial hygienists among firms of different sizes was made by the Conference Board survey. They found a general correlation between company size and employment of an industrial hygiene specialist: "Roughly three-fourths of the companies with 10,000 or more have an industrial hygiene specialist, but the ratio drops sharply to 31% in the 5000 to 9999 employee category and then ranges around 25% in the smaller ones."[64] It might be inferred, then, that if the smaller firm hires any professional, it is most often a safety engineer.

According to the study, "17% of the industrial hygiene specialists report to the company medical director. More typically, they report to an executive charged with 'safety' functions or to a personnel, corporate, or employee relations executive."[65]

Training in industrial hygiene has traditionally been at the master's degree and the Ph.D. level (in preparation for research and teaching). With support from NIOSH training grants during fiscal 1972, the University of North Carolina School of Public Health, the University of Cincinnati's Kettering Laboratories, the University of Michigan, and Harvard University either set up or expanded their M.S. or M.P.H. programs in industrial hygiene.[66] The future of these programs has been brought into doubt by the phaseout of NIOSH training grants begun in fiscal 1973, since industrial hygiene students have in the past largely been sponsored by federal agencies or other governmental units.[67]

Safety Professionals

As mentioned earlier, the level of competency of safety professionals has traditionally varied greatly, spanning technicians who perform simple controls and Ph.D.'s who develop complex systems and programs for the analysis and control of industrial hazards.

The safety professional is concerned not only with the prevention and control of trauma arising from accidents, but also with hazard control systems based on environmental and human factor analysis. According to the American Society of Safety Engineers (ASSE), the major functions of the safety professional are contained within four basic areas:[68]

1. Identification and appraisal of accident- and loss-producing conditions and practices and evaluation of the severity of the accident problem.
2. Development of accident prevention and loss control methods, procedures, and programs.
3. Communication of accident and loss control information to those directly involved.

4. Measurement and evaluation of the effectiveness of the accident and loss control system and the modifications needed to achieve optimum results.

The number of safety personnel can be estimated by using membership figures of the ASSE since it is the only professional society in the field. The ASSE is composed of approximately 11,000 safety professionals from the government, industry, and insurance sectors of the economy.[69] This indicates that there exists roughly a 5 to 1 ratio of safety personnel to industrial hygienists in the United States. A 1973 NIOSH study estimated that approximately 100 certified safety professionals enter the field each year.[70] The study further estimated an annual production of 200 occupational "safety and health specialists," which would include uncertified or nonaccredited safety professionals and similar industrial hygiene personnel as well as those jointly trained.

According to Wayne Christiansen, Managing Director of the ASSE, approximately 50% of all "safety engineers" do not have college degrees.[71] In an effort to distinguish safety-related technicians and paraprofessionals from professionals with baccalaureate and advanced degrees, the ASSE, like the AIHA, has attempted to upgrade the level of competency in the profession and to promote use of the term "safety professional" to distinguish the more highly qualified person. The ASSE has issued qualifications guidelines for OSHA and state program personnel,[72] which it has unsuccessfully attempted to have OSHA adopt, and has established a national accreditation service for colleges and universities offering a baccalaureate or higher degree.[73]

Since the passage of the OSHAct, there has been a trend toward requiring at least baccalaureate training for safety professionals. However, there are few programs to train people at the baccalaureate level. As a result of an initial NIOSH grant, Auburn University and Indiana University of Pennsylvania set up B.S. programs in safety management with some health curricula included.[74] The Illinois Institute of Technology began a bachelor's degree program in industrial safety in 1972;[75] and Texas A&M University has established a degree in industrial engineering with specialization in industrial safety.[76] These programs are designed to prepare first-line safety practitioners. The New York University's Center for Safety, in operation since 1938, trains M.S. and Ph.D. graduate students in the advanced areas of industrial safety and occupational biomechanics.[77]

Technicians and Paraprofessionals

The manpower gap in occupational health and safety will not be closed solely by increasing the number of professionals with baccalaureate and advanced degrees. The use of competent assistants to perform routine tasks under the supervision of a professional has proven an invaluable aid in such fields as dentistry and medicine. There is a shared belief among many occupational health and safety educators that such assistance would enable the occupational health and safety professional to function more efficiently.

Although the terms "technician" and "paraprofessional" are often used interchangeably, NIOSH has established a guideline for defining the levels of capability among occupational health and safety personnel which clarifies the terms. "Technicians" are high-school graduates trained through short courses and related in-service programs in industry and government. "Paraprofessionals" are junior-college graduates with associate degrees in science or occupational health and/or safety. A "professional" is a person with at least a baccalaureate degree in engineering, science, or occupational health and safety.[78]

An estimate of the number and distribution of technicians and paraprofessionals employed in job safety and health is not available from existing data, because these personnel are trained and employed in a variety of unspecialized areas.

Prior to the OSHAct, no programs existed which offered the Associate of Science degree in occupational health and safety. Under two NIOSH contracts, Texas A&M University and the California Community Colleges developed a curriculum guide for possible two-year programs in occupational health and safety. As a result of these studies, seven community colleges now offer two-year "safety technician" programs which utilize local professional resources and adapt curricula to the problems of local industries, and also provide a wider geographical distribution of manpower.[79] The two-year program is complete in itself, but also can provide upward professional mobility by making credits transferable to a senior college B.S. program.

9.2.2 Enforcement Manpower

The enforcement of health and safety standards represents a major portion of the total effort to fulfill the congressional mandate of assuring a safe and healthful workplace. The significance of competent compliance personnel is acknowledged by the 1973 *President's Report on Occupational Safety and Health*:

> The means by which these standards are enforced, and the men and women who enforce them, their responsibilities, equipment, training, and qualifications, have a direct bearing on the success of compliance efforts.[80]

About 60% of occupational health and safety compliance officers are now employed by OSHA, with the remainder employed by the states. As of June 1975 the OSHA field enforcement staff numbered 933 compliance officers.[81] It is estimated that states with approved plans employ 2500 occupational safety and health compliance personnel,[82] with a ratio of one industrial hygienist to eight safety inspectors. The ratio of industrial hygienists to safety inspectors in OSHA is approximately 1:5.[83] However, this ratio does not take into account 163 safety officers and 53 industrial hygienists who are conducting federal inspections for OSHA as a result of the 7(c)(1) agreements between OSHA and many states.[84] (See Section 5.3.4 for a description of these agreements.)

According to the 1973 *President's Report*, all federal compliance trainees are high-school graduates, many have college backgrounds, and some have degrees in engineering.[85] However, a source within the OSHA Training Institute states that the compliance staff receiving training at the Institute is generally of poor quality: they are usually in their twenties or fifties and tend to be "dropouts" from other areas. OSHA also selects applicants who have had practical job experience in occupational health and safety. As a result of pressure from organized labor and the realization that participation on local union health and safety committees—paid or unpaid—provides valuable background, OSHA now claims to be recruiting compliance officers from the ranks of union members.[86]

Generally, training for state and federal enforcement personnel is conducted through short courses. By the summer of 1972, 50% of OSHA Training Institute trainees were state personnel.[87] All new federal compliance officers and industrial hygienists attend a four-week training course at the Occupational Safety and Health Institute in Chicago. In 1973, 17 courses were given for federal and state personnel.[88] Upon completion of Institute training, OSHA compliance officers are assigned to one of 73 OSHA offices, where they receive on-the-job training by assisting experienced compliance officers on inspections before beginning their own inspections.[89]

In an effort to provide professional development for compliance officers and to develop a corps of "generalists" with a wide range of skills for inspection, OSHA developed a three-phase training model in 1972: entrance level, specialty, and development. As part of this program, compliance officers are required to take a minimum of two technical courses at the OSHA Training Institute after their initial four-week training.[90] With the help of a management information system, OSHA says it has been able to identify training deficiencies and develop programs to offset them, thereby individualizing training requirements for each compliance officer.[91]

Despite OSHA's announced intent to develop some degree of technical proficiency and specialization through its training programs, the level of competency and efficiency of inspectors has not been considered very high, particularly in the health area. In a June 1973 preliminary study of statutory qualification requirements, OSHA estimated the following figures on state and federal safety and health compliance officers: Fourteen of the 49 submitted state plans appeared to have qualification requirements for safety inspectors which were *less* stringent than those of OSHA. Thirteen states had safety inspector qualifications which were similar to OSHA's, and 24 states had similar health personnel requirements. Twelve states had safety inspector qualifications more stringent than OSHA's, and fourteen states had more stringent health personnel qualifications.[92] Whether or not the federal government will actually require qualifications of state personnel to meet the "at least as effective as" test is as yet undetermined, despite pressure by the ASSE to upgrade both federal and state safety personnel.

The government sector is not alone in the training and hiring of personnel for inspection services. Large insurance carriers provide a consultation-inspection service for their clients that is often as extensive as the federal program. According to Roger H. Wingate, Vice-President of Liberty Mutual Insurance Company, Liberty Mutual has 600 inspectors in the field of occupational health and safety, of whom 250 are specialists.[93]

The supply of federal enforcement manpower is not currently sufficient to meet the mandate of the OSHAct. It is difficult to believe that a federal program with only one-third as many compliance officers as the total number in the states can run an effective enforcement program even for those 33 states that currently lack approved state programs. Moreover, the low level of individual competence, combined with superficial training of a nonspecialist nature, of both federal and state inspectors further retards the enforcement process.

9.2.3 Research Manpower

Among the established disciplines which have important roles in environmental health research are the biological sciences (including physiology, pathology, microbiology, and molecular biology); chemistry (including analytical, biological, organic, and physical chemistry); engineering (particularly industrial engineering, environmental engineering, and bioengineering); and the medical sciences (including pharmacology, toxicology, epidemiology, oncology, mutagenesis, and occupational medicine). Fields in which a research role is becoming significant include environmental biophysics (to study role of heat, cold, noise, and vibration in the workplace), biomathematics, systems analysis, and the social and behavioral sciences (to study the relationship between human factors and environmental stress).[94] Occupational safety research draws heavily on industrial engineering and bioengineering, and on the social and behavioral sciences.

This section will focus on the problems of occupational health research manpower rather than safety research manpower for two reasons. First, health problems in the workplace about which little is known are emerging at an increasing rate and on an extremely large scale. Second, safety research, while still exploring new areas and problems, has become fairly well established, with the result that the scope of safety problems still in need of investigation is considerably narrower than that of current health problems. The area of safety research that is presently most in need of investigation is that of interactions between health and safety factors. The effects of noise (as a stressor) on visual acuity, synergistic effects between stress and toxic materials, and the effects of vibration on paralysis of the limbs are examples.

Estimates of the number of scientists involved in occupational health and safety research are extremely difficult to evaluate, and no attempt to do so will be made here. One reason for this difficulty lies in the mobility of scientists between different fields for which their skills are applicable. A second reason is that while some estimates for general health research and environmental health research manpower have been made, no breakdown of the occupational manpower has been done, nor has the omission been acknowledged. There seems to be a tacit assumption that occupational health research is not sufficiently different from general health research. Thus, our estimates must be generated from non-occupational-specific figures.

The National Institute of Health has defined scientific and professional health manpower engaged in research to include M.D.'s, Ph.D.'s, and "others with less than doctoral training who functioned as principal investigators and collaborators. In general, it does not include persons with such training who performed as research assistants; it also excludes technicians and all other supporting personnel."[95]

Because scientists involved in medical and non-job-related health research have a closer link to the occupational health field than scientists in other areas, it is useful to describe the supply of this manpower from which occupational health research programs can easily draw. Estimates for 1968 indicate that more than 70,000 scientists (as defined by NIH) were engaged in health-related research. The 1968 figure included 19,300 M.D.'s, dentists, and veterinarians; 35,600 research doctors with Ph.D.'s and Sc.D.'s; and 15,600 persons with master's or baccalaureate degrees.[96] Of the total number of health-related research scientists in 1968, 60% were employed by universities and research institutes, and the others were almost equally divided between government and industry.[97]

In May 1970 testimony before the Senate Subcommittee on Labor regarding the proposed OSHAct, the American Public Health Association presented an undated statistic attributed to Mr. C. C. Johnson of the Environmental Health Service, U.S. Public Health Service, which estimated that there were 1600 *occupational health scientists.*[98] This figure is roughly 2% of the above estimates of health-related research manpower.

The distribution of occupational health research scientists among government, industry, insurance carriers, and universities may be different than the distribution for other health research personnel. This is primarily because the industrial and insurance sectors have a more direct interest in occupational illness than in nonoccupational illness and may therefore be more supportive of such research activities. Within NIOSH, there are less than 600 research scientists (600 is the total staff size), including biochemists, physical and organic chemists, toxicologists, pathologists, physiologists, pharmacologists, biologists, physicians, nurses, physicists, psychologists, epidemiologists, engineers, bioengineers, and industrial hygienists.[99] However, NIOSH has actually lost personnel because (1) budget cuts have caused a reduction in the number of positions and (2) the agency is unable to offer competitive salaries due to a low average GS level of 7.6.[100]

An indication of the significance NIOSH attaches to training scientists with research capability can be seen in the distribution of NIOSH training grants among different occupational health and safety disciplines during fiscal 1972. One-third of the 21 grants to universities funded Ph.D. programs with specific research objectives.[101] These programs were in the fields of environmental health, industrial hygiene, occupational biomechanics, systems safety engineering, postmedical occupational health, and accident prevention. In a 1970 HEW Task Force Report on Research Planning in Environmental Health Sciences, it was estimated that between 1800 and 2000 trainees were being supported in various environmental health programs, and one-half to two-thirds of these were in programs oriented toward research careers.[102]

In sum, the supply of occupational health research manpower seems to vary greatly as old problems are solved and new ones emerge to be investigated. Until a survey of occupational health research is undertaken, a reliable estimate of the current supply will not be possible.

9.2.4 Teaching Manpower

In order to increase the rate at which qualified occupational health and safety personnel can become available, there must be an expansion of the stock of college-level instructors who are well-informed on the technical, management, and broader economic, legal, and human issues.

While occupational safety programs tend to be located in departments of industrial engineering or safety sciences, occupational health programs are most often located in medical schools, schools of public health, or departments of environmental health sciences. Some joint occupational health *and* safety programs, such as the B.S. program at Auburn University and the M.S. program in industrial hygiene at the University of Michigan, have been established in departments of industrial engineering, a trend that might possibly be due to the relatively more stable funding characteristics of engineering schools.[103] Because of this diversity in program sponsorship, and because many academic programs draw faculty from other departments to teach job safety or health courses, it is difficult to estimate either the number of academic programs in environmental and health services or the number of college instructors who teach in the field. For these reasons, I have not attempted to calculate the exact number of programs, curricula, or faculty available in occupational health and safety.

However, estimates have been made which illustrate the extent to which the supply of teaching manpower is deficient. The American Public Health Association estimates that in the 18 APHA-accredited public health schools during the 1969-1970 academic year 40 full-time and 12 part-time faculty and staff taught occupational health subjects, including industrial hygiene, occupational medicine, and nursing.[104] Of the 40 full-time faculty, 13 were physicians, 10 were engineers, and

7 were chemists.[105] In the University of North Carolina School of Public Health's Department of Environmental Sciences and Engineering, there are 8 full-time faculty in the industrial hygiene program.[106] In a report for the Harvard School of Public Health's Panel on Health Services, the Committee on Manpower and Curriculum acknowledged that in undertaking a proposed change in their occupational health curriculum, the shortage of faculty having "the necessary health services orientation" was likely to be a problem.[107] The University of Cincinnati's Department of Environmental Health has at least a 60-member faculty, but from the department's announcement it is not possible to determine the number of faculty involved in any given area.[108]

Because medical schools have traditionally lacked an interest in occupational health, their occupational medicine faculties have often been underemphasized and underfunded. As mentioned above, there exist only 12 medical schools which teach occupational medicine and 8 which provide occupational medicine residency programs. According to a December 1973 Industrial Medical Association survey, only four occupational physicians listed "medical teaching" as a professional activity, as compared with 465 who listed "administration" as a professional activity.[109]

While most Ph.D. graduates have traditionally sought careers in research or academic teaching, many are now going back into industry because of the shortage of specialists in the field and because of recent improvements in industrial salary levels. According to John Grimaldi, Director of New York University's Center for Safety, "Probably more people now interested in the occupational safety and health curricula at the doctorate level plan to go back into industry than plan to go into the academic world."[110]

A new effort to train college professors to teach associate- and baccalaureate-level courses in occupational health and safety began in 1972 with a NIOSH-funded Summer Training Institute at Texas A&M University.[111] The five-credit graduate program (three credits in safety management and two credits in industrial hygiene) concentrated on recognition, evaluation, and control of occupational health and safety exposures and included the preparation of a model overview course by the attendees.

Instructors for short courses covering aspects of voluntary compliance for construction and general industries are trained at the OSHA Training Center. Two weeklong courses prepare trainees to instruct first-line supervisors on their responsibilities for job health and safety; they are open, free of charge, to university educators, unions, insurance companies, local safety councils, local professional societies, and any other interested parties.[112]

9.3 ESTIMATES OF MANPOWER REQUIREMENTS

Estimates of occupational health and safety manpower needs are necessarily subjective in that they depend on the evaluator's perception of how safe or healthy the workplace should be. There are several clearly definable, though less clearly measurable, benchmarks which require manpower—

- to bring all firms up to the "best-practice" level, i.e., to the level of the safest firms in each industry;
- to bring all firms into compliance with the present (and future) requirements of the Act through adherence to existing (and expected future) standards;
- to match the best foreign practices;
- to achieve a certain target level of safety and health which goes beyond the requirements of the Act (reflected in existing standards) and complies with the "general duty" obligation;
- to achieve a proper balance between safety and health and between different industries.

It is important to keep in mind, in any discussion of manpower requirements, that the benchmark chosen can greatly affect the estimated requirements.

9.3.1 Health and Environmental Services

Although a full-scale manpower survey and model has not yet been undertaken as to manpower requirements,* a NIOSH working group has recently generated estimates of the existing stock, annual production, market demand, and perceived need for manpower in occupational health and safety.[113] Jeremiah R. Lynch, of NIOSH's Health Services and Mental Health Administration, presented an explanation of the results of the NIOSH studies in a June 1973 speech (see Table 9.1). Lynch stated that

> the needs figures shown are the levels of manpower necessary to assure to the extent feasible that each worker is covered by services which assure that his working environment is safe from hazards and that he is adequately protected from the injurious effects of hazards that cannot be eliminated. These figures imply the development of occupational safety and health programs which would make available a uniform set of services to all workers, and would take into consideration the types and numbers of personnel needed to provide the defined services in each class of industry and size of each establishment.[114]

The methodology of the study was explained as follows:

> While the precise modes of delivery are not spelled out, a structure of ratios of professionals or manpower required by category for each SIC code and establishment size, was constructed. These ratios were then applied to worker population and the need figures derived. In making the projection, the following assumptions were made:

*A NIOSH-sponsored feasibility study is being currently undertaken by a private consulting group.

1. that minimum standards would be set for occupational safety and health programs in industry;
2. that these standards would define services that are necessary and the qualifications of personnel performing these services;
3. that the personnel required would vary among industries based upon the size of the industry and the types of hazards present;
4. that maximum use would be made of paraprofessionals and technicians in both medical and environmental services;
5. that new categories of personnel would be developed with competencies in both industrial hygiene and safety;
6. that independent OSH groups would be developed to serve small industry, making standard services economically feasible.[115]

The figures in Table 9.1 imply that the total shortage of environmental and health services manpower is approximately 57,000. This is almost three times the 1972 NIOSH estimate of a 22,000-person shortage of health and environmental services manpower, broken down as follows: 1000 industrial hygienists, 8000 occupational nurses, 3000 industrial physicians, and 10,000 safety and health professionals.[116] According to Lynch, although all the figures in Table 9.1 are to some extent uncertain, "in every instance the deficit between what is presently available and what is needed is greater than 100%."[117]

According to Robert Copland, of the OSHA Public Information Office, another total manpower shortage estimate of 40,000 has been made by the American Society of Safety Engineers.[118]

Estimates made by professionals in the field are revealing complements to the NIOSH study figures and deserve mention. Peter Wolkonsky, Medical Director of Standard Oil of Indiana, in an April 1972 speech before the Industrial Medical Association revealed the following data resulting from a 1971 survey of 29 leading U.S. corporations:

Our study indicates that the medical requirements alone will be for *17,000 full-time occupational physicians* and the annual financial burden will be on the order of $2 billion.

Given the present shortage of physicians and money, it is doubtful that full implementation of this legislation is possible in any realistic or timely way.[119]

Several estimates regarding the need for industrial hygienists suggest that the NIOSH estimate may be somewhat low. In a *Fortune* magazine article, Dan Cordtz quotes an OSHA official as saying that 8000 to 9000 industrial hygienists are needed in the field in order to do the job well,[120] as contrasted to the 5500 industrial hygienists estimated in the recent NIOSH studies. According to another estimate, the states' needs alone amount to between 4000 and 5000 industrial hygienists.[121] Joseph Wagoner, Chief of the Division of Field Studies and Epidemiology, NIOSH, has estimated the market need for industrial hygienists to be distributed between industry, 60%, and government, 40%. [122] The Harvard School of Public

Table 9.1
Occupational Safety and Health Personnel

Category	Presently Employed	Production per Year[1]	Demand per Year	Need[2]	Shortage[3]
Occupational physician[4]	<500	5	15	1750	>1250
Industrial physician	2700[6]	50	100	6900	4200
Occupational nurse[4]	<1000	75	100	9400	>8400
Industrial nurse	17,000	200	900	36,700	19,700
Safety engineer[4]	2000	100	150	8700	6700
Industrial hygienist[4]	500	75	300	5500	5000
OSH specialist[5]	15,000	200	1000	26,700	11,700
Total	<38,700	705	2565	95,650	>56,950

1. Includes formal degree programs, short-course training, self-study, and practical, professionally supervised experience.
2. Levels of manpower to provide uniform services to each worker to ensure that to the extent feasible, the working environment is safe from hazards and the worker is protected from the effects of hazards that cannot be eliminated.
3. Number needed minus number presently employed.
4. Qualified by some accreditation or board certification process.
5. Includes professionals moving toward qualified levels as safety engineers and industrial hygienists (i.e., noncertified), and technicians.
6. Other estimates which include nonindustry physicians are between 6000 and 8000 (see Section 9.2.1).

Source: Jeremiah R. Lynch, "Governmental Occupational Safety and Health Training Programs," speech presented before the National Symposium on Occupational Safety and Health, Washington, D.C. June 6, 1973, p. 9.

Health's 1971 description of its graduate program in industrial hygiene made the following observation on the market demand for industrial hygienists: "To meet the needs of large industry alone, the U.S. Public Health Service has estimated that almost 2000 additional full-time industrial hygienists will have to be trained and on the job by 1975."[123]

The benchmarks used in all the above estimates have not been clearly indicated. However, these are certainly *under*estimates of the manpower necessary to achieve a technologically feasible level of hazard reduction necessary for full compliance with the spirit of the OSHAct.

9.3.2 Enforcement Needs

With over 4 million workplaces covered by the OSHAct, the need for enforcement personnel could be very high, were the thrust of implementation of the Act to lie in enforcing compliance. However, inspection of all U.S. workplaces is neither a goal of the Act nor a numerically feasible possibility. Nevertheless, an answer to the question of how much inspection should occur if "voluntary compliance" is to be effective—based on actual numbers of disasters, fatalities, imminent dangers, and complaints, and a random sampling of hazards and industries—would yield significant input to help determine the appropriate size of state and federal compliance staffs.

During the first nineteen months of the OSHAct, inspections of workplaces employing 14% of the workers covered under the Act took place.[124] Even with the anticipated increase in the number of compliance officers, enforcement sufficient to cover all possible hazardous situations cannot be achieved. Because the demands on OSHA exceed its manpower resources, a set of priorities has been established which emphasizes the inspection of workplaces where there have been catastrophes and fatalities.

In late 1973 George Taylor of the AFL-CIO estimated that about 2000 inspectors were needed to implement the Act, preferably under federal jurisdiction.[125] In June 1975 OSHA projected a need for a total of 4000 federal inspectors and 2500 state inspectors if the law is to be adequately enforced.[126] This projection, which exceeds that of the AFL-CIO, was based on ratings of the degree of hazardousness within industries. An additional 300 compliance officers have been authorized from funds previously earmarked for state plans that have recently been withdrawn, but OSHA did not request any additional inspectors in its fiscal 1976 budget request.[127]

I am of the opinion that "voluntary compliance" cannot really work in the United States unless a "critical mass" of highly qualified enforcement personnel is utilized. The inspector/worker ratio is significantly lower in the United States than in the European countries that have established moderately successful programs (see Chapter 11). The return of the enforcement function to the states is not likely to result in significant improvements, even though there are three times as many state as federal personnel. It is likely that occupational health hazards will be even further neglected, and it appears that no new mechanisms will be introduced to change this situation.

9.3.3 Research Needs

Researchers with academic training in the biological and physical sciences comprise only one type of manpower engaged in occupational health and safety research. Because of the growing awareness of problems relating to stress, mental fatigue, and the behavior patterns they produce, as well as organizational, social, and cultural factors, it is necessary that occupational health and safety investigators also be recruited from the social and behavioral sciences. Broad training and a multidisciplinary approach are key factors in assuring an adequate attack on the major problems.

The issue of training to meet environmental health research needs was addressed by the HEW Task Force on Research Planning in Environmental Health Science in its 1970 report.[128] The task force described what it felt to be three necessary elements in environmental health research training:

1. rigorous disciplinary study in preparation for original research;
2. broad integrated awareness of the total environmental health area;
3. research experience on an environmental health problem of demonstrable value.

The report placed further emphasis on training high-level personnel representing a variety of disciplines, both as participants in research work per se and as program administrators.

To meet the research needs spelled out in the report and to be sure the above elements are incorporated into research training programs, the task force made several significant manpower policy recommendations:[129]

1. A vigorous program of expansion of support for training should be undertaken in a wide variety of disciplines relevant to environmental health.
2. A systematic, planned program of recruiting capable students in the area of environmental health research should be initiated. This program should include provision of seminars and courses at the undergraduate level to acquaint the student with and attract him to the field. Secondly, the program should include a stipend level realistically appropriate to the cost of living and ancillary support.
3. There should be developed and widely disseminated in the scientific community a statement of the health relevance of graduate and post-doctoral training programs.

Although the task force noted that manpower estimates were too imprecise to permit reliable estimates for training needs in specific research areas, the group did attempt to respond to an aggregate need for training funds based on forecasts of research program needs. They estimated that "there must be a substantial—probably sevenfold–increase in the support of training in order to ensure an adequate supply of professional, trained investigators," and further that "training efforts will have to be approximately doubled in the next five years."[130]

NIOSH estimates that the shortage of occupational health scientists is approximately 8000.[131] Irving Selikoff of the Mt. Sinai School of Medicine has estimated that ten times the current level of occupational health and safety *research* is needed, but that even if the money were available for such research, there is simply not enough trained *manpower* to carry it out.[132] Thus, the lack of occupational health researchers appears to be the limiting factor for critical occupational health research, even if the funds for research were allocated. This is not to say, however, that some manpower could not be induced to do research relevant to occupational health and safety if research funds in this area were increased.

9.4 ATTEMPTS OF THE MARKET TO RESPOND

Given the foregoing discussions of the *existing* stock of occupational health and safety manpower and estimates of manpower *requirements*, this section describes briefly the difficulties that have been encountered in attempts to bridge this gap. The difficulties are discussed in terms of problems presented by market imperfections in general, and in terms of their present exacerbation by inappropriate government policies and actions.

9.4.1 Market Imperfections

Market imperfections seriously impede the development and utilization of adequate manpower to meet our occupational health and safety needs. Properly trained manpower in sufficient numbers is essential if we are to have the medical and technical expertise to bring about a socially desirable reduction in the level of workplace injury and disease. Moreover, the problems of insufficient numbers and inadequate training are compounded by a poor distribution of existing manpower. The safety bias that pervades this field is especially serious in the case of manpower, as manifest in a serious imbalance in the mix of safety versus health professionals at the present time, and in the types of facilities available for their training. This imbalance is in large measure attributable to the rational response of firms and insurance companies to economic forces and government regulations. Workmen's compensation, insofar as it acts as an incentive at all, primarily influences firms' demand for *safety* as opposed to *health* manpower, as few diseases are recognized as being of occupational origin for compensation purposes. The OSHAct with its emphasis on safety has, if anything, worsened the manpower imbalance, as firms demand additional safety personnel to comply with the Act and government demands additional safety personnel to enforce it.

Labor faces a serious shortage of the skilled health and safety manpower needed to perform the essential function of serving as an independent check on management. In view of the limited resources available to it for a full-time professional health and safety staff, organized labor is in urgent need of manpower on a part-time or consulting basis, especially for the purpose of conducting training programs which could begin to provide working people with the skills necessary to monitor their own job environment.

Industry faces a variety of market imperfections which are impeding the development and utilization of occupational health and safety manpower. Firms are reluctant to pay training costs for health and safety professionals, both because the professional, once trained, can sell his services to another employer, and because, even while he is employed by the firm which paid for his training, the firm does not capture the full social value of his services. The overall result is that firms devote insufficient resources to the training of health and safety manpower. Moreover, the interest of the hiring firm may diverge from the public interest, as the case of industry's safety bias illustrates. In the absence of sound public policy, there is little basis for hope that the market will induce private industry to utilize properly the health and safety manpower which it employs.

Small firms face special problems with regard to the development and utilization of occupational health and safety manpower. Often, the large fixed cost of setting up a full-time in-house manpower capability is prohibitive for the smaller firm. Yet the market has proven extremely sluggish in providing the kind of ancillary consulting services which could be a useful substitute for full-time staff to the smaller firm. (The great uncertainty of the future of federal and state occupational health and safety programs certainly accounts for some lack of market response.)

Government has not been playing its proper role in the occupational health and safety manpower field, and it has even made some decisions that have resulted in staffing cutbacks. Government has not been sufficiently active in promoting the training and development of health and safety manpower, nor has public policy been geared toward achieving the objective of proper manpower utilization by the private sector. Indeed, the activities of government have served to worsen certain imbalances in the supply and demand for health and safety manpower, directly through insufficient government demand for occupational *health* manpower (especially at the state level) and indirectly through the structuring of incentives on businesses to make them respond in a similar manner.

Likewise, universities and nonprofit research institutions have not been playing their potential effective role in the training of occupational health and safety manpower. In view of the low level of support for health and safety manpower development by government and business, universities and nonprofit research institutions have lacked the funds to place sufficiently high priority on this important objective. As a result, our society is currently underutilizing a potentially important resource—our universities and research institutions—in the development of occupational health and safety manpower.

In conclusion, market imperfections have been partly responsible for a serious shortage of qualified occupational health and safety manpower. Insufficient resources are devoted to the training of such manpower, and the cost-effectiveness of presently committed training resources is shamefully low.

9.4.2 Safety Bias in the Implementation of the OSHAct

One unfortunate consequence of the OSHAct has been to perpetuate the safety bias which pervades the occupational health and safety manpower field. OSHA compliance officers are primarily trained in safety-related rather than health-related disciplines. As noted earlier, as of June 1975, OSHA's enforcement staff included 759 compliance officers, but only 174 industrial hygienists.[133]

Though OSHA recognizes that the present 1:5 ratio of health to safety personnel is inadequate, its future staffing plans do not begin to address this serious imbalance. Though in principle OSHA accepts the need for hiring a higher proportion of health manpower, the projected staff level for fiscal 1976 includes no additional inspectors and remains 3000 short of the projected staffing goal—and it is unlikely that OSHA will in fact be able to hire a sufficient number of industrial hygienists to bring about even this modest improvement in its manpower mix. Ironically, one reason OSHA will have difficulty in reaching its projected manpower target is that the Office of Management and Budget has phased out health manpower training, which NIOSH had just begun to fund as mandated by the OSHAct.

Another significant source of bias is the Bureau of Labor Statistics' overemphasis on safety relative to health in the area of record-keeping and statistical reporting. Prior to enactment of the OSHAct, there was no systematic attempt to collect data on the incidence of occupational illness and disease—as opposed to injury—in the United States. One outcome of the OSHAct has been the initiation of new procedures under which firms are required to record separately every occurrence of occupational disease among their employees.

The results of the first 18 months of data collection under the new statistical reporting systems indicate that the BLS has been seriously deficient in carrying out its congressional mandate to institute accurate reporting practices with regard to the incidence of occupational disease. According to BLS statistics, occupational *illness* accounted for under 5% of the total incidence of occupational *injury* and *illness* during the 18-month period between July 1, 1971, and December 31, 1972. Moreover, only 133,000 cases of occupational illness, a large proportion of which were relatively nonserious (dermatitis accounted for 33.9% of the reports), and only 600 deaths from occupational disease were officially recorded during the last 6 months of 1971.[134] Yet other estimates indicate that the true incidence of the more serious types of occupational disease is substantially greater, and that as many as 100,000 deaths occur each year as a result of occupational disease.[135] The present director of the BLS has commented as follows on the 1971 statistics: "We strongly suspect that it is an understatement of the occurrence of occupational illness."[136] (The BLS traditionally prefers to stick with "hard" data and hence excludes from its statistics many areas in which there is good occasion to believe that disease and injury are work-related.) One result of the underrepresentation of occupational disease in the BLS record-keeping system is a further distortion of the information available to employers regarding the relative magnitudes of their occupational safety and health problems.[137]

In addition to the problems posed by OSHA's manpower mix and the BLS's statistical reporting practices, there is a set of problems related to the OSHAct itself and its enforcement methods. The number of safety standards contained in or promulgated under the OSHAct is far in excess of the number of health standards. Similarly, the number of safety violations cited and fines levied by compliance officers is far in excess of citations and fines for occupational *health* violations. The experience of one major Eastern university cited by Fred Foulkes is instructive in this regard:

> A student fatality led to an OSHA inspection of a large Eastern engineering school, [and] the inspectors found 1650 instances of 66 different types of violations (e.g., broken stepladders; poor housekeeping; unguarded vacuum pumps, fans, and high-voltage areas; and aisles blocked by bicycles chained to posts).[138]

Of the 1650 violations cited, *not a single one* dealt with occupational *health* hazards. The OSHA inspector routinely overlooked serious health hazards such as open containers of benzene in his zeal to cite relatively trivial safety violations.

The inappropriate incentives set up by OSHA have been seized upon by self-serving management consultants and insurance companies and diffused by these groups throughout the entire economy. Capitalizing on OSHA's safety bias, some consulting firms utilize high-pressure advertising and scare tactics to transmit these wrong incentives to private employers, particularly small ones, in an attempt to convince them of the necessity of augmenting their safety manpower through consulting services. Indeed, the activities of consultants and insurance companies in this regard have been blamed by one OSHA official for the management hysteria which has resulted in pressure for a small-business exemption,[139] although it is doubtful that the meager assistance offered to small business by consulting groups has played a large part in creating the anti-OSHA sentiment.

Insurance companies concerned to reduce workmen's compensation claims do, however, influence firms toward increased emphasis on problems of occupational *safety* relative to health. Thus, our system of workmen's compensation, as presently structured, leads to a further distortion of the safety-health manpower mix in the private sector. It remains to be seen whether proposed reforms of the workmen's compensation system will result in a sufficient increase in the recognition and compensability of occupational illness to overcome this long-standing bias.

A recent survey conducted by the National Association of Manufacturers reveals that, of companies surveyed, 61% (and a larger proportion of firms with over 100 employees) indicated that some individual within the firm is in fact vested with total responsibility for employee safety and health. Moreover, the individual in charge of overall health and safety, while often untrained in either area, is at present far more likely to be a qualified safety engineer than an industrial hygienist (see Table 9.2). The proportion of time spent by these individuals on occupational safety and health according to firm size is given in Table 9.3. Again the bias toward safety is evident.

In sum, the underemphasis of health by government and industry has helped to distort the occupational health and safety manpower market, which was exceedingly imperfect to begin with. Government policy, therefore, must specifically address this imbalance as an important part of any broader program to correct the overall deficiencies in occupational safety and health manpower.

9.5 THE ROLE AND RESPONSE OF GOVERNMENT*

It is ironic that at a time when occupational health and safety manpower needs are so enormous, due in large measure to the OSHAct, the federal government is drastically curtailing its activities in the field of health manpower development. (It

Table 9.2
Backgrounds of Company Personnel with Responsibility for Occupational Health and Safety (Percentage of Total Sample)

Company Size (No. of Employees)	Qualified Industrial Hygienist	Qualified Safety Engineer
1-100	.5%	6.8%
101-500	15.1%	15.9%
501-1000	12.7%	38.2%
1001-2000	1.8%	55.6%
2001-5000	4.9%	70.7%
over 5000	23.6%	78.2%
Total	4.2%	26.6%

Source: National Association of Manufacturers, "Where We Stand," p. 11.

*The federal government. Although the states are increasing their use of occupational health and safety manpower, they are not involved in manpower development.

Table 9.3
Proportion of Time Spent on Safety and Health by Company Personnel in Charge of Occupational Health and Safety

Company Size (No. of Employees)	Safety	Health
0-100	to 10%	to 10%
101-500	to 10%	to 10%
501-1000	30-40%	to 10%
1001-2000	40-50%	10-20%
2001-5000	70-80%	10-20%
over 5000	40-50%	to 10%

Source: National Association of Manufacturers, "Where We Stand," p. 10.

should be emphasized that this curtailment is not the result of OSHA or NIOSH decisions, but rather of more general cutbacks ordered by the Office of Management and Budget.)

This unfortunate decision may have significant adverse consequences for the ability of the private sector to comply with the mandates of the OSHAct, as well as for the ability of OSHA to enforce it, and it may seriously hinder urgently needed improvements in the level of safety and health in our workplaces. In the words of Jeremiah Lynch:

> As a chain is as strong as its weakest link, our ability to provide a safe and healthy workplace will be determined by our most critical deficiency. . . . [It] should be recognized when less than half of the required manpower is available to apply the research and standards and to enable the employer to comply, the impact of all of these activities will necessarily be less than half of what it could be if adequate manpower were available. *No single factor has a greater detrimental effect on the prevention of occupational illness, injury, and death than the manpower deficit crisis.*[140]

Among the specific occupational health manpower development programs which are being eliminated or cut back are the following:

- Training grants to educational institutions, which provide training in health specialties, are being phased out over a three-year period ending in 1976.
- Direct short-course training in HEW will no longer be federally subsidized, but will be put on a cost-covering basis.

These programs, though inadequate, have been an important source of occupational safety and health manpower development since the passage of the OSHAct. The prior level of effort in regard to occupational safety and health manpower development has been aptly characterized as follows:

> In the years prior to the passage of the Federal Coal Mine Health and Safety Act of

1969, occupational safety and health training was a very minimal effort conducted by a few graduate schools and a number of institutions offering short courses of one sort or another. Most of the practitioners in occupational safety and health were, if you will, self-trained. This can be seen from the membership of the AIHA, ASSE, and from the ranks of industrial physicians and nurses who are presently fulfilling occupational safety and health roles in industry. . . . [This] dearth of fully qualified manpower led inevitably to the very problems which were identified in the legislative hearings on the Occupational Safety and Health Act of 1970.[141]

One result of the dearth of formal training programs in this area has been forced reliance upon improperly or moderately trained manpower:

In the industrial hygiene and safety engineering fields, demand is met by internal conversion. That is to say, an engineer or a chemist who is employed by a company is designated as a safety engineer or an industrial hygienist and told to go and learn his trade.[142]

NIOSH training grants have been especially significant in providing support for training of new and badly needed categories of occupational safety and health professionals and paraprofessionals:

Significant support has been provided for the creation of innovative programs designed to produce *combined* occupational safety *and* health professionals at the baccalaureate level and for the training of paraprofessionals.[143]

Without the stimulus provided by federal subsidy, the development of such innovative programs in educational institutions may be seriously impeded.

NIOSH activities have also been significant in the development of more traditional categories of occupational safety and health manpower, with an emphasis on promoting the training of direct practitioners rather than researchers in this field:

In 1972 the NIOSH-funded training grants supported 223 trainees. An additional 972 trainees benefited from these programs, but were not supported with stipends.

Direct operations in training were concerned with increasing the number and quality of short-term courses for upgrading of skills and skill conversions, continuing education, and the training of trainers. In FY 1972, 1264 individuals received short-term training through the NIOSH training courses.[144]

Clearly, the reduction of NIOSH and OSHA involvement in occupational safety and health manpower development and training, both direct and indirect, will seriously hinder these agencies from carrying out their mandate under the OSHAct. It is difficult to understand how the following duties and responsibilities of NIOSH and OSHA can now be implemented:

The sections of the Act imposing upon the National Institute for Occupational Safety and Health the responsibility to train require that the Secretary of HEW consult with the Secretary of Labor and other appropriate departments and agencies. Since the Secretary of Labor has independent responsibilities which appear to overlap those of the Secretary of Health, Education, and Welfare, an interagency agreement between NIOSH and OSHA was negotiated. This agreement states in general that the Secretary of Labor has the responsibility to train employers and employees in the recognition, avoidance, and prevention of unsafe and unhealthful working conditions in employments covered by the Act. This has been interpreted by the agencies to mean the training of employers and workers in general in their collateral responsibilities in the field of occupational safety and health. That is to say, everyone who works in industry, regardless of his job, has some responsibility for the safety and health of himself and others and needs some training in order to be able to carry out that responsibility. The training of direct operating safety and health personnel, that is, those whose sole or major responsibilities are in occupational safety and health, is the responsibility of the Secretary of Health, Education, and Welfare. The training of government employees, state and federal, who carry out the responsibilities of the Secretary of Labor under the Act, is to be conducted by OSHA with the assistance of NIOSH under the terms of the agreement. Although this agreement appears to leave some degree of overlap, the very close, harmonious relationship between the Division of Training and other manpower activities of NIOSH and the Office of Training and Manpower Development of OSHA has enabled us to resolve specific questions without conflict.[145]

The result of federal cutbacks in this area has been to give rise to an "impending manpower crisis," according to Lynch. For example, "The employer will have the dilemma of being required to eliminate hazards while the people who could tell him how to eliminate them will not be available."[146]

Fortunately, NIOSH will continue to perform at least some important functions in the manpower training and development area. NIOSH will, for example, undertake or sponsor the following major activities:

- A census of occupational safety and health manpower.
- Standards development for those standards that require certain medical and environmental services (such as the marketing of lung-function tests or the monitoring of air quality), and for voluntary accreditation of occupational safety and health programs.
- Reimbursible training.
- Encouragement of training by others by providing educational materials, a directory of resources, and a mechanism of cooperation between different institutions and organizations.

Activities to define more closely the national needs for occupational safety and health manpower include a contract with the Division of Nursing, Health Services and Mental Health Administration, in which data were gathered on the number and qualifications of nurses presently engaged in occupational safety and health, and a contract to study the feasibility of conducting an industrywide sample survey on current staffing trends and projected needs based on industry's experience under the Occupational Safety and Health Act.[147]

The latter contract was awarded to the Salt Lake City-based consulting firm of John Short & Associates. This study will hopefully produce some sorely needed progress in regard to the job description and classification of occupational safety and health manpower, an area which is in an extremely rudimentary state at present. As part of the study, a survey questionnaire instrument was developed to form the basis for comprehensive national occupational health and safety manpower planning.*

In spite of these few valuable activities, the government effort in manpower development is critically inadequate. Inasmuch as the private sector is not likely to take up the task in the near future, for reasons already explained, occupational health and safety manpower development will remain the limiting factor in improving occupational health and safety for the American worker.

9.6 POSSIBLE SOLUTIONS

As indicated at the beginning of this chapter, before a comprehensive manpower policy can be formulated with confidence, more detailed analysis must be undertaken to confirm the observations made here.

A hindrance to analysis is the imprecision in current job descriptions and classifications, which makes assessment of the existing manpower stock, as well as projection of future requirements, a tricky and uncertain task. The source of this imprecision is the pervasive lack of specific, formal training and the enormous variety in backgrounds and skills found among those assigned to fill occupational health and safety roles. In addition, the types of expertise *required* in various situations do and should span a large number of disciplines. Thus, it is difficult to compare different health and safety professionals.

Nonetheless, it is useful to examine possible programs and solutions. These are of two types: (1) those which can easily be effected within the present political environment, and (2) those which would require a significant shift in the operating philosophies of government, industry, and labor.

At the beginning of the chapter, manpower needs were classified into four categories: enforcement, health and environmental services, research, and teaching. While research is critically needed to fill the gaps in our scientific and technical knowledge, it is *not* at this time the limiting factor in improving the work environment.** Great improvements could be brought about by applying the knowledge we already have. In addition, the manpower effort required to create a firm data base

*This study was completed and submitted to NIOSH after this chapter had been prepared;[148] hence, no evaluation of it was undertaken.

**I am aware of the tremendous lack of knowledge that exists with regard to the toxicity of the more than 12,000 chemicals now in commercial use. It is precisely because of this lack of knowledge that the amount of research that is needed cannot be achieved in a realistic sense—nor performed for the 500 new chemicals entering the environment each year. Solutions other than those relying on a firm data base will have to be made. In this sense, research capability is not the bottleneck—or, if we insist that it is, then it always will be.

for all suspect materials would be more efficiently utilized in designing new processes and equipment which could be generally applied to many materials. If additional funds were made available for occupational health and safety research, competent researchers would shift their attention to this area.

Similarly, were adequate resources devoted to the training of occupational health and safety professionals as they have traditionally been trained, the shortage of competent teaching manpower would not be a serious limitation. Occupational health and safety training is by its nature multidisciplinary, requiring engineers, chemists, physiologists, physicians, etc., who are already teaching and who could be readily shifted into relevant training activities.

However, if we are to train occupational health and safety professionals as I believe they should be trained, not only do we face a serious teaching manpower problem, but there also exists the serious problem of overcoming the opposition of the "teaching establishment" in this area to a different view of the role of the professional. Traditional programs are deficient in that they emphasize treatment rather than prevention. Professionals' attitudes toward workers and what constitutes a socially acceptable level of harm will have to be markedly changed if we are to obtain the maximum possible reduction in occupational injury and disease.

The most serious problems of manpower development can be summarized as follows:

1. Market demand for occupational safety and health manpower services does not reflect true social needs, in regard to either the level or the composition of professional services demanded. There is a severe overall shortage of qualified occupational safety and health professionals relative to the numbers which would be required to bring about a significant improvement in the level of workplace health and safety. Currently available statistics, though they tend to indicate a considerable shortage, significantly understate the true magnitude of the problem in view of the widespread tendency among many firms, especially smaller ones, to designate untrained or poorly trained individuals as company "health and safety director." Moreover, there are a number of particularly acute imbalances in the *mix* of available occupational safety and health manpower services, relative to the social need. In particular, a greater allocation of manpower to occupational health services, to smaller firms, and to enforcement agencies is needed.

2. The bulk of occupational safety and health practitioners and their professional associations have traditionally been "industry captives." That is, they have accepted the view that their primary loyalty is to their employer. In some cases, improvement in physical working conditions is, in fact, consistent with overall corporate goals, so that no conflict between loyalty to the employer and professional ethics need arise. More typically, however, it is not possible for the business enterprise to appropriate fully the benefits of reduced workplace hazard. In other cases, the cheapest method for the firm to bring about occupational safety and health im-

provements is not the least costly for society as a whole (as may be the case when the firm is faced with a choice between dismissing an older or sick employee and correcting a hazardous condition).

In many instances, the objectives of the firm are inconsistent with reducing the level of workplace hazard. In cases where corporate policy rules out elimination of a health hazard, for example, the professional may be faced with the unfortunate choice between recommending dismissal for an affected employee or allowing him to continue to work under conditions harmful to his health. Moreover, occupational health and safety professionals have often been complicit in concealing information from workers concerning the nature of health hazards they face on the job. Too many professionals accept the company's view that expenditures on the improvement of working conditions must be justified in terms of standard cost-effectiveness criteria—though from a *social* point of view, such expenditures may be warranted even when they are not for the firm.

Thus, it is not going to be enough simply to increase the *number* of occupational safety and health professionals, or even to bring about needed improvements in the professional "mix." These important and necessary changes must be coupled with an infusion of a new sense of professionalism and new standards of professional ethics. Practitioners in the occupational safety and health manpower field must be educated to the viewpoint that improvement in workers' health and safety, rather than maximum contribution to corporate profit, is their overriding professional objective. This can only come about if the management of firms regard good health and safety practices as standard operating procedure—and do not insist that "safety pay."

It is incumbent upon us to begin to specify policy alternatives which address these and other problems of manpower development and utilization. Perhaps the highest-priority area is that of enforcement manpower. Government at both the federal and state levels should greatly increase the number and quality of compliance officers for the purpose of vigorously enforcing the OSHAct. Recruitment of competent enforcement personnel in sufficient numbers may require substantial increases in the pay and status associated with this position, as well as more direct and indirect government involvement in training (in marked contrast to the current policy of cutting back the government's training effort). Compliance officers must be more well-rounded, with greater sensitivity to the broader issues of physical and mental occupational *health* than has been the case in the past.

Tuition subsidies and living expenses during training could be made available to individuals who agree to enter government service as compliance officers. Union safety stewards and committeemen and unemployed or underemployed chemists, physicists, and aerospace engineers represent a large, untapped potential source of such trainees.

It is also necessary to bring about a significant expansion in the demand for professional services in the private sector. Improved enforcement of the OSHAct would indirectly achieve part of the desired result in this regard, and it is of particular importance in view of this "multiplier effect." A shift in emphasis in enforcement, with greatly increased attention devoted to occupational *health* problems, would also help bring about a desirable alteration in the mix of professional services demanded by the private sector, which is at this time heavily safety-biased.

Another indirect method of stimulating private-sector demand for occupational safety and health manpower services would be to encourage and promote the efforts of labor unions to bring about needed improvements in working conditions. Goals should include an expansion of access to information and of training programs that help workers know their rights under the OSHAct and monitor their own job environment. As workers and their unions become more assertive of their right to a safe and healthful workplace, it can be expected that pressure on employers to reduce the level of workplace hazard will mount.

More direct approaches could also be taken to stimulate the private-sector demand for occupational safety and health professional services. The OSHAct could be amended to require that firms adhere to mandatory staffing norms. Larger firms and unions could be required to maintain a full-time, in-house staff; smaller firms and unions could be required to contract with consultants for an appropriate level of services. Consultative services could be provided by the government or by approved university or private programs.

Alternatively, or in concert with mandatory staffing requirements, private-sector demand could be stimulated by means of wage subsidies or tax incentives to firms and unions which directly employ professional safety and health manpower of various types, or which utilize approved consultative services. (Such wage subsidies have been used with moderate success in the past as a component of manpower programs to facilitate the employment of disadvantaged workers.) Subsidies of this type could be justified in economic terms on the basis of standard externality criteria: the firm is unable to appropriate all of the social benefits created by the occupational safety and health manpower which it employs; hence, the value of such professionals to the firm is below their value to society. In order to stimulate the private sector to hire a number of professionals that is sufficient from society's point of view, their wages must be subsidized, either directly or by means of tax concessions to their employer.

A range of programs are also needed on the "supply" side of the manpower problem.* Students could be encouraged by means of tuition rebates, loans, and stipends to enter occupational safety and health professional training programs. The large amount of paramedical talent among Vietnam veterans, currently unutilized,

*Incentives encouraging new entrants into occupational health and safety have in many instances a negative aspect in that they draw manpower away from other badly underserviced areas. Manpower policies should, ideally, be implemented on a much broader scale.

indicates one potential source of trainees. Special programs could also be devised to retrain unemployed or underemployed scientists and aerospace engineers laid off during the early 1970s. Union safety stewards and committeemen represent another important potential source of midcareer candidates for intensive short-course retraining. Such individuals typically have family responsibilities which would necessitate substantial financial assistance while they undertake retraining, but this could be an investment with a very large payoff. Particularly in the case of scientists and engineers, subsidized retraining would give a large group of alienated, disappointed, middle-aged individuals an excellent opportunity to redirect and utilize their expertise in a socially useful and necessary endeavor. An expansion of training opportunities for those entering the field of occupational nursing is also vitally needed; indeed, many of the professionals interviewed in the course of this study felt that investment in this area would produce the greatest return in terms of increased manpower.

Aside from individual subsidies, supply could be facilitated by means of grants to universities and other educational institutions which undertake or expand training programs for occupational safety and health professionals. Such subsidies are likely to prove helpful in overcoming the "indivisibilities" and risk characteristics which make establishment of new educational programs such a difficult undertaking. They should be used especially to encourage the establishment or expansion of innovative programs to train new categories of *combined* health and safety professionals and paraprofessionals. Perhaps the best way to produce a proper mix of concern for health and safety is to create a professional trained equally well in both areas.

Of course, once people have been trained, it is important that they be kept up to date in this rapidly developing field. Opportunities for continuing education and for professional interaction on a national or international scale should be provided so that the level of effectiveness of health and safety personnel is not only sustained but continually raised.

The problem of the "captive professional" is less easily addressed. To the extent that better enforcement of the OSHAct results in more nearly complete "internalization" of the social costs of occupational illness and injury, the interests of firms and of society at large in bringing about needed improvements in working conditions may begin to converge. To the extent that better enforcement achieves this result, the "captive-professional" problem may be diminished. Similarly, expanded union activities will necessarily result in a more proper reorientation of the activities and objectives of at least some occupational safety and health professionals.

A more direct method of overcoming this problem would be the mandatory establishment of joint labor-management health and safety committees in every enterprise. If occupational safety and health professionals employed by the firm were made accountable to such committees (as is often the case in Europe) rather than to management alone, conflict between organizational objectives and professional ethics would be less likely to arise. (See Chapter 11 for an extended discussion of the lessons to be learned from the Western European experience with works' councils.) Workers can be more effectively utilized in improving their own occupational health and safety situations and should be trained to monitor their work environment. As discussed in the next chapter, this is an indispensable supplement to a meaningful manpower policy.

In sum, there is an urgent need for a coordinated manpower policy in occupational health and safety. Occupational health and safety problems require the expertise of many disciplines, the needs are not being presently recognized by the market, and the occupational health and safety issue is fraught with political as well as scientific-technical conflicts. A massive and coordinated effort is required, since manpower is the limiting *resource* factor in the improvement of job health and safety and since there exists a gap between what is necessary and what the market response has been to this challenge.

REFERENCES

1. Hearings before the Subcommittee on Environmental Problems Affecting Small Business of the H.R. Select Committee on Small Business, 92d Congress, 2d Session, on Small Business and the Occupational Safety and Health Act of 1970, June 22, 1972, p. 299.

2. Susan Mazzocchi, "Rx for Updating American Medical Education," unpublished paper, 1973, pp. 3, 8.

3. Industrial Medical Association, *U.S. Distribution of Physicians* (Fisher-Stevens, Inc., Clifton, N.J., December 1973).

4. Marcus Key, "Research Criteria for Safety and Health Standards, Manpower Development and Training," speech presented at the American Chemical Society's Symposium on Occupational Safety and Health, Washington, D.C., June 4, 1973, p. 8.

5. Charles Culhane, "Labor, Business Press Administration to Change Safety and Health Program," *National Journal* (Washington, D.C.), June 1, 1972, p. 1094.

6. *Occupational Safety and Health Reporter* (the Bureau of National Affairs, Inc.), Vol. 3, No. 27 (December 12, 1973), p. 859. [Hereinafter cited as *OSHR*, BNA]

7. *The President's Report on Occupational Safety and Health*, G.P.O. Document No. 4000-00288 (December 1973), p. 33.

8. Mary-Win O'Brien, Solicitor's Office, Department of Labor, interview with author, Washington, D.C., August 7, 1973.

9. Jessie L. Steinfeld, "Occupational Health and Safety Programs of HEW–A Progress Report," speech presented at the Western Industrial Health Conference, Newport Beach, California, October 13, 1972, p. 11.

10. Dr. Ralph Vernon, Texas A&M University, interview with author, Dallas, Texas, March 31, 1973.

11. "Point of View: An Interview with Dr. John Grimaldi, Director, The Center for Safety, New York University," *Job Safety and Health*, Vol. 1, No. 10 (September 1973), pp. 22-23.

12. Parliamentary Committee on Safety and Health at Work, *Report of the Committee, 1970-1972 (London, 1972)*, p. 14.

13. *Ibid.*, pp. 16-17.

14. Audrey A. Brenny, "Have Head and Heart–Will Travel: Nurses' Message to U.S. Businessmen," speech presented before the Annual Convention of the American Public Health Association, San Francisco, California, November 7, 1973, p. 6.

15. Jane A. Lee, "The Plant Nurse and PL91-596, The Occupational Safety and

Health Act of 1970," speech presented before the National Symposium on Occupational Safety and Health, Washington, D.C., June 6, 1973, p. 7.

16. *President's Report*, 1973, p. 145; Jeremiah R. Lynch, "Governmental Occupational Safety and Health Training Programs," speech presented before the National Symposium on Occupational Safety and Health, Washington, D.C., June 6, 1973, p. 3.

17. *President's Report*, 1973, p. 144.

18. Mazzocchi, "Rx for Updating American Medical Education," p. 8.

19. *Ibid.*

20. "Point of View: John Grimaldi," p. 22.

21. *OSHR*, BNA, Vol. 2, No. 46 (April 19, 1973), p. 1347.

22. *New York Times*, May 3, 1973, p. 54.

23. *OSHR*, BNA, Vol. 2, No. 48 (May 3, 1973), p. 1418.

24. Lorin E. Kerr, "Occupational Health: A Discipline in Search of a Mission," speech presented at the Annual Meeting of the American Public Health Association, November 14, 1972, Atlantic City, New Jersey. Health Policy Advisory Center, *Health PAC Bulletin*, No. 44 (September 1972).

25. *U.S. Distribution of Physicians.*

26. Henry F. Howe, "Distribution of Occupational Physicians Among Industries," *Journal of Occupational Medicine*, Vol. 11 (April 1969), pp. 191-192.

27. Mazzocchi, "Rx for Updating American Medical Education," p. 11.

28. Seymour Lusterman, *Industry Roles in Health Care* (Conference Board, New York, 1974), p. 21.

29. *Ibid.*, p. 22.

30. *Ibid.*

31. Howe, "Distribution of Occupational Physicians," p. 191.

32. Lusterman, *Industry Roles*, p. 22.

33. *Ibid.*, pp. 23-24.

34. *Ibid.*, p. 24.

35. Leonard Arling, "Industrial Clinics and Occupational Medicine," *Archives of Environmental Health* (American Medical Association), Vol. 12 (May 1966), p. 644.

36. Joseph A. Page and Mary-Win O'Brien, *Bitter Wages* (Grossman, New York, 1973), p. 152.

37. Dr. Henry F. Howe, Department of Environmental, Public, and Occupational Health, American Medical Association, interview with author, June 21, 1972.

38. Mazzocchi, "Rx for Updating American Medical Education," p. 8.

39. Page and O'Brien, *Bitter Wages*, p. 152.

40. Mazzocchi, "Rx for Updating American Medical Education," p. 6.

41. *Ibid.*, pp. 14-15.

42. Lee, "The Plant Nurse and PL91-596," p. 1.

43. Lynch, "Governmental Occupational Safety and Health Training Programs," p. 7.

44. Mary Lou Bauer and Mary Louise Brown, *Occupational Health Nurses: An Initial Survey*, U.S. Public Health Service, Publication No. 1470 (May 1966).

45. Lusterman, *Industry Roles*, p. 29.

46. Brenny, "Have Head and Heart–Will Travel," p. 6.

47. Lee, "The Plant Nurse and PL91-596," p. 1.

48. Brenny, "Have Head and Heart–Will Travel," p. 1.

49. Bauer and Brown, *Occupational Health Nurses*, p. xii.

50. Mary Louise Brown, "Occupational Health Nursing in the United States: Training and Education," *Occupational Health Nursing*, Vol. 17 (November 1969), p. 13.

51. Brenny, "Have Head and Heart–Will Travel," p. 13.

52. Marjorie J. Keller and W. Theodore May, *Occupational Health Content in Baccalaureate Nursing Education*, NIOSH, Publication No. 2176 (1971).

53. *Health Resources Statistics*, Health Serv-

ices and Mental Health Administration, HEW, Publication No. (HSM) 72-1509 (1971), p. 106.

54. *Ibid.*, p. 107.

55. *OSHR*, BNA, Vol. 2, No. 39 (March 1, 1973), p. 1134.

56. Statement of Clyde M. Berry, President of the American Industrial Hygiene Association, before hearings of the House Subcommittee on Bill 14816. See also *Health Resources Statistics*, p. 107.

57. Henry Howe, interview with author.

58. *Health Resources Statistics*, p. 107.

59. Lynch, "Governmental Occupational Safety and Health Training Programs," p. 6.

60. Henry Howe, interview with author.

61. University of Michigan Survey of Industrial Health Alumni, January 17, 1973 (mimeograph).

62. *President's Report*, 1973, p. 33.

63. Roger Wingate, Vice-President, Liberty Mutual Insurance Co., interview with author, Boston, Mass, July 11, 1972.

64. Lusterman, *Industry Roles*, p. 27.

65. *Ibid.*

66. *Occupational Safety and Health Training Grants*, NIOSH, G.P.O. Document No. 759-569/1303 (1972).

67. Dr. Bertram D. Dinman, Institute of Environmental and Industrial Health, School of Public Health, University of Michigan, letter to author, February 26, 1973.

68. "Scope and Functions of the Professional Safety Position," American Society of Safety Engineers, Park Ridge, Illinois (1972).

69. B. Gawain Bonner, "A Pro's Analysis," *Trial*, Vol. 9, No. 4 (July-August 1973), p. 25.

70. Lynch, "Governmental Occupational Safety and Health Training Programs," p. 6.

71. Wayne Christensen, Managing Director, American Society of Safety Engineers, interview with author, Chicago, Illinois, February 5, 1973.

72. *OSHR*, BNA, Vol. 2, No. 39 (March 1, 1973), p. 1134.

73. *OSHR*, BNA, Vol. 2, No. 36 (February 8, 1973), p. 1044.

74. *Occupational Safety and Health Training Grants*, pp. 3, 14.

75. Robert J. Firenze, "A Bachelor's Degree in Industrial Safety," *National Safety News* (June 1972), p. 55.

76. "Engineering at Texas A&M University," College of Engineering, Texas A&M University, p. 26.

77. *Statement of Purpose and Programs*, Center for Safety, New York University (October 1970), pp. 1, 5.

78. Edgar F. Seagle, "Manpower for Occupational Safety and Health," *Occupational Health Nursing*, Vol. 20, No. 4, p. 10.

79. "Health and Safety Technician," *Job Health and Safety*, Vol. 1, No. 2 (February 1973), pp. 15-16.

80. *President's Report*, 1973, p. 33.

81. *OSHR*, BNA, Vol. 5, No. 4 (June 26, 1975), p. 125.

82. *Ibid.*

83. *President's Report*, 1973, p. 33.

84. *OSHR*, BNA, Vol. 3, No. 6 (July 12, 1973), p. 169.

85. *Ibid.*

86. "Needed: More Compliance Officers," *Job Safety and Health*, Vol. 1, No. 13 (December 1973), p. 14.

87. Robert L. McKeand, Director, OSHA Training Institute, interview with author.

88. *OSHR*, BNA, Vol. 2, No. 43 (March 29, 1973), p. 1246.

89. *President's Report*, 1973, p. 33.

90. *Ibid.*, p. 34.

91. *President's Report*, 1973, p. 33.

92. *OSHR*, BNA, Vol. 3, No. 1 (June 7, 1973), p. 9.

93. Roger Wingate, interview with author.

94. *Man's Health and the Environment–Some Research Needs*, Report of the Task Force on Research Planning in Environmental Health Sciences, National Institute of Environmental

Health Sciences (March 1970), pp. 218-219.

95. *Health Resources Statistics*, p. 42.

96. *Ibid.*, p. 41.

97. *Ibid.*, p. 42.

98. Testimony by the American Public Health Association before the Senate Subcommittee on Labor of the Committee on Labor and Public Welfare, 91st Congress, 2d Session, May 5, 1970, p. 1090.

99. Herbert M. Hohn, "Research to Determine What's Dangerous," *Monthly Labor Review*, Vol. 96, No. 8 (August 1973), p. 49. *The Advisor* (NIOSH), No. 4 (December 1972) and No. 5 (January 1973).

100. *OSHR*, BNA, Vol. 2, No. 47 (April 26, 1973), p. 1371.

101. *Occupational Safety and Health Training Grants*, pp. 3-44.

102. *Man's Health and the Environment*, p. 221.

103. Dr. Bertram Dinman, letter to author, February 26, 1973.

104. "Schools of Public Health in the United States and Canada," American Public Health Association, New York (June 1970), p. 16 (mimeograph).

105. *Ibid.*, p. 21.

106. "Programs of Graduate Study and Research," Department of Environmental Sciences and Engineering, University of North Carolina, School of Public Health, p. 3.

107. Report of the Committee on Manpower and Curriculum, Panel on Health Services, Harvard School of Public Health (October 24-25, 1972), p. 12.

108. Announcement of the Department of Environmental Health, University of Cincinnati, 1971-1972, pp. 1-3.

109. *U.S. Distribution of Physicians.*

110. "Point of View: John Grimaldi," p. 24.

111. Ralph J. Vernon and Richard B. Konzen, "Occupational Safety and Health Training Institute for College Professors," *ASSE Journal* (March 1973).

112. *Daily Labor Report* (December 28, 1973), pp. A2-3.

113. "Meeting Health Manpower Requirements of Legislatively Mandated Occupational and Environmental Health Responsibilities," NIOSH memorandum (undated mimeograph).

114. Lynch, "Governmental Occupational Safety and Health Training Programs," pp. 6-7.

115. *Ibid.*

116. Preprint of *President's Report*, 1973, p. 143 (unpublished draft).

117. Lynch, "Governmental Occupational Safety and Health Training Programs," p. 7.

118. Robert Copland, OSHA Public Information Office, interview with author, June 30, 1972.

119. "Occupational Health and Safety Letter" (Gershon W. Fishbein, Publisher), Vol. 2, No. 8 (April 22, 1972), p. 1; emphasis added.

120. Dan Cordtz, "Safety on the Job Becomes a Major Job for Management," *Fortune* (November 1972), p. 116.

121. Sharon Morris, NIOSH, interview with author, Cincinnati, Ohio, February 12, 1973.

122. Dr. Joseph Wagoner, NIOSH, interview with author, Cincinnati, Ohio, February 12, 1973.

123. "Graduate Program in Industrial Hygiene at the School of Public Health," Harvard University, 1971.

124. *Monthly Labor Review* (August 1973), p. 41.

125. *Spotlight on Health and Safety* (Industrial Union Department, AFL-CIO), Vol. 2, No. 4 (fourth quarter 1973), pp. 1, 3.

126. *OSHR*, BNA, Vol. 5, No. 4 (June 26, 1975), p. 125.

127. *Ibid.*

128. *Man's Health and the Environment*, pp. 217-221.

129. *Ibid.*, p. 222.

130. *Ibid.*, p. 221.

131. "Manpower Development for a Safe Workplace," *HSMHA World* (March-April 1972).

132. Dr. Irving Selikoff, Mt. Sinai School of Medicine, interview with author, New York, June 15, 1972.

133. *OSHR*, BNA, Vol. 5, No. 4 (June 26, 1975), p. 125.

134. "Occupational Injuries and Illnesses by Industry, July 1-December 31, 1971," Bureau of Labor Statistics Bulletin No. 1798, Table 4c, p. 14.

135. *The President's Report on Occupational Safety and Health*, G.P.O. Document No. 2915-0011 (May 1972), p. 111.

136. Maurice F. Bresnahan, "Availability of Occupational Injury and Illness Data and Future Plans," speech presented before the National Symposium on Occupational Safety and Health, Washington, D.C., June 4, 1973.

137. *OSHR*, BNA, Vol. 3, No. 34 (January 24, 1974), p. 1067.

138. Fred K. Foulkes, "Learning to Live with OSHA," *Harvard Business Review* (November-December 1973), p. 64.

139. Robert Copland, interview with author.

140. Lynch, "Governmental Occupational Safety and Health Training Programs," p. 10; emphasis added.

141. *Ibid.*, p. 1.

142. *Ibid.*, p. 6.

143. *Ibid.*, pp. 2-3; emphasis added.

144. *Ibid.*, p. 3.

145. *Ibid.*, p. 2.

146. *Ibid.*, p. 7.

147. *Ibid.*, p. 3.

148. John Short & Associates, Inc., *Feasibility Study Covering Nationwide Occupational Safety and Health Manpower Survey*, NIOSH Publication No. 75-138 (February 1975).

10 Key Activities in the Private Sector

10.1 INTRODUCTION

The purpose of this chapter is to describe and evaluate the key activities of the various forces active in occupational safety and health so that sensitive leverage points for the design of policy can be identified. Many of the assertions of the chapter are more fully referenced and explained in other parts of the text. More detailed documentation is given in the unpublished appendices, constructed from interviews as well as from published and unpublished documents describing the various organizations.* The chapter is but a brief abstract of these appendices. The design of policy for the private sector will be discussed in Chapter 13.

10.2 LEGAL ACTION TAKEN TO FORCE IMPLEMENTATION OF THE OSHAct

The legal action taken by public-interest groups and unions has played an important role in the implementation of the OSHAct. Many of the positive steps taken by OSHA have been in response to this legal action or fear of future legal action. In the cases of the three health standards adopted through the procedure established by Section 6(b) of the OSHAct, petitions or litigation were filed by unions, public-interest groups, and employer groups.

The asbestos standard was adopted by OSHA as a result of a petition from the Industrial Union Department of the AFL-CIO. The final OSHA standard was viewed as weak, and a suit was subsequently brought by the IUD and five affiliated unions: Asbestos Workers; Pressmen; Paperworkers; Oil, Chemical, and Atomic Workers; and Textile Workers. This suit charged that the June 1972 OSHA standard of five fibers per cubic centimeter failed to protect workers and that OSHA was improperly considering the economic impact of industry's compliance with the standards. The suit was based upon the belief that health is a right for which there is no trade-off on any grounds. The court of appeals decision, *Industrial Union Department v. Hodgson* (499 F.2d 467, D.C. Cir. 1974), upheld the standard taking into account economic feasibility, remanding it only on the reasons for the industrywide delay and the short record retention period.

The carcinogen standards promulgated by OSHA were the result of legal action on the part of the Health Research Group and the Oil, Chemical, and Atomic Workers Union. These groups first petitioned OSHA in December of 1972 to adopt temporary emergency standards for ten carcinogens. In April 1973 HRG and OCAW formally filed a complaint in the federal courts alleging that OSHA had unlawfully withheld action on the petition. In response to this court suit, OSHA issued temporary standards. Still another court suit was required to force the promulgation of permanent standards in January 1974.

*The detailed studies involved eight major union groups, six management groups, six major corporations, nine public-interest groups, professional societies, private organizations, insurance carriers, and consulting groups.

Similar legal action was required in order to get a temporary emergency standard issued for pesticides. On January 15, 1974, the Organized Migrants in Community Action, Inc., the RAZA Association of Spanish Surnamed Americans, and a farmworker from Florida filed suit against OSHA seeking adoption of a permanent pesticide standard (see Sections 4.4.2 and 12.3).

Although the case for a vinyl chloride standard never reached the courts prior to its issuance, the IUD petitioned OSHA to issue a temporary emergency standard for vinyl chloride in early March 1974, after the discovery of liver cancer deaths among rubberworkers exposed to the chemical. The IUD stated that information from local union leaders and a limited investigation of plants using vinyl chloride continued to support the need for an emergency standard since little or inadequate action was being taken voluntarily by management. Once issued, the permanent vinyl chloride standard was challenged in the courts from all sides.

Legal action has been aimed not only at the issuance of health standards, but also at the policies of the Department of Labor in implementing the OSHAct. As noted in Chapter 5, the unions are deeply concerned about the return of occupational health and safety enforcement to the states. A recent suit by the IUD and the AFL-CIO seeks to prohibit OSHA from approving plans that do not meet the statutory requirements of the OSHAct.[1] * The suit alleges that OSHA is illegally allowing states to operate occupational health and safety programs on OSHA-approved plans which are based on unsecured promises of future state performance and which do not meet the "at least as effective as" requirement of the Act (particularly for standards development and enforcement, personnel qualifications, and adequate funding).

In general, the labor unions have been reticent to pursue litigation against individual employers; historically, they have not had much success in the courts. The filing of OSHA complaints and Review Commission briefs, contract bargaining, and the threat of litigation have further served to minimize the incidence of court action. It should be pointed out that the expense of court action, while high even for labor unions who have their own legal departments, is prohibitive for unorganized workers.

When suits have been brought against employers, it has been in serious cases and in cases seeking to test the OSHAct. For example, court action was taken on the issue of walkaround pay, not specified in the Act as part of the walkaround right which enables employees to accompany federal compliance officers during inspections. After an inspection of a Mobil Oil Company plant in which members of the Oil, Chemical, and Atomic Workers Union participated, the company refused to pay employees for time spent in the walkaround. OCAW filed a formal complaint alleging discrimination with the Department of Labor, but it was dismissed on the grounds that no violation of the discrimination provisions of Section 11(c) of the

*The District Court for the District of Columbia dismissed the suit, upholding the challenged state plan's regulations; the case is now on appeal.

Act had occurred. The union then filed suit against Mobil Oil, but the court also upheld the company's position (*Frank Leone v. Mobil Oil*, July 1974). The issue of compensation is important because it will have a significant effect on employees' decisions concerning the exercise of the walkaround right—especially in unorganized workplaces.

The first case applying strict tort liability principles to suits involving occupational disease was brought (and won) by an individual employee against asbestos product manufacturing firms who were charged with failure to provide warning of the dangers involved in handling asbestos products (see Section 4.3.2).

Class-action suits against employers, which as yet have gone virtually untried on occupational health and safety issues, may be a very persuasive method of implementing the OSHAct. A class-action suit for $100 million was filed in 1974 in federal court on behalf of former employees of the Pittsburgh Corning Corporation asbestos plant in Tyler, Texas (*Yandel v. PPG Industries, Inc.*). The suit charges negligence by the company and a trade association in knowingly exposing workers to extremely high concentrations of asbestos fibers, failing to correct the hazards, and not warning employees of the dangers.[2]* The *Yandle* case may pave the way for more class actions on behalf of exposed workers suffering from the effects of occupational hazards (see Section 4.3.2). The possibility of a similar class action against Goodrich Rubber for exposure to vinyl chloride has no doubt influenced the company to cooperate with the Rubberworkers Union and the government in detection and control of the substance.

Legal assistance is relatively unavailable to unorganized workers. To date, the Health Research Group and the Environmental Defense Fund have provided valuable legal advice to unions and their members and assistance in filing complaints, petitions, and lawsuits. Some unions and public-interest groups are educating union members on their legal rights under the OSHAct, as will be discussed below. However, while the unorganized worker does benefit generally from the precedents set by union lawsuits, his individual problems rarely reach either OSHA or the courts. The unorganized worker is unlikely to know of the existence of the OSHAct, let alone have information about the legal assistance available from the public-interest groups.

The impact of court action and legal assistance on the establishment of health standards, on the testing of the OSHAct (particularly with regard to employee rights), and on the determination of liability regarding occupational disease has been very strong. The continuation of similar legal action would appear to be necessary for improving worker health and safety since it appears that the OSHAct is being implemented reluctantly by the present administration.

*The *Yandle* case is accompanied by Federal Torts Claims Act complaints against government officials and agencies for failure to warn the workers at Tyler, Texas, or to take enforcement action.

10.3 LOBBYING, TESTIFYING, AND RELATED ACTIVITIES

Legal action has been crucial in forcing the implementation of the OSHAct. Lobbying, testifying, and other political activities have also helped to shape the course of federal and state occupational health and safety efforts. Various groups representing management, labor, the public interest, and health and safety professionals have brought pressure to bear from all directions on the federal government through legislative and administrative agency testimony and publicity efforts.

On the management side, industry pressures have attempted to weaken enforcement of the Act. The U.S. Chamber of Commerce has testified in support of the removal of first-instance citations and sanctions, removal of the prohibition of advance notice of inspections, and decentralization of the federal authority through state takeover. In testimony on the Toxic Substances Control Act, the Manufacturing Chemists Association opposed mandatory premarket clearance and testing of new chemical substances, which would have indirect effects on workers manufacturing these products. In January 1973 the National Association of Manufacturers recommended adoption of numerous amendments to the Act, specifically amendments which would provide for on-site consultation, advance notice of inspection, and the abolition of mandatory first-instance sanctions. The small business exemption, vetoed by President Nixon, would have excluded from inspection firms with fifteen or fewer employees.

The labor movement, whose lobbying efforts contributed significantly to the passage of the OSHAct in 1970, has continued to speak out for its implementation. Since the OSHAct was passed, the labor movement, spearheaded by the AFL-CIO and its Industrial Union Department, has become increasingly disillusioned with the implementation of the Act and with the government's inadequate commitment to protecting worker health. As Franklin Wallick has noted, instead of experiencing a collaboration with the Department of Labor, as they had expected, AFL-CIO officials found themselves thrust into an adversary position.[3] As a result, the AFL-CIO has spent a great deal of its occupational health and safety effort in lobbying against a weakening of the federal health and safety program and in playing watchdog over the Labor Department. The IUD has applied steady pressure for recognition of the worker's right to health. In its policy statements submitted to the Department of Labor and in its frequent press conferences, the IUD has strongly opposed state takeover of occupational health and safety, on-site consultation, and the low funding and manpower provided both OSHA and NIOSH.

The AFL-CIO and the IUD have been joined by several large, politically aware unions on job health and safety issues. The United Steelworkers of America has vigorously opposed return of enforcement to the states. The United Mine Workers has lobbyed strongly to correct "fiscally discriminatory and punitive administrative decisions" regarding benefits under the black lung program. The Oil, Chemical, and

Atomic Workers Union has been active in lobbying on issues that directly affect workers; it has criticized the Department of Labor for denying workers the information found by OSHA on inspections, and it has endorsed the use-permit system for inclusion in the carcinogen standards.

Though pitifully underfunded, public-interest and health activist organizations such as the Health Research Group have used their legal and professional staffs to lobby and provide testimony in favor of expanded government occupational health and safety programs. In addition to filing petitions and lawsuits, HRG has testified before the House Subcommittee on Commerce and Finance on the Toxic Substances Control Act and at the 1972 oversight hearings on the OSHAct.

On the state and local levels, activist groups have attempted to influence the development and implementation of state plans and the organization of local occupational health and safety assistance programs. PACOSH, a Pittsburgh area group, participated in the legislative hearings on the Pennsylvania plan (which was subsequently withdrawn). The Massachusetts-based Urban Planning Aid prepared detailed critiques of the Massachusetts occupational health and safety laws and their enforcement. The proposed Massachusetts state plan was found to be far inferior to existing federal efforts.

While some professional societies (such as the American Chemical Society) have hesitated to become involved in political activity, others have raised the occupational health and safety banner and applied their expertise both in support of and in opposition to federal efforts. The American Conference of Governmental Industrial Hygienists has urged OSHA to make distinctions between known human carcinogens and those found in animals in the establishment of standards. Both the American Society of Safety Engineers and the American Industrial Hygiene Association have expressed concern over the number and competency of occupational health and safety personnel associated with the implementation of the OSHAct and state plans under the Act. The Industrial Medical Association indicated in testimony its support of a strong federal involvement with a broader role for NIOSH in the standard-setting process. One of the most active societies, the American Public Health Association, has exhibited a high level of concern over workers' health. In its testimony on the OSHAct, the APHA expressed its strong belief that employees' health must never be sacrificed for economics. The APHA also actively supported the OCAW strike of the Shell Oil Company over health and safety issues. The recently formed Society for Occupational and Environmental Health promises to be a valuable, well-balanced group whose viewpoints may influence future government activity in occupational and environmental health.

Management and labor have always been active in lobbying efforts. However, management has traditionally had more ready access to expert testimony on technical, medical, or scientific matters to present before congressional committees or administrative agencies. Recently, a small but effective group of experts has become available to workers through public-interest and some professional groups, but their number remains inadequate to counterbalance industry expertise.

In sum, the continued voicing of all points of view is necessary for a resolution of occupational health and safety issues. The interests of working people have not been adequately represented by the scientific and medical experts. Since many decisions turn on technical arguments, both scientific-medical expertise and information should be made as available for labor as it is for management.

10.4 EDUCATIONAL AND INFORMATIONAL ACTIVITIES

The dissemination of information about workplace hazards is crucial for improving the work environment. This includes information on how the hazards can be recognized, monitored, and controlled and on how the OSHAct can be used to eliminate occupational health and safety problems. Only by creating a heightened awareness on the part of workers, employers, health and safety professionals, and the general public can it be made clear that the mandate for a safe and healthful workplace is everyone's responsibility. (See Section 2.5 for a discussion of some relevant popular books.)

10.4.1 For Workers

Educational activities have increased, particularly for the worker. In addition to an emphasis on workplace hazards, reeducation efforts are now being directed toward changing workers' traditional acceptance of hazardous conditions and fear of job insecurity.

The budgetary constraints on unions and public-interest groups have forced them to be selective in their educational and informational efforts for workers. Such specialization brings with it a need for close interaction among interested parties to pool and organize information and educational activities. Exchanges of information do occur through the IUD and on an informal basis.

The activities of unions and public-interest groups can be categorized into the following areas: (1) education of workers concerning the law and workers' legal rights under the OSHAct; (2) dissemination of technical information on occupational disease and potential workplace hazards; (3) training of workers to identify and monitor workplace hazards and to establish self-help programs; and (4) establishment of ongoing national or regional training centers and programs.

Not all groups have developed the same approach to occupational health and safety educational activities. The differences between union programs can be traced in large part to factors of size, structure, and emphasis by union officials. Several of the large labor groups, including the AFL-CIO and the Steelworkers, have devoted most of their efforts to influencing national policy and fighting the return of jurisdiction over occupational health and safety to the states. Other unions, such as the Textile Workers and Auto Workers, have devoted most of their efforts to local problems. Similarly, the public-interest groups differ in their emphasis and their visibility to workers. The Health Research Group is quite effective in its efforts to influence the federal bureaucracy but is not well known to local unions and workers. On the other hand, the Urban Planning Aid group is well known for its close interaction with workers in eastern Massachusetts. In some local areas, unions have combined funds and donated manpower to coalition groups and activist groups such as CACOSH and PACOSH.

The Use of the Law

Education concerning the law and workers' legal rights under the OSHAct has been an increasingly important activity. Several union and public-interest groups have sponsored courses for workers to apprise them of their rights under the OSHAct and to explain the procedures followed by OSHA. The AFL-CIO in conjunction with the IUD has sponsored a series of special courses on the OSHAct at ten universities around the country. These basic four-day courses, attended by over 250 union representatives, covered five areas: explanation of the OSHAct; inspection procedures and employee rights; industrial hygiene (including threshold limit value concepts and sampling and monitoring techniques); job safety; and the role of the unions, particularly in bargaining for safety and health clauses. OCAW is conducting programs with local and district councils on the uses of the OSHAct, workers' rights during inspections, and the filing of complaints. The public-interest group PACOSH, with the cooperation of several Pittsburgh-area unions, has developed a comprehensive course on occupational health and safety which includes an emphasis on the law itself.

In addition to providing coursework on the OSHAct and workers' rights, unions and public-interest groups have collected some basic information to disseminate to union members. Several pamphlets of the Textile Workers Union of America and the United Auto Workers Union are aimed at familiarizing workers with the OSHAct and its potential uses by local unions. The Teamsters have published an excellent manual, *Job Health, Safety and You*, which includes sample letters workers can send to request an OSHA inspection.

While current educational and informational efforts in the legal area are by no means adequate, they are critical for arming workers with the knowledge that there exist legal mechanisms to help solve workplace hazard problems. These activities are political in nature and are to be distinguished from parallel efforts to educate workers on technical matters.

Technical-Medical Information

The IUD has conducted a series of courses and seminars for union members on technical-medical aspects of occupational disease. For example, in 1972 the IUD, in conjunction with the AFL-CIO and the Mt. Sinai School of Medicine, held a three-day seminar on asbestos. In cooperation with its union affiliates (steelworkers, machinists, and electrical workers), the IUD has also held a series of courses on such topics as cotton dust and carcinogens.

OCAW has been in the forefront in implementing the philosophy that scientific and technical information need not and should not be a mystery to the worker. A fourteen-week training course conducted by members of an ad hoc Scientists' Committee on Occupational Health was established by OCAW at Rutgers University in the late 1960s.

Health-PAC, a New York-based public-interest group, conducts workshops and supplies technical assistance and speakers for community, worker, and student groups. In the fall of 1973, a five-week course on occupational health and safety was taught by the professional Health-PAC staff for New Jersey workers in the United Electrical Workers Union.

All workers cannot attend courses, seminars, and conferences in order to become educated about the hazards found in the workplace. The dissemination of technical information, particularly on occupational disease and potential workplace hazards, is a broad-based activity in which increasing numbers of unions are participating. The IUD publishes an excellent quarterly newsletter, *Spotlight*, devoted to occupational health and safety and emphasizing short descriptions of key occupational hazards.

Many of the large unions either publish a separate newsletter on occupational health and safety issues or devote sections of their regular journals to the subject. As early as 1966, OCAW conducted meetings with local unions to determine the extent of the occupational health hazards facing its membership. Based on this and other work, OCAW publishes a self-help newsletter, *Lifelines*, which provides workers with the latest information on health hazards and their abatement. In addition, Dr. Jeanne Stellman and her staff at OCAW, along with various consultants, have written booklets on mercury and on benzene that describe in nontechnical but not oversimplified language the uses, proper handling, dangers, and effects of these substances. Forthcoming are similar booklets on asbestos, lead, and carbon disulfide. The United Mine Workers devotes a full column in its monthly journal to pneumoconiosis and other diseases related to coal mining. In addition, the UMW has distributed 25,000 copies of its booklet *Black Lung* which describes the disease and suggests measures to control it. The Medical Committee for Human Rights distributes a *Catalogue of Publications* to persons on its mailing list, which lists books, pamphlets, and fact sheets prepared by MCHR, Urban Planning Aid, the United Auto Workers, Teamsters Local 688, the United Electrical Workers, the Massachusetts Department of Labor and Industries, The University of Wisconsin's School for Workers, and CACOSH. As part of its role as a clearinghouse for occupational health and safety literature, MCHR has proposed that a "Work Information Center" be established. In addition to providing information, the center would give aid in utilizing existing government services, maintain a file on model collective-bargaining approaches to the problems of occupational safety and health, give seminars on specific problems, and seek media support for the groups associated with the program.

Training in Self-Help Techniques

Efforts to make workers more self-sufficient in recognizing hazards and taking appropriate action to improve the workplace have been an integral part of educational activities. Urban Planning Aid focuses much of its efforts as on teaching the worker how to identify and correct hazards, with an emphasis on training the worker to be his own on-the-job inspector. UPA helps the worker to develop self-confidence in taking action, in addition to teaching him how to identify specific hazards and how to use technical instrumentation. The OCAW Health and Safety Office is training selected workers to administer lung-function and hearing tests to members in their locals. This joint research-education program attempts to place responsibility for monitoring and control of hazards in the hands of workers. By training local health and safety representatives in industrial hygiene and teaching techniques, OCAW hopes that a multiplier effect will spread awareness of the hazards to a broad base of workers.

Institutional Training Programs

To date, most of the educational and informational activities which have been described have been fostered by small groups of dedicated people in public-interest organizations, local unions, or the small health and safety departments of international unions. They have had inadequate resources to do the task at hand, and to a large extent their efforts have been necessarily ad hoc or in response to day-to-day developments in the field. Recently, more established institutions have addressed occupational health and safety problems on a national basis. For example, the AFL-CIO established the Labor Studies Center in Silver Spring, Maryland, in 1969. The center sponsors one- to four-week courses for full-time elected officials and staff of national and international unions in such areas as collective bargaining, arbitration, fringe benefits, labor law, and the OSHAct of 1970. (The center has so far been underutilized, reflecting perhaps the emphasis of the national leadership on political rather than educational activities.)

The University of Wisconsin School for Workers, in operation since 1925, has recently placed more emphasis on occupational health and safety. The School for Workers, a department within the University Extension, provides services to local union members, stewards, officers, and staff representatives through extension classes, weekend and one-day institutes, staff training programs, research projects, and union program consultation. As part of its program of instruction, the school gives courses and seminars in occupational safety and health problems. Other fields covered by the school include collective bargaining, labor law, arbitration, labor economics, and union administration. The faculty and staff engage in research on occupational safety and health problems and participate in numerous conferences with local unions. Recent emphasis has included research into the causes of occupational injuries; study of the background and training of Wisconsin safety and health specialists; investigation of the administration of Wisconsin health and safety laws; a study of the inspection effectiveness of health and safety specialists; and technical assistance to local unions. The school is conducting research on injuries in Wisconsin to test the widely held hypothesis that accidents are due 90% to a worker's unsafe act and 10% to unsafe conditions. Early research findings suggest that injuries are related 59% to unsafe conditions and 31% to unsafe acts by workers (see Section 3.5.1).

Under a $300,000 OSHA-supported demonstration project running through mid-1976, the School for Workers will develop procedures for establishing worker safety committees and joint management-labor programs, and for processing worker complaints. The school is also working on a catalogue of resources available to help workers recognize and control hazards.[4] The School for Workers has already prepared a series of pamphlets on OSHA: *Inspection Procedure; Making Best Use of the OSHA Walkaround; Employee Access to Information; Worker Rights and Responsibilities Under OSHA;* and *Coping with Mechanical Hazards*, which the IUD is recommending for distribution to its affiliated unions.

According to Ed Estkowski, OSHA Director for Region V, OSHA complaints in Region V went up 900% in the three quarters preceding June 1972, giving the region seven times the number of complaints of any other region.[5] Eight months later he attributed this large number of complaints to the activities of the University of Wisconsin School for Workers and other groups whose efforts in worker education have been successful.[6]

Another university-based group involved in the development and delivery of education programs is the Center for Labor Research and Education at the University of California, Berkeley. Its Occupational Health and Safety Project involves preparing and disseminating materials concerning health and safety hazards and rendering technical assistance to unions and labor-management groups. The emphasis of the program is on organized workers and the collective-bargaining process. The center held a successful conference on occupational safety and health in November of 1973 and plans to develop a newsletter and other information-dissemination mechanisms in order to involve workers in the process of hazard control.

There are indications that as a result of the efforts of these groups, workers are becoming more aware of some hazards that exist in the workplace. However, much more activity is needed in this area. In the minds of many workers, health and safety issues are still secondary to the problems of wages and benefits.

The problem of getting information to the unorganized worker is particularly serious. Since OSHA does not actively disseminate information and since health and safety has not yet become an important organizing issue (except among farmworkers), the unorganized worker is relatively uninformed (see Section 7.7).

10.4.2 For Employers

Inasmuch as consultation and direct technical assistance are the mechanisms by which most employers become "informed" on the issues, the discussion of information dissemination per se will be brief. A discussion of technical assistance will be found in the following section. Management groups, trade associations, and insurance carriers have been the main providers of information and education for employers on their responsibilities under the OSHAct.

The Conference Board and the Industrial Health Foundation disseminate the results of their research findings among their membership, comprised primarily of companies. In 1971 the National Association of Manufacturers sponsored a nationwide closed-circuit teleconference on OSHAct's requirements which reached more than 10,000 executives.

In 1974, the National Safety Council received a $3 million contract from OSHA to train teaching teams from fifty local safety councils at the OSHA Training Institute. These teams will then conduct twelve courses for local employers and employees—ranging from an introduction to the OSHAct to recognition and control of specific hazards. Although the NSC has recently become somewhat more involved in occupational health issues, it still retains a safety bias and may be expected to provide employers with more information on safety issues than on the more critical health issues.

10.4.3 For Professionals*

The rapid increase in the number of health hazards has created a large information transfer problem for occupational health and safety professionals. The American Public Health Association, led by its president, an occupational physician, has given occupational health issues visibility in its newspaper and monthly journal and at its annual meeting. The AMA's Department of Environmental, Public, and Occupational Health has held 34 annual conferences to help occupational physicians keep abreast of the field. In accordance with their educational role, the AMA publishes numerous pamphlets and articles on occupational health issues ranging from pneumoconiosis, radiation, and pesticides to plant surveys, record-keeping, and physi-

*Inasmuch as a large number of professionals are employed by industry, informational and educational activities directed toward them also benefit the employer.

cian responsibilities. The Industrial Medical Association and the American Society of Safety Engineers each publish a monthly journal on significant issues in their areas of expertise.

On the health advocacy front, Health-PAC, whose premise is that adequate health care is a basic human right, publishes a monthly bulletin and a series of special reports on the power and politics of U.S. health care. The subscription list to Health-PAC's publications consists largely of professionals, university libraries, community groups, and HEW personnel—those already involved in the health field.

An increased awareness among health and safety professionals of the legal and technical problems in the field is certainly an important goal. Some professional societies and health advocacy groups are taking up the problems of education and information dissemination. However, as with activities to educate and train workers, the success of these efforts is difficult to measure.

10.4.4 For the Public

The public is relatively uninformed about either occupational health and safety problems or the OSHAct. OSHA does not utilize public-service announcements on television to advertise occupational health and safety problems. Recently, occupational health problems such as asbestos and vinyl chloride have started to attract the attention of the news media, which was previously inactive in this area.

10.5 TECHNICAL ASSISTANCE

Employers receive technical consultative services on occupational health and safety problems from three sources: private consulting firms, trade associations, and insurance carriers. In general, these services are designed to minimize the financial burden on employers in their efforts to comply with the OSHAct. Large corporations are often able to obtain sufficient technical services from their in-house safety, industrial hygiene, and medical staffs (see Section 9.2.1). Law firms are also a potential source of both legal and technical advice, but their services are usually sought only after an OSHA inspection has occurred.

Private consulting firms vary only slightly in their approach; the emphasis of their services is on making health and safety economically beneficial, or at least not detrimental, to the employer. For example, Occupational Safety and Health Associates, Inc., indicates that their services are not directed merely at safety and health, but rather at good management, increased profits, and "getting a safety and health dividend on the side." OSHCA uses a hard-sell technique based to a great extent upon the employers' fears of heavy penalties resulting from inspection. Their service package focuses on the training and supervision of employees, and on reducing the liability of employers to OSHA citations through early hazard identification.

M&M Protection Consultants of Marsh and McLennan conducts seminars to educate company personnel, performs general physical audits from which occupational health and safety priorities are developed, and assists clients to evolve an effective self-inspection program for "continuous compliance" with the OSHAct. It also evaluates the effectiveness of existing loss-prevention programs, and provides general consultation concerning industrial hygiene requirements of the Act. M&M emphasizes the need for recognition of corporate responsibility in occupational health and safety by encouraging its clients to do more than the minimum required by the letter of the law.

Tabershaw-Cooper Associates refers to itself as an occupational health agency and a medical consultant. TCA views the private consultant as a referee between two adversaries: management and labor. In response to an asbestos problem, TCA suggested both a shifting of workers to limit excessive exposure to a few workers and a pretesting of workers to single out hypersusceptible individuals. In this regard, TCA is typical of firms who claim to be holding a middle course on occupational health problems in order to satisfy both management and labor, but who tend in fact to favor management.

Some management consultants mirror the inappropriate emphases set up by OSHA and thus diffuse OSHA's philosophy throughout the entire economy. Capitalizing on OSHA's safety bias, some consulting firms utilize high-pressure advertising and scare tactics to transmit inappropriate incentives to private employers, particularly small ones, in an attempt to convince them of the necessity of augmenting their safety manpower through consulting services.

Most industry and employer groups do not provide members with direct consulting services, but rather act as forums for the exchange of information and views and as a lobby for member interests. Some groups do, however, function in a research capacity or in an information collection and dissemination role.

A group which does provide direct services is the Industrial Health Foundation, a nonprofit research association for the advancement of healthful working conditions in industry, founded in 1935. Its membership has traditionally been limited to companies, partnerships, or corporations. The Industrial Health Foundation provides members with free technical services in four categories: (1) engineering services, including a plant survey or audit; (2) occupational medicine services in the form of a medical survey; (3) short-term toxicology studies conducted by the IHF research laboratories; (4) informational services through its journals and newsletters.

The larger insurance carriers generally provide employers with assistance in the recognition and evaluation of risks, and in loss prevention and reduction. They are motivated to provide such services to the extent that their profits are increased by a reduction in compensable injury and disease. For example, the loss-control staffs of many insurance companies have the dual role of (1) determining premium levels based upon the condition of the workplace, and (2) assisting the policyholder to reduce hazards. Because insurance services and policies are most responsive to workmen's compensation claims, where the emphasis is on injury rather than disease, extensive occupational *health* services are generally not provided by carriers. In fact, the American Insurance Association is opposed to changes in workmen's compensation laws which would extend their coverage of occupational disease.

The American Insurance Association, a trade association of stock insurance companies, has exhibited a strong safety bias in its activities. This is significant in that stock insurance companies write policies collecting 69% of all workmen's compensation premiums. On the other hand, the American Mutual Insurance Alliance, an association of mutual insurance companies which provides 25% of the nation's private workmen's compensation insurance coverage, has presented a more balanced approach in its promotion of the application of both sound safety engineering and industrial hygiene principles. Like the AIA, however, the AMIA resists the extension of workmen's compensation to occupational disease.

One of the most progressive carriers is the Liberty Mutual Insurance Company, which has been the largest writer of workmen's compensation insurance since 1936. Liberty Mutual operates several rehabilitation centers and a research laboratory; it runs training programs for its clients; and one of its vice-presidents participates in the National Advisory Committee for Occupational Safety and Health. The company offers three main services: (1) a merit rating system designed to give policyholders percentage reductions in premium rates when warranted by superior physical conditions of the policyholder's plant; (2) a loss-prevention service, the amount of which is "based upon what is needed to adequately evaluate and provide assistance on the safety problems of that risk";* and (3) a claims medical service which both helps the policyholder obtain the proper mix of medical services and procedures, and reviews the treatment afforded to employees.

Liberty Mutual has often been cited as the exception to the generally accepted image of the insurance carrier. Through television commercials and the involvement of top executives (Roger Wingate and Melvin Bradshaw) in national occupational health and safety affairs, Liberty Mutual has maintained an image of social concern. Occupational *health* issues are certainly more emphasized by Liberty Mutual than by other insurance organizations, but it remains to be seen how willing the company will be to include chronic occupational diseases in its compensation structure. (See Chapter 8 for an economic analysis of workmen's compensation.)

Technical assistance to workers, while on the increase, is still less available than it is for management, in part because only management is legally responsible under

*In 1971 the Liberty Mutual Loss Service had more technical personnel than OSHA had compliance officers. Further, in 1970 Liberty Mutual Loss Prevention personnel made 132,829 visits to policyholders, versus only 14,800 inspections made by OSHA *in its first two years.* Eighteen industrial hygienists made 1300 surveys of toxic explosives and analyzed over 6000 samples in 1970. Loss-prevention services include technical assistance and "training institute" courses.

the OSHAct for making the workplace safe. Of the seven major unions studied for this report, only one has full-time medical expertise, and only three—the United Auto Workers, United Rubberworkers and OCAW—have full-time industrial hygiene or scientific expertise. The UAW's four-man health and safety team of professionals conducted a workplace pollution survey in 1970, utilizing information obtained from 400 plants in a comprehensive Environmental Occupational Health and Safety Questionnaire. The Textile Workers Union has actively utilized private or university scientific-medical consultants and NIOSH hazard investigations in local crises. OCAW is doing biological monitoring of its members for lead and mercury exposure and testing their lung function and hearing.

Generally, health activist groups have not been able to provide a great deal of technical assistance to unions. Barriers to such assistance include the workers' traditional lack of responsibility for and control of the workplace and the unavailability of resources to hire sufficient expertise and equipment. While *biological* monitoring of employees can be done to detect the presence of some hazards, unions and interested parties other than the employers and the government do not generally have access to the workplace and thus cannot do *environmental* monitoring. The legal issues surrounding an employee's right to monitor his work environment are far from resolved. Most interested unions have preferred to perform biological monitoring away from the plant, to train workers in hazard recognition, to establish worker committees to put pressure on employers, and to bargain for occupational health and safety items in their contracts in order to obtain environmental monitoring.

Both the small firm and the unorganized worker are sorely in need of new sources of information and technical-medical assistance (see Sections 7.6 and 7.7). One avenue for providing such services would be to combine occupational and general health care, perhaps delivering both *at the workplace*, as is the normal practice in Eastern Europe. A unique program in Philadelphia is now testing the feasibility of linking medical care, industrial hygiene, and safety engineering services to both employers and employees. The Franklin-Hahnemann Institute of Occupational and Environmental Safety and Health cooperates with the Health Service Plan of Pennsylvania, a nonprofit, prepaid group practice Health Maintenance Organization, in a joint effort to develop an efficient and economical health-care delivery system integrating occupational and family health services. The Health Service Plan is based upon the Kaiser Foundation Health Plan model.

The Franklin-Hahnemann Institute has three goals for its occupational health and safety work: research, education, and direct services to industry and unions. The program is an interesting experiment which should be examined closely as it progresses. The program has three distinct advantages. First, it attempts to provide a total health-care program, dealing with problems encountered on and off the job in a unified fashion. Second, the overall emphasis is on prevention rather than treatment. Third, the direct service component has been aimed at companies with under

100 employees, who are not able to run a comprehensive in-house program. These direct services are provided on a per capita subscription basis which reduces the overall expense and risk to a small company. According to Franklin-Hahnemann, the per capita service could conceivably "include basic safety monitoring and possibly the health maintenance tasks now inadequately performed via workmen's compensation, leaving the purely insurance functions of compensation for lost time and survivor benefits to insurance carriers."[7]

It would be possible for unions to contract, independently or jointly with management, with a group like Franklin-Hahnemann for preventive occupational health and safety services. While the use of such preventive services would help unions identify the health problems of their members, it is doubtful whether private consultants, while performing services solely for a union, could enter a business establishment to monitor the working conditions of union members. In this instance, the "preventive services" would most likely find health hazards only after workers have been affected (e.g., after lead or mercury has been detected in the body).

There has been mounting pressure from management for the government to provide consultative services to employers (see Sections 4.3.2 and 4.4.1). It is important to examine the motivations behind this pressure. While government dissemination of information about the OSHAct has been poor, there are indications that *consultative services would not be used even if provided by the government.* Staff members of NIOSH have been available for consultation with employers and employees through the Hazard Evaluation Program, but this service has been used only rarely—principally because management objects to government employees coming in for any purpose. Congressman William Steiger's unsuccessful amendment to provide consultative services was supported by business partly because it would essentially have converted inspectors to consultants and thereby weakened the OSHAct. States with state plans can provide consulting services. In states without state plans, a new program of agreements under Section 7(c)(1) of the OSHAct provides for the use of state personnel to act in a federal capacity as consultants; an allotment of $5 million has been made for this effort.[8]

There *is* a need for technical assistance to employers and employees—but both groups are unaware of the magnitude of the need, especially in the health area. It is not clear that private consulting groups have felt any increased demand for occupational health and safety services as a result of the OSHAct. Were the Act to be rigorously enforced and employers and employees informed of the extent of workplace hazards, a large demand for these services would arise naturally.

10.6 COLLECTIVE BARGAINING AND STRIKES*

Historically, most labor unions have not emphasized health and safety in collective

*See Section 4.5 on the relationship between the OSHAct, existing labor law, arbitration, and collective bargaining.

bargaining for several reasons. Worker concern with inflation and economic problems has often preempted health and safety considerations. The complex nature of the collective-bargaining process and the often coincident distress of a strike have limited the number of issues that can be realistically negotiated. Most importantly, the fear that jobs would be lost if strict safety and health standards were introduced, or if a worker filed a complaint with the state authorities, has often deterred unions from raising this issue except in extremely serious situations. Further, it was not until 1966 that the National Labor Relations Board established the principle that health and safety issues are mandatory subjects for bargaining. With the advent of the OSHAct, labor hoped for a strong federal role which they now realize is not forthcoming.

The present trend is toward contract bargaining as *the* union mechanism for guaranteeing improved workplace conditions. Unions are working to adopt specific clauses in their collective-bargaining agreements which provide coverage in the following areas:[9]

- *Funding:* A joint labor-management trust fund specifying a per employee-hour contribution by the employer. This fund would provide a financial base for conducting industrywide studies into the causes and prevention of occupational injuries and diseases.
- *Expanded protection:* Incorporation of OSHA regulations and standards into labor contracts, as well as health and safety provisions based upon information presented by the union.
- *Arbitration:* Provisions for an expedited arbitration procedure when occupational health and safety disputes occur. Clauses guaranteeing the right to strike over alleged dangerous conditions are also being sought by the unions. These provisions place full responsibility for resolving occupational health and safety problems directly on the parties most concerned.
- *Union inspection rights:* Provisions for unlimited union inspection rights.
- *Use of impartial experts:* Provisions for the use and funding of impartial health and safety experts when disputes arise or when conditions become dangerous.
- *Training of union health and safety stewards:* Provisions for on-the-job health and safety training for selected union members to aid in the enforcement of standards.
- *Multiphasic screening tests for employees:* Periodic medical examinations to help identify health hazards to which union members are exposed.
- *Establishment or expansion of safety and health committees:* To conduct inspections and investigations, and to participate in the setting of general policy.

It should be pointed out that many of these goals have yet to be incorporated into existing collective-bargaining agreements. As these items suggest, the emphasis in collective bargaining is shifting from compensation of workers for injuries or diseases sustained during the course of employment to direct attempts at hazard control in order to make the workplace a safe and healthful environment (in which the need to compensate workers is minimized).

Federal mediation officials have noted that in 1973, for the first time in major industrial bargaining, safety and health demands formed a major issue in contract stalemates.[10] In 1973 the main advances in negotiating strong health and safety clauses into collective-bargaining agreements occurred in the oil and automotive industries. OCAW threatened strike action until safety and health issues were included in the contracts: Shell Oil was, in fact, struck. The union gained major concessions from American Oil, Atlantic Richfield, and Standard Oil of Ohio, and was successful in having its four-point program adopted in collective-bargaining agreements. The negotiated plan included: (1) joint labor-management health and safety committees with power over workplace conditions; (2) periodic health surveys of plants by independent health consultants approved by both labor and management; (3) medical examinations of workers at company expense; and (4) the provision of annual employee morbidity and mortality data by the company.

The OCAW joint labor-management committee is a major advance over current joint committees. The contract clause outlines a detailed procedure to be followed, including provisions for third-party arbitration. In contrast, the existing contract of the United Rubberworkers leaves to management decisions regarding specific steps to be taken once an independent university expert has made recommendations. Under the OCAW contract, the means by which the company takes affirmative action to correct dangerous conditions must be proposed by the joint committee. Although under the OCAW contract the management representatives on the joint committee can deadlock the committee's progress, the arbitration clause does provide further protection for the union.

The United Auto Workers has won a six-point safety and health program in their negotiations with the automobile industry. While some of these provisions were part of prior programs, they were negotiated explicitly for the first time in 1973. The UAW contract includes: (1) the establishment of a new hierarchy of full-time company-paid UAW safety and health representatives; (2) continuation of the prior union right to strike over safety and health issues; (3) provisions for regular weekly or biweekly inspections by UAW representatives or joint labor-management plant committees; (4) UAW rights to sampling and monitoring equipment for noise and gas, to be provided by the companies; (5) supporting equipment for other environmental surveillance conducted by the union; (6) union access to detailed data from company and government records on employee exposure to dangerous physical agents and chemicals; (7) union access to detailed data on harmful physical agents and chemicals; and (8) company-paid physicals and other examinations and tests for employees exposed to harmful physical and chemical agents.[11]

Strikes and the threat of strikes were powerful weapons in the 1973 collective-bargaining negotiations. OCAW's first success on safety and health provisions was a result of management's realization that the union would in fact strike over these issues. The decision of the Shell Oil Company not to accept the OCAW contract proposals resulted in a four-month strike and a worldwide consumer boycott campaign. Wildcat strikes and threatened union strikes likewise played an important role in the UAW success.[12] Although not related to master contract bargaining, a strike by members of a Textile Workers local over a specific nerve poison found in their Columbus Coated Fabrics plant resulted in a negotiated occupational health program. Included were provisions for a new ventilation system, company-provided protective clothing, rotation of workers on hazardous jobs, and mandatory use of respirators in hazardous areas.

Collective bargaining has the potential to go far beyond the mandates of the OSHAct, by obligating employers to interact closely with workers and not merely comply with government standards. The negotiation process enables different local and industrywide needs to be met, particularly where hazards are extensive. Further, it can move the responsibility for occupational health and safety out of the sole hands of management, and it thus encourages the process of industrial democracy. To the extent that collective bargaining changes the way of doing business with regard to health and safety on the job, indirect benefits may accrue to unorganized workers, although very slowly.

Finally, it should be noted that occupational health and safety has not only been one of the main bargaining issues for the United Farm Workers, it has also been *the* main organizing issue (see Chapter 12). It remains to be seen whether occupational health and safety in general will give rise to new organizing efforts on the part of the industrial unions.

10.7 RESEARCH*

There is an insufficient scientific data base upon which to make decisions concerning the risk of harmful effects as a result of exposure to suspect hazards. Research on environmental health hazards is performed in a more or less uncoordinated way at more than eight government agencies, at about twenty schools of public health or environmental medicine, and at private industry-supported foundations such as the Industrial Health Foundation.

There is no union-sponsored foundation similar to the Industrial Health Foundation which conducts research into occupational health and safety problems, although suggestions for such a facility have been repeatedly made. Individual unions do undertake or sponsor some practical research of relevance to their workers' particular job tasks. For example, OCAW is investigating the effects of lead and mer-

*The state of the art in safety and health research was discussed at length in Section 3.5. The specific assessment of the research conducted for or by NIOSH is to be found in Chapter 6. A comprehensive HEW study to provide guidelines for research planning in environmental health science has resulted in a valuable report entitled *Man's Health and the Environment–Some Research Needs* (see Section 3.5.2).

cury and is also attempting to establish a correlation between stress-related disease and high levels of noise. The Amalgamated Clothing Workers are actually paying the salary of a NIOSH scientist to investigate their workers' problems.[13] The United Rubberworkers cosponsor, with industry, research at two public health schools, Harvard and the University of North Carolina. The Textile Workers Union uses scientists as consultants and cooperates on research projects involving cotton dust at universities. The small amount of union-sponsored research has importance beyond the value of the resulting information or data. It represents a realization on the part of the unions that access to information and expertise is critical for arguing and bargaining intelligently with management and for actually gaining control of certain technology in the workplace. Even though the payoff from supporting such research would be great, unions by themselves simply do not have the resources to become major sponsors or to balance management's control of information and expertise. Only a few unions have occupational health and safety professionals, and the demands made on them are enormous.

The National Academies of Science and Engineering have expressed the necessity of recognizing the "total body burden"[14] in evaluating the effects of an environment on the human organism. The three factors they consider crucial are: (1) multiple sources of a particular toxicant; (2) the additive or synergistic effects of all the toxic substances acting on the body; and (3) the additional effects of environmental stressors, such as poor nutrition or overcrowding. (Noise, perhaps the most prevalent stressor, and psychosocial stressors were omitted, but could well have been included.) Because research is fragmented among different governmental and private institutions and because each project tends to concentrate on a single hazard, most research is not of much direct, practical use. The exception to this generalization is epidemiological research to detect abnormal disease patterns in various populations.

As was made clear in Chapter 3, there can never be enough research done with the present or anticipated resources to address all possible hazards and their effects in combination with one another and in combination with stressors. One cannot disagree with the usefulness of more epidemiological studies of workers in suspected hazardous occupations and demonstrations of the generality or frequency of synergistic effects of multiple hazards, including the effects of stress. Indeed, such research is needed not merely to improve the scientific data base, but to demonstrate how *pervasive* the problems are in order (1) to stimulate new technology and methods of production,* (2) to build a case for shifting the burden of proof as to the safety of a material to the employer, and (3) to force the government to develop a

*As was pointed out in Chapter 3, there is abundant technology presently available which has not been applied to improve job safety and health. It would appear that the mere application of presently known technology, such as ventilation systems, would significantly improve many industrial workplaces.

coordinated policy instead of responding only to crises.

The kind of research chosen should have a multiplier effect. Demonstration projects, not mechanistic studies, are what is needed. Scientists will argue for the next twenty years on the applicability of animal studies to humans, on the concept of the "zero threshold" for carcinogens, and on the specific causes of disease. The increasing problems of job injury and disease cannot wait for the agreement of experts. Strict regulation in the pharmaceutical industry has led to fewer new products and fewer new hazards, a longer use period for accepted drugs, and increased—not decreased—research and development. Like pharmaceuticals, the rapid proliferation of chemical substances might also be checked without harm to either the economy or society. Regulation would stimulate more industry-sponsored research on possible harmful effects.

In sum, research is needed to demonstrate that the present course of research is not the best alternative use of resources and manpower, that merely expanding the data base is not sufficient, that new technology process design is required, and that the lack of a firm data base should not be an excuse for continuing to proliferate health hazards in both the general and work environments.

10.8 THE SETTING OF STANDARDS

The OSHAct relies very heavily on the development and enforcement of occupational health and safety standards. Historically, standards have been developed in two distinct ways. As the name implies, the "consensus method" sought to bring together representatives of industry, labor, government, and the public interest to agree to the "proper degree of safety" required for a particular material, process, or piece of equipment. Since many of the standards agreed to were not mandatory, but merely guidelines, management usually dominated the consensus activities, which were conducted more for the purpose of coordinating industrial practice than for coming to some agreement with labor. Labor had neither the expertise nor the ability to counterbalance industry representatives on the consensus committees. Most standards adopted under the consensus method were safety standards rather than health standards. In the initial period of the implementation of the OSHAct, OSHA adopted a whole set of safety standards which had their origins in these so-called consensus standards.

The second method for establishing standards is to generate them through panels of "experts" and then use an administrative agency's procedure to promulgate them. If any groups are not represented in the recommendation process, their views can be given public exposure through an open hearing before the standards are adopted as final. This has been the method most often used by OSHA since the adoption of the initial batch of standards.

Both unions and the public-interest groups participate actively in the hearings on new occupational standards, but very often the die has already been cast by the panel of experts who have recommended a particular standard. What the open-hearing mechanism provides is information for the public and a record upon which an interested party can find grounds to sue the Department of Labor for abuse of discretionary authority in adopting a standard which insufficiently protects the worker.

Section 6(b)(5) of the OSHAct mandates the setting of standards for toxic materials or harmful physical agents by which "to the extent feasible . . . *no* employee will suffer material impairment of health or functional capacity even if such employee has regular exposure to the hazard dealt with by such standard for the period of his working life." Study of the legislative history of the OSHAct shows clearly that standards were not intended to be the result of any balance or compromise between health and economics.

Thus, the consensus process—a compromise system—is not suited and is no longer used for the adoption of health standards. The consensus organizations continue to function, however, as "advisory committees" and sources of "expert information."

Although OSHA uses committees of experts and public hearings to develop standards, it has been sharply criticized for compromising by taking *economic* feasibility into account in the setting of health standards, in apparent violation of the spirit of the Act. In the decision regarding an asbestos standard, the issue of economic feasibility was legitimized in the setting of standards. It is difficult to understand why society should make greater economic compromises in regulating the work environment than it does in regulating the general environment, food consumption, or drug use. In any event, the adoption of occupational health standards by OSHA is proceeding so slowly that to a large extent workers remain unprotected from significant health hazards, and the enforcement of standards already adopted has been grossly inadequate. (See Chapter 6 for an extensive discussion of standard adoption and enforcement.)

10.9 SELF-POLICING AND VOLUNTARY COMPLIANCE

OSHA has referred repeatedly to its dependence on "voluntary compliance" in enforcing the OSHAct. However, in the more than five years since the passage of the OSHAct, the available statistics do not indicate any improvement in the incidence of job injury and disease (see Section 3.4.1).

The "voluntary compliance" which OSHA hopes to encourage on the part of industry is ineffective because (1) regulations and standards do *not* exist for most serious health hazards; (2) the inspection force is *not* large enough to present a serious possibility of random inspection; (3) significant penalties are *not* being levied; (4) the inspectors are poorly qualified; and (5) employees are *not* well informed on their rights. Furthermore, the return of enforcement of health and safety standards to states that submit plans will make the situation worse, not better (perhaps with a few exceptions, such as California). (See Section 6.3.1 for a further discussion of the prerequisites for a successful program of "voluntary compliance.")

Insurance carriers do exercise some persuasive power on their industrial clients to adopt good safety practices, but obviously, with a 29% rise in the industrial injury rate in the last decade, not enough pressure has been exerted. Moreover, insurance carriers are interested predominantly in acute safety problems, which show up in compensation payments, and have shown little if any interest in chronic occupational disease problems (see Chapter 8).

REFERENCES

1. *Occupational Safety and Health Reporter* (the Bureau of National Affairs, Inc.), Vol. 3, No. 44 (April 4, 1974), p. 1387.

2. *Occupational Safety and Health Reporter*, Vol. 3, No. 32 (January 10, 1974), pp. 995-996.

3. Franklin Wallick, *The American Worker: An Endangered Species* (Ballantine Books, New York, 1972).

4. Phyllis Lehmann, "Safety and Health Go to College–and Back Home," *Job Health and Safety* (OSHA), Vol. 1, No. 9 (August 1973), p. 6.

5. Edward Estkowski, OSHA Director for Region V, interview with author, Chicago, Illinois, June 30, 1972.

6. Edward Estkowski, interview with author, February 5, 1973.

7. Franklin-Hahnemann Institute of Occupational and Environmental Safety and Health, "The Development of Comprehensive Safety and Health Service Concepts," report presented at the Consultation on Occupational Health in Small Industries, Geneva, Switzerland, October 1970.

8. *Occupational Safety and Health Reporter*, Vol. 4, No. 34 (January 23, 1975), p. 1028.

9. Report on the Proceedings of the Conference on Occupational Health and Safety held at the Jack Tar Hotel, San Francisco, November 8, 1973, sponsored by the California Labor Federation, AFL-CIO, and the Center for Labor Research and Education, University of California, Berkeley, pp. 22-23.

10. *Daily Labor Report* (the Bureau of National Affairs, Inc.), No. 2 (January 3, 1974), p. C-1.

11. *Ibid.*

12. *Ibid.*

13. *Spotlight on Health and Safety* (Industrial Union Department, AFL-CIO), Vol. 3, No. 1 (first quarter 1974), pp. 1-2.

14. *Man, Materials and Environment*, A Report for the National Commission on Materials Policy by the National Academy of Engineering and the National Academy of Sciences (March 1973), p. 72.

11 Foreign Experience and Its Relevance for the United States

This chapter, which is meant to be indicative only, will focus on three topics: (1) differences between American and Western European views and practices concerning worker health and safety; (2) a 1972 British parliamentary study of health and safety at work; and (3) the importance of international cooperation in occupational health and safety.

11.1 THE WESTERN EUROPEAN VIEW*

There are inherent difficulties in making comparisons between national systems, due in part to historical differences in social, cultural, and political values. A comparative discussion is further complicated by difficult-to-know subtleties of each system. For example, although a country may have protective legislation regarding worker health and safety, it is necessary to know how extensively the legislation is enforced, what the qualifications and roles of the inspectors are, and how effective compliance is at the local level. The *effectiveness* of legislation, then, is difficult to measure. Nevertheless, there are significant differences among Western European countries, in legislation, in enforcement, and in the patterns of health and safety concern and activity within the private sector, that are worthy of consideration by U.S. policy-makers.

No attempt is being made to represent the situation in Europe as ideal. There are problems in each of these countries, both in terms of the amount of job injury and disease and in terms of systemic problems preventing the further reduction of injury and disease. For example, lack of a reliable scientific data base and lack of services and protection in small firms are problems affecting many countries. Nor is the purpose of this discussion to make comparative judgments of the various European approaches. Rather, its goal is to draw the best examples from the European occupational health and safety approaches in order to show that there are ways in which other productivity-oriented economies pay more attention to worker protection, and especially to health problems, than does the United States.

Table 11.1 is an attempt to view Western European occupational health and safety approaches in broad terms by spotlighting unique legislative features (where known) among the countries visited.

11.1.1 The Role of the National Government

In many European countries, the heads of the agencies and departments involved

*The observations made in this chapter are based on a ten-week investigation of eight countries in Western Europe (Belgium, the Federal Republic of Germany, Finland, France, Italy, Sweden, the Netherlands, and the United Kingdom). I am aware of the shortness of this study, and I make no claims to expertise on each of the systems studied. However, a strong effort was made to be selective and balanced in the persons and institutions visited, and some clear and definite patterns did emerge.

Table 11.1
Characteristics of Western European Occupational Safety and Health Programs

Country	Coverage of Legislation	Administrative Responsibility	Penalties	Occupational Health Services
Belgium	All workers except those in family businesses	Ministry of Labor: Industrial Safety Admin., Health and Industrial Med. Admin.	• Fines fixed by the courts • Employer can be charged with criminal neglect	Mandatory for all firms at employer's expense
France	All workers except some in transportation	• Ministry of Labor, Labor Inspectorate • Ministry of Health (Social Security Caisse)	Fines, increased tenfold for second violation; third violation can close firm	• Mandatory for all firms at employer's expense • Factory doctor required in firms with over 200 employees
Netherlands	All workers except the self-employed	Ministry of Social Affairs and Public Health (Labor Inspectorate)	• Fines fixed by the courts • Provision for closure of firms	Mandatory for all firms with 750 or more employees and for all firms with special health hazards
Sweden	All workers except those who work at home	• Ministry of Social Affairs: Labor Protection Board (Labor Inspectorate and special inspectorates) • Local inspectors for workplaces with 10 or fewer workers and without machines	• First-instance orders to improve • Fines • Provision for closure of firms • No penalty for failure to appoint safety rep. or committee	• Encouraged by labor-management agreement at the federation level • Firms with over 1500 workers have internal services • Smaller firms have joint services • Unique building industry program
Finland	All workers except the self-employed, household workers, and family businesses	Ministry of Social Affairs, Labor Inspectorate	• Fines and/or imprisonment (after warning)	Not mandated by law
Federal Republic of Germany		• Ministry of Labor and Social Affairs • Enforcement on state level • Industrywide insurance and accident-prevention institutes (BGS)	• Improvement orders • Fines • BGS premium increases	• Not mandated by law • Proposed bill would make physician mandatory in firms with more than 800 workers • Occupational health services available from BGS

Health/Safety or Works' Council Committees in the Workplace	Recognition of Occupational Disease	Workmen's Compensation	Most Serious Problems	Research Capability
Mandatory in firms with 50 or more employees			• silicosis • noise	Centered in schools of public health
• Mandatory in factories with 50 or more employees, and in commercial firms with over 500 employees • Factory doctor is responsible to the committee	• 48 listed as compensable • If not compensable, employee can sue at common law	• Mandatory • Rate based on firm size and risk • Rate increases serve as important economic lever	• silicosis	One main government-sponsored institute, funded by workmen's compensation payments
• Mandatory for firms with 25 or more employees • Shipping, construction, mining, and railways exempt		• Mandatory • Paid primarily by the employer • Benefits regardless of cause	• silicosis • asbestosis • noise	• Many independent and semi-independent institutes • Concern with mental stress
• Mandatory works' council and safety committee in firms with 50 or more employees • Mandatory safety rep. in firms with more than 5 employees	Recognized and compensable under health-insurance scheme		• ergonomic errors in design of work methods • noise • mental load • asbestosis • vibration	• National Institute of Occupational Health (government-sponsored) • Private institutions
• Mandatory in firms with more than 10 employees • Safety advisor mandatory in firms with more than 20 employees (law pending)	• Recognized by law in 1967 • Compensable under workmen's compensation	• Mandatory • Premium increase for noncompliance with safety recommendations	• silicosis • asbestosis • vibration and nerve disorders • lead • inorganic solvents	• Institute for Occupational Health (private) • Special concern for heart disease, vibration effects
• Mandatory works' council in firms with more than 5 employees • Council has OHS responsibility	• Compensable for diseases arising from certain substances	• Mandatory through BGS • Premium based on % of wage bill, risk level, and performance record	• noise • organic dusts	• Federal research coordinating institute • BGS-sponsored institutes

Table 11.1 (continued)

Country	Coverage of Legislation	Administrative Responsibility	Penalties	Occupational Health Services
United Kingdom*	All workers except household workers and transportation workers	Safety and Health Commission (autonomous, responsible to Sec. of State for Employment) • Inspectorates • Employment medical advisory service	• Warning notices • Provision for closure of firms • Fines	Not mandated by law
Italy	All workers except those in transportation and family businesses	Ministry of Labor and Social Security: Labor Inspectorate, Medical Inspectorate	• Apply to employers *and* workers • Fines and imprisonment (up to 3 months) • Harsher for mines, quarries, tampering with safety apparatus	• Medical exams required by law for some hazards

*Characteristics shown are those of the proposed Safety and Health at Work Bill.

with occupational health and safety have more direct experience in relevant fields than do their American counterparts. For example, in Belgium, the counterpart of the Secretary of Labor comes from the labor movement. In Britain, Mr. Brian Harvey, Chief of the Factory Inspectorate, is a public health professional.

The European counterpart of NIOSH either is located higher in the government hierarchy or is an independent organization. In France, the National Institute for Safety and Research (INRS) is managed autonomously, with its budget and directives coming from the National Caisse (the powerful national social security and workmen's compensation system). Similarly, the Italian National Institute for Accident Prevention (ENPI) is an autonomous public board, responsible directly to the Minister of Labor and Social Security, with 40% of its support coming from workmen's compensation premiums. The Finnish Institute of Occupational Health, an independent organization, has a staff of 409 persons, as compared to NIOSH's staff of 600 for 25 times the working population.

In most of the countries under discussion, the government does more consulting with employers on methods for improving health and safety conditions than does OSHA. In Sweden, the United Kingdom, West Germany, and Finland, a citation is

Health/Safety or Works' Council Committees in the Workplace	Recognition of Occupational Disease	Workmen's Compensation	Most Serious Problems	Research Capability
Joint consultation (including safety committees) voluntary on an individual basis	• Reportable under Factories Act • Some diseases (e.g., pneumoconiosis, byssinosis) are compensable	• Mandatory • Flat-rate premium paid by employer and worker • Common-law suit for damages	• pneumoconiosis • byssinosis	• Government, industry, and university • No coordinating agency • Wide concerns (dusts, cancer, toxic substances)
• Not required by law • Worker constitution	Recognized by law Legal controls required			• Central organization (ENPI) research, exams, and medical hygiene inspections • Schools and departments of industrial medicine • Special interest in ergonomics

issued only after advice and warning has been given. Government enforcement is looked upon as a last resort in Sweden. In France, first-instance sanctions are available; however, labor inspectors rarely cite on first instance.

Standards are generally regarded as guidelines, although they may have the force of law. Discretionary power lies with the inspector, who can issue a citation for unsafe or unhealthy conditions even if no specific standard exists, a procedure similar to citation for violation of the "general duty" clause in the United States. In West Germany, Finland, and Sweden, improvement orders to correct hazards can be issued on "general duty" grounds and are not uncommon.

European inspectors tend to be better trained and more specialized than U.S. inspectors at the federal and state levels. British occupational health and safety inspectors are mostly college graduates in engineering or chemistry, averaging over 26 years of age; they receive a two-year apprenticeship on the job.[1] Of the 650 factory inspectors, over 400 are specialists. Inspectors take seriously the fact that they are "civil servants" and are far less susceptible to bribery than were the state inspectors in the United States prior to the passage of the OSHAct. In West Germany, there are three distinct levels of inspectors: high, intermediate, and junior, with qualifications of advanced degrees, technical college degrees, and technician skills, respectively.

In the United States, there are 1000 federal and 1500-1800 state inspectors to serve 57 million workers covered under the OSHAct (1:20,000). There is no doubt that it is difficult to make international comparisons of inspector-to-worker ratios—due, in part, to differences in occupations covered, local and state inspector capability, and number of man-hours spent. Nonetheless, some general statistics are illustrative of the extent to which the European countries utilize inspection manpower, regardless of whether that manpower consults or cites.

In Finland, on the local level alone (excluding federal factory inspectors), there are 500 inspectors for 2.2 million workers (1:4400). In the German state of North Rhine-Westphalia, there are 710 state inspectors (the *total* German working population numbers 22 million). There are 650-700 British factory inspectors to serve 8.5 million workers (1:12,000) covered under the Factories Act (which excludes commercial firms, mines, agriculture, and transportation).

Legal requirements for occupational health services in the workplace are frequent among the Western European countries studied. In the case of small firms, the law usually requires participation in a joint medical service. In Belgium, France, and the Netherlands, occupational health services, headed by a factory physician, are mandatory in large firms. The minimum size of "large" firms varies among the countries studied from 200 to 1500. In Belgium and the Netherlands, firms with high risk and serious hazards, irrespective of size, must also have occupational health services. Joint occupational health services for small firms are established in France, Belgium, and Sweden (which mandates services not by law but by labor-management agreement). The construction industry in Sweden has developed an industry-supported foundation to provide medical, industrial hygiene, and safety services to building sites via mobile treatment units and to collect data for research purposes. In Germany, industrywide insurance institutes (Berufsgenossenschaften) provide occupational health services for member employers.

Providing occupational health services for small firms is a problem in many of these countries because of the large number of such firms and their wide geographic distribution, especially in remote areas. Problems hampering improvement of such services have been investigated by the World Health Organization in the United Kingdom, France, and Sweden. The Swedish government has made a strong effort to develop regional occupational health services and has about sixty currently in operation.

It may very well be that job safety, as opposed to health, is more advanced in the United States. Indeed, much of the safety equipment used in Europe is made in the United States. Inasmuch as all statistics on occupational injury and disease represent gross underreporting, statistical comparisons between countries are not very revealing.

Occupational disease has been widely recognized in Europe for years. Much of the data on asbestosis comes from British studies conducted in the 1930s. The programs for reporting, controlling, and compensating occupational disease are all far superior to those of the United States. In Sweden, occupational illness has been compensable for twenty years, and a steadily increasing number of diseases have been added to the list. This is also true in Finland, where occupational disease has been reportable and compensable since 1939, with an updated list provided in 1967. Under the German BGS insurance scheme, 47 occupational diseases are compensable. In France, there are 48 compensable occupational diseases, and if a disease is not included as compensable, an employee can sue at common law. In France and Britain, a physician is required by law to report and diagnose properly all cases of occupational disease. In Britain, diseases caused by twenty chemical substances or their derivatives as well as pneumoconiosis and byssinosis must be reported to the Factory Inspectorate. In Italy, control of 47 hazardous substances has been required by law since 1952 (amended 1967). In many cases, an illness may be classified as occupational if a hazardous substance which can cause a variety of illnesses is found in the work environment, even if no direct connection is known.

In addition to requiring occupational health services and recognizing and compensating occupational disease as well as industrial injury, it is not uncommon for European countries to have separate medical inspectorates, generally staffed by doctors, to conduct industrial hygiene inspections and medical examinations. This is true in Italy, France, Belgium, and Germany.

Workmen's compensation insurance, while mandatory in many of the countries studied, varies in the form it takes and the extent to which it affects accident and illness prevention. In France, Finland, and Germany, premiums (which are paid by the employers) are based on a percentage of the total wage bill, risk level, and firm performance record. In France, the threat of rate increases serves as an important economic lever in controlling accident and illness, particularly since the Regional Caisses have an inspection role. The British system utilizes a flat-rate premium and offsets the lack of individual firm incentive (caused by spreading the risk) by enabling employees to sue their employers for negligence to cover any damages not paid for by workmen's compensation insurance.

11.1.2 The Role of Labor and Management

A higher percentage of the labor force in Europe is organized than in the United States. Approximately 27% of all American workers and 50% of the manufacturing workers are unionized. In France, 31% of the workforce is unionized; in the Netherlands, 40-50%; in Belgium, 64% (90% in the industrial sector); in Finland, 70-80%; and in Sweden, 95% (of the industrial workforce). These percentages, however, do not by themselves guarantee a greater impact of trade unions on each country's political and economic establishment.

In all countries studied except the United Kingdom and Italy, either health and safety committees or works' councils with occupational health and safety responsibility are required in workplaces by law. In Belgium, France, and Sweden (where requirements are set by a labor-management agreement), safety committees are mandatory in firms with 50 or more employees. In the Netherlands, the necessary number of employees is 25; in Finland, 10. In Germany, workers' councils are required in firms with more than five employees. Italy has a Worker Constitution which provides worker protection on the plant floor.

On the national level, powerful government departments run jointly by equal numbers of workers and employer representatives are found in the Swedish Worker Protection Board, and the French Regional Caisses and National Institute of Safety and Research. Collective bargaining in Sweden is conducted on the federation rather than the plant level. Agreements on occupational health and safety issues have been negotiated, including the establishment and subsequent expansion of the Joint Industrial Safety Council.

It appears, then, that responsibility for occupational health and safety at both the plant and national level is shared by employers and employees in most European countries. Management does not consider health and safety to be its sole prerogative, as is so often the case in the United States. The interaction between labor and management on safety and health issues is therefore characterized by a greater sense of fair play in a less adversary setting than in the United States.

11.1.3 Characteristics of Research and Services

Health and safety are given more equal emphasis in many of the countries studied than in the United States. In general, government requirements for occupational health services, medical and industrial hygiene inspections, and recognition and compensation of occupational disease are more extensive in these countries. This observation is reinforced by the level and type of interest in occupational health and safety research and services.

In many of the countries visited, preventive medicine plays a significant role in job health and safety. In France, the Netherlands, Belgium, and Sweden, the physician's role in the occupational health service is almost exclusively preventive; treatment may be given only in the case of an emergency. In addition, industrial hygiene monitoring and preventive measures, directed by the physician, form an integral part of the occupational health services. In Italy, there is a move to broaden the scope of occupational physicians' training to include a better understanding of the work environment.

In Italy, France, the Netherlands, and Belgium, the combination of medicine and industrial hygiene in the program of preventive medicine has, in fact, resulted in underdevelopment of the field of industrial hygiene. Only recently has a realization been made of the distinct contribution that industrial hygienists can make to the work environment. In Sweden, safety engineers are employed extensively in industry and often share responsibilities with the physician in both accident and disease prevention.

In the research area, mental health in the workplace receives considerable attention in Belgium, the Netherlands, and Sweden, where stress and mental load are accepted as serious job health problems. The connection between vibration and nervous system damage are being studied in Finland and Sweden, where lumberjacks using chainsaws are a large population at risk. In France, a jackhammer without vibration has been developed at a cost only slightly higher than those in current use. Ergonomics—the study of the design of operations and equipment to take into account human physical and mental needs—has been influential in Europe for many years. Ergonomic research is actively underway in Italy, France, Sweden, Finland, and Germany.

Although research on toxicants and airborne dust diseases is still being done, much of the available data on these problems was collected years ago in Europe. A new and widespread interest in occupational mental stress and ergonomics—subjects still almost wholly neglected in the United States—is an indication of how much farther along many European countries are in tackling occupational health problems.

11.1.4 International Cooperation

The transfer of ideas across national boundaries is a common occurrence among Western European countries. These countries are, of course, knitted together by close economic and geographic bonds (through the EEC and OECD), but it is also true that European countries have utilized the World Health Organization (WHO) and the International Labour Organization (ILO) much more than has the United States as a vehicle for the exchange of ideas on occupational health and safety.

Great importance is attached by the Europeans to international cooperation in achieving common standards and work practices. The traditional nonadversary nature of labor-management relations and the willingness of management to accept government regulation in the area of health and safety also contribute to a stronger national commitment to a safe and healthful workplace and a greater sense of fair play.

11.1.5 Summary: Characteristics of the Western European Approach

The following are the major characteristics noted in this section which differentiate the Western European from the American approach to occupational health and safety:

- It is more often the case that the leadership in the various agencies responsible for occupational health and safety has the necessary health and safety expertise.

- The counterpart of NIOSH is at a higher organizational level within the government—or is an independent organization.
- Where the NIOSH counterpart exists, there is relatively greater capability and the research is more relevant to critical occupational health and safety problems.
- The government consults more with the employer, and first-instance citations are infrequent. Warnings and improvement notices are the major enforcement mechanisms.
- Standards are generally regarded as guidelines, and the inspector has discretionary power.
- The inspectors are more specialized and better trained.
- Physicians are very often part of the inspectorate system.
- There are legal requirements for occupational health services in the workplace or, in the case of small firms, for participation in a joint medical service.
- Occupational disease has been recognized longer and more extensively for both compensation and control purposes.
- Safety and health have a more equal emphasis.
- Insofar as occupational disease is recognized by the compensation system, economic incentives are more effective for removing health hazards on the job.
- A larger proportion of workers are organized.
- Either works' councils or joint health and safety committees for even small firms are generally mandated by law.
- It is accepted that health and safety is the joint responsibility of employer and employee.
- Employers and employees are more often equally represented in the agencies or organizations responsible for occupational health and safety.
- Collective bargaining is often done at the federation rather than at the plant level.
- There is a much greater national commitment to provide a safe and healthy workplace and a greater sense of fair play.
- Preventive medicine plays a more significant role in job health and safety.
- Industrial hygiene is undeveloped as a profession and falls within the duties of the plant physician or the safety engineer.
- Mental health and ergonomic factors in the workplace receive more attention.
- The transfer of ideas across national boundaries occurs much more readily.
- Much more importance is given to international cooperation and the need to agree on standards and work practices.

11.2 THE ROBENS REPORT: A BRITISH STUDY OF SAFETY AND HEALTH AT WORK

In May 1970 a Parliamentary Committee headed by Lord Robens was appointed by the British Secretary of State for Employment and Productivity to review the scope of safety and health legislation and make recommendations regarding changes in existing law and its implementation. The resulting report, *Safety and Health at Work*, also known as the Robens Report, has created a great deal of controversy, particularly because of its emphasis on "self-regulation" in enforcement and in the setting of standards. The report has apparently influenced some American officials, so that the following discussion is quite relevant to the U.S. situation.

One of the report's initial conclusions is that the existence of "too much law" hampers the molding of a comprehensive program. British safety and health legislation has been developed in a piecemeal fashion, with the result that there are nine major Acts which seek to promote health and safety among varying segments of the working population and to eliminate various hazards. Further, the report concludes that the law is "intrinsically unsatisfactory" and "written in a language and style that renders it largely unintelligible."[2] With responsibility for administering the laws diffused among five government departments and seven separate inspectorates, it is no surprise that the Committee viewed the fragmentation of administrative jurisdictions as a major systemic problem. The Committee states its conclusion as follows:

> In sum, the excessive fragmentation of the legislation and of its administration is a serious obstacle to the creation of a more modern code of law, to its effective implementation, and to the development of a clear and comprehensive strategy for the promotion of safety and health at work. It is apparent that tinkering with the present statutory provisions and administrative arrangements will not cure the situation. A major exercise in unification is needed to rationalise the state's contribution to safety and health at work and to maximise its efficiency.[3]

The Committee's recommendation for unification was, understandably, the enactment of a single comprehensive Act which would bring relevant statutes under one central, self-contained, and autonomous administration.

While the primary duty of British inspectors is to secure compliance with the law, their main mode of action is advice to employers rather than legal sanctions. The Robens Report agreed with this policy and stated its belief that a rigorous enforcement policy is "misconceived" and "undesirable."[4] This belief is based on the Committee's principal thesis that "self-regulation" on the part of those directly involved in the workplace is the best mechanism for achieving a safe and healthy work environment:

> A principal theme of this report is the need for greater acceptance of shared responsibility and more reliance on self-inspection and self-regulation and less on state regulation. This calls for a greater degree of real participation in the process of decision-making at all levels.[5]

On the national level, the policy of self-policing is manifested in a greater reliance on consultation rather than legal coercion in enforcement and in the setting and use of standards. The Committee believes that such voluntary standards and codes of practice will provide more flexibility and greater discrimination.[6]

"Self-regulation" also forms the backbone of the Committee's recommendations on sanctions for job health and safety violations. In Britain, unlike the United States, there is no administrative justice system for law enforcement. Violations of the health and safety statutes are prosecuted in the courts as *criminal offenses.* While philosophically this is perhaps a stronger deterrent than the threat of civil prosecution or fines, in practice few violations are ever prosecuted. In 1970, 300,000 inspections produced fewer than 3000 prosecutions.[7] The Committee concluded that the traditional sanction of criminal prosecution is neither respected, threatening, nor appropriate, that fines should be increased, and that a system of administrative sanctions should be adopted. They recommended that inspectors issue an Improvement Notice, which is a warning to comply with the requirements of the law, administrative regulations, or nonstatutory codes and standards. In cases where serious hazards or imminent dangers exist, the inspector could issue a Prohibition Notice ordering the operation or establishment to be closed down, either immediately or within a specified period of time.[8] The question of Improvement Notices versus first-instance sanctions has been a focal point of criticism of the report.

Self-regulation at the plant level hinges on more voluntary management-labor interaction. The only legal requirement the Committee recommended to encourage this interaction was "that there be a statutory duty on every employer to consult with his employees or their representatives at the workplace on measures for promoting safety and health at work, and to provide arrangements for the participation of employees in the development of such measures."[9] In principle, this *general duty to consult* has persuasive force; however, the lack of specifications as to what constitutes "consultation" provides considerable leeway for only minimal fulfillment of this legal obligation.

The Committee's recommendations at the plant level are aimed at creating better management performance. They include requirements for explicit policy objectives in written form and an effective health and safety organization with responsibilities clearly defined.[10] Although full worker participation must be encouraged by the appointment of workers' safety representatives and/or joint safety committees, this option is outside the domain of state control.

The Committee's belief in limited state intervention and increased self-regulation is summed up in the following statement:

> One main objective of reform of the statutory arrangements should be the creation of a more unified and integrated system to increase the effectiveness of the state's contribution to safety and health at work. This calls for the acceptance and exercise of appropriate responsibilities at all levels within industry and commerce. It calls for better systems of safety organisation, for more management initiatives, and for more involvement of workpeople themselves.[11]

Criticism of the Robens Report in Britain has centered around the laissez-faire concept of self-regulation and the recommendations that derive from it. The Public Interest Research Centre Limited, a Ralph Nader group located in London, while agreeing with recommendations on some specific issues (e.g., unification of law and administration), believes the report to be little more than a "statement of faith in existing principles." The group disagrees strongly with the general philosophy of self-regulation, for, as they see it, "the main responsibility for reducing the level of industrial accidents has *always* rested with management—yet their contribution has typically been minimal."[12]

In addition, the Centre calls the philosophy "wishful thinking." Their skepticism derives from their belief that

> the value of such ideals *cannot* be appreciated in individual works, in the way in which the Committee would wish. The problem is this:
>
> The Robens Committee has for two years been preoccupied with an *aggregation* of accident statistics. The perspective of plant management is totally different: not only is it preoccupied with the cutting of costs, but it has only to contend with relatively occasional accidents.[13]

Dr. W. T. Jones, of the London School of Hygiene and Tropical Medicine, has also criticized self-policing as naive, given the past history of accident control:

> Self-inspection and self-regulation of hazards at the workplace by cooperative effort of management and employees, which Robens puts forward as an alternative to enforcement, is too hopeful an ideal. It assumes that the present search for individual recognition and participation carries with it the corollary of the acceptance of increased individual responsibility. This puts too much faith in human nature on present evidence. It divides responsibility for safety amongst everyone and so it becomes no-one's business. In any case, the Factory Inspectorate has been coaxing along this idea for two decades now with only limited success. The accident figures remain stubbornly uninfluenced.[14]

He notes further the paradox between the Committee's conclusion that "the most important single reason for accidents at work is apathy"[15] and the recommendation that only voluntary, self-generating efforts by "those who create the risks and those who work with them"[16] will solve the problem. The inconsistency is that the lack of motivation, will, and information—the presumed apathy—is the very problem which must be addressed first if self-regulation is to be a viable alternative and not merely a stripping down of state intervention. In an interview, members of the Robens Committee agreed that "low priority" was a much more apt description of the employer than "apathy." Aside from these semantic problems, though, the occupational health problems are greatly compounded by ignorance—and "apathy" is hardly an accurate description of the cause of the difficulties.

Throughout the report, no distinction is made between safety and health. And in failing to delineate the differences between the two categories, the Committee also failed to take advantage of the distinction in formulating solutions. Dr. Jones echoes this view:

> Equally important is the confusion throughout the Report between "Safety" and "Health." Both subjects are truly inter-related in this field, but "Health" is used here in the restricted sense that any improvement in "Safety" must mean a reduction in injury or illness arising from work, and thus an improvement in "Health." "Safety" as used in the Report considers the control of the total physical environment enlisting the diagnostic knowledge of occupational medicine and the controlling skills of occupational hygiene; ergonomics is given a passing pat on the head. In this sense *the Report is not about Health at all; it is concerned solely with Safety.*[17]

Interviews with members of the Robens Committee convinced me that health problems had, in fact, been almost wholly ignored, and that the members' appreciation of critical problems was seriously deficient.

Both NIOSH and OSHA spokesmen have praised the Robens Report and supported adoption of its recommendations in the United States—especially the recommendation that "self-regulation" be the operating norm.

Dr. Marcus Key, then Director of NIOSH, summarized the principles of the Robens Report in a 1972 speech before the American Public Health Association and stated that not all problems can be solved "by the strict language of a standard."[18] Dr. Key went on to recommend flexibility in the language of standards. (It should be remembered that NIOSH has been very slow in developing criteria document and would prefer that standards be described as work practice *guidelines.*)

George Guenther, former Assistant Secretary of Labor for Occupational Safety and Health agreed with the appropriateness for the United States of the following Robens Report conclusions: (1) there is too much law; (2) the law is not relevant to the worker's situation; and (3) the various administrative agencies are unnecessarily fragmented.[19] It should be remembered, though, that it is the British system that is characterized by fragmented legislation; this is *not* the case in the United States. Thus, Guenther was misusing the Robens Committee's observation that "there is too much law" to justify *not* developing regulations.

The Robens Report recommends self-regulation as a total operating philosophy. At the federal level, self-regulation as defined by the Robens Committee (i.e., warnings instead of first-instance sanctions, flexible nonstatutory work practices rather than rigid standards, and discretionary power of inspectors) is a package which must have all its parts in order if it is to be effective. It is not my purpose to analyze whether or not this proposal is workable in the United Kingdom, but rather to analyze whether it is feasible for the United States. In order to utilize a consultative inspection approach and thereby eliminate first-instance citations, a system must have a critical mass of highly qualified and uncorruptible inspectors. It must also have muscle to use if the warning is not heeded. In the United Kingdom, this clout is provided by the proposed Prohibition Notice which enables an inspector to close down an operation or a workplace completely. Any attempt to use a consultative or warning approach, or nonspecific standards, *without* (1) a sufficient number of highly qualified, uncorruptible inspectors and (2) a set of strict backup sanctions to support the inspector in the use of his discretionary powers, would of necessity doom the program to failure. Unfortunately, OSHA meets neither prerequisite: its inspection force is of low quality (which will certainly not be improved by return of enforcement to the states), and it lacks the power to close down a factory without a court order. Those in OSHA and NIOSH who believe that they can adopt only the consultative or flexible-standard portions of self-regulation would, in fact, weaken the power of the OSHAct and render its enforcement function ineffective. The conclusion is simple: under present circumstances, OSHA is incapable of utilizing self-regulation at the federal level.

OSHA spokesmen have consistently used the term "voluntary compliance" to characterize the Department of Labor's approach to improving job health and safety. The compliance with health and safety regulations is supposed to occur "voluntarily" as a result of selective enforcement of strict standards by first-instance citations in the few workplaces OSHA's small inspection force is able to visit. In Section 6.3.1, the following requisites for "voluntary compliance" were described:

1. understandable and accessible standards and regulations;
2. an inspection capability large enough to present a significant chance of the employer being randomly inspected;
3. significant penalties on being found in violation;
4. well-qualified inspectors; and
5. employees well-informed about workplace hazards and their rights of protection under the law.

It is my opinion that "voluntary compliance" does not work in the United States—and neither would its big sister, self-regulation.

11.3 THE IMPORTANCE OF INTERNATIONAL COOPERATION FOR THE UNITED STATES

11.3.1 Activities of International Organizations

The two most active international organizations in occupational health and safety are the International Labour Organization (ILO) and the World Health Organization (WHO).

The ILO has been involved in worker health and safety since 1919, when it was established by the Treaty of Versaille. Over the years the ILO has adopted 54 Conventions and 52 Recommendations directly related to occupational health and safety. Ratification of a Convention is tantamount to a statement of the intention of a government to bring its laws into conformity with that Convention. One of the most important ILO Conventions requires a state factory inspection system. Among the relevant occupational health and safety Recommendations are one concerning the establishment of occupational health services and another concerning control of workplace hazards. The setting of international standards has been greatly facilitated by the activities of the ILO. While the United States has not adopted the Conventions and Recommendations (although it contributes heavily to the ILO's support), extensive adoption in Europe is paving the way for uniform health and safety standards and work practices. The ILO's other activities include facilitating the exchange of information and operating technical assistance programs in which teams of experts study a country's problems and recommend action programs.

The World Health Organization has been involved in occupational health since 1948. Like the ILO, it believes that occupational health goes beyond the prevention of occupational disease to encompass the worker's physical, mental, and social well-being both on and off the job. WHO is also committed to the unification of occupational and general health care; hence, it differs from the ILO in that it advocates a central health authority in each government in place of separated occupational health units in departments of labor and general health units in departments of public health. WHO, like the ILO, provides direct scientific-medical assistance to member countries.

Both WHO and the ILO have performed critical roles in aiding the exchange of information and scientific-medical expertise. Although NIOSH is a correspondent of the CIS information system of the ILO and welcomes foreign visitors when they come to the United States, it remains relatively uninformed about foreign research and approaches. Throughout my travels in Europe, I found that neither NIOSH nor its leadership was highly regarded by those who had reason to know both. The recent change in leadership at NIOSH will hopefully change this perception.

Aside from WHO and the ILO, very little has been done by international organizations with regard to occupational health and safety. The European Economic Community is attempting to develop uniform standards for the general environment, but it has so far essentially ignored the work environment. The Organization for Economic Cooperation and Development is just now awakening to environmental problems and social indicators of quality-of-life issues. The United Nations Conference on the Environment followed the usual course of ignoring the work environment; only Sweden chose to submit a comprehensive document on the problems of the work environment.

11.3.2 Economic Imperatives and International Competition

Although foreign trade accounts for a smaller percentage of GNP in the United States than in the Western European countries, more concern is voiced here for the possible effects of the enforcement of occupational health and safety legislation on the international competitive position of American businesses. (It should be noted, however, that the EEC countries have now surpassed the United States in trade with the developing countries.) The National Association of Manufacturers has expressed special concern in this regard. Multinational companies with alternative production sites may, in fact, pick a particular country because the occupational health and safety of workers is less emphasized there and, hence, the product of their labor is cheaper. What this really means, of course, is that the workers' lives are cheaper there.*

The United States has, unfortunately, been as unconcerned about foreign consumers' health and safety as it is about American workers' health and safety. While the proposed Toxic Substances Control Act seeks to impose premarket testing requirements on products distributed in this country, it explicitly exempts products to be sold abroad.[20] Thus, Congress seems content to export the possible harm it will not tolerate at home. The sale of herbicide orange to South American countries echoes this willingness.[21]

Ironically, if imposing strict legislation raises the price of American products (such as asbestos) to the point where we import the product from other countries, we will, in fact, be *exporting* disease and injury. (A solution would be to levy an import tax on all *unsafely produced* goods.) This is precisely why equity demands that international standards and work practices for occupational health and safety be established. Europe is moving in this direction, but the United States seems unlikely to follow. In this case it is not a competitive disadvantage we are seeking to avoid, but rather an advantage we wish to capture—at the expense of our workers.

Moreover, the notion that the United States would suffer a competitive disadvantage internationally by enforcing strict environmental controls in the occupational health and safety area is open to much question. It may, however, change the mix of small and large firms in a hazardous industry, with smaller and marginal firms disappearing. (For a fuller discussion of this issue, see Section 7.2.)

It would seem, then, that competition in international markets does not create an economic imperative for relaxing occupational health and safety regulations. What seems to emerge is the moral imperative to adopt uniform international standards.

11.3.3 Reforms Suggested by the Foreign Experience

It should be obvious to the reader that the exchange of scientific-medical information and expertise related to occupational health and safety is both profitable and necessary. The amount of research that needs to be done—in both basic science and epidemiology—is so vast that coordination on an international scale is required. It

*In a lecture at Harvard University in March 1974, Mr. Birger Viklund, Swedish Labor Attaché in the United States, emphasized that international cooperation and uniform standard setting are essential in order to guarantee that the lives of workers in every country are equally valuable. He did not advocate the equalization of wages, since purchasing power varies greatly among countries. Viklund criticized the United States for paying too little attention to occupational health and safety and for the adversary roles played by management and labor in this area.

should not have been necessary for the United States to rediscover "black lung" or asbestosis. Foreign information officially unrecognized is foreign information ignored. (It has been noted, for example, that environmental factors may be responsible for as much as 90% of all human cancers.[22] The striking variation in cancer rates in different areas of the world provides a challenge and an opportunity to identify the environmental factors which must be responsible.)

The differences in the way Western European countries address the occupational health and safety problems suggest reforms for the U.S. system. The only "reform" advocated loudly by OSHA and NIOSH is no reform at all—which would mean a collapse of our just-beginning efforts under the laissez-faire banner of self-regulation. Reforms suggested by the foreign experience include the following:

- The chief administrators of NIOSH and OSHA should be competent health and safety professionals as well as good administrators.
- Mechanisms need to be encouraged which facilitate closer cooperation between OSHA and NIOSH in policy formulation.
- NIOSH should be either located higher within the HEW hierarchy, combined with the National Institute of Environmental Health Sciences, or removed entirely from HEW.
- Occupational health and safety research needs serious reconsideration in order to make research more relevant to critical occupational health and safety problems.
- The government should provide more information to employers, employees, and the public, but not in the form of consultative services that would detract from the presently weak enforcement capability.
- More emphasis should be placed on enforcement of health standards and the general duty clause to improve health and safety conditions.
- Better training of physicians and other health professionals in the recognition of occupational diseases and in the importance of prevention strategies should be encouraged by means of federal support.
- More and better training of inspectors, especially in occupational health, is desperately needed.
- Serious consideration should be given to mandating in-house occupational health services or, in the case of small firms, participation in a joint medical service.
- Occupational disease needs to be fully reported, through a national registry, by employers and private physicians.
- Occupational disease needs to be fully recognized for compensation purposes (this should include *any sickness caused by exposure to toxic materials*).
- Serious consideration should be given to the mandating of joint occupational health and safety committees for firms, whether organized or not.
- More equal representation of management and labor should be attempted in the staffing of the government and private agencies responsible for health and safety.
- The problems of mental health, ergonomic factors, synergistic effects, and carcin-

ogenesis should receive more attention.

• Mechanisms for the international exchange of scientific-medical information and personnel should be developed further and fully utilized.

• Attempts at the adoption of international standards should receive sympathetic attention.

Many of these reforms are possible within the present occupational health and safety system. However, they require a firm national commitment to improve job health and safety—and such a commitment does not appear to be forthcoming as of this writing.

REFERENCES

1. Mr. Brian Harvey, Chief of the Factory Inspectorate, Department of Employment, interview with author, London, July 1972.

2. Parliamentary Committee on Safety and Health at Work, *Report of the Committee 1970-1972* (London, 1972), paragraph 29.

3. *Ibid.*, para. 98.

4. *Ibid.*, para. 208.

5. *Ibid.*, para. 114.

6. *Ibid.*, para. 147-148.

7. *Ibid.*, para. 259.

8. *Ibid.*, para. 269, 276-277.

9. *Ibid.*, para. 70.

10. *Ibid.*, para. 51.

11. *Ibid.*, para. 41.

12. "Preliminary Assessment of the Report of the Robens Committee on Health and Safety at Work," Public Interest Research Centre Limited, London, 1972, p. 1 (mimeograph).

13. *Ibid.*, p. 2.

14. W. T. Jones, "The Robens Report," unpublished paper, London, 1972, p. 2.

15. Parliamentary Committee . . . , *Report*, para. 13.

16. *Ibid.*, para. 28.

17. Jones, "Robens Report," p. 3; emphasis added.

18. Marcus Key, "Report from NIOSH," speech presented before the Annual Meeting of the American Public Health Association, Atlantic City, N.J., November 14, 1972.

19. George Guenther, lecture at the Kennedy School of Government, Harvard University, April 26, 1973.

20. Toxic Substances Control Act of 1973, Section 15(a); see S. Rep. No. 93-254 (June 26, 1973), 93d Congress, 1st Session (*Report of the Senate Committee on Commerce on S. 426*).

21. "Herbicides: Agent Orange Stockpile May Go to South Americans," *Science*, Vol. 180, No. 4081 (April 6, 1973), p. 43.

22. *Man's Health and the Environment—Some Research Needs*, Report of the Task Force on Research Planning in Environmental Health Science, National Institute of Environmental Health Sciences, HEW (March 1970), p. 147.

12 Agricultural Workers: A Special Case

Throughout this report, the problems of the industrial worker have been repeatedly cited. The occupational health and safety of agricultural workers deserve special attention for several reasons: (1) their occupational health and safety problems are particularly serious; (2) they are not recognized as "industrial workers" and hence do not enjoy the protection of existing labor law other than the OSHAct; (3) they are both unorganized and economically disadvantaged; (4) their "workplace" is the natural environment; and (5) to the extent that a "crisis" atmosphere has begun to surround the question of U.S. food production, it is necessary that national policy include concern for their health and safety, which must not be traded off for lower food prices.

12.1 INTRODUCTION

There are between 3 and 3½ million persons who receive payment in the course of a year for doing some farm work. One-fifth of these (between 600,000 and 700,000) are "regular" farmworkers who perform 150 or more days of farm work yearly. The balance of about 2,750,000 are seasonal workers, half of whom work fewer than 25 days on farms each year. Of this seasonal farm work force, between 400,000 and 450,000 (about one-sixth of all seasonal farmworkers) are migrants.[1] The Department of Agriculture defines migrant laborers as "workers who cross county lines overnight to perform farm wage work with the expectation of eventually returning home."[2]

It is difficult, however, to cite meaningful statistics on the subject, as unemployment and underemployment are pervasive problems among agricultural workers. L. W. Knapp, Jr., has arrived at a different set of statistics from extrapolation of the 1970 census data. He estimates that "of the approximate 4,500,000 employed in agriculture, only 1,165,000 were hired, and of the hired, 172,000 (16%) were migratory."[3] Although the working and living conditions of farmworkers are generally very poor, those of migrant workers are substantially worse. The special plight of migrants was pointed out as long ago as 1960 by Edward R. Murrow in the television documentary "Harvest of Shame."

The agricultural workforce in this country suffers a disproportionate number of occupational injuries and illnesses. National Safety Council estimates for 1971 indicate that although farm employees make up only about 4.4% of the workforce, they appear to suffer some 16% of the occupational deaths and 9% of the disabling injuries.[4] According to Gerard F. Scannell, former Chief of the OSHA Office of Standards, "the job illness and injury rate in agriculture is only exceeded by those in mining and construction."[5] The National Safety Council estimates the death rate of agricultural workers to be 67 per 100,000 workers. Only mining, with 100 deaths per 100,000 workers, and construction, with 72 deaths per 100,000 workers, exceed the figures for agriculture. (The overall national death rate is 18 per

100,000.) Moreover, according to the National Safety Council, the number of farm deaths has been *increasing* in recent years. Although the total number of farmworkers decreased by 50% between 1953 and 1973, in 1971 alone there were 2300 farm deaths in the United States, a 10% increase over 1961.[6] Despite this disproportionate number of injuries and deaths, however, agriculture was not included in OSHA's Target Industries Program.

Statistics on occupational health are not readily available. Since many states do not cover agricultural workers under their workmen's compensation laws, few records or statistics are kept, and few individuals are responsible for agricultural health. The situation is further complicated by the fact that, more than in any other industry, illnesses go unreported. The seasonal, transient nature of the work, the inaccessibility to medical facilities, and the lack of transportation are also contributing factors.[7] This data problem has not gone unrecognized by OSHA. In the summer of 1971, record-keeping forms and regulations were mailed to 650,000 farm employers, and the Bureau of Labor Statistics was ordered to begin surveying the group.[8] As of this writing, though, the results of the survey have not been made public.

A good deal of information regarding farmworkers can be found in the transcript of the Senate Hearings on Migrant and Seasonal Farmworker Powerlessness. There one learns that the migrant farmworker has a life expectancy twenty years less than that of the average American. The migrant's infant and maternal mortality rate is 125% higher than the national average. The death rate from influenza and pneumonia is 200% higher than the national rate, and that from tuberculosis it is 250% higher than the national rate.[9]

Agricultural employees traditionally have been little protected by either state or federal labor legislation in the United States. For example, only 28 states and Puerto Rico have included agricultural workers under their workmen's compensation laws.[10] Only 16 of these states provide agricultural workers with approximately the same type of coverage as other workers—a reform recommended in the report of the National Commission on State Workmen's Compensation Laws, set up by the OSHAct. In no states are farmworkers covered by unemployment insurance, and only in some states are they covered by disability insurance. Farmworkers are covered by minimum wage laws, but the minimum rate is lower and there are more accompanying restrictions for agricultural workers.

Since farmworkers were excluded from the National Labor Relations Act of 1935, today they must continue to strike for the right to organize. Presently only about 1% of the nation's farmworkers are unionized.[11]

The OSHAct is the first piece of federal legislation which on its face treats agricultural workers on an equal basis with those in industry. Unfortunately, OSHA does not seem to have rendered much help yet to agricultural workers. The Act included only four standards which loosely apply to agriculture,[12] and only one of general application. An Agricultural Advisory Committee to OSHA was formed, and several relevant subcommittees, such as those on farm equipment, housing and sanitation, and pesticides, have also been set up.[13] An emergency pesticide standard was adopted by OSHA but then suspended; a housing standard has also been promulgated, but the degree of protection it offers is not significant. It is not surprising, then, that the United Farm Workers Union prefers to write health and safety clauses into its contracts, rather than rely on the government for enforcement. In fact, health and safety is one of the UFW's biggest organizing issues.[11]

12.2 MAJOR OCCUPATIONAL HEALTH AND SAFETY PROBLEMS OF AGRICULTURAL WORKERS

The most widely publicized health and safety problems of agricultural workers have been tractor and machinery accidents and pesticide exposure. Trauma arising from the accidents is said to be "the largest occupational health problem facing the agricultural worker."[14] Exposure to noise, vibration, and animal medicinals also present significant problems.[15] Arthritis remains an important malady, complicated by the intense physical nature of farm work.[16] As with other work situations, insufficient record-keeping and inadequate or nonexistent diagnosis of occupationally related disease obscures the severity of the health situation.

12.2.1 Pesticide Exposure

One of the most serious health hazards currently faced by agricultural workers and one of the most controversial problems faced by OSHA is that of pesticides. Pesticides are applied by spraying. These chemicals are essentially poisons, and there is danger from both inhalation and skin exposure. The inhalation danger is limited to the time of spraying, while direct contact presents a continuous, though lessening, danger as time elapses after the spraying ceases.

Wind velocity, rainfall, and probably sunlight all serve to decrease the pesticide remaining on the foliage. Since these factors differ from region to region, the safe "reentry time" will also differ. ("Reentry time," the time that must elapse after spraying before workers can safely reenter a field, is the commonly used surrogate standard for pesticide concentration on the foliage.) Different reentry times in different regions could be used to guarantee the same minimum exposure. Thus, the fact that OSHA demands a national standard rather than regional standards greatly complicates the problem. (It has been suggested, however, that proposals for the development of separate regional standards are intended to *impede* the development of protective standards, since more time will be needed and local farm-interest pressure may be more easily brought to bear.)

Statistics on exposure are difficult to obtain. An HEW official stated during the

Senate Hearings on Migratory Labor that an estimated 800 persons are killed each year as a result of improper use of pesticides, and another 80,000 are injured.[17] This figure has been widely attacked and disputed as having been a national extrapolation from a figure for only one county in Florida. Others claim the high death rates represent suicides unrelated to the pesticides. However, it should be noted that one class of widely used pesticides, organophosphate insecticides, are virtually nerve gases, affecting the level of red cells and plasma cholinesterase, which control the body's involuntary nerve responses. In August of 1963, in California, over 90 peach pickers became sufficiently ill with parathion poisoning to seek medical attention.[18] There were about 600 pickers in the area of whom about 70 apparently well pickers were tested at random. Fifty percent of the workers tested showed a significant reduction of red cell and plasma cholinesterase levels but were either asymptomatic or had not sought medical attention for their symptoms. Although most of the cases were mild or moderate, about one-third were hospitalized and one died. Spray schedules for the fields were obtained, and unusually heavy spraying with parathion during the spring and summer was found. In all cases, spray schedules showed that over two weeks had elapsed between the last parathion application and picking. However, because of a greater surface area on the leaves, they had collected more parathion than the fruit.

In a paper presented before the APHA meeting in November 1973, William Serat states that

> in the light of modest data currently available . . . spraymen in agricultural operations and particularly in fogging operations can avoid insult from pesticide exposure by wearing protective clothing including a mask. Moreover, in normal agricultural spray operations, in the absence of a wind or faulty equipment which could broadcast spray in the vicinity of the sprayman, the intake of pesticides by inhalation is minimal. Even so, there is always danger to individuals who carry out such operations for several hours day after day.[19]

Unfortunately, protective equipment is not always available, and individuals do carry out such operations for several hours day after day. Even when available, protective equipment is often too hot and impractical under conditions of hard labor and extreme heat.

At present, six crops (tobacco, cotton, citrus fruits, grapes, peaches, and apples) and approximately 21 pesticides (the majority of them organophosphates) are responsible for most of the reentry poisonings and have been the focus of OSHA's attempts at setting standards.[20] * Although the effects of all pesticides used are unknown, several pesticides which have been banned or severely restricted in use in

*The crops were chosen because they all involve extensive hand labor. The pesticides were chosen first by an arbitrary decision to limit the study to organophosphates. Then, the most dangerous of these based on toxicity, persistence, and human experience were selected. Dermatitis and similar conditions were not considered.

other countries are still being used in the United States. For example, mevinphos, TEPP, and demeton have been banned from use in the Soviet Union because of their toxicity, and use of parathion and OMPA is permitted only to an extremely limited extent.[21] Moreover, no new compound may be used in agriculture in the Soviet Union until it has been subjected to detailed toxicological examination and been approved by the nation's chief medical officer. (It must be noted, however, that in the Soviet Union, the meaning of the words "ban" and "enforcement," as well as the criteria for toxicity, are different than in the United States.)

The lack of a firm data base on which to base scientifically rigorous standards is the cause of much controversy.[22] Here, as with other occupational situations where ill effects are prevalent, we are faced with the question of the extent to which we should limit the use of a suspect substance until it is proven safe or harmful. There has been a substantial amount of pressure upon OSHA by Washington-based public-interest groups to set emergency standards immediately for a number of pesticides while pursuing the long process of developing permanent standards. There has also been strong opposition to the adoption of such emergency standards by farm owners, growers, and several scientists in the field. In each case, the opponents of the standards assert that no emergency situation exists and that more data is needed before standards of any kind can be adopted. A proponent of the emergency pesticide standards, Dr. Dwight Culver of the University of California at Irvine, has put the problem this way: "For a change, I'd like to protect the health of the worker first. Let's protect his health, then find out what the exact standards should be."[23] The members of the group proposing a standard were evenly split as to whether the proposed standards were overly protective or implied a false sense of safety and, therefore, were insufficiently protective.

12.2.2 Farm Equipment

Although public attention has recently been focused on the health hazards caused by pesticides, agricultural workers face extreme danger from farm equipment as well. From 1968 to 1970 the state of Wisconsin averaged 82 fatal farm accidents per year. Over 39% of these fatalities involved tractors; an additional 15.7% resulted from other types of farm machinery.[24] A study by the New York State Health Department's Office of Biostatistics indicates that from 1967 to 1971, 145 deaths in upstate New York resulted from agricultural accidents. More than half of these involved tractors. Forty-nine deaths resulted from fractures, and 44 deaths from crushing injuries.[25] As of this writing, OSHA has issued only one farm equipment standard—a requirement for rollover protective structures on agricultural tractors.[26]

L. W. Knapp, Jr., has recently reviewed accidents on the farm, and his conclusions bear repeating here:

Mechanization, which has reduced the number of people exposed to the body insults of agriculture, is also producing an unprecedented number of injuries. Such machines as agriculture has needed to maintain its productivity are usually designed to grasp, pull, cut, pound, or shake agricultural products in such a manner as to render them usable by the farmer, and as they are not selective in their dealing with living protoplasm, one can expect that corn stalks or fingers, logs or legs, feed or feet, will be treated similarly. Such trauma is and should be one of our major concerns for such accidents usually involve the human extremities so important to farm work.[27]

Trauma is the largest occupational health problem facing the agricultural worker and his family, and machinery in one form or another produces nearly one-third of all injuries on the farm. The devices which are designed to reduce labor input and aid the farmer in his endeavors are capable of producing serious injury because they themselves are, or are connected to, power sources of considerable energy. This is particularly true if inadvertent or improper contact is made with moving parts by which the farm produce enters the machine to be processed.

The farm tractor, which produces the greatest number of fatalities from a machinery point of view, isn't necessarily the most dangerous machine on the farm from an exposure point of view, according to studies conducted in Michigan and Ohio. However, the seriousness of injuries and fatalities with its use places it high on the safety problems list, and the concerned efforts of industry, educators and government to resolve this problem has borne fruit. The development of Roll Over Protective Structures (ROPS) to prevent injury in case of overturn, and their mandatory use (an OSHA requirement being prepared for promulgation), will hopefully change this picture, as will similar efforts affecting power take-off injuries. Unfortunately, such devices on tractors, even though mandatory, will not immediately change the picture for two basic reasons: first, the existing tractors on today's farms have an average operative life in excess of 11 years, and, as less than 150,000 new units are added annually, the dilution factor for reducing the accident potential will be small. Secondly, the hired labor force involved in agriculture is only roughly one-third of the total labor force, thus resulting in further dilution. Lastly, new studies indicate that a further dilution is evident, for when comparing tractor accident rates versus age on a per million hours of use basis, operators between age 10-14 years and those over 65 years had rates of ten times and seven times higher, respectively, than the average, and as these two groups are largely made up of family members, they will thus not be subject to any mandatory regulation. Yet, there will be a considerable spin-off effect in spite of the aforementioned reasons, for many farmers will adopt such safety devices as good farming practices. This is particularly true when such devices are a part of an environmental enclosure (cab) which the farmer desires for health and comfort reasons.[28]

12.3 THE PESTICIDE PROBLEM

The history of the development of a pesticide standard is a good example of the interweaving of various conflicts involved in the setting of any occupational health standard.

OSHA's initial involvement in the pesticide area was through the Subcommittee on Pesticides of OSHA's Agricultural Advisory Committee. Using data collected by Auburn University, the Subcommittee made a report to the Advisory Committee in March 1973 stating that it would not recommend field reentry times because there was as yet insufficient data.[29] However, while the Subcommittee was preparing this report, the Migrant Legal Action Program, Inc., of Washington, D.C., filed a petition against OSHA charging that that body had failed in its duty to provide farmworkers with safe and healthful working conditions, as required by the OSHAct.[30] Under threat of suit, and despite the recommendations of the Pesticides Subcommittee, OSHA issued a temporary emergency standard on May 1, 1973.[31] The standard, which was to take effect on June 18, contained a reentry schedule more stringent than that advised by the Subcommittee and the Advisory Committee. On June 1, the Chairman of the Subcommittee on Pesticides, Dr. Frank Arant, resigned from his position in protest against the emergency standard. According to Arant, "The Pesticides Subcommittee was unanimous in its position that no emergency existed with respect to occupational exposure to organophosphorous pesticides and that there was no justification for such a standard."[32]

NIOSH had also been researching the pesticide issue. In fact, it had asked the Council on Environmental Quality (a federal agency) to set up a task force to study the effects of occupational exposure to pesticides. The task force agreed with Arant's subcommittee on the basic issues. Both groups felt that (1) not enough scientific data existed to justify the adoption of permanent standards, and (2) standards should be regional rather than national because of variations in weather factors such as wind and rain. (Keith Long, who served on the task force, felt that analytical scientific techniques to measure pesticide exposure, such as measuring the concentration on leaf surfaces, had not as yet been sufficiently developed in 1973. Long also felt that at least one or two years would be needed to generate sufficient data to set standards.)[33]

Not only did Arant and the OSHA and NIOSH committees protest the new standard, so did growers and Congress. The Florida Peach Growers Association and other associations filed suit against OSHA, and both houses of Congress made moves to subvert the emergency standards.

In the Senate, an amendment to the proposed Agriculture and Consumer Protection Act of 1973 (S. 1888 and H.R. 8860), introduced by Senator Ernest Hollings (D-South Carolina), would have required hearings on the standard and provided that no standard could take effect until Congress had had thirty legislative days to review the evidence. Senator Hollings charged that the pesticide standard was another example of "high-handed, arbitrary actions" on the part of OSHA. Had it passed, the Senate amendment would have, in effect, stayed the standard.

In the House of Representatives, an amendment was proposed (but ultimately rejected) that would have transferred from the Labor Department to the Agriculture Department the authority to establish regulations to protect farmworkers from exposure to dangerous pesticides.[34] Both Ralph Nader's Health Research Group and the Legislative Department of the Steelworkers Union strongly opposed this amendment in written position papers, citing, among other points, the fragmentation in standard-setting function among different agencies that would result from any such move, the need for an immediate emergency pesticide standard, and the lack of expertise in the Agriculture Department to deal with such problems.

Thus, once again under tremendous legal and political pressure, OSHA suspended the emergency standard on June 15, 1973, stating that it would publish a new standard shortly. Meanwhile, on June 19, a motion for a temporary restraining order to prevent suspension of the emergency standard was filed in the U.S. District Court for the District of Columbia by the Migrant Legal Action Program, Inc., and the National Health Law Program. OSHA soon thereafter announced an amended temporary emergency standard, excluding 9 of the 21 pesticides previously covered and reducing reentry times for certain crop areas, to take effect on July 13. A stay of this effective date was issued by the Fifth U.S. Circuit Court of Appeals on July 10 on procedural grounds and not on the merit of the plea.[35]

At this point, a simmering intergovernmental jurisdictional dispute came into the open. The Environmental Protection Agency, whose role is to assure the safety of pesticides in interstate commerce, had been for some time expending considerable effort and large sums of money through its Office of Pesticide Programs on the development of sophisticated programs to establish toxicity criteria. Despite statements that OSHA would coordinate its efforts in this area with the EPA,[36] the atmosphere between the two agencies seemed to be one of antagonism and rivalry rather than one of cooperation. In fact, the EPA claimed that it had "primary responsibility for establishing the standards concerning occupational safety arising from the use of pesticides" and that the EPA standards would preempt any that might be set up by OSHA.[37] (The basis of this claim was the fact that labels already contained reentry times, established under authority of the Federal Insecticide, Fungicide and Rodenticide Act. However, these times were suggested by the manufacturer and were not complete, uniform, or mandatory; hence, one could argue that they were not really standards.)

On May 10, 1974, the EPA promulgated its own standard for pesticides, and a subsequent court decision, *Organized Migrants in Community Action, Inc. v. Brennan*, upheld the EPA's preemption of standard-setting power.

12.4 THE OSHAct AND THE FARMWORKERS

Except for investigating deaths and disabling injuries on the farm, the OSHAct was used very minimally in agriculture in the first years after its passage. Though ultimately vetoed by President Nixon, Congress did pass a small business exemption which would have prohibited federal funds from being used to inspect workplaces with fewer than fifteen employees (see Section 4.3.4). This bill would have exempted from protection some 30% of the nation's workforce and, more significantly here, 98% of farmworkers. Congress may not have intended this last result; it seems, rather, to be another vivid illustration of the low visibility of farmworker interests in the area of health and safety.

Cesar Chavez, leader of the United Farm Workers Organizing Committee, has stated that "the issue of health and safety of farmworkers in California and throughout the United States is the single most important issue facing the United Farm Workers Organizing Committee. . . . [The Committee] is attempting to solve this pervasive problem by the collective bargaining process."[38] The contract clause included in many of the UFW's negotiated agreements is given at the end of this chapter. It is interesting to note the specificity of the contract language with regard to pesticide dangers.

A UFW spokesman stated in an interview that the UFW has no great attachment to the OSHAct, primarily because the union does not trust the government to enforce it. The UFW prefers to write extensive health and safety clauses into its contracts, and it uses health and safety as one of its major organizing issues.[39] Another spokesman claims that the clauses have been honored in the fields, and that a lot of leverage has been gained out of the contract itself.[40] However, other observers sympathetic to the farmworkers have informed me that the UFW contract clause is not always effective.

In any event, the record of the government to date does not leave the farmworkers very much hope. As previously mentioned, only four standards relating to agriculture existed when the OSHAct was passed, and only one new standard has since been issued. Several public-interest groups and congressmen believe this situation to be due, at least in part, to the composition of the OSHA Agricultural Advisory Committee, which has come under attack in and apart from its reluctance to recommend pesticide standards. The Migrant Labor Action Program claims that the four persons chosen by OSHA to represent farmworkers on the Advisory Committee are "either in management positions or are discredited ex-members of farmworker organizations"; it has therefore called for the reconstitution of the Agricultural Advisory Committee.[41] Congressman James G. O'Hara (D-Michigan), Chairman of the House Subcommittee on Agricultural Labor, has criticized the lack of employee representation on the Agricultural Advisory Committees of both the U.S. Department of Agriculture and OSHA.[42]

It is significant to note that while the "general duty" requirement of the OSHAct could be used to protect farmworkers while standards are being developed, OSHA chooses not to upset any political baskets by utilizing it. The result is almost the total absence of any meaningful protection for the farmworker. And even if the UFW were successful in its collective-bargaining efforts, it only represents 1% of the agricultural workforce.

12.5 FUTURE PROBLEMS, NEEDS, AND SOLUTIONS

As slowly as the federal government has moved to protect the farmworker specifically, if control for occupational health and safety is returned to the states, even less will be done. The states' records are generally poor, and the failure of OSHA to issue a pesticide standard will not encourage better action from most of the states. (One exception is California, which established the original field reentry standards for pesticides. These standards, which have been in force for several years, are generally better than the federal standards.) The EPA's assertion of control does at least provide some protection on a *federal* basis, which might be better than leaving occupational pesticide control in the hands of an inept or biased state agency. However, the EPA's enforcement authority is not geared to the workplace, and its procedures are time-consuming.

If farmworkers were treated like other industrial employees with regard to workmen's compensation coverage, there might be some economic incentive for the farm managers to reduce injury. Had the recommendations of the National Commission on State Workmen's Compensation Laws been followed by legislation, "as of July 1, 1973, it is estimated that it would have applied to some 430,000 farm employers in the 50 states."[43]

Today the basic premium per $100 of payroll ranges from $2.45 to $16.01.[44] In some cases, a large premium payment could provide an incentive to improve safety; but, as noted in the case of industry, it is harder to build incentives for improved health measures into the workmen's compensation scheme.

In sum, then, the future of legislative and administrative action to improve farmworkers' health and safety remains quite uncertain. It is unlikely that significant changes will occur. Of all members of the labor force, the farmworker's working conditions will almost certainly be the slowest to improve.

Of course, the farmworker faces other severe problems too, such as unemployment and underemployment, malnutrition, and intolerable living conditions. Unless these problems are directly attacked, the issues of health and safety will probably not rank very high among the demands of either agricultural workers themselves or their supporters. This is not to say, of course, that increased farm inspections, for example, would not have direct and positive results. The more the OSHAct is utilized in the agricultural workplace, the greater the visibility of all the farmworkers' problems will be.

Finally, innovations in farming methods can be encouraged which both conserve agricultural workers' health and safety and improve food production. The problems of the agricultural worker need to be systematically attacked before food production develops into an area of "crisis," after which clear thinking and a regard for workers' health and safety are likely to be set aside.

POSTSCRIPT: AN EXAMPLE OF A UFW HEALTH AND SAFETY CLAUSE

SECTION XVIII. *HEALTH AND SAFETY*

A. Company agrees to consult with Union's Health and Safety Committee in formulation of policies relating to Health and Safety of workers, including but not limited to the following:
   Use of economic poisons; use of protective garments, materials, tools and equipment as they may affect the health and safety of workers and sanitary conditions.

B. 2, 4-D, 2, 4-5T, DDT, DDD, ALDRIN, DIELDRIN, and ENDRIN shall not be used. Other economic poisons which under any circumstances cannot be used without extreme danger to farm workers and consumers will not be used.

C. Company agrees that it shall notify Health and Safety Committee Chairman as soon as possible prior to the application of economic poisons. Company, after consultation with Committee, shall determine the length of time during which workers will not be required to enter the treated field subsequent to said application.

   In making this decision, the Company and Committee shall be guided by the recommendations of Federal and State authorities but shall also be entitled to refer to recognized experts in the fields in making its determinations.

   The foregoing notwithstanding, it is agreed that no worker shall be required to enter a field which has been treated with parathion less than twenty-one (21) days after it has been treated with parathion. Essential work may be done in the area after five (5) days if appropriate protective clothing and respirators are supplied.

D. Company agrees to keep the following records and to make them available to Committee or such Union member as Committee shall appoint as its agent:
   1. Location of field to be treated.
   2. Pesticide and economic poison used, including brand name, registration number on label, and manufacturer's batch or lot number as shown on label.
   3. Dates and time applied or to be applied.
   4. Amount of each application
   5. Formulation and concentration in spray or dust.
   6. Method of application.
   7. Applicator's name or names.
   8. Estimated date of harvest.

E. There shall be adequate toilet facilities, separate for men and women in the field readily accessible to workers, that will be maintained in a clean and sanitary manner. These may be portable facilities and shall be maintained at the ratio of one for every forty workers.

F. Each place where there is work being performed shall be provided with suitable, cool, potable drinking water convenient to workers. Individual paper drinking cups shall be provided.

G. Workers will have a rest period of ten (10) minutes in the morning and ten (10) minutes in the afternoon which insofar as practical shall be in the middle of each work period.

H. Tools and equipment required by the Company to perform the work and protective garments necessary to safeguard the Health of, or to prevent injury to, a worker's person shall be provided, maintained, and paid for by the Company. Each worker shall be responsible for returning all equipment checked out to him except for breakage caused by normal wear and tear. Workers shall be charged actual cost of such equipment that is not broken and not returned.

I. Adequate first-aid supplies shall be provided and kept in clean and sanitary dust-proof containers, as required by the State.

J. One baseline cholinesterase test and other necessary cholinesterase tests shall be taken on those workers employed as applicators at Company's expense when organo-phosphates are used. Union shall be given results of said tests immediately.

K. Company shall not require any employee to go or be in any employment or place of employment which is not safe. "Safe" means such freedom from danger to life or safety of employee as the nature of employment reasonably permits.

REFERENCES

1. Testimony of Frank A. Potter, Director, Office of Farm Labor Service, Hearings on Migrant and Seasonal Farmworker Powerlessness Before the Senate Subcommittee on Migratory Labor, Committee on Labor and Public Welfare, 91st Congress, p. 34. [Hereinafter cited as Hearings]

2. Hearings, Testimony of Raymond M. Wheeler, Charlotte, N.C., p. 4984.

3. L. W. Knapp, Jr., Director, Institute of Agricultural Medicine, College of Medicine, University of Iowa, Oakdale, Iowa, "Agriculture in the United States–Whom Should We Be Protecting?" speech presented at the Annual Meeting of the American Public Health Association, San Francisco, November 5-8, 1973, p. 2.

4. George C. Guenther, Assistant Secretary of Labor, statement before the House Subcommittee on Agricultural Labor Regarding Occupational Safety and Health, September 25, 1972, p. 1.

5. *Occupational Safety and Health Reporter* (the Bureau of National Affairs, Inc.) Vol. 1, No. 34 (December 23, 1971), p. 675. [Hereinafter cited as *OSHR*, BNA]

6. Donald L. Ream, Chief, Division of Workmen's Compensation Standards, Employment

Standards Administration, U.S. Department of Labor, "Workmen's Compensation Coverage of Agricultural Workers," speech presented at the Annual Meeting of the American Public Health Association, San Francisco, November 5-8, 1973, p. 1.

7. Migrant Legal Action Program, Inc., et al., "Petition to the Occupational Safety and Health Administration," September 1, 1972, p. 8.

8. Guenther, statement, p. 2.

9. Hearings, Testimony of Raymond M. Wheeler, p. 4984.

10. Ream, "Workmen's Compensation," p. 4.

11. Andrew Klesco, United Farm Workers, AFL-CIO, interview with author, June 1973.

12. Guenther, statement, p. 3.

13. *Ibid.*, pp. 3, 4, and 6.

14. Knapp, "Agriculture in the United States," p. 10.

15. *Ibid.*, pp. 6, 9.

16. Ream, "Workmen's Compensation," p. 7.

17. S. Rep. No. 91-1282, 91st Congress, 2d Session, 91-1282, p. 4.

18. Irma West, Bureau of Occupational Health and Safety of the California Department of Public Health, "Occupational Experiences with Pesticides in California," lecture presented at the International Short Course on the Occupational Aspects of Pesticides, Oklahoma Center for Continuing Education, Norman, Oklahoma, November 20-22, 1963.

19. William F. Serat, State Department of Health, Berkeley, Calif., "Quantitating Health Aspects of Pesticide Exposures," speech presented at the Annual Meeting of the American Public Health Association, San Francisco, November 5-8, 1973, pp. 4, 5.

20. Knapp, "Agriculture in the U.S.," p. 6.

21. "Pesticides, Organophosphorous," *Encyclopedia of Occupational Health and Safety* (International Labour Organization, Geneva, 1972), Vol. 11, p. 1024.

22. Knapp, "Agriculture in the United States," p. 7.

23. Quoted in Richard D. Lyons, "Pesticide Safeguards on Farms Asked," *New York Times* August 30, 1973.

24. *OSHR*, BNA, Vol. 1, No. 40 (February 3, 1972), p. 820.

25. *OSHR*, BNA, Vol. 3, No. 1 (June 7, 1973), p. 32.

26. *OSHR*, BNA, Vol. 4, No. 48 (May 1, 1975), p. 1601.

27. Knapp, "Agriculture in the United States," p. 6.

28. *Ibid.*, pp. 10, 11.

29. *OSHR*, BNA, Vol. 2, No. 39 (March 1, 1973), p. 1120.

30. Migrant Legal Action Program et al., "Petition," p. 1.

31. *OSHR*, BNA, Vol. 2, No. 48 (May 3, 1973), p. 1403.

32. *OSHR*, BNA, Vol. 3, No. 1 (June 7, 1973), p. 11.

33. Keith Long, Department of Community Health, University of Iowa, telephone interview with author, June 6, 1973.

34. *OSHR*, BNA, Vol. 3, No. 7 (July 19, 1973), p. 195.

35. *Ibid.*, p. 187.

36. Guenther, statement, p. 4.

37. *OSHR*, BNA, Vol. 3, No. 24 (November 15, 1973), p. 771.

38. Statement of Cesar Chavez before the House of Representatives, *Congressional Record*, 91st Congress, 1st Session, October 3, 1969.

39. Andrew Klesco, interview with author.

40. Nick Jones, United Farm Workers, AFL-CIO, interview with author, March 15, 1973.

41. Migrant Legal Action Program et al., "Petition," p. 2.

42. *OSHR*, BNA, Vol. 2, No. 17 (September 28, 1972), p. 448.

43. Ream, "Workmen's Compensation," p. 5.

44. *Ibid.*, p. 6.

13 Epilogue: The Design of Policy

In the introduction to this report, it was stated that the "field" of occupational health and safety is not a field at all, but rather a collection of activities unconnected in any meaningful way. Further, it was noted that these activities were so characterized by conflict that a more accurate description might be occupational *politics*, health, and safety. The conflicts arise from systemic problems and from problems related to societal values; they were conveniently classified into five types:

- conflicts reflecting different self-interests
- conflicts caused by a lack of knowledge
- conflicts due to the differences in perception of what is just or fair
- conflicts due to the nonconnectedness of important institutions or agencies
- conflicts due to unidisciplinary approaches to problems.

Earlier in this century, governmental response to problems raised by the rapidly increasing occupational injury rate took the form of workmen's compensation legislation at the state level. This legislation was aimed at settling the question of responsibility and payment for injury arising out of the course of employment in a rapidly industrializing nation. In contrast, the federal Coal Mine Health and Safety Act of 1969 and Occupational Safety and Health Act of 1970 represent a very different use of the law—namely, as a means of exercising social control over technology in the workplace. This new exercise of federal control in the *work* environment was a natural parallel development to the National Environmental Policy Act focusing on *general* environmental regulation. Unlike workmen's compensation legislation, the federal occupational health and safety legislation has the potential to alter technology in the workplace and hence "internalize" the costs of industrial disease and injury. What the new legislation has in fact done is to rekindle hitherto unresolved conflicts centering around *health* in the *work environment.*

This rekindling is occurring in the midst of intense and related philosophical controversy concerning appropriate public policy with regard to health, unemployment, and environmental control. For example, it is prevention- rather than treatment-oriented medicine that is required to reduce occupational disease—and new health-care legislation is, in fact, moving in the direction of preventive strategies. Unemployment policy is inextricably related to governmental efforts to control inflation and ease the present economic crisis. Proposed toxic substances control legislation has reopened the debate on whether society should prohibit use of a new substance until it is proven safe or allow its use until it is found to be harmful. Lessons relevant to the resolution of this question are to be found in the high rates of occupational disease in the United States, where waiting for "proof" of harm has been the operating philosophy.

Another philosophical controversy directly related to technology and its effects has been generated by the study *Limits to Growth*, which assesses the desirability of slowing down technological growth and its attendant consequences.[1] Continued rapid technological growth has brought rapidly increasing problems of pollution and resource depletion. The "energy crisis," which has implications for both resources and the environment, is a recent example. Occupational disease and injury are in fact closely tied to a rapidly developing industrialized economy, and policies affecting technology will have consequences for occupational health and safety.

In the midst of hard economic times, old conflicts are renewed. But times are different. We today have the most educated population and workforce in our history. General environmentalists are finding that their interests greatly overlap those of people in the labor movement concerned with occupational disease and injury, especially since the realization that a significant portion of disease may be occupational in origin. New coalitions may be forming, and the elitism of the environmental movement is disappearing.

We may be approaching the "critical mass" of concern and activity necessary for significant changes to occur. The fact that vinyl chloride, the twenty-sixth most commonly used chemical, has been linked to liver cancer has had a remarkable effect in "legitimizing" the concern for job health and safety. What, then, are the likely public and private initiatives which may evolve to deal with health in the work environment, and how effective are these efforts likely to be?

The use of the law, in this case the OSHAct, as a mechanism for the social control of technology in the workplace has proven to be only minimally effective so far. As happens so often, insufficient manpower or political interference can turn a good idea into an ineffective program. Further, it is endemic to our system that the market mechanism alone cannot be relied upon to solve the health and safety problems of the work environment, no matter what the case may be for the general environment.

Solving the occupational health and safety problems requires both a change in the society's concept of equity (such as that suggested by John Rawls in his *Theory of Justice*)[2] and a change in the operating philosophy of the firm. A strong national leadership and a law implemented in such a way as to embody new rules of the game as to who pays the cost and who reaps the benefit of industrial technology are required. Legislation can be relatively ineffective in controlling technology unless its implementation is motivated by the same goal as its enactment—that is, to settle equity questions.

Safeguarding the health and safety of workers is not inconsistent with a growth economy, as the experiences of other countries demonstrate, but business philosophy must reflect that concern. The business philosophy of American firms does, to some extent, reflect concern with working conditions aside from their probable effect on productivity. For example, no firm today, even faced with failure, would think of closing down its toilet facilities in order to increase its production capability because the toilets "don't pay for themselves"; the firm would close down its operations before acting in this ridiculous manner. However, the European firm (at least in Sweden and the Netherlands) seems to have an operating philosophy which incorporates concern for job health and safety to a much greater extent than its American counterpart. Employers and employees work together in joint health and safety committees, while joint agreements on the federation level indicate that concern for job health and safety is becoming the normal way of doing business. A change of such fundamental rules of the game within the United States, based upon a different concept of equity than has been the norm even for the American firm motivated by enlightened self-interest, is required if the hazards to health in the workplace are to be significantly reduced.

How to change such concepts of equity is a key issue. Can we rely on the evolution of a more benevolent employer, that is, one willing to extend his paternalism to job health and safety, or must the worker gain more participation in or control of his work environment in order to improve health and safety on the job? In view of the rapidly rising industrial injury rate in the 1960s, it would be fair to say that voluntary improvement of health and safety conditions in the industrial workplace has not kept pace with the rapidly changing technology and the introduction of new materials into that workplace.

Management-labor relations being adversary in our system, worker participation or control comes either through collective-bargaining agreements that the employer will provide improved working conditions—this is the usual American mechanism—or through the acquisition by workers of the right to have access to information and to monitor and change that environment themselves. The OSHAct, for all intents and purposes, perpetuates the established management prerogative over health and safety. It imposes sanctions for violation of the duties and obligations of the employer; but while employees are "obligated to obey safety rules," the OSHAct imposes no enforceable duty on them. It gives employees important rights, but these rights are safeguards that the employer will perform his responsibilities correctly. The Act gives rights but no enforceable responsibilities to the worker.

From a practical point of view, increased *responsibility* of the worker for his own work environment can come only with a corresponding increase in both his *knowledge* and his *control* of that work environment. The struggle for job health and safety in the United States has been both a demand for worker participation or control and an effort to keep the government from weakening the effects of the Act. Thus, in a sense, the management of the Shell Oil Company was right in seeing the OCAW strike over health and safety issues as going much deeper than those issues. The extension of industrial democracy in the United States may well be more an outgrowth of the demand for improved health and safety than an outgrowth of generalized job dissatisfaction related to the meaningless nature of work. A look at the activities of management, labor, government, public-interest groups, and "professionals" since the enactment of the OSHAct lends support to the suggestion that a fundamental shift in power regarding management prerogatives is in progress. To the extent that collective bargaining is being used to settle occupational health and safety issues by extracting better *employer* performance, there is little shift in power. However, to the extent that workers are gaining participatory rights in the process, important changes may be occurring.*

No attempt has been made thus far to present policy alternatives for the important participants in occupational health and safety—i.e., for management, labor, public-interest groups, professionals, or government. This report has sought to demonstrate that occupational health and safety problems will not yield significantly to simplistic strategies based on single approaches. "Enforcing the law," as is argued by some, will not suffice; nor will "research" to determine "safe levels" of exposure or "hypersusceptibility." Rather, the design of policy to bring about improvements in occupational environments must rely upon four sets of policy instruments: (1) the law, (2) market incentives, (3) the generation, dissemination, and utilization of knowledge, and (4) the development of personnel in the various professions, in labor unions, in management, and in government with the requisite knowledge of the issues. These policy instruments can be utilized effectively only if management, labor, and government are willing to expend sufficient resources for preventing occupational disease and injury. Groups within labor and within management vary greatly as to their concern, but their concern is increasing. Government activity has been shamefully inadequate. It is a sad commentary on the lack of government responsibility that the former Assistant Secretary of Labor, while directing the OSHA efforts, saw an opportunity to encourage contributions from management for the reelection of President Nixon in exchange for treading softly in the implementation of the OSHAct. (See the memo reproduced at the end of this chapter.) More recently—especially since the controversy over vinyl chloride—the government has displayed increased awareness of the importance of reducing occupational disease, but it still has a long way to go.

While the policy instruments cited above must be further developed and aggressively wielded, they will nonetheless be insufficient even when working in concert. Workers themselves must assume a greater role in self-monitoring the quality of their own work environments. It is true that both law and tradition place ultimate

*Many employers have responded, more or less, to the problems related to job safety. Where *safety* problems have been addressed, management has assumed the "responsibility" and provided safety equipment, "trained" its employees, etc. On the other hand, most job *health* problems have gone unattended. It is not clear that management will hire occupational health professionals, monitor plants, or provide occupational health care and training for its employees since, at present, the economic and legal incentives to do so are weak. If the demands of workers to monitor and control the health hazards in their own work environment become appreciable, the responsibility for occupational health may be more fully shared than is the case for job safety.

responsibility for job health and safety on the employer; but full reliance on law or market incentives to encourage fulfillment of this obligation is impossible in any practical sense. While progress can be expected in the courts and hearings—and in the laboratory and classrooms—much more is likely if workers and their representatives are enabled to function responsibly and effectively in fruitful interaction with employers, and as a necessary check on even the best-intentioned employers.

This latter conclusion is based on the fact that too little is now known—or even, at least in the short run, knowable—about health hazards for scientific proof and definitive standards to be widely forthcoming. Rather, most decisions regarding workplace environmental quality must be worked out in other forums on the basis of existing and often underutilized knowledge. In part, these forums will be the courts or administrative agencies, and in part, they will be the arenas of politics and public opinion. Mainly, however, these issues need to be addressed in the day-to-day interaction between informed employers and employees—and in a deepening of the trend toward a higher priority for job environment issues in collective bargaining.

The choice of policy instruments and leverage points depends strongly on the philosophical basis of the decision-maker and the answers he or she would give to the following questions:

- What is an acceptable level of workplace injury and disease?
- How protective of the worker should the society be? Should workers be allowed to exercise free choice in deciding to accept a disproportionately higher amount of injury and disease if they are told of the risks on the job and if society as a whole benefits?
- Should the social goal of maintaining the viability of the small firm be achieved at the expense of the workers' health and safety in those firms?
- Should the small firm be subsidized in order to raise the standard of its workplace conditions?
- Should either society or management bear the burden of relocating or retraining workers displaced as a result of the closing of unsafe and unhealthy workplaces?
- Should technological growth per se be limited directly, or should the incentives for technological growth (like the patent system) be revised?
- Is health a basic American right like education or free speech?

My own opinions on these issues have no doubt come across very clearly throughout this report. Nevertheless, I have attempted an objective *description* of the problems and attempts at their solutions. To the extent that any philosophical base is subjective, the recommendations for reform made in this report are also subjective. It should be clear to the reader, however, how a different philosophy would give rise to different recommendations.

Governmental reforms suggested by the foreign experience were elaborated in Section 11.3.3. Recommendations for the private sector fall into two general types:

*Activities that can have a direct and early impact on occupational health and safety problems**

1. The education of workers
   - by employers
   - by universities with worker education programs or by other institutes and centers offering long- and short-term training in occupational health and safety, collective bargaining, and industrial relations
   - by health activist and public-interest groups to train workers or their representatives in their legal rights under the OSHAct, in recognizing occupational hazards, and in utilizing self-help techniques
2. The support of information dissemination activities and the development of new information sources
   - by encouraging the transfer of technical and medical information between employer and employees
   - by encouraging efforts to develop and disseminate readable materials for small employers and unorganized workers
   - by supporting joint development of audiovisual materials for use by labor, management, and public-interest groups
   - by supporting information clearinghouses of a neutral nature or designed to meet the information needs of particular groups
   - by establishing mechanisms to ensure the rapid international transfer of information from abroad
3. Cooperative efforts in the rapid development and transfer of new technology designed to assure safer and healthier working conditions
4. Encouragement of professional manpower development
   - by maintaining the viability of occupational health and safety programs in universities where they already exist but where support has been withdrawn by the government
   - by establishing or supporting college programs for training multidisciplinary professionals in combined occupational health and safety curricula
   - by supporting the development and teaching of specific occupational health safety courses in the medical, nursing, public health, law, and business schools
5. Increased utilization of occupational health and safety manpower by employers and unions
6. The support of individuals or groups within the professions who can awaken professional and public-interest groups to these problems
7. The sponsoring of conferences on occupational health and safety (there is a

*The mechanisms for supporting these activities can include direct funding, the establishment of academic chairs in occupational health and safety, scholarship aid for students, and the support of internships in labor unions and industry.

need for such conferences within various interest groups—labor, management, professional groups, etc.—and also for a variety of conferences to which all interested parties are invited)

8. The support of selective research to demonstrate the seriousness of combined effects of hazards (synergism) and the effects of stress as a cocausative factor in occupational disease
9. The support of health activist and environmental and public-interest groups in the use of legal action to force implementation of the OSHAct
10. The support of the above-mentioned activities on an immediate but a short-range basis while serious policy questions are examined

*Policy analysis of occupational and environmental health and safety to be done over the longer term**

1. The monitoring of public policy and the performance of government agencies, including governmentally produced data on the nature and extent of job health and safety problems and on efforts to alleviate the problems
2. Mechanisms for linking approaches to problems of the general environment with those of the work environment, and problems of occupational health with those of general health-care delivery (the design of toxic substances control legislation and national health insurance are important considerations)
3. Alternative institutional frameworks for implementing the OSHAct, delivering occupational health and safety services, educating employers and employees, and acquiring and disseminating information
4. Examination of the changing disease and injury profile in the United States, and the setting of priorities for curative or preventive action
5. Research policy: recognizing that there is a serious lack of a firm data base and that all the research that needs to be done cannot be done, it is important that the research that is conducted be carefully selected according to a conscious policy
6. Development of alternative technologies to decrease occupational exposure to harmful materials (including studies to test the feasibility of using closed systems for chemical processing and of using less harmful substitute materials)
7. Assessing the potential of job and workplace redesign for reducing workplace injury and disease (both ergonomic and motivational factors should be included in such studies)
8. Development of simple measurement and control techniques which can be used

*Such policy analysis should be conducted at universities, research institutes, and independent centers. It should be multidisciplinary and should engage health and safety professionals, lawyers, economists, engineers, scientists, and persons with business, managerial, and labor experience.

in the workplace by workers, their representatives, and first-line or middle-management personnel
9. Manpower development: the appropriate numbers, training, distribution, and mix of occupational health and safety professionals should be examined along with policies that determine the supply and demand for such personnel
10. Economic policy: the impact of occupational health and safety legislation on GNP, on the mix of large and small firms in various industries, on employment, and on international competition should be assessed
11. Examination of foreign experience in occupational safety and health and its relevance to the United States
12. Formulation and assessment of alternative coordinated policies for funding activity by private philanthropy and by other institutional elements.

The success or failure of these activities will depend on the manpower used to carry them out. The recommendations make use of the policy instruments previously described as "the generation, dissemination, and utilization of knowledge" and "manpower development." A word of warning: these activities are viewed by some as only properly carried out by "experts." It may not be wise to continue placing sole reliance on experts to solve our occupational health problems. In a sense, the experts got us into the trouble we are in by claiming to have a monopoly on the requisite knowledge. The fact is that all too often medical experts have not been interested in the prevention of occupational disease, and safety experts have not been interested in health. Scientific researchers seeking "objective truth" are employed principally by management and quite naturally reflect its view. And the lawyers who administer many government programs all too often act as advocates and present biased views. Ralph Nader's response to this observation is worth quoting: "If scientists think lawyers can present one-sided cases, they may wish to discover themselves."[4]

Of course, experts are needed, but the experts have not been in the forefront of social change and they are slow to alter the traditional ways of doing things as defined by the "accepted" practices of their profession. Furthermore, they are reluctant to have the mystique of science, medicine, or the law laid bare so that those outside the professions can enter into the debate. If the "tyranny of the experts" is not to slow down critically needed progress,[5] workers and employers alike need to find out for themselves what occupational disease and injury is all about. It is hoped that this report will help to speed that process.

APPENDIX

U.S. DEPARTMENT OF LABOR
Occupational Safety and Health Administration
Washington, D.C. 20210

Office of the Assistant Secretary

June 14, 1972 CONFIDENTIAL

MEMORANDUM TO THE UNDER SECRETARY [Lawrence Silberman]

Subject: OSHA program actions from now through November

In considering this subject, I have assembled an informal group of seven staff members whose loyalty and political credentials are certain. This group and I have conceived the following program actions. We plan to meet from time to time to assess progress and determine additional initiatives.

ACTION

1. *Standards setting.* While promulgation and modification activity must continue, no highly controversial standards (i.e., cotton dust, etc.) will be proposed by OSHA or by NIOSH. A thorough review with NIOSH indicates that while some criteria documents, such as on noise, will be transmitted to us during this period, neither the contents of these documents nor our handling of them here will generate any substantial controversy.
2. *Inspections.* OSHA will continue, administratively, to concentrate inspection activity on other than tiny establishments. . . .
4. *Personnel.* We are drafting an outline of OSHA's recruiting and hiring plan for the next six months. Subject to your approval, it is our intention to provide copies of this detailed plan to the Republican National Committee and the Committee to Re-elect the President. I can then consider applicants they propose. . . .
9. *Field staff.* I am having the registration of all RA's [Regional Administrators] and AD's [Assistant Directors] checked. When this reading is complete, we will consider further actions to be taken by the Republicans.

There are undoubtedly other ideas for action which we will come up with. You will be advised.

While I have discussed with Lee Nunn the great potential of OSHA as a sales point for fund raising and general support by employers, I do not believe the potential of this appeal is fully recognized. Your suggestions as to how to promote the advantages of four more years of properly managed OSHA for use in the campaign would be appreciated.

(initialed)
George C. Guenther

REFERENCES

1. Donella Meadows et al., *The Limits to Growth: A Report for the Club of Rome's Project on the Predicament of Mankind* (Universe Books, New York, 1972).

2. John Rawls, *A Theory of Justice* (Harvard University Press, Cambridge, Mass., 1971).

3. *Facts & Analysis: Occupational Safety and Health* (Industrial Union Department, AFL–CIO), No. 18 (July 22, 1974).

4. William A. Thomas, ed., *Scientists in the Legal System* (Ann Arbor Science Publishers, Ann Arbor, Mich., 1974).

5. Jethro K. Lieberman, *The Tyranny of the Experts* (Walker & Co., New York, 1970).

Appendix
The OSHAct of 1970 (Public Law 91-596)

An Act

To assure safe and healthful working conditions for working men and women; by authorizing enforcement of the standards developed under the Act; by assisting and encouraging the States in their efforts to assure safe and healthful working conditions; by providing for research, information, education, and training in the field of occupational safety and health; and for other purposes.

Be it enacted by the Senate and House of Representatives of the United States of America in Congress assembled, That this Act may be cited as the "Occupational Safety and Health Act of 1970". Occupational Safety and Health Act of 1970.

CONGRESSIONAL FINDINGS AND PURPOSE

SEC. (2) The Congress finds that personal injuries and illnesses arising out of work situations impcse a substantial burden upon, and are a hindrance to, interstate commerce in terms of lost production, wage loss, medical expenses, and disability compensation payments.

(b) The Congress declares it to be its purpose and policy, through the exercise of its powers to regulate commerce among the several States and with foreign nations and to provide for the general welfare, to assure so far as possible every working man and woman in the Nation safe and healthful working conditions and to preserve our human resources—

(1) by encouraging employers and employees in their efforts to reduce the number of occupational safety and health hazards at their places of employment, and to stimulate employers and employees to institute new and to perfect existing programs for providing safe and healthful working conditions;

(2) by providing that employers and employees have separate but dependent responsibilities and rights with respect to achieving safe and healthful working conditions;

(3) by authorizing the Secretary of Labor to set mandatory occupational safety and health standards applicable to businesses affecting interstate commerce, and by creating an Occupational Safety and Health Review Commission for carrying out adjudicatory functions under the Act;

(4) by building upon advances already made through employer and employee initiative for providing safe and healthful working conditions;

(5) by providing for research in the field of occupational safety and health, including the psychological factors involved, and by developing innovative methods, techniques, and approaches for dealing with occupational safety and health problems;

(6) by exploring ways to discover latent diseases, establishing causal connections between diseases and work in environmental conditions, and conducting other research relating to health problems, in recognition of the fact that occupational health standards present problems often different from those involved in occupational safety;

(7) by providing medical criteria which will assure insofar as practicable that no employee will suffer diminished health, functional capacity, or life expectancy as a result of his work experience;

(8) by providing for training programs to increase the number and competence of personnel engaged in the field of occupational safety and health;

(9) by providing for the development and promulgation of occupational safety and health standards;

(10) by providing an effective enforcement program which shall include a prohibition against giving advance notice of any inspection and sanctions for any individual violating this prohibition;

(11) by encouraging the States to assume the fullest responsibility for the administration and enforcement of their occupational safety and health laws by providing grants to the States to assist in identifying their needs and responsibilities in the area of occupational safety and health, to develop plans in accordance with the provisions of this Act, to improve the administration and enforcement of State occupational safety and health laws, and to conduct experimental and demonstration projects in connection therewith;

(12) by providing for appropriate reporting procedures with respect to occupational safety and health which procedures will help achieve the objectives of this Act and accurately describe the nature of the occupational safety and health problem;

(13) by encouraging joint labor-management efforts to reduce injuries and disease arising out of employment.

DEFINITIONS

SEC. 3. For the purposes of this Act—

(1) The term "Secretary" mean the Secretary of Labor.

(2) The term "Commission" means the Occupational Safety and Health Review Commission established under this Act.

(3) The term "commerce" means trade, traffic, commerce, transportation, or communication among the several States, or between a State and any place outside thereof, or within the District of Columbia, or a possession of the United States (other than the Trust Territory of the Pacific Islands), or between points in the same State but through a point outside thereof.

(4) The term "person" means one or more individuals, partnerships, associations, corporations, business trusts, legal representatives, or any organized group of persons.

(5) The term "employer" means a person engaged in a business affecting commerce who has employees, but does not include the United States or any State or political subdivision of a State.

(6) The term "employee" means an employee of an employer who is employed in a business of his employer which affects commerce.

(7) The term "State" includes a State of the United States, the District of Columbia, Puerto Rico, the Virgin Islands, American Samoa, Guam, and the Trust Territory of the Pacific Islands.

(8) The term "occupational safety and health standard" means a standard which requires conditions, or the adoption or use of one or more practices, means, methods, operations, or processes, reasonably necessary or appropriate to provide safe or healthful employment and places of employment.

(9) The term "national consensus standard" means any occupational safety and health standard or modification thereof which (1), has been adopted and promulgated by a nationally recognized standards-producing organization under procedures whereby it can be determined by the Secretary that persons interested

and affected by the scope or provisions of the standard have reached substantial agreement on its adoption, (2) was formulated in a manner which afforded an opportunity for diverse views to be considered and (3) has been designated as such a standard by the Secretary, after consultation with other appropriate Federal agencies.

(10) The term "established Federal standard" means any operative occupational safety and health standard established by any agency of the United States and presently in effect, or contained in any Act of Congress in force on the date of enactment of this Act.

(11) The term "Committee" means the National Advisory Committee on Occupational Safety and Health established under this Act.

(12) The term "Director" means the Director of the National Institute for Occupational Safety and Health.

(13) The term "Institute" means the National Institute for Occupational Safety and Health established under this Act.

(14) The term "Workmen's Compensation Commission" means the National Commission on State Workmen's Compensation Laws established under this Act.

APPLICABILITY OF THIS ACT

SEC. 4. (a) This Act shall apply with respect to employment performed in a workplace in a State, the District of Columbia, the Commonwealth of Puerto Rico, the Virgin Islands, American Samoa, Guam, the Trust Territory of the Pacific Islands, Wake Island, Outer Continental Shelf lands defined in the Outer Continental Shelf Lands Act, Johnston Island, and the Canal Zone. The Secretary of the Interior shall, by regulation, provide for judicial enforcement of this Act by the courts established for areas in which there are no United States district courts having jurisdiction.

67 Stat. 462.
43 USC 1331 note.

(b)(1) Nothing in this Act shall apply to working conditions of employees with respect to which other Federal agencies, and State agencies acting under section 274 of the Atomic Energy Act of 1954, as amended (42 U.S.C. 2021), exercise statutory authority to prescribe or enforce standards or regulations affecting occupational safety or health.

73 Stat. 688.

(2) The safety and health standards promulgated under the Act of June 30, 1936, commonly known as the Walsh-Healey Act (41 U.S.C. 35 et seq.), the Service Contract Act of 1965 (41 U.S.C. 351 et seq.), Public Law 91–54, Act of August 9, 1969 (40 U.S.C. 333), Public Law 85–742, Act of August 23, 1958 (33 U.S.C. 941), and the National Foundation on Arts and Humanities Act (20 U.S.C. 951 et seq.) are superseded on the effective date of corresponding standards, promulgated under this Act, which are determined by the Secretary to be more effective. Standards issued under the laws listed in this paragraph and in effect on or after the effective date of this Act shall be deemed to be occupational safety and health standards issued under this Act, as well as under such other Acts.

49 Stat. 2036.
79 Stat. 1034.
83 Stat. 96.
72 Stat. 835.
79 Stat. 845;
Ante, p. 443.

(3) The Secretary shall, within three years after the effective date of this Act, report to the Congress his recommendations for legislation to avoid unnecessary duplication and to achieve coordination between this Act and other Federal laws.

Report to Congress.

(4) Nothing in this Act shall be construed to supersede or in any manner affect any workmen's compensation law or to enlarge or diminish or affect in any other manner the common law or statutory rights, duties, or liabilities of employers and employees under any law with respect to injuries, diseases, or death of employees arising out of, or in the course of, employment.

DUTIES

SEC. 5. (a) Each employer—

(1) shall furnish to each of his employees employment and a place of employment which are free from recognized hazards that are causing or are likely to cause death or serious physical harm to his employees;

(2) shall comply with occupational safety and health standards promulgated under this Act.

(b) Each employee shall comply with occupational safety and health standards and all rules, regulations, and orders issued pursuant to this Act which are applicable to his own actions and conduct.

OCCUPATIONAL SAFETY AND HEALTH STANDARDS

80 Stat. 381; 81 Stat. 195. 5 USC 500.

SEC. 6. (a) Without regard to chapter 5 of title 5, United States Code, or to the other subsections of this section, the Secretary shall, as soon as practicable during the period beginning with the effective date of this Act and ending two years after such date, by rule promulgate as an occupational safety or health standard any national consensus standard, and any established Federal standard, unless he determines that the promulgation of such a standard would not result in improved safety or health for specifically designated employees. In the event of conflict among any such standards, the Secretary shall promulgate the standard which assures the greatest protection of the safety or health of the affected employees.

(b) The Secretary may by rule promulgate, modify, or revoke any occupational safety or health standard in the following manner:

Advisory committee, recommendations.

(1) Whenever the Secretary, upon the basis of information submitted to him in writing by an interested person, a representative of any organization of employers or employees, a nationally recognized standards-producing organization, the Secretary of Health, Education, and Welfare, the National Institute for Occupational Safety and Health, or a State or political subdivision, or on the basis of information developed by the Secretary or otherwise available to him, determines that a rule should be promulgated in order to serve the objectives of this Act, the Secretary may request the recommendations of an advisory committee appointed under section 7 of this Act. The Secretary shall provide such an advisory committee with any proposals of his own or of the Secretary of Health, Education, and Welfare, together with all pertinent factual information developed by the Secretary or the Secretary of Health, Education, and Welfare, or otherwise available, including the results of research, demonstrations, and experiments. An advisory committee shall submit to the Secretary its recommendations regarding the rule to be promulgated within ninety days from the date of its appointment or within such longer or shorter period as may be prescribed by the Secretary, but in no event for a period which is longer than two hundred and seventy days.

(2) The Secretary shall publish a proposed rule promulgating, modifying, or revoking an occupational safety or health standard in the Federal Register and shall afford interested persons a period of thirty days after publication to submit written data or comments. Where an advisory committee is appointed and the Secretary determines that a rule should be issued, he shall publish the proposed rule within sixty days after the submission of the advisory committee's recommendations or the expiration of the period prescribed by the Secretary for such submission. Publication in Federal Register.

(3) On or before the last day of the period provided for the submission of written data or comments under paragraph (2), any interested person may file with the Secretary written objections to the proposed rule, stating the grounds therefor and requesting a public hearing on such objections. Within thirty days after the last day for filing such objections, the Secretary shall publish in the Federal Register a notice specifying the occupational safety or health standard to which objections have been filed and a hearing requested, and specifying a time and place for such hearing. Hearing, notice. Publication in Federal Register.

(4) Within sixty days after the expiration of the period provided for the submission of written data or comments under paragraph (2), or within sixty days after the completion of any hearing held under paragraph (3), the Secretary shall issue a rule promulgating, modifying, or revoking an occupational safety or health standard or make a determination that a rule should not be issued. Such a rule may contain a provision delaying its effective date for such period (not in excess of ninety days) as the Secretary determines may be necessary to insure that affected employers and employees will be informed of the existence of the standard and of its terms and that employers affected are given an opportunity to familiarize themselves and their employees with the existence of the requirements of the standard.

(5) The Secretary, in promulgating standards dealing with toxic materials or harmful physical agents under this subsection, shall set the standard which most adequately assures, to the extent feasible, on the basis of the best available evidence, that no employee will suffer material impairment of health or functional capacity even if such employee has regular exposure to the hazard dealt with by such standard for the period of his working life. Development of standards under this subsection shall be based upon research, demonstrations, experiments, and such other information as may be appropriate. In addition to the attainment of the highest degree of health and safety protection for the employee, other considerations shall be the latest available scientific data in the field, the feasibility of the standards, and experience gained under this and other health and safety laws. Whenever practicable, the standard promulgated shall be expressed in terms of objective criteria and of the performance desired. Toxic materials.

(6) (A) Any employer may apply to the Secretary for a temporary order granting a variance from a standard or any provision thereof promulgated under this section. Such temporary order shall be granted only if the employer files an application which meets the requirements of clause (B) and establishes that (i) he is unable to comply with a standard by its effective date because of unavailability of professional or technical personnel or of materials and equipment needed to come into compliance with the standard or because necessary construction or alteration of facilities cannot be completed by the effective date, (ii) he is taking all available steps to safeguard his employees against the hazards covered by the standard, and (iii) he has an effective program for coming into compliance with the standard as quickly as Temporary variance order.

practicable. Any temporary order issued under this paragraph shall prescribe the practices, means, methods, operations, and processes which the employer must adopt and use while the order is in effect and state in detail his program for coming into compliance with the standard. Such a temporary order may be granted only after notice to employees and an opportunity for a hearing: *Provided*, That the Secretary may issue one interim order to be effective until a decision is made on the basis of the hearing. No temporary order may be in effect for longer than the period needed by the employer to achieve compliance with the standard or one year, whichever is shorter, except that such an order may be renewed not more than twice (I) so long as the requirements of this paragraph are met and (II) if an application for renewal is filed at least 90 days prior to the expiration date of the order. No interim renewal of an order may remain in effect for longer than 180 days.

Notice, hearing.

Renewal.

Time limitation.

(B) An application for a temporary order under this paragraph (6) shall contain:

(i) a specification of the standard or portion thereof from which the employer seeks a variance,

(ii) a representation by the employer, supported by representations from qualified persons having firsthand knowledge of the facts represented, that he is unable to comply with the standard or portion thereof and a detailed statement of the reasons therefor,

(iii) a statement of the steps he has taken and will take (with specific dates) to protect employees against the hazard covered by the standard,

(iv) a statement of when he expects to be able to comply with the standard and what steps he has taken and what steps he will take (with dates specified) to come into compliance with the standard, and

(v) a certification that he has informed his employees of the application by giving a copy thereof to their authorized representative, posting a statement giving a summary of the application and specifying where a copy may be examined at the place or places where notices to employees are normally posted, and by other appropriate means.

A description of how employees have been informed shall be contained in the certification. The information to employees shall also inform them of their right to petition the Secretary for a hearing.

(C) The Secretary is authorized to grant a variance from any standard or portion thereof whenever he determines, or the Secretary of Health, Education, and Welfare certifies, that such variance is necessary to permit an employer to participate in an experiment approved by him or the Secretary of Health, Education, and Welfare designed to demonstrate or validate new and improved techniques to safeguard the health or safety of workers.

Labels, etc.

(7) Any standard promulgated under this subsection shall prescribe the use of labels or other appropriate forms of warning as are necessary to insure that employees are apprised of all hazards to which they are exposed, relevant symptoms and appropriate emergency treatment, and proper conditions and precautions of safe use or exposure. Where appropriate, such standard shall also prescribe suitable protective equipment and control or technological procedures to be used in connection with such hazards and shall provide for monitoring or measuring employee exposure at such locations and intervals, and in such manner as may be necessary for the protection of employees. In

Protective equipment, etc.

addition, where appropriate, any such standard shall prescribe the type and frequency of medical examinations or other tests which shall be made available, by the employer or at his cost, to employees exposed to such hazards in order to most effectively determine whether the health of such employees is adversely affected by such exposure. In the event such medical examinations are in the nature of research, as determined by the Secretary of Health, Education, and Welfare, such examinations may be furnished at the expense of the Secretary of Health, Education, and Welfare. The results of such examinations or tests shall be furnished only to the Secretary or the Secretary of Health, Education, and Welfare, and, at the request of the employee, to his physician. The Secretary, in consultation with the Secretary of Health, Education, and Welfare, may by rule promulgated pursuant to section 553 of title 5, United States Code, make appropriate modifications in the foregoing requirements relating to the use of labels or other forms of warning, monitoring or measuring, and medical examinations, as may be warranted by experience, information, or medical or technological developments acquired subsequent to the promulgation of the relevant standard.

Medical examinations.

80 Stat. 383.

(8) Whenever a rule promulgated by the Secretary differs substantially from an existing national consensus standard, the Secretary shall, at the same time, publish in the Federal Register a statement of the reasons why the rule as adopted will better effectuate the purposes of this Act than the national consensus standard.

Publication in Federal Register.

(c)(1) The Secretary shall provide, without regard to the requirements of chapter 5, title 5, United States Code, for an emergency temporary standard to take immediate effect upon publication in the Federal Register if he determines (A) that employees are exposed to grave danger from exposure to substances or agents determined to be toxic or physically harmful or from new hazards, and (B) that such emergency standard is necessary to protect employees from such danger.

Temporary standard. Publication in Federal Register. 80 Stat. 381; 81 Stat. 195. 5 USC 500.

(2) Such standard shall be effective until superseded by a standard promulgated in accordance with the procedures prescribed in paragraph (3) of this subsection.

Time limitation.

(3) Upon publication of such standard in the Federal Register the Secretary shall commence a proceeding in accordance with section 6(b) of this Act, and the standard as published shall also serve as a proposed rule for the proceeding. The Secretary shall promulgate a standard under this paragraph no later than six months after publication of the emergency standard as provided in paragraph (2) of this subsection.

(d) Any affected employer may apply to the Secretary for a rule or order for a variance from a standard promulgated under this section. Affected employees shall be given notice of each such application and an opportunity to participate in a hearing. The Secretary shall issue such rule or order if he determines on the record, after opportunity for an inspection where appropriate and a hearing, that the proponent of the variance has demonstrated by a preponderance of the evidence that the conditions, practices, means, methods, operations, or processes used or proposed to be used by an employer will provide employment and places of employment to his employees which are as safe and healthful as those which would prevail if he complied with the standard. The rule or order so issued shall prescribe the conditions the employer must maintain, and the practices, means, methods, operations, and processes which he must adopt and utilize to the extent they

Variance rule.

differ from the standard in question. Such a rule or order may be modified or revoked upon application by an employer, employees, or by the Secretary on his own motion, in the manner prescribed for its issuance under this subsection at any time after six months from its issuance.

Publication in Federal Register.

(e) Whenever the Secretary promulgates any standard, makes any rule, order, or decision, grants any exemption or extension of time, or compromises, mitigates, or settles any penalty assessed under this Act, he shall include a statement of the reasons for such action, which shall be published in the Federal Register.

Petition for judicial review.

(f) Any person who may be adversely affected by a standard issued under this section may at any time prior to the sixtieth day after such standard is promulgated file a petition challenging the validity of such standard with the United States court of appeals for the circuit wherein such person resides or has his principal place of business, for a judicial review of such standard. A copy of the petition shall be forthwith transmitted by the clerk of the court to the Secretary. The filing of such petition shall not, unless otherwise ordered by the court, operate as a stay of the standard. The determinations of the Secretary shall be conclusive if supported by substantial evidence in the record considered as a whole.

(g) In determining the priority for establishing standards under this section, the Secretary shall give due regard to the urgency of the need for mandatory safety and health standards for particular industries, trades, crafts, occupations, businesses, workplaces or work environments. The Secretary shall also give due regard to the recommendations of the Secretary of Health, Education, and Welfare regarding the need for mandatory standards in determining the priority for establishing such standards.

ADVISORY COMMITTEES; ADMINISTRATION

Establishment; membership.

80 Stat. 378.
5 USC 101.

SEC. 7. (a)(1) There is hereby established a National Advisory Committee on Occupational Safety and Health consisting of twelve members appointed by the Secretary, four of whom are to be designated by the Secretary of Health, Education, and Welfare, without regard to the provisions of title 5, United States Code, governing appointments in the competitive service, and composed of representatives of manangement, labor, occupational safety and occupational health professions, and of the public. The Secretary shall designate one of the public members as Chairman. The members shall be selected upon the basis of their experience and competence in the field of occupational safety and health.

(2) The Committee shall advise, consult with, and make recommendations to the Secretary and the Secretary of Health, Education, and Welfare on matters relating to the administration of the Act. The Committee shall hold no fewer than two meetings during each calendar year. All meetings of the Committee shall be open to the public and a transcript shall be kept and made available for public inspection.

Public transcript.

(3) The members of the Committee shall be compensated in accordance with the provisions of section 3109 of title 5, United States Code.

80 Stat. 416.

(4) The Secretary shall furnish to the Committee an executive secretary and such secretarial, clerical, and other services as are deemed necessary to the conduct of its business.

(b) An advisory committee may be appointed by the Secretary to assist him in his standard-setting functions under section 6 of this Act. Each such committee shall consist of not more than fifteen members and shall include as a member one or more designees of the Secretary of Health, Education, and Welfare, and shall include among its members an equal number of persons qualified by experience and affiliation to present the viewpoint of the employers involved, and of persons similarly qualified to present the viewpoint of the workers involved, as well as one or more representatives of health and safety agencies of the States. An advisory committee may also include such other persons as the Secretary may appoint who are qualified by knowledge and experience to make a useful contribution to the work of such committee, including one or more representatives of professional organizations of technicians or professionals specializing in occupational safety or health, and one or more representatives of nationally recognized standards-producing organizations, but the number of persons so appointed to any such advisory committee shall not exceed the number appointed to such committee as representatives of Federal and State agencies. Persons appointed to advisory committees from private life shall be compensated in the same manner as consultants or experts under section 3109 of title 5, United States Code. The Secretary shall pay to any State which is the employer of a member of such a committee who is a representative of the health or safety agency of that State, reimbursement sufficient to cover the actual cost to the State resulting from such representative's membership on such committee. Any meeting of such committee shall be open to the public and an accurate record shall be kept and made available to the public. No member of such committee (other than representatives of employers and employees) shall have an economic interest in any proposed rule.

80 Stat. 416.

Recordkeeping.

(c) In carrying out his responsibilities under this Act, the Secretary is authorized to—

(1) use, with the consent of any Federal agency, the services, facilities, and personnel of such agency, with or without reimbursement, and with the consent of any State or political subdivision thereof, accept and use the services, facilities, and personnel of any agency of such State or subdivision with reimbursement; and

(2) employ experts and consultants or organizations thereof as authorized by section 3109 of title 5, United States Code, except that contracts for such employment may be renewed annually; compensate individuals so employed at rates not in excess of the rate specified at the time of service for grade GS–18 under section 5332 of title 5, United States Code, including traveltime, and allow them while away from their homes or regular places of business, travel expenses (including per diem in lieu of subsistence) as authorized by section 5703 of title 5, United States Code, for persons in the Government service employed intermittently, while so employed.

Ante, p. 198-1.

80 Stat. 499; 83 Stat. 190.

INSPECTIONS, INVESTIGATIONS, AND RECORDKEEPING

SEC. 8. (a) In order to carry out the purposes of this Act, the Secretary, upon presenting appropriate credentials to the owner, operator, or agent in charge, is authorized—

(1) to enter without delay and at reasonable times any factory, plant, establishment, construction site, or other area, workplace or environment where work is performed by an employee of an employer; and

(2) to inspect and investigate during regular working hours and at other reasonable times, and within reasonable limits and in a reasonable manner, any such place of employment and all pertinent conditions, structures, machines, apparatus, devices, equipment, and materials therein, and to question privately any such employer, owner, operator, agent or employee.

Subpoena power.

(b) In making his inspections and investigations under this Act the Secretary may require the attendance and testimony of witnesses and the production of evidence under oath. Witnesses shall be paid the same fees and mileage that are paid witnesses in the courts of the United States. In case of a contumacy, failure, or refusal of any person to obey such an order, any district court of the United States or the United States courts of any territory or possession, within the jurisdiction of which such person is found, or resides or transacts business, upon the application by the Secretary, shall have jurisdiction to issue to such person an order requiring such person to appear to produce evidence if, as, and when so ordered, and to give testimony relating to the matter under investigation or in question, and any failure to obey such order of the court may be punished by said court as a contempt thereof.

Recordkeeping.

(c)(1) Each employer shall make, keep and preserve, and make available to the Secretary or the Secretary of Health, Education, and Welfare, such records regarding his activities relating to this Act as the Secretary, in cooperation with the Secretary of Health, Education, and Welfare, may prescribe by regulation as necessary or appropriate for the enforcement of this Act or for developing information regarding the causes and prevention of occupational accidents and illnesses. In order to carry out the provisions of this paragraph such regulations may include provisions requiring employers to conduct periodic inspections. The Secretary shall also issue regulations requiring that employers, through posting of notices or other appropriate means, keep their employees informed of their protections and obligations under this Act, including the provisions of applicable standards.

Work-related deaths, etc.; reports.

(2) The Secretary, in cooperation with the Secretary of Health, Education, and Welfare, shall prescribe regulations requiring employers to maintain accurate records of, and to make periodic reports on, work-related deaths, injuries and illnesses other than minor injuries requiring only first aid treatment and which do not involve medical treatment, loss of consciousness, restriction of work or motion, or transfer to another job.

(3) The Secretary, in cooperation with the Secretary of Health, Education, and Welfare, shall issue regulations requiring employers to maintain accurate records of employee exposures to potentially toxic materials or harmful physical agents which are required to be monitored or measured under section 6. Such regulations shall provide employees or their representatives with an opportunity to observe such monitoring or measuring, and to have access to the records thereof. Such regulations shall also make appropriate provision for each employee or former employee to have access to such records as will indicate his own exposure to toxic materials or harmful physical agents. Each employer shall promptly notify any employee who has been or is being exposed to toxic materials or harmful physical agents in concentrations or at levels which exceed those prescribed by an applicable occupational safety and health standard promulgated under section 6, and shall inform any employee who is being thus exposed of the corrective action being taken.

(d) Any information obtained by the Secretary, the Secretary of Health, Education, and Welfare, or a State agency under this Act shall be obtained with a minimum burden upon employers, especially those operating small businesses. Unnecessary duplication of efforts in obtaining information shall be reduced to the maximum extent feasible.

(e) Subject to regulations issued by the Secretary, a representative of the employer and a representative authorized by his employees shall be given an opportunity to accompany the Secretary or his authorized representative during the physical inspection of any workplace under subsection (a) for the purpose of aiding such inspection. Where there is no authorized employee representative, the Secretary or his authorized representative shall consult with a reasonable number of employees concerning matters of health and safety in the workplace.

(f)(1) Any employees or representative of employees who believe that a violation of a safety or health standard exists that threatens physical harm, or that an imminent danger exists, may request an inspection by giving notice to the Secretary or his authorized representative of such violation or danger. Any such notice shall be reduced to writing, shall set forth with reasonable particularity the grounds for the notice, and shall be signed by the employees or representative of employees, and a copy shall be provided the employer or his agent no later than at the time of inspection, except that, upon the request of the person giving such notice, his name and the names of individual employees referred to therein shall not appear in such copy or on any record published, released, or made available pursuant to subsection (g) of this section. If upon receipt of such notification the Secretary determines there are reasonable grounds to believe that such violation or danger exists, he shall make a special inspection in accordance with the provisions of this section as soon as practicable, to determine if such violation or danger exists. If the Secretary determines there are no reasonable grounds to believe that a violation or danger exists he shall notify the employees or representative of the employees in writing of such determination.

(2) Prior to or during any inspection of a workplace, any employees or representative of employees employed in such workplace may notify the Secretary or any representative of the Secretary responsible for conducting the inspection, in writing, of any violation of this Act which they have reason to believe exists in such workplace. The Secretary shall, by regulation, establish procedures for informal review of any refusal by a representative of the Secretary to issue a citation with respect to any such alleged violation and shall furnish the employees or representative of employees requesting such review a written statement of the reasons for the Secretary's final disposition of the case.

Reports, publication.

(g)(1) The Secretary and Secretary of Health, Education, and Welfare are authorized to compile, analyze, and publish, either in summary or detailed form, all reports or information obtained under this section.

Rules and regulations.

(2) The Secretary and the Secretary of Health, Education, and Welfare shall each prescribe such rules and regulations as he may deem necessary to carry out their responsibilities under this Act, including rules and regulations dealing with the inspection of an employer's establishment.

CITATIONS

SEC. 9. (a) If, upon inspection or investigation, the Secretary or his authorized representative believes that an employer has violated a requirement of section 5 of this Act, of any standard, rule or order promulgated pursuant to section 6 of this Act, or of any regulations prescribed pursuant to this Act, he shall with reasonable promptness issue a citation to the employer. Each citation shall be in writing and shall describe with particularity the nature of the violation, including a reference to the provision of the Act, standard, rule, regulation, or order alleged to have been violated. In addition, the citation shall fix a reasonable time for the abatement of the violation. The Secretary may prescribe procedures for the issuance of a notice in lieu of a citation with respect to de minimis violations which have no direct or immediate relationship to safety or health.

(b) Each citation issued under this section, or a copy or copies thereof, shall be prominently posted, as prescribed in regulations issued by the Secretary, at or near each place a violation referred to in the citation occurred.

Limitation. (c) No citation may be issued under this section after the expiration of six months following the occurrence of any violation.

PROCEDURE FOR ENFORCEMENT

SEC. 10. (a) If, after an inspection or investigation, the Secretary issues a citation under section 9(a), he shall, within a reasonable time after the termination of such inspection or investigation, notify the employer by certified mail of the penalty, if any, proposed to be assessed under section 17 and that the employer has fifteen working days within which to notify the Secretary that he wishes to contest the citation or proposed assessment of penalty. If, within fifteen working days from the receipt of the notice issued by the Secretary the employer fails to notify the Secretary that he intends to contest the citation or proposed assessment of penalty, and no notice is filed by any employee or representative of employees under subsection (c) within such time, the citation and the assessment, as proposed, shall be deemed a final order of the Commission and not subject to review by any court or agency.

(b) If the Secretary has reason to believe that an employer has failed to correct a violation for which a citation has been issued within the period permitted for its correction (which period shall not begin to run until the entry of a final order by the Commission in the case of any review proceedings under this section initiated by the employer in good faith and not solely for delay or avoidance of penalties), the Secretary shall notify the employer by certified mail of such failure and of the penalty proposed to be assessed under section 17 by reason of such failure, and that the employer has fifteen working days within which to notify the Secretary that he wishes to contest the Secretary's notification or the proposed assessment of penalty. If, within fifteen working days from the receipt of notification issued by the Secretary, the employer fails to notify the Secretary that he intends to contest the notification or proposed assessment of penalty, the notification and assessment, as proposed, shall be deemed a final order of the Commission and not subject to review by any court or agency.

(c) If an employer notifies the Secretary that he intends to contest a citation issued under section 9(a) or notification issued under subsection (a) or (b) of this section, or if, within fifteen working days

of the issuance of a citation under section 9(a), any employee or representative of employees files a notice with the Secretary alleging that the period of time fixed in the citation for the abatement of the violation is unreasonable, the Secretary shall immediately advise the Commission of such notification, and the Commission shall afford an opportunity for a hearing (in accordance with section 554 of title 5, United States Code, but without regard to subsection (a)(3) of such 80 Stat. 384.
section). The Commission shall thereafter issue an order, based on findings of fact, affirming, modifying, or vacating the Secretary's citation or proposed penalty, or directing other appropriate relief, and such order shall become final thirty days after its issuance. Upon a showing by an employer of a good faith effort to comply with the abatement requirements of a citation, and that abatement has not been completed because of factors beyond his reasonable control, the Secretary, after an opportunity for a hearing as provided in this subsection, shall issue an order affirming or modifying the abatement requirements in such citation. The rules of procedure prescribed by the Commission shall provide affected employees or representatives of affected employees an opportunity to participate as parties to hearings under this subsection.

JUDICIAL REVIEW

SEC. 11. (a) Any person adversely affected or aggrieved by an order of the Commission issued under subsection (c) of section 10 may obtain a review of such order in any United States court of appeals for the circuit in which the violation is alleged to have occurred or where the employer has its principal office, or in the Court of Appeals for the District of Columbia Circuit, by filing in such court within sixty days following the issuance of such order a written petition praying that the order be modified or set aside. A copy of such petition shall be forthwith transmitted by the clerk of the court to the Commission and to the other parties, and thereupon the Commission shall file in the court the record in the proceeding as provided in section 2112 of title 28, United States Code. Upon such filing, the court shall have jurisdiction 72 Stat. 941;
of the proceeding and of the question determined therein, and shall 80 Stat. 1323.
have power to grant such temporary relief or restraining order as it deems just and proper, and to make and enter upon the pleadings, testimony, and proceedings set forth in such record a decree affirming, modifying, or setting aside in whole or in part, the order of the Commission and enforcing the same to the extent that such order is affirmed or modified. The commencement of proceedings under this subsection shall not, unless ordered by the court, operate as a stay of the order of the Commission. No objection that has not been urged before the Commission shall be considered by the court, unless the failure or neglect to urge such objection shall be excused because of extraordinary circumstances. The findings of the Commission with respect to questions of fact, if supported by substantial evidence on the record considered as a whole, shall be conclusive. If any party shall apply to the court for leave to adduce additional evidence and shall show to the satisfaction of the court that such additional evidence is material and that there were reasonable grounds for the failure to adduce such evidence in the hearing before the Commission, the court may order such additional evidence to be taken before the Commission and to be made a part of the record. The Commission may modify its findings as to the facts, or make new findings, by reason of additional evidence so taken and filed, and it shall file such modified or new findings, which findings with respect to questions of fact, if supported by substantial evi-

dence on the record considered as a whole, shall be conclusive, and its recommendations, if any, for the modification or setting aside of its original order. Upon the filing of the record with it, the jurisdiction of the court shall be exclusive and its judgment and decree shall be final, except that the same shall be subject to review by the Supreme Court of the United States, as provided in section 1254 of title 28,
62 Stat. 928. United States Code. Petitions filed under this subsection shall be heard expeditiously.

(b) The Secretary may also obtain review or enforcement of any final order of the Commission by filing a petition for such relief in the United States court of appeals for the circuit in which the alleged violation occurred or in which the employer has its principal office, and the provisions of subsection (a) shall govern such proceedings to the extent applicable. If no petition for review, as provided in subsection (a), is filed within sixty days after service of the Commission's order, the Commission's findings of fact and order shall be conclusive in connection with any petition for enforcement which is filed by the Secretary after the expiration of such sixty-day period. In any such case, as well as in the case of a noncontested citation or notification by the Secretary which has become a final order of the Commission under subsection (a) or (b) of section 10, the clerk of the court, unless otherwise ordered by the court, shall forthwith enter a decree enforcing the order and shall transmit a copy of such decree to the Secretary and the employer named in the petition. In any contempt proceeding brought to enforce a decree of a court of appeals entered pursuant to this subsection or subsection (a), the court of appeals may assess the penalties provided in section 17, in addition to invoking any other available remedies.

(c) (1) No person shall discharge or in any manner discriminate against any employee because such employee has filed any complaint or instituted or caused to be instituted any proceeding under or related to this Act or has testified or is about to testify in any such proceeding or because of the exercise by such employee on behalf of himself or others of any right afforded by this Act.

(2) Any employee who believes that he has been discharged or otherwise discriminated against by any person in violation of this subsection may, within thirty days after such violation occurs, file a complaint with the Secretary alleging such discrimination. Upon receipt of such complaint, the Secretary shall cause such investigation to be made as he deems appropriate. If upon such investigation, the Secretary determines that the provisions of this subsection have been violated, he shall bring an action in any appropriate United States district court against such person. In any such action the United States district courts shall have jurisdiction, for cause shown to restrain violations of paragraph (1) of this subsection and order all appropriate relief including rehiring or reinstatement of the employee to his former position with back pay.

(3) Within 90 days of the receipt of a complaint filed under this subsection the Secretary shall notify the complainant of his determination under paragraph 2 of this subsection.

THE OCCUPATIONAL SAFETY AND HEALTH REVIEW COMMISSION

Establishment; membership.

SEC. 12. (a) The Occupational Safety and Health Review Commission is hereby established. The Commission shall be composed of three members who shall be appointed by the President, by and with the advice and consent of the Senate, from among persons who by reason of training, education, or experience are qualified to carry out the functions of the Commission under this Act. The President shall designate one of the members of the Commission to serve as Chairman.

(b) The terms of members of the Commission shall be six years except that (1) the members of the Commission first taking office shall serve, as designated by the President at the time of appointment, one for a term of two years, one for a term of four years, and one for a term of six years, and (2) a vacancy caused by the death, resignation, or removal of a member prior to the expiration of the term for which he was appointed shall be filled only for the remainder of such unexpired term. A member of the Commission may be removed by the President for inefficiency, neglect of duty, or malfeasance in office. Terms.

(c)(1) Section 5314 of title 5, United States Code, is amended by adding at the end thereof the following new paragraph: 80 Stat. 460.

"(57) Chairman, Occupational Safety and Health Review Commission."

(2) Section 5315 of title 5, United States Code, is amended by adding at the end thereof the following new paragraph: Ante, p. 776.

"(94) Members, Occupational Safety and Health Review Commission."

(d) The principal office of the Commission shall be in the District of Columbia. Whenever the Commission deems that the convenience of the public or of the parties may be promoted, or delay or expense may be minimized, it may hold hearings or conduct other proceedings at any other place. Location.

(e) The Chairman shall be responsible on behalf of the Commission for the administrative operations of the Commission and shall appoint such hearing examiners and other employees as he deems necessary to assist in the performance of the Commission's functions and to fix their compensation in accordance with the provisions of chapter 51 and subchapter III of chapter 53 of title 5, United States Code, relating to classification and General Schedule pay rates: *Provided*, That assignment, removal and compensation of hearing examiners shall be in accordance with sections 3105, 3344, 5362, and 7521 of title 5, United States Code. 5 USC 5101, 5331. Ante, p. 198-1.

(f) For the purpose of carrying out its functions under this Act, two members of the Commission shall constitute a quorum and official action can be taken only on the affirmative vote of at least two members. Quorum.

(g) Every official act of the Commission shall be entered of record, and its hearings and records shall be open to the public. The Commission is authorized to make such rules as are necessary for the orderly transaction of its proceedings. Unless the Commission has adopted a different rule, its proceedings shall be in accordance with the Federal Rules of Civil Procedure. Public records. 28 USC app.

(h) The Commission may order testimony to be taken by deposition in any proceedings pending before it at any state of such proceeding. Any person may be compelled to appear and depose, and to produce books, papers, or documents, in the same manner as witnesses may be compelled to appear and testify and produce like documentary evidence before the Commission. Witnesses whose depositions are taken under this subsection, and the persons taking such depositions, shall be entitled to the same fees as are paid for like services in the courts of the United States.

(i) For the purpose of any proceeding before the Commission, the provisions of section 11 of the National Labor Relations Act (29 U.S.C. 161) are hereby made applicable to the jurisdiction and powers of the Commission. 61 Stat. 150; Ante, p. 930.

Report. (j) A hearing examiner appointed by the Commission shall hear, and make a determination upon, any proceeding instituted before the Commission and any motion in connection therewith, assigned to such hearing examiner by the Chairman of the Commission, and shall make a report of any such determination which constitutes his final disposition of the proceedings. The report of the hearing examiner shall become the final order of the Commission within thirty days after such report by the hearing examiner, unless within such period any Commission member has directed that such report shall be reviewed by the Commission.

(k) Except as otherwise provided in this Act, the hearing examiners shall be subject to the laws governing employees in the classified civil service, except that appointments shall be made without regard to section 5108 of title 5, United States Code. Each hearing examiner shall receive compensation at a rate not less than that prescribed for GS-16 under section 5332 of title 5, United States Code.

80 Stat. 453.

Ante, p. 198-1.

PROCEDURES TO COUNTERACT IMMINENT DANGERS

SEC. 13. (a) The United States district courts shall have jurisdiction, upon petition of the Secretary, to restrain any conditions or practices in any place of employment which are such that a danger exists which could reasonably be expected to cause death or serious physical harm immediately or before the imminence of such danger can be eliminated through the enforcement procedures otherwise provided by this Act. Any order issued under this section may require such steps to be taken as may be necessary to avoid, correct, or remove such imminent danger and prohibit the employment or presence of any individual in locations or under conditions where such imminent danger exists, except individuals whose presence is necessary to avoid, correct, or remove such imminent danger or to maintain the capacity of a continuous process operation to resume normal operations without a complete cessation of operations, or where a cessation of operations is necessary, to permit such to be accomplished in a safe and orderly manner.

(b) Upon the filing of any such petition the district court shall have jurisdiction to grant such injunctive relief or temporary restraining order pending the outcome of an enforcement proceeding pursuant to this Act. The proceeding shall be as provided by Rule 65 of the Federal Rules, Civil Procedure, except that no temporary restraining order issued without notice shall be effective for a period longer than five days.

28 USC app.

(c) Whenever and as soon as an inspector concludes that conditions or practices described in subsection (a) exist in any place of employment, he shall inform the affected employees and employers of the danger and that he is recommending to the Secretary that relief be sought.

(d) If the Secretary arbitrarily or capriciously fails to seek relief under this section, any employee who may be injured by reason of such failure, or the representative of such employees, might bring an action against the Secretary in the United States district court for the district in which the imminent danger is alleged to exist or the employer has its principal office, or for the District of Columbia, for a writ of mandamus to compel the Secretary to seek such an order and for such further relief as may be appropriate.

REPRESENTATION IN CIVIL LITIGATION

SEC. 14. Except as provided in section 518(a) of title 28, United States Code, relating to litigation before the Supreme Court, the Solicitor of Labor may appear for and represent the Secretary in any civil litigation brought under this Act but all such litigation shall be subject to the direction and control of the Attorney General.

80 Stat. 613.

CONFIDENTIALITY OF TRADE SECRETS

SEC. 15. All information reported to or otherwise obtained by the Secretary or his representative in connection with any inspection or proceeding under this Act which contains or which might reveal a trade secret referred to in section 1905 of title 18 of the United States Code shall be considered confidential for the purpose of that section, except that such information may be disclosed to other officers or employees concerned with carrying out this Act or when relevant in any proceeding under this Act. In any such proceeding the Secretary, the Commission, or the court shall issue such orders as may be appropriate to protect the confidentiality of trade secrets.

62 Stat. 791.

VARIATIONS, TOLERANCES, AND EXEMPTIONS

SEC. 16. The Secretary, on the record, after notice and opportunity for a hearing may provide such reasonable limitations and may make such rules and regulations allowing reasonable variations, tolerances, and exemptions to and from any or all provisions of this Act as he may find necessary and proper to avoid serious impairment of the national defense. Such action shall not be in effect for more than six months without notification to affected employees and an opportunity being afforded for a hearing.

PENALTIES

SEC. 17. (a) Any employer who willfully or repeatedly violates the requirements of section 5 of this Act, any standard, rule, or order promulgated pursuant to section 6 of this Act, or regulations prescribed pursuant to this Act, may be assessed a civil penalty of not more than $10,000 for each violation.

(b) Any employer who has received a citation for a serious violation of the requirements of section 5 of this Act, of any standard, rule, or order promulgated pursuant to section 6 of this Act, or of any regulations prescribed pursuant to this Act, shall be assessed a civil penalty of up to $1,000 for each such violation.

(c) Any employer who has received a citation for a violation of the requirements of section 5 of this Act, of any standard, rule, or order promulgated pursuant to section 6 of this Act, or of regulations prescribed pursuant to this Act, and such violation is specifically determined not to be of a serious nature, may be assessed a civil penalty of up to $1,000 for each such violation.

(d) Any employer who fails to correct a violation for which a citation has been issued under section 9(a) within the period permitted for its correction (which period shall not begin to run until the date of the final order of the Commission in the case of any review proceeding under section 10 initiated by the employer in good faith and not solely for delay or avoidance of penalties), may be assessed a civil penalty of not more than $1,000 for each day during which such failure or violation continues.

(e) Any employer who willfully violates any standard, rule, or order promulgated pursuant to section 6 of this Act, or of any regulations prescribed pursuant to this Act, and that violation caused death to any employee, shall, upon conviction, be punished by a fine of not more than $10,000 or by imprisonment for not more than six months, or by both; except that if the conviction is for a violation committed after a first conviction of such person, punishment shall be by a fine of not more than $20,000 or by imprisonment for not more than one year, or by both.

(f) Any person who gives advance notice of any inspection to be conducted under this Act, without authority from the Secretary or his designees, shall, upon conviction, be punished by a fine of not more than $1,000 or by imprisonment for not more than six months, or by both.

(g) Whoever knowingly makes any false statement, representation, or certification in any application, record, report, plan, or other document filed or required to be maintained pursuant to this Act shall, upon conviction, be punished by a fine of not more than $10,000, or by imprisonment for not more than six months, or by both.

65 Stat. 721; 79 Stat. 234.

(h)(1) Section 1114 of title 18, United States Code, is hereby amended by striking out "designated by the Secretary of Health, Education, and Welfare to conduct investigations, or inspections under the Federal Food, Drug, and Cosmetic Act" and inserting in lieu thereof "or of the Department of Labor assigned to perform investigative, inspection, or law enforcement functions".

62 Stat. 756.

(2) Notwithstanding the provisions of sections 1111 and 1114 of title 18, United States Code, whoever, in violation of the provisions of section 1114 of such title, kills a person while engaged in or on account of the performance of investigative, inspection, or law enforcement functions added to such section 1114 by paragraph (1) of this subsection, and who would otherwise be subject to the penalty provisions of such section 1111, shall be punished by imprisonment for any term of years or for life.

(i) Any employer who violates any of the posting requirements, as prescribed under the provisions of this Act, shall be assessed a civil penalty of up to $1,000 for each violation.

(j) The Commission shall have authority to assess all civil penalties provided in this section, giving due consideration to the appropriateness of the penalty with respect to the size of the business of the employer being charged, the gravity of the violation, the good faith of the employer, and the history of previous violations.

(k) For purposes of this section, a serious violation shall be deemed to exist in a place of employment if there is a substantial probability that death or serious physical harm could result from a condition which exists, or from one or more practices, means, methods, operations, or processes which have been adopted or are in use, in such place of employment unless the employer did not, and could not with the exercise of reasonable diligence, know of the presence of the violation.

(l) Civil penalties owed under this Act shall be paid to the Secretary for deposit into the Treasury of the United States and shall accrue to the United States and may be recovered in a civil action in the name of the United States brought in the United States district court for the district where the violation is alleged to have occurred or where the employer has its principal office.

STATE JURISDICTION AND STATE PLANS

SEC. 18. (a) Nothing in this Act shall prevent any State agency or court from asserting jurisdiction under State law over any occupational safety or health issue with respect to which no standard is in effect under section 6.

(b) Any State which, at any time, desires to assume responsibility for development and enforcement therein of occupational safety and health standards relating to any occupational safety or health issue with respect to which a Federal standard has been promulgated under section 6 shall submit a State plan for the development of such standards and their enforcement.

(c) The Secretary shall approve the plan submitted by a State under subsection (b), or any modification thereof, if such plan in his judgment—

(1) designates a State agency or agencies as the agency or agencies responsible for administering the plan throughout the State,

(2) provides for the development and enforcement of safety and health standards relating to one or more safety or health issues, which standards (and the enforcement of which standards) are or will be at least as effective in providing safe and healthful employment and places of employment as the standards promulgated under section 6 which relate to the same issues, and which standards, when applicable to products which are distributed or used in interstate commerce, are required by compelling local conditions and do not unduly burden interstate commerce,

(3) provides for a right of entry and inspection of all workplaces subject to the Act which is at least as effective as that provided in section 8, and includes a prohibition on advance notice of inspections,

(4) contains satisfactory assurances that such agency or agencies have or will have the legal authority and qualified personnel necessary for the enforcement of such standards,

(5) gives satisfactory assurances that such State will devote adequate funds to the administration and enforcement of such standards,

(6) contains satisfactory assurances that such State will, to the extent permitted by its law, establish and maintain an effective and comprehensive occupational safety and health program applicable to all employees of public agencies of the State and its political subdivisions, which program is as effective as the standards contained in an approved plan,

(7) requires employers in the State to make reports to the Secretary in the same manner and to the same extent as if the plan were not in effect, and

(8) provides that the State agency will make such reports to the Secretary in such form and containing such information, as the Secretary shall from time to time require.

(d) If the Secretary rejects a plan submitted under subsection (b), he shall afford the State submitting the plan due notice and opportunity for a hearing before so doing. Notice of hearing.

(e) After the Secretary approves a State plan submitted under subsection (b), he may, but shall not be required to, exercise his authority under sections 8, 9, 10, 13, and 17 with respect to comparable standards promulgated under section 6, for the period specified in the next sentence. The Secretary may exercise the authority referred to above until he determines, on the basis of actual operations under the

State plan, that the criteria set forth in subsection (c) are being applied, but he shall not make such determination for at least three years after the plan's approval under subsection (c). Upon making the determination referred to in the preceding sentence, the provisions of sections 5(a)(2), 8 (except for the purpose of carrying out subsection (f) of this section), 9, 10, 13, and 17, and standards promulgated under section 6 of this Act, shall not apply with respect to any occupational safety or health issues covered under the plan, but the Secretary may retain jurisdiction under the above provisions in any proceeding commenced under section 9 or 10 before the date of determination.

Continuing evaluation.

(f) The Secretary shall, on the basis of reports submitted by the State agency and his own inspections make a continuing evaluation of the manner in which each State having a plan approved under this section is carrying out such plan. Whenever the Secretary finds, after affording due notice and opportunity for a hearing, that in the administration of the State plan there is a failure to comply substantially with any provision of the State plan (or any assurance contained therein), he shall notify the State agency of his withdrawal of approval of such plan and upon receipt of such notice such plan shall cease to be in effect, but the State may retain jurisdiction in any case commenced before the withdrawal of the plan in order to enforce standards under the plan whenever the issues involved do not relate to the reasons for the withdrawal of the plan.

Plan rejection, review.

(g) The State may obtain a review of a decision of the Secretary withdrawing approval of or rejecting its plan by the United States court of appeals for the circuit in which the State is located by filing in such court within thirty days following receipt of notice of such decision a petition to modify or set aside in whole or in part the action of the Secretary. A copy of such petition shall forthwith be served upon the Secretary, and thereupon the Secretary shall certify and file in the court the record upon which the decision complained of was issued as provided in section 2112 of title 28, United States Code. Unless the court finds that the Secretary's decision in rejecting a proposed State plan or withdrawing his approval of such a plan is not supported by substantial evidence the court shall affirm the Secretary's decision. The judgment of the court shall be subject to review by the Supreme Court of the United States upon certiorari or certification as provided in section 1254 of title 28, United States Code.

72 Stat. 941; 80 Stat. 1323.

62 Stat. 928.

(h) The Secretary may enter into an agreement with a State under which the State will be permitted to continue to enforce one or more occupational health and safety standards in effect in such State until final action is taken by the Secretary with respect to a plan submitted by a State under subsection (b) of this section, or two years from the date of enactment of this Act, whichever is earlier.

FEDERAL AGENCY SAFETY PROGRAMS AND RESPONSIBILITIES

SEC. 19. (a) It shall be the responsibility of the head of each Federal agency to establish and maintain an effective and comprehensive occupational safety and health program which is consistent with the standards promulgated under section 6. The head of each agency shall (after consultation with representatives of the employees thereof)—

(1) provide safe and healthful places and conditions of employment, consistent with the standards set under section 6;

(2) acquire, maintain, and require the use of safety equipment, personal protective equipment, and devices reasonably necessary to protect employees;

(3) keep adequate records of all occupational accidents and illnesses for proper evaluation and necessary corrective action; Recordkeeping.

(4) consult with the Secretary with regard to the adequacy as to form and content of records kept pursuant to subsection (a)(3) of this section; and

(5) make an annual report to the Secretary with respect to occupational accidents and injuries and the agency's program under this section. Such report shall include any report submitted under section 7902(e)(2) of title 5, United States Code. Annual report. 80 Stat. 530.

(b) The Secretary shall report to the President a summary or digest of reports submitted to him under subsection (a)(5) of this section, together with his evaluations of and recommendations derived from such reports. The President shall transmit annually to the Senate and the House of Representatives a report of the activities of Federal agencies under this section. Report to President. Report to Congress.

(c) Section 7902(c)(1) of title 5, United States Code, is amended by inserting after "agencies" the following: "and of labor organizations representing employees".

(d) The Secretary shall have access to records and reports kept and filed by Federal agencies pursuant to subsections (a)(3) and (5) of this section unless those records and reports are specifically required by Executive order to be kept secret in the interest of the national defense or foreign policy, in which case the Secretary shall have access to such information as will not jeopardize national defense or foreign policy. Records, etc.; availability.

RESEARCH AND RELATED ACTIVITIES

SEC. 20. (a)(1) The Secretary of Health, Education, and Welfare, after consultation with the Secretary and with other appropriate Federal departments or agencies, shall conduct (directly or by grants or contracts) research, experiments, and demonstrations relating to occupational safety and health, including studies of psychological factors involved, and relating to innovative methods, techniques, and approaches for dealing with occupational safety and health problems.

(2) The Secretary of Health, Education, and Welfare shall from time to time consult with the Secretary in order to develop specific plans for such research, demonstrations, and experiments as are necessary to produce criteria, including criteria identifying toxic substances, enabling the Secretary to meet his responsibility for the formulation of safety and health standards under this Act; and the Secretary of Health, Education, and Welfare, on the basis of such research, demonstrations, and experiments and any other information available to him, shall develop and publish at least annually such criteria as will effectuate the purposes of this Act.

(3) The Secretary of Health, Education, and Welfare, on the basis of such research, demonstrations, and experiments, and any other information available to him, shall develop criteria dealing with toxic materials and harmful physical agents and substances which will describe exposure levels that are safe for various periods of employment, including but not limited to the exposure levels at which no employee will suffer impaired health or functional capacities or diminished life expectancy as a result of his work experience.

(4) The Secretary of Health, Education, and Welfare shall also conduct special research, experiments, and demonstrations relating to occupational safety and health as are necessary to explore new problems, including those created by new technology in occupational safety and health, which may require ameliorative action beyond that

which is otherwise provided for in the operating provisions of this Act. The Secretary of Health, Education, and Welfare shall also conduct research into the motivational and behavioral factors relating to the field of occupational safety and health.

Toxic substances, records.

(5) The Secretary of Health, Education, and Welfare, in order to comply with his responsibilities under paragraph (2), and in order to develop needed information regarding potentially toxic substances or harmful physical agents, may prescribe regulations requiring employers to measure, record, and make reports on the exposure of employees to substances or physical agents which the Secretary of Health, Education, and Welfare reasonably believes may endanger the health or safety of employees. The Secretary of Health, Education, and Welfare also is authorized to establish such programs of medical examinations and tests as may be necessary for determining the incidence of occupational illnesses and the susceptibility of employees to such illnesses. Nothing in this or any other provision of this Act shall be deemed to authorize or require medical examination, immunization, or treatment for those who object thereto on religious grounds, except where such is necessary for the protection of the health or safety of others. Upon the request of any employer who is required to measure and record exposure of employees to substances or physical agents as provided under this subsection, the Secretary of Health, Education, and Welfare shall furnish full financial or other assistance to such employer for the purpose of defraying any additional expense incurred by him in carrying out the measuring and recording as provided in this subsection.

Medical examinations.

Toxic substances, publication.

(6) The Secretary of Health, Education, and Welfare shall publish within six months of enactment of this Act and thereafter as needed but at least annually a list of all known toxic substances by generic family or other useful grouping, and the concentrations at which such toxicity is known to occur. He shall determine following a written request by any employer or authorized representative of employees, specifying with reasonable particularity the grounds on which the request is made, whether any substance normally found in the place of employment has potentially toxic effects in such concentrations as used or found; and shall submit such determination both to employers and affected employees as soon as possible. If the Secretary of Health, Education, and Welfare determines that any substance is potentially toxic at the concentrations in which it is used or found in a place of employment, and such substance is not covered by an occupational safety or health standard promulgated under section 6, the Secretary of Health, Education, and Welfare shall immediately submit such determination to the Secretary, together with all pertinent criteria.

Annual studies.

(7) Within two years of enactment of this Act, and annually thereafter the Secretary of Health, Education, and Welfare shall conduct and publish industrywide studies of the effect of chronic or low-level exposure to industrial materials, processes, and stresses on the potential for illness, disease, or loss of functional capacity in aging adults.

Inspections.

(b) The Secretary of Health, Education, and Welfare is authorized to make inspections and question employers and employees as provided in section 8 of this Act in order to carry out his functions and responsibilities under this section.

Contract authority.

(c) The Secretary is authorized to enter into contracts, agreements, or other arrangements with appropriate public agencies or private organizations for the purpose of conducting studies relating to his responsibilities under this Act. In carrying out his responsibilities under this subsection, the Secretary shall cooperate with the Secretary of Health, Education, and Welfare in order to avoid any duplication of efforts under this section.

(d) Information obtained by the Secretary and the Secretary of Health, Education, and Welfare under this section shall be disseminated by the Secretary to employers and employees and organizations thereof.

(e) The functions of the Secretary of Health, Education, and Welfare under this Act shall, to the extent feasible, be delegated to the Director of the National Institute for Occupational Safety and Health established by section 22 of this Act. Delegation of functions.

TRAINING AND EMPLOYEE EDUCATION

SEC. 21. (a) The Secretary of Health, Education, and Welfare, after consultation with the Secretary and with other appropriate Federal departments and agencies, shall conduct, directly or by grants or contracts (1) education programs to provide an adequate supply of qualified personnel to carry out the purposes of this Act, and (2) informational programs on the importance of and proper use of adequate safety and health equipment.

(b) The Secretary is also authorized to conduct, directly or by grants or contracts, short-term training of personnel engaged in work related to his responsibilities under this Act.

(c) The Secretary, in consultation with the Secretary of Health, Education, and Welfare, shall (1) provide for the establishment and supervision of programs for the education and training of employers and employees in the recognition, avoidance, and prevention of unsafe or unhealthful working conditions in employments covered by this Act, and (2) consult with and advise employers and employees, and organizations representing employers and employees as to effective means of preventing occupational injuries and illnesses.

NATIONAL INSTITUTE FOR OCCUPATIONAL SAFETY AND HEALTH

SEC. 22. (a) It is the purpose of this section to establish a National Institute for Occupational Safety and Health in the Department of Health, Education, and Welfare in order to carry out the policy set forth in section 2 of this Act and to perform the functions of the Secretary of Health, Education, and Welfare under sections 20 and 21 of this Act. Establishment.

(b) There is hereby established in the Department of Health, Education, and Welfare a National Institute for Occupational Safety and Health. The Institute shall be headed by a Director who shall be appointed by the Secretary of Health, Education, and Welfare, and who shall serve for a term of six years unless previously removed by the Secretary of Health, Education, and Welfare. Director, appointment, term.

(c) The Institute is authorized to—

(1) develop and establish recommended occupational safety and health standards; and

(2) perform all functions of the Secretary of Health, Education, and Welfare under sections 20 and 21 of this Act.

(d) Upon his own initiative, or upon the request of the Secretary or the Secretary of Health, Education, and Welfare, the Director is authorized (1) to conduct such research and experimental programs as he determines are necessary for the development of criteria for new and improved occupational safety and health standards, and (2) after consideration of the results of such research and experimental programs make recommendations concerning new or improved occupational safety and health standards. Any occupational safety and health standard recommended pursuant to this section shall immediately be forwarded to the Secretary of Labor, and to the Secretary of Health, Education, and Welfare.

(e) In addition to any authority vested in the Institute by other provisions of this section, the Director, in carrying out the functions of the Institute, is authorized to—

(1) prescribe such regulations as he deems necessary governing the manner in which its functions shall be carried out;

(2) receive money and other property donated, bequeathed, or devised, without condition or restriction other than that it be used for the purposes of the Institute and to use, sell, or otherwise dispose of such property for the purpose of carrying out its functions;

(3) receive (and use, sell, or otherwise dispose of, in accordance with paragraph (2)), money and other property donated, bequeathed, or devised to the Institute with a condition or restriction, including a condition that the Institute use other funds of the Institute for the purposes of the gift;

(4) in accordance with the civil service laws, appoint and fix the compensation of such personnel as may be necessary to carry out the provisions of this section;

(5) obtain the services of experts and consultants in accordance with the provisions of section 3109 of title 5, United States
80 Stat. 416. Code;

(6) accept and utilize the services of voluntary and noncompensated personnel and reimburse them for travel expenses, including per diem, as authorized by section 5703 of title 5, United
83 Stat. 190. States Code;

(7) enter into contracts, grants or other arrangements, or modifications thereof to carry out the provisions of this section, and such contracts or modifications thereof may be entered into without performance or other bonds, and without regard to section 3709 of the Revised Statutes, as amended (41 U.S.C. 5), or any other provision of law relating to competitive bidding;

(8) make advance, progress, and other payments which the Director deems necessary under this title without regard to the provisions of section 3648 of the Revised Statutes, as amended (31 U.S.C. 529); and

(9) make other necessary expenditures.

Annual report to HEW, President, and Congress.

(f) The Director shall submit to the Secretary of Health, Education, and Welfare, to the President, and to the Congress an annual report of the operations of the Institute under this Act, which shall include a detailed statement of all private and public funds received and expended by it, and such recommendations as he deems appropriate.

GRANTS TO THE STATES

Sec. 23. (a) The Secretary is authorized, during the fiscal year ending June 30, 1971, and the two succeeding fiscal years, to make grants to the States which have designated a State agency under section 18 to assist them—

(1) in identifying their needs and responsibilities in the area of occupational safety and health,

(2) in developing State plans under section 18, or

(3) in developing plans for—

(A) establishing systems for the collection of information concerning the nature and frequency of occupational injuries and diseases;

(B) increasing the expertise and enforcement capabilities of their personnel engaged in occupational safety and health programs; or

(C) otherwise improving the administration and enforcement of State occupational safety and health laws, including standards thereunder, consistent with the objectives of this Act.

(b) The Secretary is authorized, during the fiscal year ending June 30, 1971, and the two succeeding fiscal years, to make grants to the States for experimental and demonstration projects consistent with the objectives set forth in subsection (a) of this section.

(c) The Governor of the State shall designate the appropriate State agency for receipt of any grant made by the Secretary under this section.

(d) Any State agency designated by the Governor of the State desiring a grant under this section shall submit an application therefor to the Secretary.

(e) The Secretary shall review the application, and shall, after consultation with the Secretary of Health, Education, and Welfare, approve or reject such application.

(f) The Federal share for each State grant under subsection (a) or (b) of this section may not exceed 90 per centum of the total cost of the application. In the event the Federal share for all States under either such subsection is not the same, the differences among the States shall be established on the basis of objective criteria.

(g) The Secretary is authorized to make grants to the States to assist them in administering and enforcing programs for occupational safety and health contained in State plans approved by the Secretary pursuant to section 18 of this Act. The Federal share for each State grant under this subsection may not exceed 50 per centum of the total cost to the State of such a program. The last sentence of subsection (f) shall be applicable in determining the Federal share under this subsection.

(h) Prior to June 30, 1973, the Secretary shall, after consultation with the Secretary of Health, Education, and Welfare, transmit a report to the President and to the Congress, describing the experience under the grant programs authorized by this section and making any recommendations he may deem appropriate. Report to President and Congress.

STATISTICS

SEC. 24. (a) In order to further the purposes of this Act, the Secretary, in consultation with the Secretary of Health, Education, and Welfare, shall develop and maintain an effective program of collection, compilation, and analysis of occupational safety and health statistics. Such program may cover all employments whether or not subject to any other provisions of this Act but shall not cover employments excluded by section 4 of the Act. The Secretary shall compile accurate statistics on work injuries and illnesses which shall include all disabling, serious, or significant injuries and illnesses, whether or not involving loss of time from work, other than minor injuries requiring only first aid treatment and which do not involve medical treatment, loss of consciousness, restriction of work or motion, or transfer to another job.

(b) To carry out his duties under subsection (a) of this section, the Secretary may—

(1) promote, encourage, or directly engage in programs of studies, information and communication concerning occupational safety and health statistics;

(2) make grants to States or political subdivisions thereof in order to assist them in developing and administering programs dealing with occupational safety and health statistics; and

(3) arrange, through grants or contracts, for the conduct of such research and investigations as give promise of furthering the objectives of this section.

(c) The Federal share for each grant under subsection (b) of this section may be up to 50 per centum of the State's total cost.

(d) The Secretary may, with the consent of any State or political subdivision thereof, accept and use the services, facilities, and employees of the agencies of such State or political subdivision, with or without reimbursement, in order to assist him in carrying out his functions under this section.

Reports. (e) On the basis of the records made and kept pursuant to section 8(c) of this Act, employers shall file such reports with the Secretary as he shall prescribe by regulation, as necessary to carry out his functions under this Act.

(f) Agreements between the Department of Labor and States pertaining to the collection of occupational safety and health statistics already in effect on the effective date of this Act shall remain in effect until superseded by grants or contracts made under this Act.

AUDITS

SEC. 25. (a) Each recipient of a grant under this Act shall keep such records as the Secretary or the Secretary of Health, Education, and Welfare shall prescribe, including records which fully disclose the amount and disposition by such recipient of the proceeds of such grant, the total cost of the project or undertaking in connection with which such grant is made or used, and the amount of that portion of the cost of the project or undertaking supplied by other sources, and such other records as will facilitate an effective audit.

(b) The Secretary or the Secretary of Health, Education, and Welfare, and the Comptroller General of the United States, or any of their duly authorized representatives, shall have access for the purpose of audit and examination to any books, documents, papers, and records of the recipients of any grant under this Act that are pertinent to any such grant.

ANNUAL REPORT

SEC. 26. Within one hundred and twenty days following the convening of each regular session of each Congress, the Secretary and the Secretary of Health, Education, and Welfare shall each prepare and submit to the President for transmittal to the Congress a report upon the subject matter of this Act, the progress toward achievement of the purpose of this Act, the needs and requirements in the field of occupational safety and health, and any other relevant information. Such reports shall include information regarding occupational safety and health standards, and criteria for such standards, developed during the preceding year; evaluation of standards and criteria previously developed under this Act, defining areas of emphasis for new criteria and standards; an evaluation of the degree of observance of applicable occupational safety and health standards, and a summary of inspection and enforcement activity undertaken; analysis and evaluation of research activities for which results have been obtained under governmental and nongovernmental sponsorship; an analysis of major occupational diseases; evaluation of available control and measurement technology for hazards for which standards or criteria have been developed during the preceding year; description of cooperative efforts undertaken between Government agencies and other interested parties in the implementation of this Act during the preceding year; a progress report on the development of an adequate supply of trained manpower in the field of occupational safety and health, including estimates of future needs and the efforts being made by Government and others to meet those needs; listing of all toxic substances in industrial usage for which labeling requirements, criteria, or standards have not yet been established; and such recommendations for additional legislation as are deemed necessary to protect the safety and health of the worker and improve the administration of this Act.

NATIONAL COMMISSION ON STATE WORKMEN'S COMPENSATION LAWS

SEC. 27. (a) (1) The Congress hereby finds and declares that—

(A) the vast majority of American workers, and their families, are dependent on workmen's compensation for their basic economic security in the event such workers suffer disabling injury or death in the course of their employment; and that the full protection of American workers from job-related injury or death requires an adequate, prompt, and equitable system of workmen's compensation as well as an effective program of occupational health and safety regulation; and

(B) in recent years serious questions have been raised concerning the fairness and adequacy of present workmen's compensation laws in the light of the growth of the economy, the changing nature of the labor force, increases in medical knowledge, changes in the hazards associated with various types of employment, new technology creating new risks to health and safety, and increases in the general level of wages and the cost of living.

(2) The purpose of this section is to authorize an effective study and objective evaluation of State workmen's compensation laws in order to determine if such laws provide an adequate, prompt, and equitable system of compensation for injury or death arising out of or in the course of employment.

(b) There is hereby established a National Commission on State Workmen's Compensation Laws. Establishment.

(c) (1) The Workmen's Compensation Commission shall be composed of fifteen members to be appointed by the President from among members of State workmen's compensation boards, representatives of insurance carriers, business, labor, members of the medical profession having experience in industrial medicine or in workmen's compensation cases, educators having special expertise in the field of workmen's compensation, and representatives of the general public. The Secretary, the Secretary of Commerce, and the Secretary of Health, Education, and Welfare shall be ex officio members of the Workmen's Compensation Commission: Membership.

(2) Any vacancy in the Workmen's Compensation Commission shall not affect its powers.

(3) The President shall designate one of the members to serve as Chairman and one to serve as Vice Chairman of the Workmen's Compensation Commission.

Quorum.

(4) Eight members of the Workmen's Compensation Commission shall constitute a quorum.

Study.

(d) (1) The Workmen's Compensation Commission shall undertake a comprehensive study and evaluation of State workmen's compensation laws in order to determine if such laws provide an adequate, prompt, and equitable system of compensation. Such study and evaluation shall include, without being limited to, the following subjects: (A) the amount and duration of permanent and temporary disability benefits and the criteria for determining the maximum limitations thereon, (B) the amount and duration of medical benefits and provisions insuring adequate medical care and free choice of physician, (C) the extent of coverage of workers, including exemptions based on numbers or type of employment, (D) standards for determining which injuries or diseases should be deemed compensable, (E) rehabilitation, (F) coverage under second or subsequent injury funds, (G) time limits on filing claims, (H) waiting periods, (I) compulsory or elective coverage, (J) administration, (K) legal expenses, (L) the feasibility and desirability of a uniform system of reporting information concerning job-related injuries and diseases and the operation of workmen's compensation laws, (M) the resolution of conflict of laws, extraterritoriality and similar problems arising from claims with multistate aspects, (N) the extent to which private insurance carriers are excluded from supplying workmen's compensation coverage and the desirability of such exclusionary practices, to the extent they are found to exist, (O) the relationship between workmen's compensation on the one hand, and old-age, disability, and survivors insurance and other types of insurance, public or private, on the other hand, (P) methods of implementing the recommendations of the Commission.

Report to President and Congress.

(2) The Workmen's Compensation Commission shall transmit to the President and to the Congress not later than July 31, 1972, a final report containing a detailed statement of the findings and conclusions of the Commission, together with such recommendations as it deems advisable.

Hearings.

(e) (1) The Workmen's Compensation Commission or, on the authorization of the Workmen's Compensation Commission, any subcommittee or members thereof, may, for the purpose of carrying out the provisions of this title, hold such hearings, take such testimony, and sit and act at such times and places as the Workmen's Compensation Commission deems advisable. Any member authorized by the Workmen's Compensation Commission may administer oaths or affirmations to witnesses appearing before the Workmen's Compensation Commission or any subcommittee or members thereof.

(2) Each department, agency, and instrumentality of the executive branch of the Government, including independent agencies, is authorized and directed to furnish to the Workmen's Compensation Commission, upon request made by the Chairman or Vice Chairman, such information as the Workmen's Compensation Commission deems necessary to carry out its functions under this section.

(f) Subject to such rules and regulations as may be adopted by the Workmen's Compensation Commission, the Chairman shall have the power to—

80 Stat. 378. 5 USC 101.

5 USC 5101, 5331.

(1) appoint and fix the compensation of an executive director, and such additional staff personnel as he deems necessary, without regard to the provisions of title 5, United States Code, governing appointments in the competitive service, and without regard to the provisions of chapter 51 and subchapter III of chapter 53 of such title relating to classification and General Schedule

pay rates, but at rates not in excess of the maximum rate for GS–18 of the General Schedule under section 5332 of such title, and Ante, p. 198-1.

(2) procure temporary and intermittent services to the same extent as is authorized by section 3109 of title 5, United States Code. 80 Stat. 416.

(g) The Workmen's Compensation Commission is authorized to enter into contracts with Federal or State agencies, private firms, institutions, and individuals for the conduct of research or surveys, the preparation of reports, and other activities necessary to the discharge of its duties. Contract authorization.

(h) Members of the Workmen's Compensation Commission shall receive compensation for each day they are engaged in the performance of their duties as members of the Workmen's Compensation Commission at the daily rate prescribed for GS–18 under section 5332 of title 5, United States Code, and shall be entitled to reimbursement for travel, subsistence, and other necessary expenses incurred by them in the performance of their duties as members of the Workmen's Compensation Commission. Compensation; travel expenses.

(i) There are hereby authorized to be appropriated such sums as may be necessary to carry out the provisions of this section. Appropriation.

(j) On the ninetieth day after the date of submission of its final report to the President, the Workmen's Compensation Commission shall cease to exist. Termination.

ECONOMIC ASSISTANCE TO SMALL BUSINESSES

SEC. 28. (a) Section 7(b) of the Small Business Act, as amended, is amended— 72 Stat. 387; 83 Stat. 802. 15 USC 636.

(1) by striking out the period at the end of "paragraph (5)" and inserting in lieu thereof "; and"; and

(2) by adding after paragraph (5) a new paragraph as follows:

"(6) to make such loans (either directly or in cooperation with banks or other lending institutions through agreements to participate on an immediate or deferred basis) as the Administration may determine to be necessary or appropriate to assist any small business concern in effecting additions to or alterations in the equipment, facilities, or methods of operation of such business in order to comply with the applicable standards promulgated pursuant to section 6 of the Occupational Safety and Health Act of 1970 or standards adopted by a State pursuant to a plan approved under section 18 of the Occupational Safety and Health Act of 1970, if the Administration determines that such concern is likely to suffer substantial economic injury without assistance under this paragraph."

(b) The third sentence of section 7(b) of he Small Business Act, as amended, is amended by striking out "or (5)" after "paragraph (3)" and inserting a comma followed by "(5) or (6)".

(c) Section 4(c)(1) of the Small Business Act, as amended, is amended by inserting "7(b)(6)," after "7(b)(5),". 80 Stat. 132. 15 USC 633.

(d) Loans may also be made or guaranteed for the purposes set forth in section 7(b)(6) of the Small Business Act, as amended, pursuant to the provisions of section 202 of the Public Works and Economic Development Act of 1965, as amended. 79 Stat. 556. 42 USC 3142.

ADDITIONAL ASSISTANT SECRETARY OF LABOR

SEC. 29. (a) Section 2 of the Act of April 17, 1946 (60 Stat. 91) as amended (29 U.S.C. 553) is amended by— 75 Stat. 338.

(1) striking out "four" in the first sentence of such section and inserting in lieu thereof "five"; and

(2) adding at the end thereof the following new sentence, "One of such Assistant Secretaries shall be an Assistant Secretary of Labor for Occupational Safety and Health.".

80 Stat. 462. (b) Paragraph (20) of section 5315 of title 5, United States Code, is amended by striking out "(4)" and inserting in lieu thereof "(5)".

ADDITIONAL POSITIONS

Sec. 30. Section 5108(c) of title 5, United States Code, is amended by—

(1) striking out the word "and" at the end of paragraph (8);

(2) striking out the period at the end of paragraph (9) and inserting in lieu thereof a semicolon and the word "and"; and

(3) by adding immediately after paragraph (9) the following new paragraph:

"(10)(A) the Secretary of Labor, subject to the standards and procedures prescribed by this chapter, may place an additional twenty-five positions in the Department of Labor in GS–16, 17, and 18 for the purposes of carrying out his responsibilities under the Occupational Safety and Health Act of 1970;

"(B) the Occupational Safety and Health Review Commission, subject to the standards and procedures prescribed by this chapter, may place ten positions in GS–16, 17, and 18 in carrying out its functions under the Occupational Safety and Health Act of 1970."

EMERGENCY LOCATOR BEACONS

72 Stat. 775. Sec. 31. Section 601 of the Federal Aviation Act of 1958 is amended
49 USC 1421. by inserting at the end thereof a new subsection as follows:

"EMERGENCY LOCATOR BEACONS

"(d)(1) Except with respect to aircraft described in paragraph (2) of this subsection, minimum standards pursuant to this section shall include a requirement that emergency locator beacons shall be installed—

"(A) on any fixed-wing, powered aircraft for use in air commerce the manufacture of which is completed, or which is imported into the United States, after one year following the date of enactment of this subsection; and

"(B) on any fixed-wing, powered aircraft used in air commerce after three years following such date.

"(2) The provisions of this subsection shall not apply to jet-powered aircraft; aircraft used in air transportation (other than air taxis and charter aircraft); military aircraft; aircraft used solely for training purposes not involving flights more than twenty miles from its base; and aircraft used for the aerial application of chemicals."

SEPARABILITY

Sec. 32. If any provision of this Act, or the application of such provision to any person or circumstance, shall be held invalid, the remainder of this Act, or the application of such provision to persons or circumstances other than those as to which it is held invalid, shall not be affected thereby.

APPROPRIATIONS

SEC. 33. There are authorized to be appropriated to carry out this Act for each fiscal year such sums as the Congress shall deem necessary.

EFFECTIVE DATE

SEC. 34. This Act shall take effect one hundred and twenty days after the date of its enactment.

Approved December 29, 1970.

LEGISLATIVE HISTORY:

HOUSE REPORTS: No. 91-1291 accompanying H.R. 16785 (Comm. on Education and Labor) and No. 91-1765 (Comm. of Conference).
SENATE REPORT No. 91-1282 (Comm. on Labor and Public Welfare).
CONGRESSIONAL RECORD, Vol. 116 (1970):
Oct. 13, Nov. 16, 17, considered and passed Senate.
Nov. 23, 24, considered and passed House, amended, in lieu of H.R. 16785.
Dec. 16, Senate agreed to conference report.
Dec. 17, House agreed to conference report.

Selected Bibliography on the OSHAct and the Foreign Experience

I. The OSHAct

E.F. Brady, "The New Occupational Safety and Health Act—Its Impact on Contractors and Sureties," *The Forum*, Vol. 8 (1972), p. 164.

M.E. Gross, "The Occupational Safety and Health Act: Much Ado About Something," *Loyola Law Journal*, Vol. 3 (1972), p. 246.

George C. Guenther, "The Significance of the Occupational Safety and Health Act to the Worker in the United States," *International Labour Review*, Vol. 195, No. 1 (January 1972).

Earl D. Heath, "The Implementation and Philosophy of the Williams-Steiger Occupational Safety and Health Act of 1970," *University of Florida Law Review*, Vol. 25, No. 2 (winter 1973).

Horneberger, "Occupational Safety and Health Act of 1970," *Cleveland State Law Review*, Vol. 21 (1972), p. 1.

J. Lazarus and J. Redeker, "A Guide to Employer Alternatives Within the Occupational Safety and Health Act of 1970," U.S. Chamber of Commerce pamphlet (1972).

R. L. Moore, "The Federal Occupational Safety and Health Act of 1970," *Wisconsin Bar Bulletin* (February 1972), p. 46.

R. D. Moran, "A Critique of the Occupational Safety and Health Act of 1970," *Northwestern University Law Review*, Vol. 67 (1972), p. 200.

R. D. Moran, "Address of Chairman Moran of the Review Commission to the ABA," Labor Law Section, August 15, 1972.

R. S. Morey, "The General Duty Clause of the Occupational Safety and Health Act of 1970," *Harvard Law Review*, Vol. 86 (1973), p. 988.

D. M. Neill, "Federal Focus: Occupational Safety and Health Act," *Journal of the Missouri Bar* (August and September 1971).

Periman, "Landmark Job Safety Law Wins Final Congress Okay," *AFL-CIO News*, December 19, 1970, p. 1.

W. L. White and J. T. Carney, "OSHA Comes of Age: The Law of Work Place Environment," *The Business Lawyer* (July 1973), p. 1309.

II. The Foreign Experience*

International Labour Organization

International Directory of Occupational Safety and Health Services (International Labour Organization, Geneva, 1969).

"In-depth Review of Occupational Safety and Health Programs (Further Considerations)," Document No. GB.181/FA/8/9 (November 1970).

"Report of the Ad Hoc Committee on the Occupational Safety and Health Program, Geneva, May 8-12, 1972," Document No. GB.18b/5/11.

"Report of the Meeting of Experts on Control and Prevention of Occupational Cancer," Geneva, January 10-17, 1972.

Marcel Roberts and Luigi Parmeggiani, "Fifty Years of International Cooperation in Occupa-

*This section gives some of the documents on which the analysis of Chapter 11 (and the more extensive analysis contained in the unpublished appendices) is based. While the listing is by no means all-inclusive, it should help the reader to get into the literature.

tional Safety and Health," *International Labour Review*, Vol. 99, No. 1 (January 1969).

World Health Organization

World Health (World Health Organization, Geneva, March 1969).

International Agency for Research on Cancer, *IARC Monographs on the Evaluation of Carcinogenic Risk of Chemicals to Man* (World Health Organization, Geneva, 1972).

"Occupational Health Programmes," *WHO Chronicle*, Vol. 26, No. 12 (December 1972).

"Resolution of the World Health Assembly Regarding Occupational Health Programmes" (mimeograph, May 26, 1972).

M. A. El Batawi, "Occupational Health in Small Industries in Developing Countries," Document No. 71.6.

Belgium

Recueil des Exposés et des Rapports relatifs à la Journée Nationale de la Médicine du Travail (Ministère de l'Emploi et du Travail, Administration de l'Hygiène et de la Médicine du Travail, 1970).

Manpower Policy in Belgium (Organization for Economic Cooperation and Development, Paris, 1971).

Federal Republic of Germany

Detlev Zoellner, *Social Legislation in the Federal Republic of Germany* (Adgard-Verlag GMBH, Bonn, 1970).

Ernst-Gerhard Erdmann, *The Myth of Codetermination* (Bundesvereinigung der Deutschen Arbeitgeberverbande, Cologne, 1966).

"German Works Council Act, 1972" (Commerce Clearing House, Inc., Chicago, 1972).

"The Industrial Injuries Insurance Institute (die Berufsgenossenschaften)" (information brochure).

"Leaflet about Statutory Accident Insurance" (Berufsgenossenschaften, Bonn, July 1971).

"Betriebs Wacht 1972" (Berufsgenossenschaften, Bonn, 1972).

Finland

Kaj Ahlman, "Occupational Diseases in Finland," in *Annual Report, 1970* (The Occupational Health Foundation and the Institute of Occupational Health, Helsinki, 1972).

S. Hernberg and Anneli Kikkonen, "Current Research Projects of the Institute of Occupational Health" (Institute of Occupational Health, Helsinki, 1971).

"Social Insurance" (Ministry of Social Affairs and Health, Helsinki, 1969).

"Social Services in Finland" (Ministry of Social Affairs and Health, Helsinki, December 1964).

France

G. H. Camerlynck and G. Lyon-Caen, *Précis de Droit du Travail* (Librairie Dalloz, Paris, 1972).

Jean-Jacques Dupeyroux, *Sécurité Sociale* (Librairie Dalloz, Paris, 1969).

J. Pluyette, *Hygiène et Sécurité* (Technique et Documentation, Paris, 1971).

Les Comités Techniques des Organismes de Sécurité Sociale (Institut National de Recherche et de Sécurité, Paris, 1972).

Recherches en vue d'une Organisation plus Humaine du Travail Industriel (La Documentation Francaise, Paris, 1972).

Statistiques Nationales d'Accidents du Travail (Caisse National de l'Assurance Maladie de Travailleurs Salries, Paris, 1970).

Textes Relatifs aux Comités d'Hygiène et de Sécurité (INRS, Paris, 1971).

Travail et Sécurité (INRS, Paris, March, June, July 1972).

Italy

Angelo Iannaccone, "Malattie Professionali," *Trattato di Patalogia Medica*, Vol. 3, No. 2 (1971), pp. 3073-3074.

The Netherlands

Environmental Health in the Netherlands (Ministry of Social Affairs and Public Health, the Hague, 1968).

Verslag (Netherlands Institute for Preventive Medicine, TNO, Leiden, 1971).

R. L. Zielhuis, "Statement on Education and Training in Occupational Health in the Netherlands" (Coronel Laboratory, University of Amsterdam, 1972).

"An Introduction to TNO" (Industrial Liaison Department, TNO, Delft, n.d.).

Industrial Medicine Act (February 19, 1959) (International Labour Organization, Geneva, Legislative Series 1959 [Neth.2], p. 3).

Sweden

Sven Forssman, *Occupational Health in Sweden* (The Swedish Institute, Stockholm, 1971).

Eric Ostlund and Anders Englund, "Occupational Safety and Health in the Swedish Construction Industry," *Work-Environment-Health*, Vol. 8, No. 1 (1971).

LO Executive Board, "Working Environment" (Swedish Trade Union Confederation [LO], Stockholm, 1971).

"Promoting Mutual Interests on Sweden's Labour Market" (Swedish Employers' Confederation and Confederation of Swedish Trade Unions, Stockholm, 1961).

"Safety and Health at the Building Site,"*Byggnads Industrin*, No. 19.

United Kingdom

Parliamentary Committee on Safety and Health at Work, *Report of the Committee, 1970-72* (London, 1972). Volume 2 of the report gives selected evidence.

Index

Cases Cited